Edition 3

Seeing Sociology

AN INTRODUCTION

Joan Ferrante
Northern Kentucky University

With contributions from
Chris Caldeira
University of California, Davis

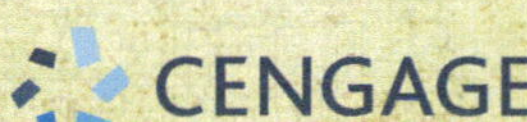

Australia • Brazil • Mexico • Singapore • United Kingdom • United States

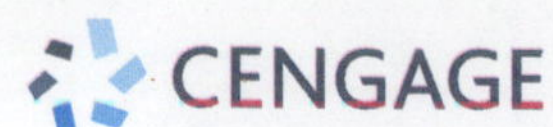

Seeing Sociology: An Introduction,
Third Edition
Joan Ferrante

Product Director: Marta Lee-Perriard

Product Manager: Jennifer Harrison

Content Developer: John Chell

Product Assistant: Julia Catalano

Media Developer: John Chell

Product Development Manager: Erik Fortier

Marketing Manager: Kara Kindstrom

Content Project Manager: Cheri Palmer

Senior Art Director: Michelle Kunkler

Manufacturing Planner: Judy Inouye

IP Analyst: Deanna Ettinger

IP Project Manager: Amber Hosea

Production Service/Project Manager: Jill Traut, MPS Limited

Photo Researcher: Dharanivel Baskar, Lumina Datamatics Ltd.

Text Researcher: Nandhini Srinivasagopalan, Lumina Datamatics Ltd.

Copy Editor: David Heath

Text Designer: Diane Beasley

Cover Designer: Chris Miller

Cover Images: S, Courtesy of Joan Ferrante; E, © Artens/Shutterstock.com; E, © XiXinXing/Shutterstock.com; I, © karelnoppe/Shutterstock.com; N, © MSPhotographic/Shutterstock.com; G, © Volodymyr Kyrylyuk/Shutterstock.com; S, © Chris Alcock/Shutterstock.com; O, © hxdbzxy/Shutterstock.com; C, © Perfect Gui/Shutterstock.com; I, © Shots Studio/Shutterstock.com; O, © Vasin Lee/Shutterstock.com; L, © blvdone/Shutterstock.com; O, © TTstudio/Shutterstock.com; G, © Angelo Giampiccolo/Shutterstock.com; Y, Courtesy of Joan Ferrante.

Compositor: MPS Limited

For product information and technology assistance, contact us at
Cengage Customer & Sales Support, 1-800-354-9706
For permission to use material from this text or product,
submit all requests online at **www.cengage.com/permissions**
Further permissions questions can be e-mailed to
permissionrequest@cengage.com

Library of Congress Control Number: 2014942029

Student Edition: 978-1-305-09436-9

Cengage
20 Channel Center Street
Boston, MA 02210
USA

Cengage is a leading provider of customized learning solutions with office locations around the globe, including Singapore, the United Kingdom, Australia, Mexico, Brazil, and Japan. Locate your local office at **www.cengage.com/global**

Cengage products are represented in Canada by Nelson Education, Ltd.

To learn more about Cengage platforms and services, register or access your online learning solution, or purchase materials for your course, visit **www.cengage.com**.

Printed at CLDPC, USA, 02-19

About the Author

JOAN FERRANTE is a professor of sociology at Northern Kentucky University (NKU). She received her PhD from the University of Cincinnati in 1984. Joan decided early in her career that she wanted to focus her publishing efforts on introducing students to the discipline of sociology. She believes it is important for that introduction to cultivate an appreciation for the methods of social research and for sociological theory beyond the three major perspectives. As a professor, she teaches sociology from an applied perspective so that students come to understand the various career options that the serious student of sociology can pursue. Joan is the author of "Careers in Sociology" (a Wadsworth sociology module), a guide to making the most of an undergraduate degree in sociology. She also teaches sociology in a way that emphasizes the value and power of the sociological framework for making a difference in the world. With the support of the Mayerson Family Foundations, Joan designed the curriculum for a student philanthropy project at NKU in which students, as part of their course work, must decide how to use $4,000 in a way that addresses some community need. That curriculum has now been adopted by dozens of universities across the United States. For the past decade, she has also supported a study abroad scholarship called "Beyond the Classroom" for which any NKU student who has used her sociology texts (new or used) can apply. Joan's university has twice recognized her as an outstanding professor with the Frank Milburn Sinton Outstanding Professor Award and the Outstanding Junior Faculty Award.

Doug Hume

To Dr. Horatio C Wood, IV, MD

Brief Contents

Contents

NKU Philosophy, Rudy Garns

Chris Caldeira

Chris Caldeira

NKU Sociology, Missy Gish

Chris Caldeira

Aleena Ferrante

Chris Caldeira

Chris Caldeira

Photo by Chris Caldeira: Mural by Paul Ygartua/www.ygartua.com

Chris Caldeira

Chris Caldeira

NKU Sociology, Missy Gish

Chris Caldeira

Preface

This is the third edition of *Seeing Sociology*—my third attempt at writing a visually oriented introduction to the discipline. As I went about doing the work of this revision, I gave much thought to students who are new to sociology—the primary audience for this book. I thought about all the conversations over the course of my teaching career where I overheard students proclaiming they can get by without reading a textbook. And, of course, being an author, I don't want this to be my book's fate. But I also know reading is challenging; in fact, it is very hard work. As Professor Jeffrey Davis (2014) at Wheaton University describes it, "Reading takes effort: you have to be alert and force your eyes across a page, back and forth, back and forth, hundreds of times. It is tiring. It is brain-draining. It is tough on the eyes and neck. It requires concentration, recall, and synthesis. You can't zone out." Reading is made even more challenging because we live in an age where the discipline and dedication required to read seems out of place as digital technologies distract us and lure us to move on and forget what we have just read. Recognizing these pressures, I tried to write in a clear and concise way to support the challenges of reading. In addition, *Seeing Sociology* is structured to support the reading experience. Specifically, the book has two signature qualities:

- *Seeing Sociology* contains 13 chapters, each broken into about seven self-contained modules, four to eight pages in length. The modular format gives readers focused and manageable "chunks" of reading. It also gives instructors the flexibility to assign all or selected modules within a chapter. Each module begins by posing a question that prompts readers to recall an experience or that elicits a reaction priming them mentally for the material to come.
- *Seeing Sociology* capitalizes on the instructional value of photographs as a tool for provoking thought, clarifying abstract concepts, and conveying sociology's significance as a perspective. Photographs, seamlessly integrated into the flow of the surrounding text, are presented as objects of analysis to demonstrate how sociologists see, interpret, and analyze all that is going on in the world around them.

Major Changes to This Edition

In revising *Seeing Sociology*, my strategy was simple: I tried to make the third edition better than the second. As I revised, I read each word, sentence, paragraph, module, and chapter. I imagined the eyes of the first-year student looking at the text. I think back to a time when I was a first-year student reading my textbooks and simply lost about what I should take away from the reading. I also remember being overwhelmed with so much to know and wondering why I should know it. I became sidetracked and frustrated when I could not understand exactly what the author was seeking to convey. Often, instead of reading on, I stopped. While some may argue that writing with the first-year student in mind means dumbing down the discipline, that is not how I see it. I believe writing guided by these kinds of remembered experiences actually helped strengthen the way I present the discipline and convey its conceptual power. These remembered experiences motivated me to carefully choose words, to evaluate the effectiveness of each sentence, and to pay close attention to flow. At first glance, someone comparing this third edition to the second or first may conclude that large sections are "the same." A close reading shows that my revisions are aimed not simply at updating—adding new and removing outdated examples and concepts. My revisions involve revising words so they say what I need them to convey. For those readers who make the effort to concentrate on my words, I want to deliver a polished product.

I also reviewed every photograph used in the second edition, asking: "Can I find a more effective photograph to represent a given sociological idea?" In the end, I replaced 439 of 711 photos with new, hopefully more effective ones. The adage "a picture is worth a thousand words" helps to explain the photograph's pedagogical value. Let me be clear about that value: while pictures offer a subject for analysis and a tool to illustrate social dynamics, we must also realize that a photograph by itself does not convey some fixed meaning. Meaning changes depending on the point of view of the person viewing it. The point of view this book cultivates is sociological and the accompanying captions explain how sociologists "see" the freeze-frames of activity. Placing such high pedagogical value on photographs is especially appropriate given that we live in an age in which photographs have assumed a tsunami-like force as billions of photographs (most notably selfies) travel the Internet. Sociology offers tools for thinking about what photographs capture about the society in which we live; how they shape our sense of self, our interactions and relationships; and the ways an event occurring locally is shared on a potentially global scale.

In addition to revising my writing and photographic choices, I made a number of other major changes. Earlier editions of *Seeing Sociology* included an Applying Theory module at the close of each chapter. Initially, those modules were written with the purpose of highlighting and applying a variety of micro-, meso-, and macro-level theories in sociology including, but also going beyond, the big three—functionalism, conflict, and symbolic interaction. Applying Theory modules in the first and second editions focused on global society theories, phenomenology, post-structural theories, critical theory, world system theory, and more. For this edition I took seriously reviewer suggestions that this was too much for new sociology students to absorb. Thus, for the third edition I limited the Applying Theory modules to the big three plus one. That one is the feminist perspective.

The Applying Theory modules focus on comparing how sociologists inspired by each of the four perspectives might think about targeted issues related to each chapter topic. As a result, the target of the comparative analysis is something very specific. To illustrate further, instead of addressing how sociologists inspired by each of the four perspectives present an impossibly large and abstract force (e.g., culture, social structure, race, or socialization), I address how each perspective describes the sociological significance of something very specific such as blue jeans as an item of material culture, the social structure of Vietnamese nail salons, race categories as identity-building tools, and interactive games as agents of socialization. I believe this kind of focused and comparative analysis sends the message that the four perspectives are powerful conceptual tools that can be used to think about any area of life and not just the most abstract forces in our lives, important as they are.

The Write a Caption feature that was at the end of each module in the first and second editions has been dropped and replaced with a new feature: What Do Sociologists See? As the reviewers noted, the Write a Caption features seemed too difficult for students to do and too difficult for professors to implement and assess. Now each module still ends with a photo, but there is an accompanying description detailing what sociologists might see when they gaze at the photograph. This new feature serves to further demonstrate and reinforce the sociological perspective as an interpretive guide to routine, and sometimes extraordinary, happenings in our lives and world around us.

Of course, there are also other changes. Those that I consider most significant are listed below by chapter and module.

Chapter 1: The Sociological Perspective

Module 1.1 (What Is Sociology?)—the module that opens the book—is for all practical purposes new. I revised this critical module thinking about the power of first impressions and that these first pages have the potential to set the tone for the class and the reading to come. Module 1.5 (Sociological Perspectives) and Module 1.6 (Methods of Research) are now organized around social robotics, or robots programmed to interact with humans. Most of us have encountered social robots when we have tried to carry on a conversation with a robotic telemarketer or phone tree operator. This topic is used as a vehicle to demonstrate how sociologists inspired by each of the four perspectives think about social robots (Module 1.5) and also to demonstrate how sociologists design a research study, in this case a research study about a cutting-edge societal transformation integrating robots in into the workplace.

Chapter 2: Culture

Module 2.6 (Applying Theory: Blue Jeans as Material Culture) focuses on how sociologists inspired by each of the four perspectives analyze what is arguably the most popular item of clothing on the planet.

Chapter 3: Socialization

Module 3.6 (Applying Sociology: Interactive Technologies as Agents of Socialization) considers what sociologists from each of the four perspectives would make of interactive digital technologies designed to allow children to imitate animated characters, to role play, and to engage in games. What lessons do these digital technologies convey about the self, its relationships to others, and its place in the world?

Chapter 4: Social Structures

Social structure is arguably sociology's signature concept. The concept directs sociologists to think about the largely invisible system that coordinates human activity. This chapter has been revised to showcase the power of social structure to shape and constrain interactions, relationships, and experiences but also to showcase the power of human agency to change social structures. Module 4.1 (Institutions and Social Structures), Module 4.2 (Levels of Social Structure), and Module 4.3 (Social Structure and Human Agency) seek to present this invisible system in all its levels and complexity and to showcase the analytic power of this concept to assess and change how human activity is organized. Module 4.9 (Applying Theory: The Social Structure of Nail Salons) applies the four perspectives to an analysis of the social structure of nail salons in the United States and how Vietnamese immigrants came to dominate this industry and shape the experience of going to a nail salon.

Chapter 5: The Social Construction of Reality

Module 5.1 (Definition of the Situation) has been revised to systematically describe the social dynamics that influence how people see and make sense of what is going on around them. This module places emphasis on the shared and learned "knowledge" people draw upon to create a reality upon which they act. Module 5.6 (Applying Theory: Language and Reality Construction) compares how each of the four perspectives presents the power of language to both constrain and empower thinking about our selves, others, and the larger society.

Chapter 6: Deviance

Module 6.8 (Applying Theory: Laws) reminds us that people who violate laws are not always "criminals." Each of the four perspectives alerts us to situations in which laws are enacted to control behavior that by any definition cannot be called criminal (at least in the popular way we think of criminals). As one example, feminists direct our attention toward laws that maintain and perpetuate gender ideals and inequalities and that regulate behavior and opportunities based on gender.

Chapter 7: Social Inequalities

Module 7.1 (Assigning Social Worth) and Module 7.4 (Unearned "Failures") now include discussions on unearned failures and unearned successes, both of which are derived from sources unrelated to individual merit or effort. Taken together these modules examine how people can become unemployed and earn poverty-level wages through no fault of their own (e.g., economic restructuring, a capitalist system that destroys as it creates, and an economy that depends on poverty-wage labor). Module 7.6 (Applying Theory: The World's Billionaires) asks: What does it mean to be one of the world's billionaires—one of 1,645 people in the world with this vast amount of wealth? We consider the answers sociologists from each of the four perspectives offer as explanations.

Chapter 8: Race and Ethnicity

Module 8.8 (Applying Theory: Racial Classification) examines how each of the four theoretical perspectives helps us think about the meaning and purpose of racial classification.

Chapter 9: Gender and Sexualities

Module 9.7 (Applying Theory: Sex Testing) gives attention to sex testing *in utero* and presents how sociologists from each of the theoretical perspectives see the purpose of sex testing.

Chapter 10: Economics and Politics

Chapter 10 places greater emphasis on how the economy and politics intertwine to shape job opportunities and income. Simply consider that governments enact tens of thousands of laws that affect income and wealth. Module 10.2 (The U.S. Economy and Jobs) includes a heavily revised analysis of the two-tier labor market, broadly polarized into privileged and disadvantaged workers. This module is organized around the long-standing forces supporting the two-tier system. Module 10.7 (Applying Theory: The Power and Reach of the U.S. Military) considers the U.S. military—the largest military in the world—from the point of view of each of the four perspectives.

Chapter 11: Families

Every module of the family chapter was revised with a focus on understanding why new forms of family and intimate relationships once thought of as odd, dysfunctional, or deviant are now experiencing some acceptance and even becoming accepted as "normal." In particular, it places new emphasis on social movements and demographic changes as vehicles of change ushering in the rise and increased visibility of new family forms. These social movements and demographic changes are responses to the challenges of our time (e.g., increased life expectancy, lower fertility, female empowerment, economic restructuring). From a sociological perspective, family is not unchanging or "static"; it is a dynamic response to shifting relational contexts.

Chapter 12: Education and Religion

Module 12.3 (Education in a Knowledge Economy) examines education in the context of revolutionary changes to the U.S. economy and the global economy of which it is a part. The new economy is knowledge-dominated and symbolized by smart technologies. This transformation raises questions about which school systems are best at preparing their students to compete in a knowledge economy. Module 12.7 (Applying Theory: Private Schools) considers how sociologists inspired by each of the four perspectives view private school education, specifically in what ways, if any, this educational experience advantages or disadvantages its students.

Chapter 13: Social Change and the Pressing Issues of Our Time

This chapter covers a variety of interrelated issues that are among the most compelling of our time, shaping the lives of every individual on the planet. Those issues relate to technology (Module 13.2), globalization (Module 13.3), social movements (Module 13.4), aging societies (Modules 13.5 and 13.6), the changing environment (Module 13.7), and health care (module 13.8). The Applying Theory module focuses attention on the criteria sociologists from each the four perspectives employ to evaluate a major change.

MindTap™: The Personal Learning Experience

MindTap Sociology for Ferrante's Seeing Sociology: An Introduction Powered by Knewton from Cengage Learning represents a new approach to a highly personalized, online learning platform. MindTap combines all of a student's learning tools—readings, multimedia, activities, and assessments—into a Learning Path that guides the student through the Introduction to Sociology course. Instructors personalize the experience by customizing the presentation of these learning tools to their students, even seamlessly introducing their own content into the Learning Path. Learn more at www.cengage.com/mindtap.

MindTap Sociology for Ferrante's Seeing Sociology: An Introduction Powered by Knewton is easy to use and saves instructors time by allowing them to:

- Customize the course—from tools to text—and make adjustments "on the fly," making it possible to intertwine breaking news into their lessons and incorporate today's teachable moments;
- Promote personalization by segmenting course content into movable objects, encourage interactivity, and ensure student engagement;
- Integrate multimedia assets, in-context exercises, and supplements; student engagement will increase, leading to better student outcomes;
- Track students' use, activities, and comprehension in real time, providing opportunities for early intervention to influence progress and outcomes;
- Assess knowledge throughout each section: after readings, during activities, homework, and quizzes; and
- Automatically grade homework and quizzes.

Ancillary Materials

I believe that a textbook is only as good as its supplements. For this reason, I have written the Test Bank, PowerPoint Slides, and Instructor's Manual, with assistance from Kristie Vise, my colleague at NKU, to accompany *Seeing Sociology*. We have tried to create ancillary materials that support the vision of this textbook.

Instructor's Manual

The Instructor's Manual includes standard offerings such as Learning Objectives video recommendations and activity suggestions a Key Terms glossary. It also includes sample answers to Critical Thinking questions: each module ends with a Critical Thinking question, the purpose of which is to get students to reflect on key ideas, concepts, and theories covered. Typically, the questions can be answered in 250 to 400 words. The Instructor's Manual includes a sample answer from an actual sociology student to each Critical Thinking question.The sample answer can serve as an example to share with the students as a way of stimulating thoughts about how to answer these questions. Instructors may also want to read sample answers as a way to prepare for questions students may have about them.

Test Bank

Like most textbooks, the ancillary materials for instructors include a Test Bank with multiple-choice and true-false questions. In addition to test questions about the textbook material, there are several multiple-choice questions relating to the short film clips. These questions can be found at the end of the multiple-choice questions for each chapter and are labeled by topic ("TOP").

PowerPoint® Lecture Slides

These slides highlight key ideas and points covered in each module. They are useful if instructors want to give students a quick overview of material covered or post online as a review.

Cengage Learning Testing Powered by Cognero®

Cengage Learning Testing Powered by Cognero is a flexible online system that allows instructors to write, edit, and manage Test Bank content and quickly create multiple test versions. You can deliver tests from your LMS, your classroom—or wherever you want.

Acknowledgments

The acknowledgment section—the place to recognize and give credit to those who have influenced the ideas in this book and its creation—is the most difficult part of the book to write. I have always struggled to find the words to capture the essence and depth of the various relationships that are special to my intellectual and, by extension, personal life. I find myself using clichés like "This book could not have been written without . . . ," "I wish to extend my deepest gratitude to . . . ," "I acknowledge the profound influence of . . . ," and so on. I am never satisfied with the words I use to convey the colloboration I value so highly. Here I will simply state the names and collaborations for which I am most thankful and leave it at that.

Chris Caldeira, my former editor and now a graduate student at the University of California–Davis, conceived the book's structure and approach. She is the lead photographer, contributing 120 photographs to this edition, and she is the person with whom I talk most about this book. Her role is so large that her intellectual and photographic contributions are acknowledged on the title page of this text.

Phillip (deceased) and Annalee Ferrante, my parents, whom I most admire for their work ethic, their optimism and perseverance in the face of difficulties, and their belief that the best effort matters.

Missy Gish, who manages the overwhelming number of details associated with writing a textbook and preparing it for production, including taking 67 photographs.

Robert K. Wallace, my husband and colleague, who offers unwavering support.

There are also the colleagues and students (former and current) who contributed one or more photographs to this edition. They include Leslie Ackerson (5), Tabitha Adams (1), Prince Brown, Jr. (1), Katie Caputo (2), Rachel Ellison (27), Katie Englert (2), Jeremiah Evans (2), Rudy Garns (20), Mark Gish (1), Aleena Ferrante (5), Sharyn Jones (21), Boni Li (5), Tony Rotundo (14), Billy Santos (1), Terra Schultz (1), Lisa Southwick (25), Tom Zaniello (1) Jibril McCaster (2), and the Asia and Pacific Transgender Network (1). It is important to note that photos set in Fiji taken by Sharyn Jones are based upon work supported by the National Science Foundation under Grant No. 1156479 awarded to Dr. Sharyn Jones. I must also mention that photos taken by Rachel Ellison were funded by a Northern Kentucky University Undergraduate Research Grant. I would like to thank four students who have written many of the sample critical thinking question responses that are included in the instructor's manual. Those students are Joshua Blackaby, Caitlin Harrah, Jibril McCaster, and Meredith Sparkes.

Behind the scenes there is a team of people who worked to make this book a reality. You can find their names listed in an unassuming manner on the copyright page of this book. As one measure of the human effort expended, consider that there were dozens of people reading, copyediting, designing, and proofing the pages of the book for at least six months before it reached the market. All of this human effort was coordinated by Jill Traut, the project manager, who I have had the pleasure of working with on this and other editions of my text. I stand in awe of the seeming ease by which she manages all the details to guide a book through production to press.

I dedicate this book to Dr. Horatio C. Wood IV, MD. Our relationship goes back to my days in graduate school. Over the decades I have always made a point of formally acknowledging the tremendous influence he has had on my intellectual life, academic career, and philosophy of education. Dr. Wood died on May 28, 2009, but his influence remains as important and strong as ever today.

Reviewers

I am grateful to the many colleagues from universities and colleges across the United States who reviewed one or more editions of this text and provided insightful comments: Annette Allen, Troy University; Andre Arceneaux, Saint Louis University; Arnold Arluke, Northeastern University; Aurora Bautista, Bunker Hill Community College; David P. Caddell, Mount Vernon Nazarene University; Gregg Carter, Bryant University; Andrew Cho, Tacoma Community College; Mirelle Cohen, Olympic College; Pamela Cooper, Santa Monica College; Janet Cosbey, Eastern University; Gayle D'Andrea, J. Sargeant Reynolds Community College; Karen Dawes, Wake Technical Community College; Kay Decker, Northwestern Oklahoma State University; Melanie Deffendall, Delgado Community College; David Dickens, University of Nevada–Las Vegas; Dennis Downey, University of Utah; Angela Durante, Lewis University; Keith Durkin, Ohio Northern University; Murray A. Fortner, Tarrant County College; Matt Gregory, Tufts University and Boston College; Derrick Griffey, Gadsden State Community College; Laura Gruntmeir, Redlands Community College; Kellie J. Hagewen, University of Nebraska; Anna Hall, Delgado Community College;

Laura Hansen, University of Massachusetts; James Harris, Dallas Community College; Garrison Henderson, Tarrant County College; Melissa Holtzman, Ball State University; Xuemei Hu, Union County College; Jeanne Humble, Bluegrass Community and Technical College; Allan Hunchuk, Thiel College; Hui M. Huo, Highline Community College; Faye Jones, Mississippi Gulf Coast Community College; Rachel Tolbert Kimbro, Rice University; Philip Lewis, Queens College; Carolyn Liebler, University of Minnesota; Beth Mabry, Indiana University of Pennsylvania; Gerardo Marti, Davidson College; Tina Martinez, Blue Mountain Community College; Donna Maurer, University of Maryland, University College; Marcella Mazzarelli, Massachusetts Bay Community College; Jeff McAlpin, Northwestern Oklahoma State University; Douglas McConatha, West Chester University of Pennsylvania; Janis McCoy, Itawamba Community College; Elizabeth McEneaney, California State University–Long Beach; Melinda Messineo, Ball State University; Arman Mgeryan, Los Angeles Pierce College; Cathy Miller, Minneapolis Community and Technical College; Krista Lynn Minnotte, Utah State University; Lisa Speicher Muñoz, Hawkeye Community College; Elizabeth Pare, Wayne State University; Denise Reiling, Eastern Michigan University; Robert Reynolds, Weber State University; Judith Richlin-Klonsky, Santa Rosa Junior College; Lisa Riley, Creighton University; Luis Salinas, University of Houston; Alan Spector, Purdue University–Calumet; Rose A. Suggett, Southeast Community College; Don Stewart, University of Nevada–Las Vegas; Toby Ten Eyck, Michigan State University; Katherine Trelstad, Bellevue College; Sharon Wettengel, Tarrant Community College; Robert Wood, Rutgers University; and James L. Wright, Chattanooga State Technical Community College.

I also thank the focus group participants whose valuable feedback helped us shape this book: Ghyasuddin Ahmed, Virginia State University; Rob Benford, Southern Illinois University; Ralph Brown, Brigham Young University; Tawny Brown-Warren, Columbia College Online; Kay Coder, Richland College; Jodi Cohen, Bridgewater State College; Sharon Cullity, California State University, San Marcos; Anne Eisenberg, State University of New York–Geneseo; Dana Fenton, Lehman College; Rhonda Fisher, Drake University; Lara Foley, University of Tulsa; Glenn Goodwin, University of La Verne; Rebecca Hatch, Mt. San Antonio College; Idolina Hernandez, Lone Star College–CyFair; Joceyln Hollander, University of Oregon; Jennifer Holsinger, Whitworth University; Amy Holzgang, Cerritos College; Michelle Inderbitzin, Oregon State University; Mike Itashiki, Collin County Community College; Greg Jacobs, Dallas Community College District; Kevin Lamarr James, Indiana University–South Bend; Krista Jenkins, Fontbonne University; Elizabeth Jenner, Gustavus Adolphus College; Art Jipson, University of Dayton; Maksim Kokushkin, University of Missouri–Columbia; Amy Lane, University of Missouri–Columbia; Marci Littlefield, Indiana University–Purdue University, Indianapolis; Belinda Lum, University of San Diego; Ali Akbar Mahdi, Ohio Wesleyan University; Tina Martinez, Blue Mountain Community College; Lori Maida, State University of New York, Westchester Community College; Linda McCarthy, Greenfield Community College; Richard McCarthy and Elizabeth McEneaney, California State University–Long Beach; Julianne McNalley, Pacific Lutheran University; Krista McQueeney, Salem College; Angela Mertig, Middle Tennessee State University; Melinda Miceli, University of Hartford; Anna Muraco, Loyola Marymount University; Aurea Osgood, Winona State University; Rebecca Plante, Ithaca College; Dwaine Plaza, Oregon State University; Jennifer Raymond, Bridgewater State College; Cynthia Reed, Tarrant County College–Northeast; Nicholas Rowland, Pennsylvania

State University; Michael Ryan, Dodge City Community College; Martin Sheumaker, Southern Illinois University–Carbondale; Carlene Sipma-Dysico, North Central College; Juyeon Son, University of Wisconsin–Oshkosh; Kathy Stolley, Virginia Wesleyan College; Stacie Stoutmeyer, North Central Texas College, Corinth Campus; Ann Strahm, California State University–Stanislaus; Carrie Summers-Nomura, Clackamas Community College; Zaynep Tufekci, University of Maryland; Deidre Tyler, Salt Lake Community College; Georgie Ann Weatherby, Gonzaga College; Sharon Wettangel, Tarrant County College; Rowan Wolf, Portland Community College; James Wood, Dallas County Community College; LaQueta Wright, Dallas Community College District; and Lori Zottarelli, Texas Woman's University.

I am grateful to the student focus group participants who provided helpful input: Greg Arney, Pauline Barr, Wesley Chiu, John Liolos, Jessica St. Louis, Fengyi (Andy) Tang, and Rojay Wagner.

I also appreciate the useful feedback of the survey participants: Mike Abel, Brigham Young University– Idaho; Wed Abercrombie, Midlands Technical College–Airport; Dwight Adams, Salt Lake Community College; Chris Adamski-Mietus, Western Illinois University; Isaac Addaii, Lansing Community College; Pat Allen, Los Angeles Valley College; Robert Aponte, Indiana University–Purdue University, Indianapolis; David Arizmendi, South Texas College–McAllen; Yvonne Barry, John Tyler Community College; Nancy Bartkowski, Kalamazoo Oakley, University of Cincinnati; Kirsten Olsen, Anoka Ramsey Community College; Roby Page, Radford University; Richard Perry, Wake Technical Community College–Raleigh; Kenya Pierce, College of Southern Nevada–Cheyenne; Sarah Pitcher, San Diego City College; Cynthia Reed, Tarrant County College–Northeast; Paul Renger, Saint Philip's College; Melissa Rifino-Juarez, Rio Hondo College; Barbara Ryan, Widener University; Rita Sakitt, Suffolk Community College–Ammerman; Luis Salinas, University of Houston; Robert Saute, William Paterson University; Terri Slonaker, San Antonio College; Kay Snyder, Indiana University of Pennsylvania; Julie Song, Chaffey College; Andrew Spivak, University of Nevada–Las Vegas; Kathleen Stanley, Oregon State University; Rachel Stehle, Cuyahoga Community College–Western; Susan St. John-Jarvis, Corning Community College; Tanja St. Pierre, Pennsylvania State University–State College; Rose Suggett, Southeast Community College–Lincoln; Becky Trigg, University of Alabama–Birmingham; Tim Tuinstra, Kalamazoo Valley Community College; David Van Aken, Hudson Valley Community College; Vu Duc Vuong, De Anza College; Tricia Lynn Wachtendorf, University of Delaware; Kristen Wallingford, Wake Technical Community College–Raleigh; Margaret Weinberg, Bowling Green State University; Robyn White, Cuyahoga Community College; Beate Wilson, Western Illinois University; Sue Wortmann, University of Nebraska–Lincoln; Bonnie Wright, Ferris State University; Sue Wright, Eastern Washington University; Delores Wunder, College of DuPage; and Meifang Zhang, Midlands Technical College–Beltline.

Theme Index

INDEX OF GLOBALIZATION/GLOBAL INTERDEPENDENCE COVERAGE

INDEX OF DIVERSITY

INDEX OF TECHNOLOGY COVERAGE

Seeing Sociology

AN INTRODUCTION

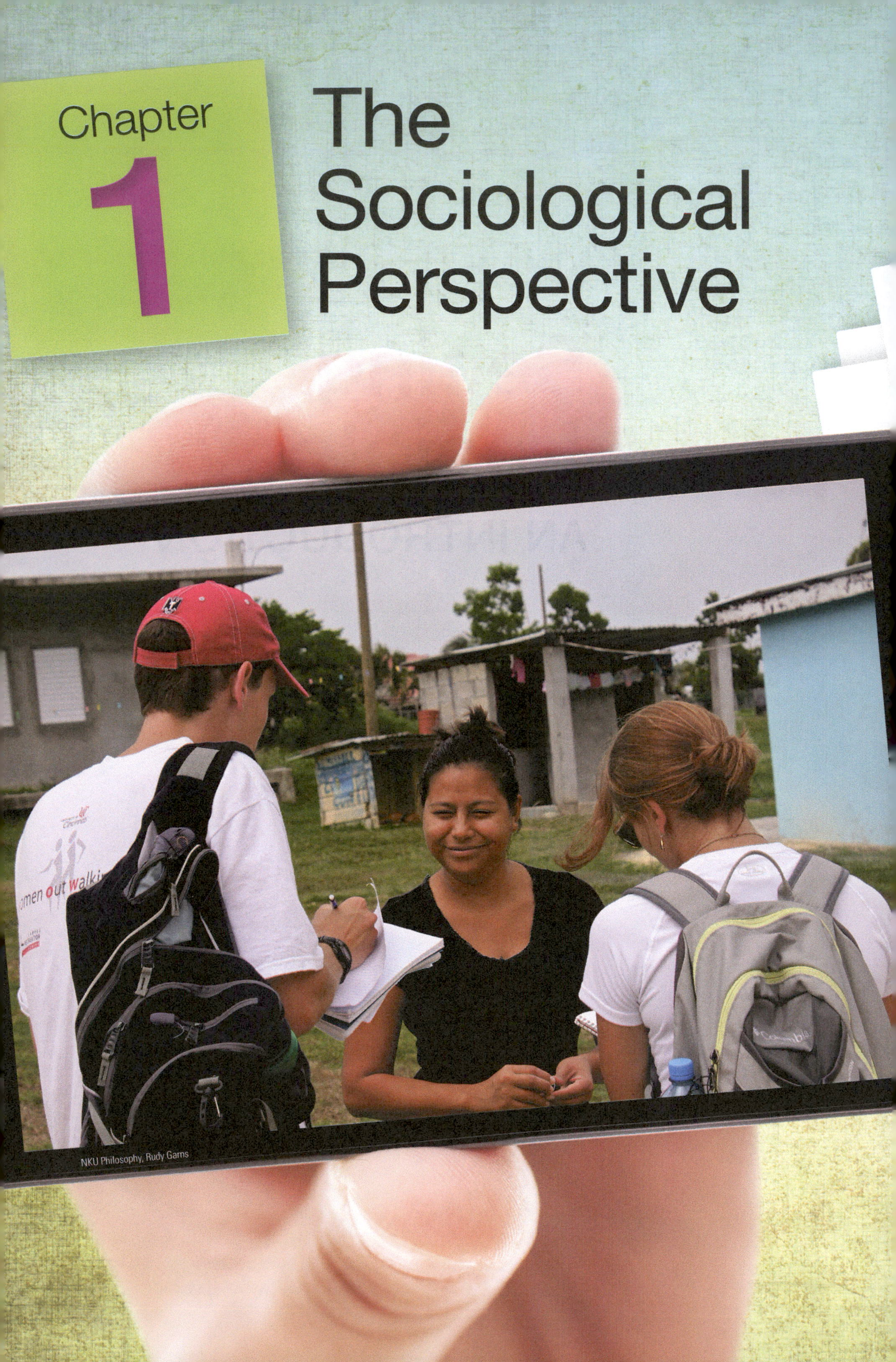

NKU Philosophy, Rudy Garns

Sociology is a field of study that invites you to see the world around you in new ways, to be open to new experiences, to be curious about what is taking place around you, and to wonder and care about those who live nearby and beyond. The sociological perspective allows you to see how the time you are born in history, the place you live, and how countless numbers of people known and unknown, living and dead, profoundly shape what you think and do. As a student of sociology, you will come to understand that "things are not what they seem" (Berger 1963, 21).

Module
1.1 What Is Sociology?

Objective

You will learn that sociologists focus their attention on the social forces that shape the way people think, interact, and organize activities.

Do you ever wonder what kind of person you would be if you grew up in another place or at another time? If you answered yes to this question, then you will most certainly appreciate the sociological perspective.

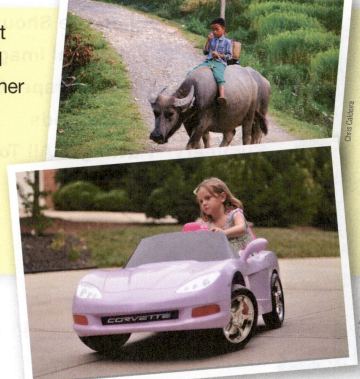

Chris Caldeira

Lisa Southwick

Sociology: A Definition

Sociology is the systematic and scientific study of society. To put it another way, sociologists work to understand how human activities, including social interactions and relationships, are organized, with a goal of analyzing how that organization affects people's lives, thinking, and responses to others and to the world around them. Simply consider the two photos above. Each depicts one of many ways children's play can be organized. The little boy on the water buffalo lives in rural Vietnam and the little girl in the motorized car lives in a suburb in North Carolina. There is no doubt that each child's sense of self, thought, and behavior is profoundly shaped in different ways by the way each is playing.

As a second example, consider that the way people organize food production affects what people eat, the way they eat, how they think about food, and the ways they relate to others to secure meals.

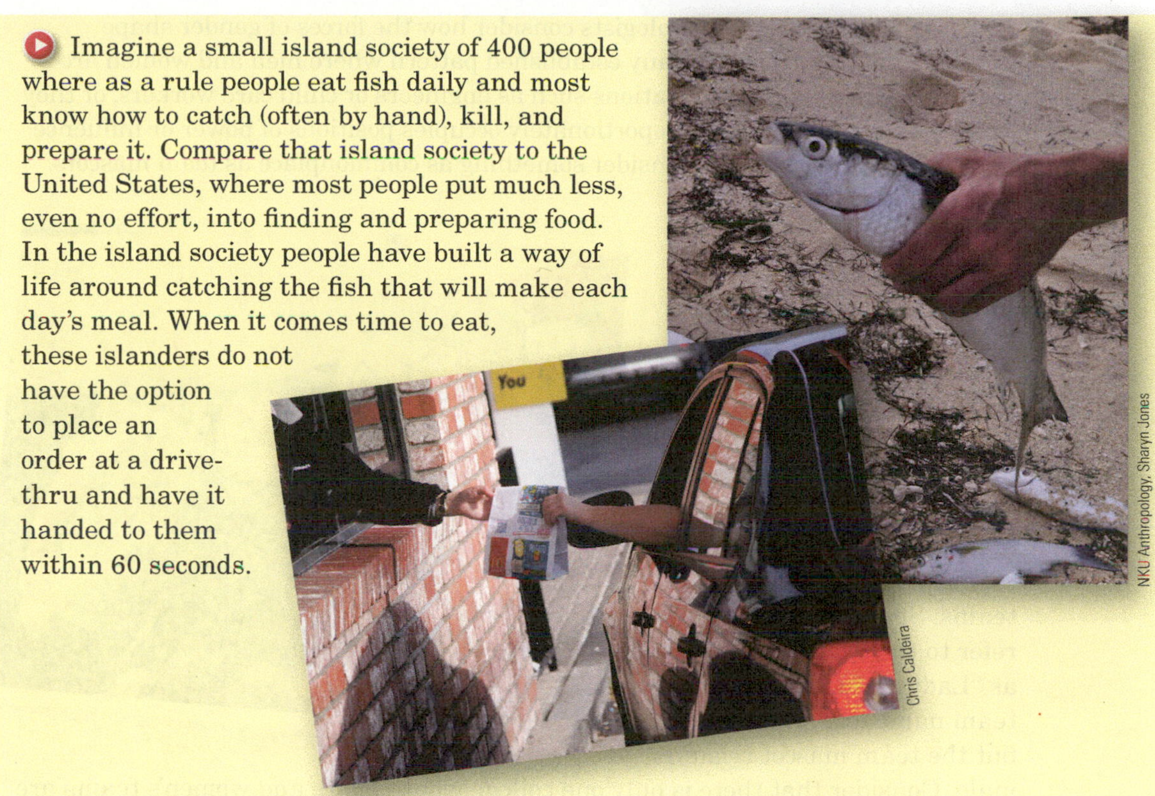

Imagine a small island society of 400 people where as a rule people eat fish daily and most know how to catch (often by hand), kill, and prepare it. Compare that island society to the United States, where most people put much less, even no effort, into finding and preparing food. In the island society people have built a way of life around catching the fish that will make each day's meal. When it comes time to eat, these islanders do not have the option to place an order at a drive-thru and have it handed to them within 60 seconds.

NKU Anthropology, Sharyn Jones

Chris Caldeira

Sociologists are especially interested in identifying and understanding the social forces that shape the ways people organize activities, whether it be to secure food, engage in play, form friendships, earn a livelihood, or anything else. For example, in explaining how an island lifestyle is organized around catching fish, sociologists would certainly consider as critical forces the geographic remoteness of the island, the surrounding waters where fish are plentiful, and small islands' marginalized status in the world economy.

This textbook introduces readers to some of the social forces that shape our lives and relationships to people around us and beyond. Those forces relate to culture, socialization, family, technology, religion, education, race, gender, class, the economy, and much more.

The culture of which we are a part gives us a language that acts as a social force broadly shaping what we think and how we convey meaning. Most English-speaking people living in the United States tend to think of this bird as singing, whereas most Korean-speaking people living in Korea think of it as weeping. For those who are Korean, the language elicits feelings of sadness; for English-speaking Americans, the feeling elicited is joy.

NKU Philosophy, Rudy Garns

As a final example, when sociologists consider how the forces of gender shape human activity, they look for any established pattern where men and women are segregated into distinct occupations such as engineers or child care workers, or they look to see if one gender disproportionately occupies positions of power or influence relative to another gender. Consider something as commonplace as team mascots.

NKU Philosophy, Rudy Garns

Sociologists maintain it is no accident that these little girls are posing with a mascot considered male. Virtually every high school, college, and professional team in the United States employs a male-appearing mascot to represent its teams, even the women's teams. Some schools may refer to their female athletes as "Lady . . ." followed by the team name (e.g., Lady Norse), but the team mascot remains male. Consider that there is only one college whose men's and women's teams are represented by a mascot considered female (the Fightin' Blue Hens of Delaware).

Why Study Sociology

In time you will learn that sociology offers a framework for analyzing the social forces that shape the way life around us is organized. That framework allows sociologists to think about that organization in terms of (1) the shared, often competing, meanings held by those involved; (2) anticipated and unanticipated consequences on thought, interaction, and relationships; and (3) patterns of inequality. This sociological framework is especially relevant today, if only because the most pressing issues we face are rooted in the ways in which activities have been organized. Change agents understand that there are many ways to organize human activities and that each way has advantages and disadvantages. Change agents also recognize when the way something is organized needs to change and they take the necessary steps to make change.

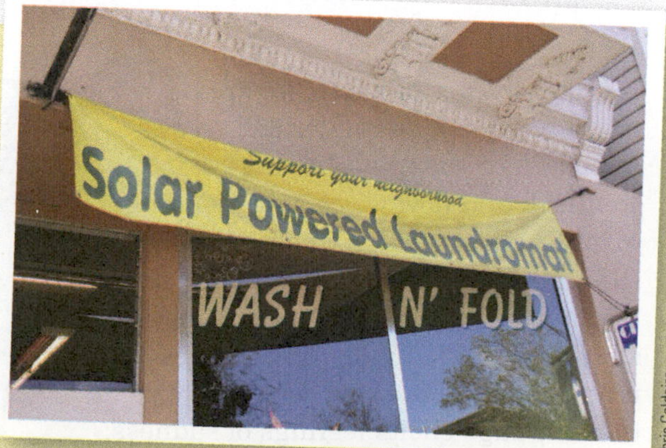

Chris Caldeira

One might see the owners of this Laundromat as change agents because they have created a solar-powered, instead of a fossil fuel-powered service. No doubt those millions of people worldwide who earn a living extracting, refining, delivering, selling, and profiting from fossil fuels will find fault with solar power. Regardless, the owners recognize that the current way we use energy is unsustainable and has to change, and they have chosen to be part of that change.

The sociological perspective supports social innovation. To be a social innovator one must (1) grasp how current ways of organizing our lives are problematic, (2) recognize barriers to change, (3) think carefully about how to make change, and (4) be ready for change by anticipating the advantages and disadvantages change can bring. Two examples of how social innovators address organizational shortcomings follow.

- Many high-skilled women who care for children, elderly, and disabled need to work but only have time to work part-time. Part-time work, however, tends to be low-skilled and low-paying, with few benefits. As a result, many high-skilled women accept employment below their ability and worth. In doing so they unwittingly fill the only kinds of jobs low-skilled applicants are qualified for. A nonprofit, Women Like Us, partners with employers and recruiting agencies to match this hidden pool of high-skilled talent with employers who need employees to fill higher-paying and skilled part-time positions. Each day, Women Like Us posts 180 such jobs for women to review (European Social Innovation Competition 2014).
- At least 22 cities in the United States have instituted innovation in transportation by offering fare-free public transportation. The key to successful fare-free transportation is to offer high-quality service to attract riders of all social classes. Chapel Hill, North Carolina, did just that. In the 10 years following a decision to make public transportation fare-free for all, ridership increased from 3 million to 7 million passengers a year. As a result, the city's highways are less crowded, pollution was reduced, and all workers (but especially low-income workers) have reliable transportation to work (Jaffe 2013).

What Do Sociologists See?

Sociologists see this advertisement as one business owner's effort to respond to, even resist, the social forces ushering a dramatic change in how people share information. As the percentage of the population using digital forms of communication increases, a corresponding decrease in customers who print copies of photos and other documents is inevitable. This sign represents a desperate effort to preserve paper in an increasingly paperless environment.

Chris Caldeira

Critical Thinking

Write about some social force that shapes your life and relationships to people around you and beyond.

Key Term

sociology

Module 1.2 The Emergence of Sociology

Objective

You will learn about a major historical event that triggered the birth of sociology.

NKU Anthropology, Sharyn Jones

Chris Caldeira

What would it be like to live in a society where there are no machines to carry loads or power boats and other vehicles?

Throughout much of history, human and animal muscle were the key sources of power. The key source of power changed to fossil fuels with the Industrial Revolution, the name given to the dramatic changes in manufacturing, agriculture, transportation, and mining that transformed virtually every aspect of society from the 1300s on. The defining feature of the Industrial Revolution was mechanization, the process of replacing human and animal muscle with machines powered by burning wood and fossil fuels (e.g., coal, oil, and natural gas). The new energy sources eventually replaced hand tools with power tools, sailboats with steamships and then freighters, and horse-drawn carriages with trains. Mechanization changed how goods were produced and how people worked. It turned workshops into factories, skilled workers into machine operators, and handmade goods into machine-made products.

Library of Congress Prints and Photographs Division[LC-DIG-nclc-05247]

Consider the effort required to make bread before mechanization. Bakers plunged their fists into gluey dough and massaged it with their fingers until their muscles hurt (Zuboff 1988).

People also took their dough to small local bakeries, where it was shaped and baked in wood- or coal-heated brick ovens. This baker and his apprentice used long-handled wooden shovels to move bread in and out of the oven (Advameg, Inc. 2007).

Library of Congress Prints and Photographs Division [LC-DIG-nclc-05178]

Library of Congress Prints and Photographs Division[LC-USF34-035069-D]

With mechanization, the effort workers once exerted to make bread was largely eliminated. Moreover, bakers no longer spent seven or more years in apprenticeships. Now people with little or no skill could do the skilled baker's work, but at a faster pace. Before mechanization, customers knew the person who baked their bread. With mechanization, they came to depend on "strangers" to sustain them.

Changes to Society

Bread baking eventually moved out of the home and small bakery shops, and by the 1940s commercial bakeries were stocking grocery shelves. While this may seem unimportant, it is just one example of the way the Industrial Revolution weakened people's ties to others in their community, their workplace, and their home. Other innovations (social forces) that changed the ways people related to each other included the railroad, the steamship, running water, central heating, electricity, and the telegraph. Month-long trips by stagecoach, for example, became daylong trips by coal-powered trains. These trains permitted people and goods to travel day and night—in rain, snow, or sleet—and to previously unconnected areas.

The railroad and other subsequent innovations in transportation (airplane, freighter, trucks) facilitated economic competition and interdependence. Now people in one area could be priced out of a livelihood if people in another area could provide lower-cost labor, goods, and/or materials (Gordon 1989). Bread, for example, could now be made outside the community and then shipped long distances for sale to strangers. The nature of bread also changed to accommodate this new reality; over time, dozens of additives gave commercial bread a standard texture, shape, taste, and most importantly, a shelf life that allowed it to be shipped long distances and sit on a store shelf for weeks.

The Industrial Revolution, centered in Europe and the United States, also pulled together, often by force, people from even the most remote parts of the planet into a global-scale division of labor. People who did not know one another became interconnected and dependent upon one another. (Tire makers in the United States depended on rubber tappers in Brazil to supply the material they needed to produce tires.) In sum, the Industrial Revolution changed everything including how goods were produced, relationships, the kinds of job people held, the density of human populations, and the importance and influence of the home. It ushered in a consumption-driven economy and made the accumulation of wealth a valued pursuit. In *The Wealth of Nations,* Adam Smith (1776) argued that the invisible hand of the free market, via private ownership and self-interested competition, held the key to progress and prosperity. These unprecedented changes caught the attention of the early sociologists. In fact, sociology emerged out of their effort to understand the effects of the Industrial Revolution on society.

What Do Sociologists See?

Sociologists see mechanization—specifically, they see a power saw replacing axes "powered" by human muscle. Mechanization substantially reduced the labor once needed to cut wood (saving cost and time). Now more trees could be sawed into wood at a faster pace. In addition, mechanization brought new health risks (severed fingers and limbs) and it also brought an "uncomplaining" machine that could run 24/7 without need of lunch or bathroom breaks.

Library of Congress Prints and Photographs Division[LC-USE6-D-001121]

Critical Thinking

Look around your home and take an inventory of some of the products in your life made by the labor of those working in a foreign country.

Module 1.3 Standing on the Shoulders of Giants

Objective

You will learn about the ideas of six early theorists who are considered the giants of the discipline.

Auguste Comte — © DIZ Muenchen GmbH, Sueddeutsche Zeitung Photo/Alamy

Karl Marx — Prints & Photographs Division, Library of Congress. LC-USZ62-16530

Émile Durkheim — French Photographer/The London Art Archive/Alamy

Max Weber — akg-images/Newscom

W.E.B. DuBois — George Grantham Bain Collection, Prints & Photographs Division, Library of Congress, LC-DIG-ggbain-07435

Jane Addams — George Grantham Bain Collection, Prints & Photographs Division, Library of Congress, LC-DIG-ggbain-12065

Have you ever read something that changed the way you look at the world? If so you will understand why the six people pictured are celebrated as "giants" in the field of sociology.

In this module we consider the transforming ideas of six early theorists. Three of the six—Karl Marx, Émile Durkheim, and Max Weber—are nicknamed the "big three" because their writings form the heart of the discipline. We also consider three other central figures: Auguste Comte because he gave sociology its name, Jane Addams for her efforts to apply sociological knowledge to change people's lives, and W.E.B. DuBois for his thoughts on the color line.

Auguste Comte (1798–1857)

The French philosopher Auguste Comte gave sociology its name in 1839. Comte argued that sociology is a science and that only those sociologists who follow the scientific method should expect to have a voice in guiding human affairs. The scientific method (also known as **positivism**) rejects personal opinion and political agendas as a basis for analysis and encourages disciplined thought and objective research. Comte identified four methods to guide analysis of human activities: (1) direct observation, (2) experimental design, (3) comparative analysis, and (4) historical research (looking at the past to understand the present).

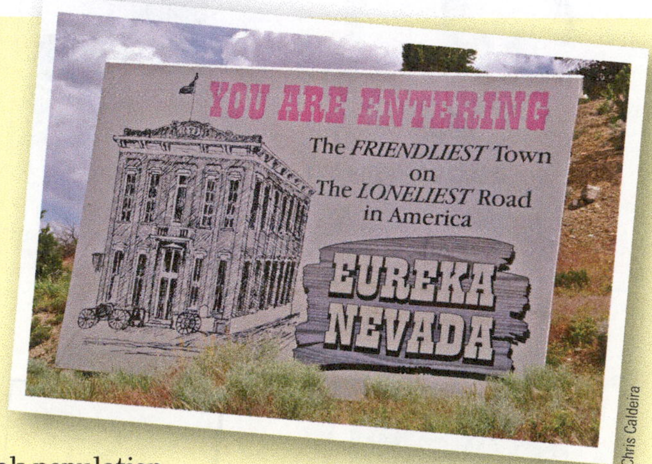

Historical research, one of the methods of analysis Comte recommended, can be applied to Eureka, Nevada. This method of inquiry directs us to consider how Eureka's past informs the town's present and future. Eureka is the friendliest town on the loneliest road for a reason. Eureka was established as a mining town in 1864. It reached its peak population in 1878 (10,000 pop.), and today is home to about a thousand people. From the beginning, the town's identity centered on global demand for its gold, silver, and lead. As the value of those ores declined in the marketplace, the residents in Eureka and surrounding towns moved away. Eureka's identity still centers on its now-closed mines as tourist attractions and its location along the "Loneliest Road in America."

Karl Marx (1818–1883)

Karl Marx was born in Germany but spent much of his professional life in London working and writing in collaboration with Friedrich Engels. He is best known for *The Communist Manifesto*, a 23-page pamphlet issued in 1848 and translated into more than 30 languages (Marcus 1998). Upon reading it today, one is "struck by the eerie way in which its 1848 description of capitalism resembles the restless, anxious and competitive world of today's global economy" (Lewis 1998, A17).

Marx made it his mission to analyze and explain **conflict**, the major force that drives social change. Specifically, Marx saw **class conflict** as the vehicle that propelled society from one historical epoch to another. He described class conflict as an antagonism growing out of the opposing interests held by exploiting and exploited classes. The nature of that conflict is shaped by the relationship of each class to the **means of production**—the resources such as land, tools, equipment, factories, transportation, and labor that are essential to the production and distribution of goods and services. To illustrate: Marx maintained that the Industrial Revolution was accompanied by the rise of two distinct classes: the **bourgeoisie**, the owners of the means of production; and the **proletariat**, those individuals who must sell their labor to the bourgeoisie. The bourgeoisie have an interest in making a profit. To maximize profit, the bourgeoisie seek to employ the lowest-cost workers, to find the cheapest materials to make products, and to create technologies that replace human labor. The bourgeoisie's profit-making goals clash with the great interest of the proletariat, which is to increase their wages.

Marx believed that this drive to make a profit drove the explosion of technological innovation and the production for which the Industrial Revolution is known. Marx felt that capitalism was the first economic system capable of maximizing human ingenuity and productive potential. He also maintained that capitalism ignored too many human needs, that too many workers could not afford to buy the products of their labor, and that relentless efforts to reduce labor costs left the worker vulnerable and insecure. Marx called the drive for profit a "boundless thirst—a werewolf-like hunger—that takes no account of the health and the length of life of the worker unless society forces it to do so" (Marx and Engels 1887, 142). Marx argued that if this economic system were in the right hands—the hands of the workers, or the proletariat—public wealth would be abundant and distributed according to need.

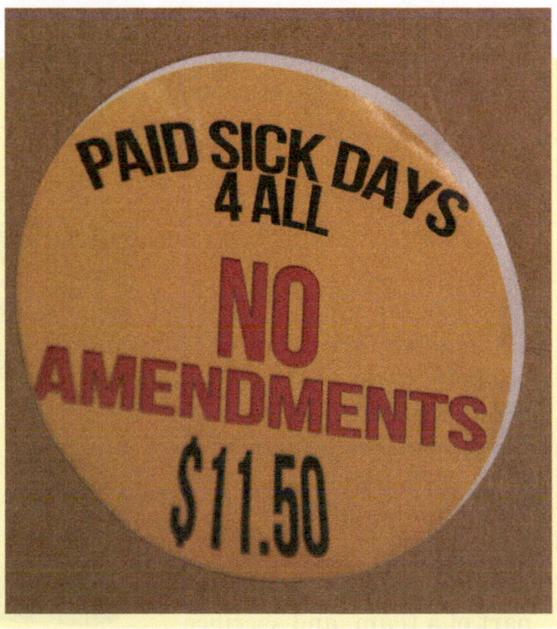

The source of class conflict for Marx was between employers' interests in keeping wages low and the workers' interests in increasing wages. This campaign button speaks to the clash between the two classes: workers' demand for paid sick days and at least $11.50 per hour clashes with employers' drive to increase profits.

Émile Durkheim (1858–1918)

To describe the Industrial Revolution and its effects, the Frenchman Émile Durkheim focused on the division of labor and solidarity. Durkheim noticed that the Industrial Revolution had a profound effect on the division of labor, or the way a society divided up tasks needed to make a product or to accomplish some other goal. Durkheim was particularly interested in how this change affected

solidarity, the system of social ties that acts as a cement bonding people to one another and to the wider society. Durkheim observed that the nature of solidarity had been changed from mechanical to organic.

Preindustrial societies are characterized by **mechanical solidarity**, a system of social ties based on uniform thinking and behavior. Durkheim believed that uniformity is common in societies with a simple division of labor, where just about everyone knows how to perform tasks critical to support the group's way of life. This shared knowledge gives rise to common experiences, skills, and beliefs. In preindustrial societies, religion and family are extremely important, and the social ties that bind people to each other are grounded in tradition, obligation, and duty. These societies do not have the technology or resources to mass produce a variety of products that people can buy to distinguish themselves from others.

The Industrial Revolution ushered in a complex division of labor where the workers needed to manufacture a product did not have to know or live near one another; in fact, they could live in different parts of a country or the world. In addition, the materials needed to make products came from many locations around the globe. This new way of dividing labor gave rise to **organic solidarity**, a system of social ties in which people became known for their specialized role in the division of labor. Under this new arrangement, few people possessed the knowledge, skills, and resources to be self-sufficient. Consequently, social ties are still strong, not because people know one another, but because they came to depend on strangers to survive. In industrial societies most day-to-day interactions are short-lived, impersonal, and instrumental (i.e., we interact with strangers for a specific reason). For example, we interact with others as "customers" or "clients." Customers can buy tires from any dealer, travel in airplanes flown by a pilot they might never see, and drink coffee made from beans harvested and roasted by people they've never met.

NKU Philosophy, Rudy Garns

The solidarity that binds team members together is mechanical—the players are united by a shared purpose, know each other on and off the court very well, wear uniforms to show they are part of a team, and sacrifice individual desires to achieve success. Still the players are connected to a global division of labor because they depend on strangers to, among other things, build the stadiums in which they play and to make their uniforms and shoes. Simply consider that by one estimate it takes 200 pairs of hands to make and deliver shoes to players' feet (Chang 2009, 98). The solidarity that binds the players to these "strangers" is organic.

Max Weber (1864–1920)

The German scholar Max Weber made it his task to analyze and explain how the Industrial Revolution affected **social action**—actions people take in response to others—with emphasis on the larger forces and settings that motivate people to act in certain ways. Weber suggested that sociologists should focus on the meanings guiding thought and action. He believed that social action is motivated in one of four ways. In reality, motives are not so clear-cut but involve some mixture of the four.

1. *Traditional*—a goal is pursued because it was pursued in the past (i.e., "that is the way it has always been").
2. *Affectional*—a goal is pursued in response to an emotion such as revenge, love, or loyalty (a soldier throws him- or herself on a grenade out of love and sense of duty for those in the unit).
3. *Value-rational*—a desired goal is pursued with a deep and abiding awareness that there can be no shortcuts or compromises made in reaching it. Instead, the actions taken to reach a desired goal are guided by a set of standards or codes of conduct (Weintraub and Soares 2005).
4. *Instrumental rational*—a valued goal is pursued by the most efficient means, irrespective of the consequences. Since the Industrial Revolution ushered in a system with profit-making as THE valued goal, instrumental rational action was rewarded. Instrumental rational action, with its focus on efficiency at any cost, leaves no time to adhere to a code of conduct, to adhere to tradition, or to feel affection. One might equate instrumental rational action with the behavior of an addict who will seek a drug at any cost to self or to others. There is an inevitable self-destructive quality to this form of action (Henrik 2000).

Weber believed that instrumental rational action could lead to **disenchantment**, a great spiritual void accompanied by a crisis of meaning in which the natural world becomes less mysterious and revered and becomes the object of human control and manipulation. The industrial model for raising chickens for human consumption applies. Note that disenchantment results when the goal of profit outweighs any moral responsibility to treat animals with kindness and when the means used to turn a profit are such that "we no longer recognize the animals in a factory farm as living creatures capable of feeling pain and fear" (Angier 2002, 9).

Regis Lefebure, USDA

Disenchantment is an outcome of instrumental rational action. This kind of action is embodied in the way factory farms raise pigs for eventual slaughter. Pigs are destined to live in a space where there may not be enough room to turn around. This treatment is the result of the most extreme cost-containing measures, with the goal of making a profit. It ignores any code of good animal husbandry involving an obligation to care for animals' well-being by providing food, protection, and shelter and the chance to be "pigs" that can move around freely (Scully 2002).

W.E.B. DuBois (1868–1963)

A voice that was initially ignored but later "discovered" as important to sociology is that of American-born W.E.B. DuBois. In trying to describe the Industrial Revolution and its effects on society, DuBois offered the concept of the **color line**, a barrier supported by customs and laws separating nonwhites from whites, especially with regard to their roles in the division of labor. That is, the positions considered high status were reserved for whites. DuBois (1919 [1970]) traced the color line's origin to the scramble for Africa's resources, beginning with the slave trade, upon which the British Empire and American Republic were built, costing black Africa "no less than 100,000,000 souls" (246). The end of the slave trade was followed by the colonial expansion that accompanied the Industrial Revolution. That expansion involved rival European powers competing to secure labor and natural resources. By 1914 virtually all of Africa had been divided into European colonies. DuBois maintained that the world was able "to endure this horrible tragedy by deliberately stopping its ears and changing the subject in conversation" (246). He felt that an honest review of Africa's history could only show that Western governments and corporations coveted Africa for its natural resources and for the cheap labor needed to extract them. The color line reflects the deep social divisions between Europeans and Africans that were solidified by the slave trade and by colonization.

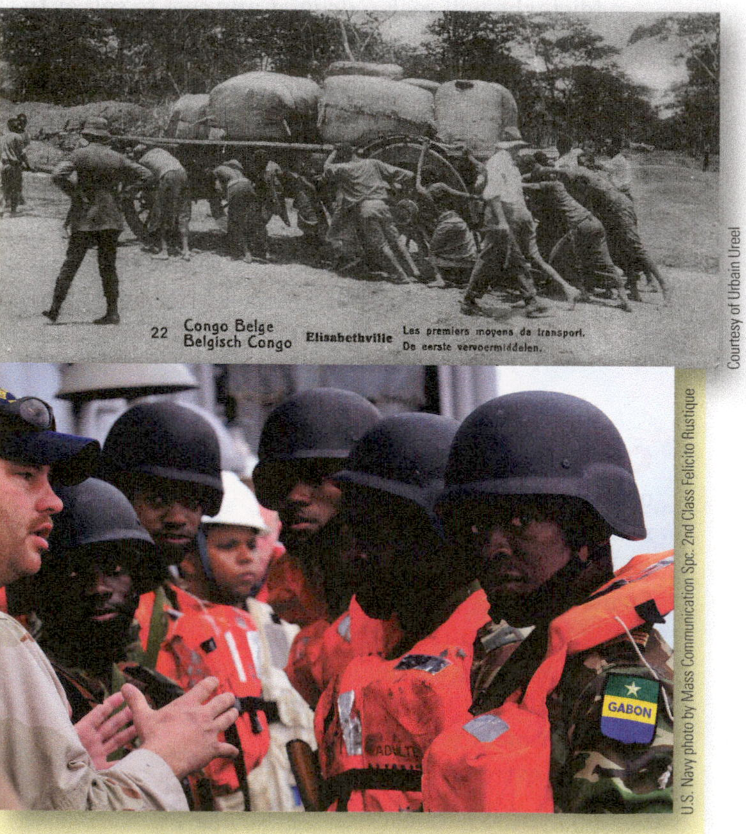

Courtesy of Urbain Ureel

U.S. Navy photo by Mass Communication Spc. 2nd Class Felicito Rustique

DuBois' observations are supported by scenes of African labor (top) moving resources out of the continent for export to Europe and the United States around the turn of the 20th century. If DuBois were alive today, he would call attention to U.S. military presence around the Horn of Africa to ensure the safe passage of oil, uranium, cobalt, titanium, diamonds, gold, bauxite, copper, and other minerals considered strategic to U.S. economic security. Here members of Gabon's navy are aboard a U.S. ship as part of joint military exercises aimed at securing the movement of goods out of the Atlantic oil- and mineral-bearing basin.

Jane Addams (1860–1935)

In 1889 Jane Addams (with Ellen Gates Starr) cofounded one of the first settlement houses in the United States, Hull House. Settlement houses were community centers that served the poor and other marginalized populations. Wealthy donors supported them, and university faculty and college students lived with the clients, serving and learning from them. The Chicago Hull House was one of the largest and most influential settlements in the United States. At the time of its founding, immigrants constituted almost half of Chicago's population. In addition, the city was industrializing and experiencing unprecedented population growth. These dramatic changes were accompanied by a variety of social problems, including homelessness, substandard housing, and unemployment.

Hull House facilities contained a school, boys' and girls' clubs, recreation facilities, a library, and much more. Hull House had strong ties with the University of Chicago School of Sociology. Jane Addams was a forerunner of what is today called public sociology because she demonstrated an "unwavering commitment to social improvement," whether it be aimed at child labor, worker safety, or other social concerns (Hamington 2007). Addams maintained that the settlement houses were equivalent to an applied university where theories about how to change people's situations could be tested. Addams advocated for **sympathetic knowledge**, firsthand knowledge gained by living and working among those being studied, because knowing others increases the "potential for caring and empathetic moral actions" (Addams 1912, 7). Addams made a point of never addressing an audience about the clients Hull House served without bringing a member who knew their conditions more intimately than she "to act as an auditor" of her words (Addams 1910, 80). Addams believed that firsthand, lived experiences acted as a test of theory and that the voices of those marginalized—the people sociologists and other theorists write about—must be included in the community of ideas (Hamington 2010).

Jane Addams believed that anyone who theorizes about or makes decisions that affect the lives of marginalized groups, including low-wage workers, has the responsibility to actively know the people they are trying to help—to establish sympathetic understanding so they know what it means to be among the ranks of the lowest paid.

Lisa Southwick

What Do Sociologists See?

DoD photo by Senior Master Sgt. David H. Lipp, North Dakota National Guard

The interpretation a sociologist makes depends on which theorist inspires their analysis. Sociologists inspired by Durkheim see organ donation as a way to establish deep ties between a deceased donor and as many as 50 patients. Organ donation is also a way to deepen ties between living donors and patients (who may be relatives, friends, or strangers). Sociologists inspired by Marx see organ donation (as practiced in the United States) as a process whereby a living donor or the family of deceased donor earn no income for organs while hospitals and surgeons profit. Sociologists inspired by Weber would consider motives that shape policies governing organ procurement. Given the revenue-generating potential associated with transplants ($1.2 million for intestine transplants, $977,000 for heart transplant), medical facilities may be motivated to secure organs at any cost (instrumental rational action).

Critical Thinking

Which one of the six theorists' ideas best reflects how you tend to think? Explain.

Key Terms

bourgeoisie
class conflict
color line
conflict
disenchantment
means of production
mechanical solidarity
organic solidarity
positivism
proletariat
social action
solidarity
sympathetic knowledge

Module 1.4

The Sociological Imagination

Objective

You will learn that the sociological imagination allows us to see how human life—even the most intimate details—is shaped by the time and place in which we live.

When you shower do you leave the water running while you lather and scrub your hair or do you turn it off?

NKU Sociology, Missy Gish

If you live in the United States, you would likely keep the water running as you shampoo your hair. If you live in Germany, you would likely wet your hair, then turn the water off while you lather and scrub your hair. Sociologists ask, "What is it about the United States society that encourages water *consumption*?" "What is it about German society that encourages water *conservation*?" If these kinds of questions interest you, then you will appreciate sociologist C. Wright Mills' (1963) writings on the **sociological imagination**, a perspective that allows us to consider how outside forces, especially the time and place we were born and live, shape our life story or biography. A **biography** consists of all the significant, and even seemingly insignificant, events and day-to-day interactions from birth to death that make up a person's life. If you live in Germany, your biography includes personal efforts to conserve resources—that is, to not let water run while brushing teeth or until it gets warm. If you live in Germany, you are surrounded by national, local, and personal efforts to achieve sustainability and to make renewable resources THE source of power. "Going Green" in Germany isn't defined as simply buying "green" products; it is a serious and long-standing policy. The Germans are working to become a world leader in conservation and sustainability.

As this example suggests, the forces of history and society affect our most personal experiences. We should not be surprised, then, to learn that the forces of time and place shape expressions of affection and closeness as well. It seems that the rules governing touch in the United States dictate that "one must be in a romantic relationship to get much touch, that touch has sexual connotations, and that daily interpersonal interaction tends not to involve touch" (Traina 2005).

Katie Englert (left), Chris Caldeira (right)

Many societies such as Papua New Guinea (left) allow men and women to walk arm-in-arm in public without onlookers assuming romantic involvement. Not so in the United States, where for the most part such physical contact assumes sexual involvement (right).

The sociological imagination is empowering because it allows those who possess it to distinguish between what sociologist C. Wright Mills (1959) called **troubles** and issues. Mills (1959) defined troubles in individual terms, as caused by personal shortcomings related to motivation, attitude, ability, character, or bad judgment. The resolution of a trouble, if it can indeed be resolved, lies in changing the person in some way. For example, Mills states that when only one person is unemployed in a city of 100,000, that situation is likely a trouble. For its relief, we can focus on that person's shortcomings—"She is lazy," "He has a bad attitude," "He didn't try very hard in school," or "She had the opportunity but didn't take it."

An **issue**, on the other hand, is a societal matter that affects many people and that can only be explained by larger social forces that are bigger than those affected. When 24 million men and women are unemployed or underemployed in a nation with a workforce of 156 million, that situation is an issue. Clearly, we cannot hope to solve widespread unemployment by focusing on the character flaws of 24 million individuals. A constructive assessment of this crisis requires us to think beyond personal shortcomings and to consider the underlying social forces that created it. For example, the economy is structured so that corporate success is measured by ever-increasing profit margins. Under such an arrangement, profits are increased by lowering labor costs, which can be achieved through laying off employees, downsizing, transferring jobs from high-wage to low-wage areas, and otherwise reducing, even eliminating, human labor needed to produce a product or deliver a service.

When you see a "going out of business" sign, do you think about the technological innovations that made renting videos and DVDs go out of fashion? Do you think about the people who will lose their jobs not because they were lazy or unmotivated but because they were caught up in forces beyond their control? If you see job loss in these terms, you are framing unemployment as an issue.

NKU Sociology, Missy Gish

The ability to distinguish between troubles and issues allows us to think more deeply about the cause of and potential solutions to problems that seem, on the surface, to be entirely personal. Arguably the best-known effort to connect personal troubles to larger social issues was that of sociologist Émile Durkheim, who wrote *Suicide* in 1897 and who is still regarded as an authority on that subject today.

Suicide

When we think about who commits suicide, we often think of people who are deeply and personally troubled. In *Suicide*, Durkheim argued that to understand this act, it is futile to think in uniquely personal terms about the circumstances that lead people to kill themselves. For example, one person may kill herself in the midst of newly acquired wealth, whereas another kills herself in the midst of poverty. One person may kill himself because he is unhappy in his marriage and feels trapped, whereas another kills himself because his unhappy marriage has just ended in divorce. We can find cases of people who kill themselves after losing a business; in other cases a lottery winner kills herself because she cannot tolerate family and friends fighting one another to share in the newfound fortune. Because almost any personal circumstance served as a context for suicide, we must look beyond the personal.

Durkheim offered a sociological vision of suicide that goes beyond its popular meaning (the act of intentionally killing oneself). He drew attention to suicide as an act that ultimately severs relationships. Durkheim's vision takes the spotlight off the victim and points it outward toward relationships or the social ties severed. To make his case, Durkheim argued that every group has a greater or lesser propensity for suicide. The suicide rates by age, sex, and race in the

United States show that suicide is more prevalent for males in general, and especially males considered white, age 65 and older. Durkheim believed that comparing suicide rates across groups yields important insights about the larger social forces that push people to take their own lives. In this case Durkheim would ask: What do the different rates suggest about men's and women's social ties to one another and the society of which they are a part?

Chart 1.4a: Male-Female Differences in Suicide Rates (per 100,000), United States
The chart shows the annual number of suicides per 100,000 people for 16 age and sex groups. Note that males age 85 and over have the highest suicide rate—each year 45 of every 100,000 men commit suicide. Females in that age category have a suicide rate of 4 of every 100,000 suicides. Is there any age category where females have the higher rate relative to males?

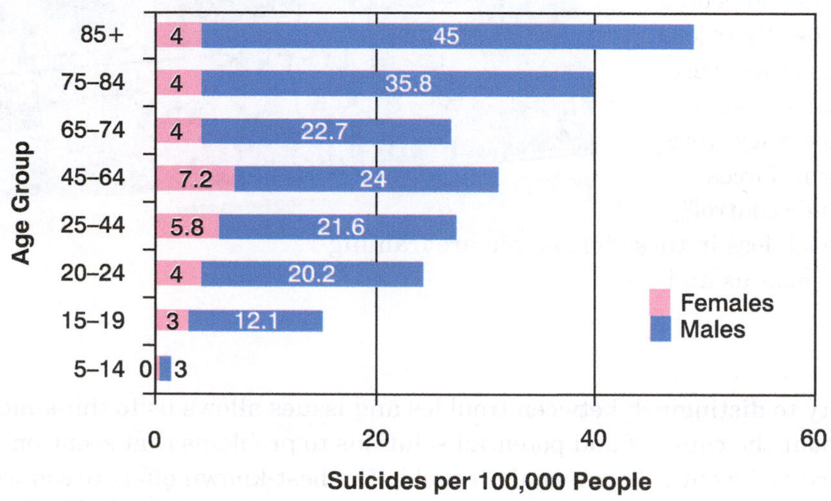

Source: Centers for Disease Control and Prevention (2013).

In reviewing these rates, it is important to point out that women attempt suicide about three times more often than men. In addition, the most common method of suicide among women is poisoning with drugs or other chemicals, a method that is more likely to fail; suicide among men most commonly involves firearms (Centers for Disease Control 2013). Durkheim would maintain that the different attempted and successful suicide rates reflect the pressures a society places on men to succeed at suicide and not use it as a cry for help as women are allowed to do. The rates also reflect males' greater access to guns and knowledge of how to use them (guns are viewed as an expression of masculinity and as something males, and not females, are expected to possess and know how to use).

In thinking about the character of relationships, Durkheim identified four types of relationships or social ties that bind the individual too weakly or strongly to others. Two types relate to problems of integration—too much (altruistic) and too little (egoistic). Another two types describe problems of regulation—too little (anomic) or too much (egoistic).

Tony Rotundo

Tony Rotundo

Durkheim maintained that the key to understanding suicide rates lay with issues of integration and regulation. Individuals connected to a group such as bikers are protected from suicide by their ties to this group. In other words, the group acts as an integrating force in members' lives. Those individuals whose social ties to a partner have been broken by a decision to divorce or a breakup lose a source of regulation or stability in their life and are thus more vulnerable to suicide. In this sense marriage acts as a regulatory force protecting the married from taking their own lives.

Egoistic describes a state in which the ties integrating the individual to others in society are weak. When individuals are weakly integrated, they encounter less resistance to suicide. Relative to men, society offers women more opportunities to form meaningful relationships with others; women are disproportionately assigned nurturing roles and men are disproportionately assigned roles that connect them less directly to others. These differences in the ties that bind men and women to others, and by extension to society, offer insights about why suicide rates differ for males and females.

Altruistic describes a state in which the individual is excessively integrated into the group. In other words, the ties attaching the individual to the group are such that a person's sense of self is lost to the group. When such people commit suicide, it is on behalf of a group they love more than themselves. The classic example is soldiers willing to sacrifice their lives to advance the ideals and cause of their unit.

Anomic describes a state in which the forces that regulate social ties are disrupted by dramatic changes in circumstances. Durkheim gave particular emphasis to the economy, which functions as a regulatory force. When the economy is turbulent—that is, in recession, depression, or economic boom—the social ties that bind the individual to the group are disrupted. To put it another way, the turbulence casts many individuals into lower or higher statuses. Those who lose jobs or money are cast into a lower status, forcing them to reduce their desires, restrain their needs, and practice self-control. Those who find themselves suddenly at a higher status after winning a lottery or other financial windfall must adjust to increased prosperity. Aspirations and desires may be unleashed, and an insatiable thirst to acquire more goods and services, and even to feel that one needs still more, may arise.

Fatalistic describes a state in which ties regulate lives too tightly. The ties attaching the individual to the group are so oppressive that there is no hope of release. Under such conditions, individuals see their futures as permanently blocked. Durkheim asked, "Do not the suicides of slaves, said to be frequent under certain conditions, belong to this type?" (1897, 276).

What Do Sociologists See?

Sociologists see the sign announcing crisis counseling and emergency phone as an effort to prevent suicidal people from jumping off this bridge to their death. Sociologists would recommend special vigilance when unemployment rates are dramatically high or low, when a major employer in the area downsizes, or when new laws tightening illegal drug use go into effect such that the addicted have harder time fueling their habit.

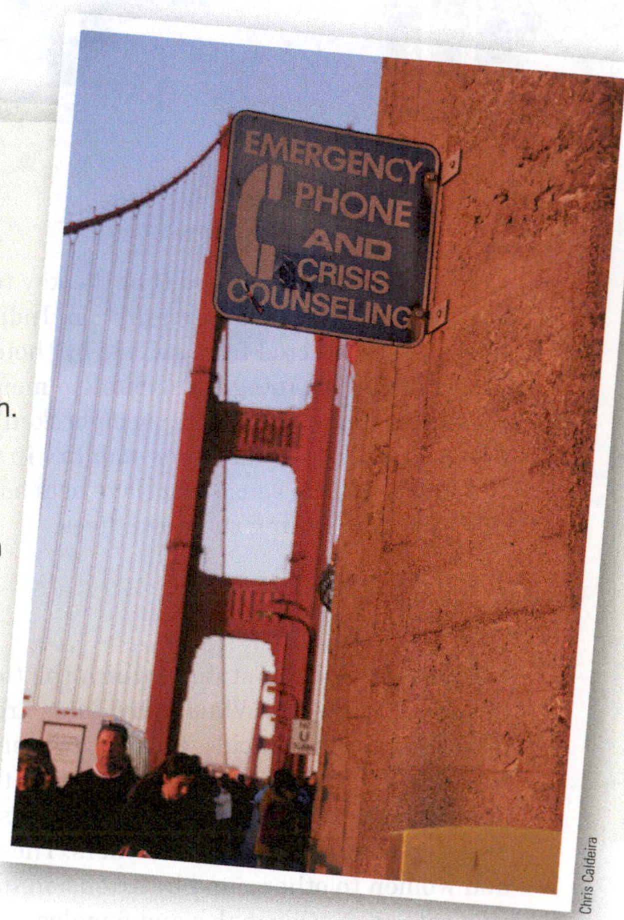

Chris Caldeira

Critical Thinking

Describe a time in your life when you were aware that your social ties to some group or person were problematic—that is, they were too weak, too strong, broken, or oppressively restrictive.

Key Terms

altruistic
anomic
biography
egoistic
fatalistic
issue
sociological imagination
troubles

Module 1.5 Sociological Perspectives

Objective

You will learn how sociological perspectives provide questions to guide analysis of any topic.

How will humans relate to social robots in workplaces, classrooms, and other settings?

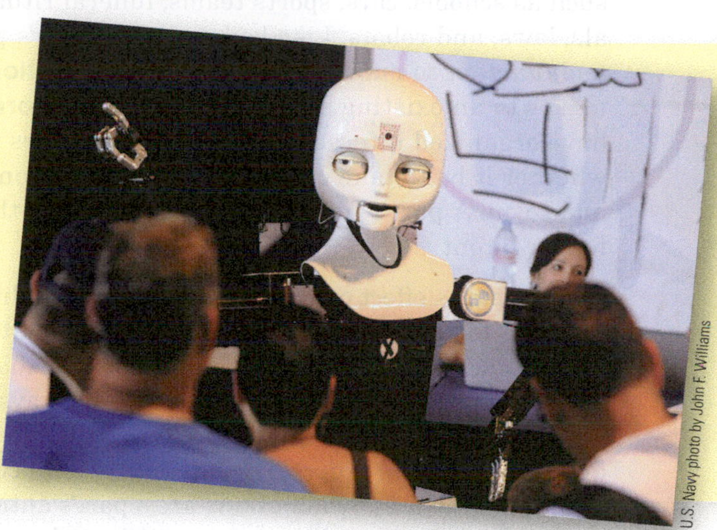

U.S. Navy photo by John F. Williams

Imagine you decided to do research on the topic of social robots, the kind of robots capable of interacting with humans. The search engine Google Scholar yields about *200,000 hits* for "social robots." As you browse though the titles, you wonder how to select and organize all that information. This research can be less complicated and stressful if it is informed by at least one of the four major sociological perspectives. This is because each perspective offers a vision of society, key questions that guide readings and analysis, as well as a vocabulary to answer those questions.

A **sociological perspective** is a framework for thinking about, describing, and explaining how human activities are organized and how people relate to one another and respond to their surroundings. In sociology there are four major sociological perspectives, each of which focuses our attention on different slices of reality. The perspectives are functionalist, conflict, symbolic interaction, and feminist. Sociologists use the perspectives to guide analysis. Each offers a central question to direct thinking and key concepts to organize answers. Keep in mind that, taken alone, no single perspective can offer a complete picture of a situation. But we can acquire the most complete picture by applying all four.

Functionalist Perspective

Functionalists are known for the focus they give to order and stability. They define society as a system of interrelated, interdependent parts. To illustrate this vision, functionalists use the human body as an analogy for society. The human body is composed of parts including bones, cartilage, ligaments, muscles, a brain, a spinal cord, hormones, blood, blood vessels, a heart, kidneys, and lungs. All of these body parts work together in impressive harmony. Each part functions in a unique way to maintain the entire body, but it cannot be separated from other body parts that it affects and that in turn help it function.

Society, like the human body, is made up of an incalculable number of parts such as schools, cars, sports teams, funeral rituals, holidays, religious rituals, laws, and robots. Like body parts, society's parts are interdependent and *function* to maintain social order. A **function** is the contribution a part makes to an existing social order. The **social order** refers to the way people have organized interaction and other activities to achieve some valued goal—whether it be to take care of the sick, to pass on knowledge, to raise a family, and so on. A part's function or contribution to the social order can be manifest or latent.

When a part's effect on the social order is something that is expected, anticipated, or intended, that effect is a **manifest function**. When that part's effect is unintended, not anticipated, or unexpected, that effect is a **latent function**. Functionalists recognize that parts do not always contribute to order and stability—they can cultivate *dysfunctions*; that is, they can have disruptive consequences to a social order. Like functions, dysfunctions can be either manifest or latent. **Manifest dysfunctions** are a part's anticipated disruptions to a social order. **Latent dysfunctions** are unanticipated or unintended disruptions to a social order. The application that follows clarifies how a part can serve both to maintain order and stability and to disrupt it.

The Functionalist Perspective Applied to Social Robots

To analyze social robots functionalists ask: How do social robots contribute to social order and stability? In what ways might social robots disrupt order and stability? Three of the most often-stated reasons for developing social robots speak to their manifest or anticipated functions: (1) to provide assistive care to those with disabilities, including children with autism and the growing number of elderly in need of care; (2) to do jobs that are repetitive, boring, dangerous, and prone to human error; and (3) to reduce labor costs.

What might be some possible unanticipated (latent) functions associated with social robots? For one, if social robots are developed to the point where they can assume caregiving roles, risks of inappropriate sexual contact and abuse of children and other clients are eliminated. Second, social robots will not suffer from caregiver burnout or become impatient with their clients. Robots are also even-tempered, predictable, nonjudgmental, and always available and responsive.

In addition to a part's functions, functionalists also consider the expected or anticipated disruptions (manifest dysfunctions) a part may have on the stability of an existing social order. One obvious dysfunction relates to employment. Programmers are writing artificial intelligence (AI) algorithms that direct social robots in roles as housekeepers, butlers, bartenders, receptionists, prison guards, and pharmacy assistants (Heater 2012). So we might predict that the number of people working in such jobs will decline.

Josh Ellingson, Courtesy of Chris Caldeira

A latent or unexpected dysfunction relates to the emotional attachment humans may come to feel toward devoted and uncomplaining social robots, such that people may actually prefer the company of robots over that of humans.

Conflict Perspective

In contrast to functionalists, who focus on social order, conflict theorists focus on conflict as an inevitable fact of social life and as the most important agent for social change. Conflict can take many forms including physical confrontations, exploitation, disagreements, and direct competition. In any society, advantaged and less disadvantaged groups compete for scarce and valued resources (access to material wealth, education, health care, well-paying jobs, and so on). Those who gain control of and access to these resources strive to protect their own interests against the competing interests of others. Conflict theorists ask this basic question: Who benefits from a particular social arrangement, and at whose expense? In answering this question, conflict theorists seek to describe the social arrangements that advantaged groups have established, consciously or unconsciously, to promote and protect their privileged position. Exposing these practices helps explain inequalities that exist in society.

Conflict theorists work to expose the **façade of legitimacy**—an explanation to justify the existing social arrangement that downplays or dismisses charges that the arrangement advantages some and disadvantages others. On close analysis, the justifications are often based on unsupported assertions and faulty assumptions but nevertheless are presented as credible explanations (Carver 1987). In general, conflict theorists dismiss any altruistic justifications as camouflaging the real motives, most notably earning a profit.

The Conflict Perspective on Social Robots

In analyzing social robots, conflict theorists ask: Who benefits from social robots, and at whose expense? In answering this question, they would point out an obvious fact: The creation of social robots is driven by a desire to create a "social arrangement" that maximizes profit by eliminating human labor and its associated costs (wages, benefits packages).

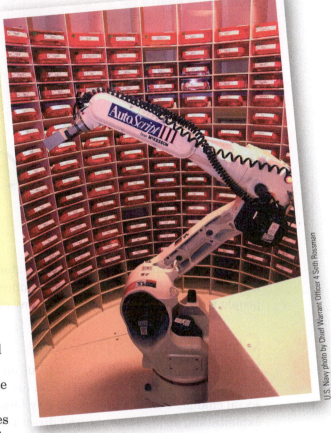

Autoscript III is prescription-filling robot. From a conflict point of view, this robot can replace the labor of pharmacy technicians who prepare medications under the direction of a pharmacist. The average salary of pharmacy technicians working full-time is $30,500 (U.S. Bureau of Labor Statistics 2014b). Assuming the robot costs $50,000, it would pay for itself in less than four months because it can work 24/7 shifts and it can do the work of five or six technicians.

U.S. Navy photo by Chief Warrant Officer 4 Seth Rossman

Conflict theorists view any professed altruistic motives for developing social robots such as to assist the growing numbers of disabled elderly as a façade to cover up the real motives. From a conflict point of view, which emphasizes profit-making as the ultimate force driving innovation, the elderly represent a huge market to sell social robots. One can easily envision the endless number of add-on apps to be offered at extra cost beyond the basic social robot—it might be an app for moving around in cluttered indoor environments, or an app for specific types of physical assistance such as opening doors, turning water in showers on and off, carrying laundry, and so on. "Incrementally new" versions of social robots could be released every six months (Blackman 2013).

Conflict theorists predict that social robot technologies will usher in a new divide separating advantaged populations with the money to acquire social robots from those disadvantaged populations without the financial resources to do so. The divide will create advantaged and disadvantaged individuals, households, and communities. That divide will also exist on a global scale, separating the richest countries with disproportionately greater access to social robots from the poorest countries with significantly less or no access.

Symbolic Interaction Perspective

Symbolic interactionists focus on **social interaction**, everyday encounters in which people communicate, interpret, and respond to one another's words and actions. These theorists ask: How, when interacting, do people "take account of what [the] other is doing or is about to do" and then direct their conduct accordingly (Blumer 1969)? The process depends on (1) self-awareness, (2) shared symbols, and (3) negotiated order.

Self-awareness occurs when a person is able to observe and evaluate the self from another's viewpoint. People are self-aware when they imagine how others are viewing, evaluating, and interpreting their words and actions. Through this imaginative process, people become objects to themselves; they come to recognize that others see them, for instance, "as being a man, young in age, a student, in debt, trying to become a doctor, coming from an undistinguished family, and so forth" (Blumer 1969, 172). In imagining others' reactions, people respond and make adjustments (apologize, change facial expressions, lash out, and so on).

A **symbol** is any kind of object to which people assign a name, meaning, or value (Blumer 1969). Objects can be classified as physical (smartphones, cars, a color, a facial expression), social (a friend, a parent, a celebrity, a bus driver), or abstract (freedom, greed, justice, empathy). Objects can take on different meanings depending on audience and context: a tree can have different meanings to an urban dweller, a farmer, a poet, a homebuilder, an environmentalist, or a lumberjack (Blumer 1969). Freedom can have different meanings to prisoners, teenagers, and older adults. People learn shared meanings that their culture attaches to objects. That is, they learn that a wave of the hand means good-bye, that letters of the alphabet can be selected and arranged to make countless words, and that dogs are considered as pets but crickets are not.

Look at the expression on the two children's faces. Their seemingly blank expressions suggest that they have no idea what to do even with prompting from their mothers. That is because they lack self-awareness—the ability to think about how others see them and what others expect of them in a particular situation.

Robert K. Wallace

When we enter into interaction with others—whether it is with a store clerk, a professor, friends, or colleagues—we take for granted that a system of expected behaviors and shared meanings is already in place to guide the course of interaction. Still we are generally aware that we should behave and talk in a certain way. In most interactions, room for negotiation exists; that is, the parties involved have the option of negotiating other expectations and meanings. For example, students know that class is to end at set time but they can often negotiate by five minutes or so the official time class is to end by packing up and looking toward the door.

The Symbolic Interactionist Perspective on Social Robots

The symbolic interaction focus on self-awareness, shared symbols, and negotiated order is especially relevant to describing the skills social robots must possess if they are to interact successfully with humans. For robots to be truly social, they must be self-aware; attach shared meanings to words, gestures, and facial expressions; and then respond accordingly and negotiate social order.

Because the success of social robots is dependent on their ability to interact successfully with people, the symbolic interactionist perspective is especially useful for thinking about factors that shape the success of human–robot interaction. To what extent does successful interaction depend on the robot's appearance? For example, does it matter if the robot is anthropomorphic (human-like), zoomorphic (animal-like), or machine-like in appearance? To be even more specific, should a robot charged with babysitting be adult-like in appearance so that parents perceive it as responsible? Should it appear gender-neutral or have traits that evoke associations of masculinity or femininity? Apart from appearance, there are questions about how a robot should convey disagreement. Should the robot express disagreement in explicit ways ("I think this choice is not correct") or in implicit ways ("Are you sure?")? Research suggests that German clients and Chinese clients tend to be more comfortable with explicit styles, and Americans with implicit (Li et al. 2010).

Feminist Perspective

The feminist perspective is considered a variation on conflict theory because it gives central focus to the unequal distribution of power and other valued resources as it relates to gender. While feminists acknowledge that there are unequal divides based on such things as nation, race, class, and age, they define the gender divide as the most basic, persistent, prevalent, and resistant to change. In fact, the minute babies are born, a glance at the genitals identifies them as male or female, and that glance profoundly affects babies' life chances in unequal ways (Epstein 2006).

Feminists seek to understand why females as a group tend to disproportionately occupy subordinate statuses relative to men. Simply consider that thrones and executive offices are overwhelmingly occupied by men and that "judges in courtrooms; priests, rabbis, and mullahs; leaders and members of unions and clubs" (Epstein 2006, 3) are most likely males who make policies and decisions that affect the lives of women and girls. How is it that males as a group are economically and politically advantaged relative to women virtually everywhere in the world? While the answers to these questions are complex, feminists do not accept essentialism as an answer. **Essentialism** is the belief that men and women are different by divine design and/or nature, and that inequalities between them are natural. By this logic, men are by nature dominant and women are submissive. Instead, feminists seek to uncover gender inequalities, understand how they are human-constructed, and push for change.

Feminist Perspective on Social Robots

In analyzing social robots, feminists ask to what extent gender inequalities are reflected in the way engineers and programmers design robots. In other words, do social robots (nanny bots) built to teach young children have female bodies and voices, while social robots built to do difficult and complicated tasks appear to be males? Science and technology—the disciplines that seek power and control over nature and the environment—are areas in which males dominate. Given that the programmers and designers are men, how will they design female robots? Will these "created bodies contribute to the reproduction of traditional gender stereotypes, especially in the tasks they are assigned and the way they respond?" (García-Ordaz et al. 2013). These are just some of the questions feminists ask. We might anticipate that feminists will challenge scientists to create robotic bodies that defy stereotypes and that recognize women's skills and contributions.

Feminists question why most, if not all, NASA robots have male bodies when when female astronauts have traveled and worked in space since 1963 and became commonplace in the 1980s.

Critique of Four Sociological Perspectives

Each perspective has its strengths and weaknesses. The functionalist perspective's strength lies in the balanced overview that comes with considering a part's anticipated and unanticipated consequences for the existing social order. One weakness is that functionalists leave us wondering about a part's *overall* effect on that order. So with regard to social robots, we are left asking: Do the functions performed by social robots outweigh the associated dysfunctions?

One strength of the conflict perspective is that it forces us to look beyond popular justifications for why humans structure social activity in the ways they do and to ask questions about whose interests are protected and promoted and at whose expense. The conflict perspective is criticized for presenting advantaged groups as driven only by profit and desire to protect their interests. The most advantaged are portrayed as all-powerful and capable of imposing their will without resistance; the disadvantaged are portrayed as exploited victims. But there are likely many among the advantaged who think beyond the bottom line and self-interest. In addition, conflict theorists fail to recognize that the drive to make a profit is responsible for innovation. Without the profit motive, how much innovation would take place?

The strength of the symbolic interactionist theory is that it focuses on up-close and personal factors that shape the course of interaction and relationships. A symbolic interactionist, however, can get caught up in interaction dynamics and lose sight of the larger structural issues in which that interaction is embedded, including the existing social order and a profit-driven economy. The strength of the feminist perspective is that it focuses attention on the experiences of women and advocates for gender equality. The weakness of the perspective is that it places too much emphasis on gender. Gender is just one of many critical statuses that shape a person's opportunities and challenges.

What Do Sociologists See?

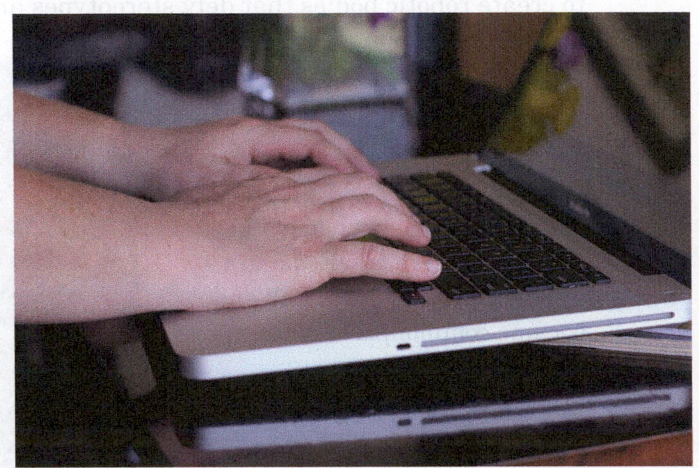

Chris Caldeira

Drawing on all four perspectives, sociologists see a technology that allows people to connect to others and to information 24/7 (function), and that also serves as a distraction from tasks at hand (dysfunction). They see a person who can afford access to this technology and wonder how it disadvantages those who cannot (conflict). Sociologists might consider how the kind of computer one uses sends a message to onlookers about status and technological abilities (symbolic interaction). Sociologists might also question the ways computers support gender inequalities. Because computers facilitate 24/7 access and apps that allow users to check in at home with surveillance cameras, are women's caregiving responsibilities extended accordingly?

Critical Thinking

Which of the four perspectives presents a question that you have found yourself asking at one time or another? Explain.

Key Terms

essentialism	latent functions	social order
façade of legitimacy	manifest dysfunctions	sociological perspective
function	manifest functions	symbol
latent dysfunctions	social interaction	

Module 1.6

Research Methods

Objective

You will learn the process by which sociologists research answers to the questions they ask.

Can humans and robots become friends and coworkers? How might you design a research study to answer this question?

NASA

In this module we will feature a research project of sociological significance—a National Science Foundation-funded study of a social robot integrated into a workplace—conducted by Min Kyung Lee, Sara Kiesler, Jodi Forlizzi, and Paul Rybski (2012a, 2012b) at the Human-Computer Interaction Institute and the Robotics Institute, and hereafter referred to as Lee.

Research Methods

Research methods are the various strategies sociologists and other scientists use to formulate and answer meaningful research questions and to collect, analyze, and interpret data gathered. Here, the term *data* is used in the broadest sense; it applies to observations recorded, responses to survey and interview questions, and much more. The four sociological perspectives (see Module 1.5) inspire the questions sociologists ask when they do research, and the perspectives offer frameworks for interpreting the findings. In turn, research findings can offer support for or challenge the sociological perspective(s).

Sociologists adhere to the **scientific method**, a carefully planned research process with the goal of generating observations and data that can be verified by others. The research process involves at least six interdependent steps:

- determining the topic or research question,
- reviewing the literature,
- choosing a research design,

- identifying variables and specifying hypotheses,
- collecting and analyzing the data, and
- drawing conclusions.

It is important to know that researchers do not always follow the six steps in sequence. For example, they may not decide on a specific research question until they have familiarized themselves with the literature. Sometimes an opportunity arises to collect specific data, and a research question is defined to fit that opportunity. Although the six steps need not be followed in any particular order, all must be completed at some point in the research process to ensure the quality of the project.

Choosing a Topic/Reviewing the Literature

It is impossible to list all the topics that sociologists study because almost any subject involving human activity represents a potential subject for research. Sociology is distinguished from other disciplines not by the topics it investigates but by the perspectives it takes in framing, studying, and drawing conclusions. The sociological significance of Lee's study lies in its focus on meanings employees assign to a social robot and the effect the robot's presence has on the workplace. Specifically, the questions driving Lee's research were: How will employees relate to and interact with a social robot? How will the robot affect the workplace? What implications might the findings of this study have for designing social robots in the workplace?

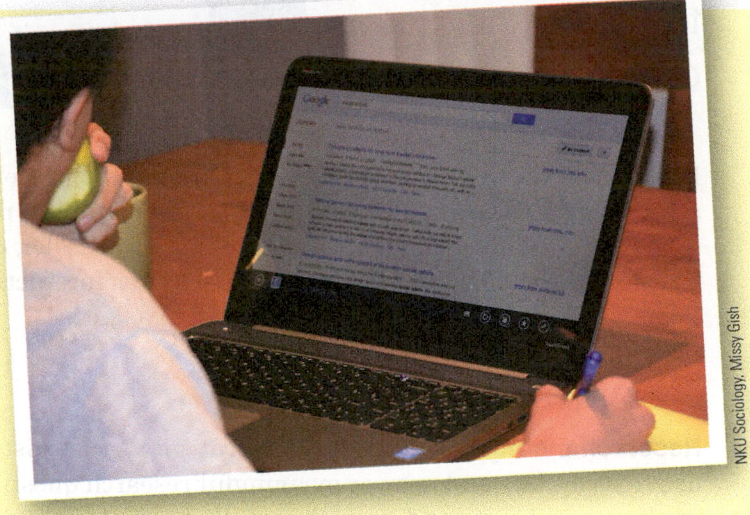

When doing research, sociologists consider what knowledgeable authorities have written on the chosen topic, if only to avoid reinventing the wheel. Reading the relevant literature can also generate insights that researchers may not have considered.

NKU Sociology, Missy Gish

In writing up an account of their research study, sociologists cite those who have influenced their work. Lee cited about 50 references to scholarly books and articles written by credentialed authors who carefully documented their sources of information and/or who actually conducted research studies. Ideally, scholarly writings are reviewed by qualified experts who assess quality, recommend whether the writings deserve to be published, and offer suggestions for improvement.

In choosing the literature to review, researchers certainly read the most current materials, but the publication date should not be the only criterion. Classic and groundbreaking articles written decades or even centuries ago can offer inspiration. In reviewing the literature, researchers learn not only what is known about the topic of interest but also about things not yet known. By identifying what is not known, researchers establish what they can do to advance knowledge on the chosen topic.

After reviewing the existing literature, Lee learned that people are most likely to accept, trust, and engage with social robots when the robots are capable of making small talk and displaying signs of empathy. She noticed that much of the research on how humans respond to social robots had been conducted in labs or in public settings involving one-time encounters. To her knowledge, "no studies have followed the same employees over an extended period." In other words, there seemed to be no research about how social robots fit into the "culture of a real workplace." Lee realized that the literature could benefit from a research study that focused on human responses to social robots over extended and repeated encounters to offset the novelty effect.

Choosing a Research Design

After determining the research question, researchers typically decide upon a **research design**, which involves deciding who or what to study and the method of gathering data. Lee's research design involved (1) observing employees as they interacted with a robot, (2) doing face-to-face interviews with employees about their experiences, and (3) distributing surveys. For this study, Lee and her colleagues designed Snackbot, a robot capable of delivering snacks, and decided to try it out on people who worked in a university. Lee hung flyers and sent out postcards to people working on a floor of a computer science building to attract customers. In the end, 21 people signed up—eight women and 13 men ranging in age from 22 to 51. Eleven of the customers were graduate students, eight were staff, and two were faculty. Only one customer had some experience with robots. Customers ordered desired snacks from a website and Snackbot, a 4.5-foot-tall anthropomorphic robot with a male voice, made deliveries between 2:30 and 4:00 p.m. Snackbot could smile, frown, or show a neutral expression. Snackbot was fitted with a Web camera and microphone to record his interactions with customers. Later, Lee and her colleagues reviewed these video recordings in search of meaningful patterns.

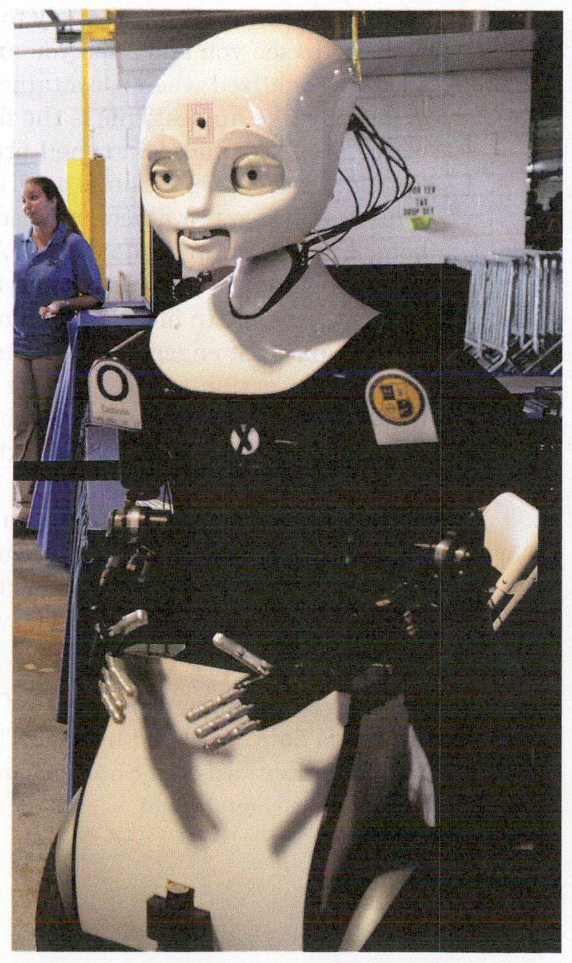

Snackbot looks something like the robot you see pictured here. Lee and her colleagues controlled Snackbot's speech and facial expressions using a wireless network. This "Wizard of Oz"—or "behind-the-curtain"—setup allowed operators working in a room offsite to see and hear customers as they interacted with Snackbot.

U.S. Navy photo by John F. Williams

Altogether, Lee recorded 346 interactions over a four-month period. The snacks were free and given as compensation to those participating in the study who agreed to do three things: allow their interactions with Snackbot to be recorded, complete a satisfaction survey, and do a 30- to 60-minute face-to-face interview with Lee when the study ended.

Specifying Variables, Operational Definitions, and Hypotheses

A **variable** is any behavior or characteristic that consists of more than one category. Age is a variable, as people can range in age from seconds old (at birth) to 100-plus years old. Grade point average is a variable ranging from 0 to 4.0. All variables used in a research study must be **operationalized**; that is, the researcher must give clear, precise instructions about how they observed or measured them.

Three variables important to Lee's study were (1) presence or absence of personalized conversation with the Snackbot, (2) whether or not customers cooperated with Snackbot's requests, and (3) whether or not customers assigned or did not assign human-like qualities to Snackbot. The operational definition of each follows:

Personalized conversation involves dialogue in which one party makes reference to things specific to the other party. Conversation was considered impersonal when the comments could apply to anyone, such as "It is a nice day today. I am glad to see you and hope you are doing well." Conversation was considered personalized when it contained statements like "It seems as though you really like apples. This is the fifth time you have ordered that," or "You were out of the office last time I brought you an apple. I missed seeing you." Personalized conversation is a variable because Snackbot followed either an impersonal script or a personalized one. For the first four deliveries, Snackbot stuck to an impersonal script. After that, customers were randomly assigned to one of two groups. For those assigned to one group, Snackbot continued to engage in impersonal conversation; for those assigned to the second group, Snackbot engaged in personalized conversation.

Cooperation is action taken to assist or meet a request someone makes. To measure cooperation with the robot, Lee recorded whether or not customers (1) gave suggestions to Snackbot when it asked for names of places on campus that might be of interest to tourists, (2) complied when the robot suggested to a customer to take a break from work and join him in doing a "neck stretch," and (3) agreed to try a special healthy mystery snack Snackbot recommended. Cooperation is a variable because customers can cooperate on all three requests, two requests, one request, or none of the requests.

Anthropomorphizing means the act of ascribing human characteristics to something that is not considered human. This is a variable in Lee's study because customers either assigned Snackbot human characteristics such as feelings and motivations or they did not.

Do you think of this robot as having human qualities? If you answered yes, you have *anthropomorphized* the robot.

Courtesy of Andrea Thomaz

Operational definitions must be reliable and valid. An operational definition is considered reliable if someone using the operational definition as described repeats the measure and gets essentially the same results. Consider the **reliability** of Lee's measure of cooperation. If customers answered Snackbot's question about good locations in the building to take tourists, took a break from work to join the robot in doing a "neck stretch," and agreed to try the mystery snack, it seems likely that anyone would classify those customers as fully cooperative. So in this sense Lee's measure is reliable. Issues of reliability arise, however, when it is not clear whether someone cooperated with Snackbot. What if someone did the neck stretch but only half-heartedly stretched while still working? What if the person rolled her eyes while doing the neck stretch? Two people trying to classify the customers' reactions as cooperative or not may not reach the same conclusion, raising questions about the reliability of the operational definition of cooperation with Snackbot. Unless clear instructions guide how to classify "half-hearted" efforts or eye rolling, the measure for cooperation cannot be reliable.

An operational definition must also be assessed for its **validity**, or the extent to which the described measure actually measures what it claims to measure. Does Lee's measure of cooperation *really* measure willingness to cooperate with a robot? One might argue that agreeing to do a neck stretch or answering a question about the location of an event are trivial tests of cooperation and therefore not valid measures. Perhaps a stronger measure of cooperation with a robot might involve giving the robot a different, more serious role in the workplace such as a health coach promoting employee fitness. As health coach, the robot could request that customers eat a healthier snack or take a brisk walk on their lunch break. If customers cooperated by doing something that really mattered, then we have a valid measure of cooperation with a robot in workplace settings.

INDEPENDENT AND DEPENDENT VARIABLES. Researchers do studies because they want to explain some behavior. The behavior to be explained or predicted is the **dependent variable**; the variable that explains or predicts the dependent variable is the **independent variable**. In Lee's study the researchers sought to determine whether customers' exposure to personalized conversation with Snackbot (the independent variable) increased the likelihood they would anthropomorphize Snackbot and cooperate with his requests (the dependent variables). The relationship between independent and dependent variables is described in a **hypothesis**, or a prediction about the relationship between the independent and dependent variables. Two of the hypotheses Lee tested are:

Personalized conversation ⟶ anthropomorphize

Hypothesis 1: Customers who engage in personalized conversations with Snackbot are more likely to anthropomorphize Snackbot than those who engage in impersonal scripted conversation.

Personalized conversation ⟶ cooperation

Hypothesis 2: Customers who experience personalized conversations with Snackbot are more likely to cooperate with Snackbot's requests than those who experience impersonal scripted conversation.

If a hypothesis is supported by the data, then researchers can claim that if they know the value of an independent variable (whether a customer has engaged in personalized conversation), then they can predict the independent variable (anthropomorphize and/or cooperate with Snackbot). In addition to identifying independent and dependent variables, researchers also identify **control variables**, other variables that may affect the dependent and independent variable and that researchers hold constant. Think about it this way—there are any number of intervening variables that both relate to how someone might respond to personalized versus impersonalized conversation AND a decision to cooperate. One such variable might be gender. Perhaps female customers are more likely to value personalized conversation than are males, and for that reason personalized conversation is a significant factor in females' decisions to cooperate.

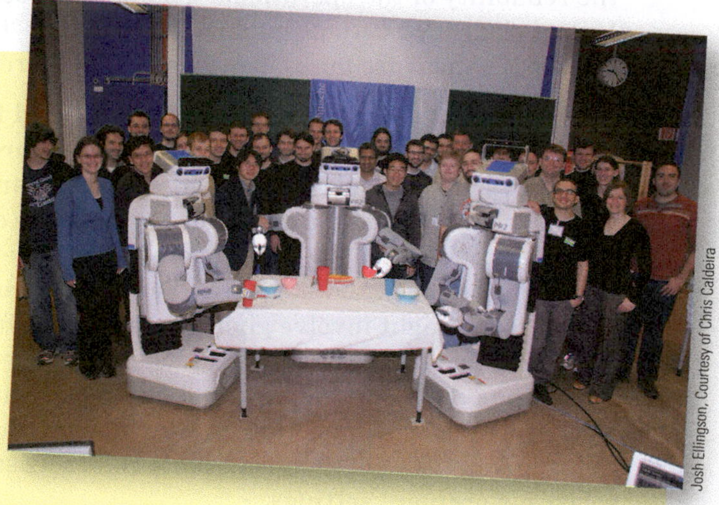

Josh Ellingson, Courtesy of Chris Caldeira

How do researchers hold variables such as gender constant? Imagine the employees in this photo were research participants in Lee's study. To control for gender, the researcher would divide the group into four groups (1) males, personalized contact with robot; (2) males, impersonalized contact; (3) females, personalized contact; and (4) females, impersonalized contact.

Then researchers compare the males' level of cooperation in Group 1 (impersonalized) with the males' level of cooperation in Group 2 (personalized). If males' levels of cooperation are the same in both groups, we can say that the type of conversation had no bearing on males' decisions to cooperate. Likewise, if we compare females' levels of cooperation in Group 1 (impersonalized) versus Group 2 (personalized) and we find that females who have engaged in personalized conversation are more cooperative than those who have not, then we can claim that personalized conversation increases females' propensity to cooperate. We do the same comparisons for transgender customers. Controlling for gender allows researchers to be more precise about for whom personalized conversation matters (or does not matter) in securing cooperation. If the level of cooperation is no different across the three gender categories—male, female, transgender—then we can say that in this study gender does not play a role in explaining cooperation. We could control for a number of social characteristics other than gender such as a customer's age, race, and status in the workplace. Controlling helps researchers be very clear about which kinds of customers are more likely to cooperate.

Collecting and Analyzing the Data

Researchers collect data that they then analyze to see if there is support for their hypotheses. When researchers analyze the collected data, they search for common themes and meaningful patterns. In presenting their findings, researchers may use graphs, tables, photos, statistical data, quotes from interviews, and so on. Lee and her colleagues presented data in the form of bar charts. Figures 1.6a and 1.6b summarize data collected which reveal the percentage of customers exposed to personalized and impersonalized conversation who cooperated with Snackbot's request to do a neck stretch and who agreed to try a recommended snack. We can see from the charts that personalized conversation dramatically increased the likelihood that customers would cooperate with Snackbot's requests to do neck stretches (Fig. 1.6a) and try a mystery snack (Fig. 1.6b).

Figure 1.6a: Percentage of Customers in Personalized and Impersonalized Conversation Groups Who Cooperated with Snackbot's Request to Do Neck Stretches

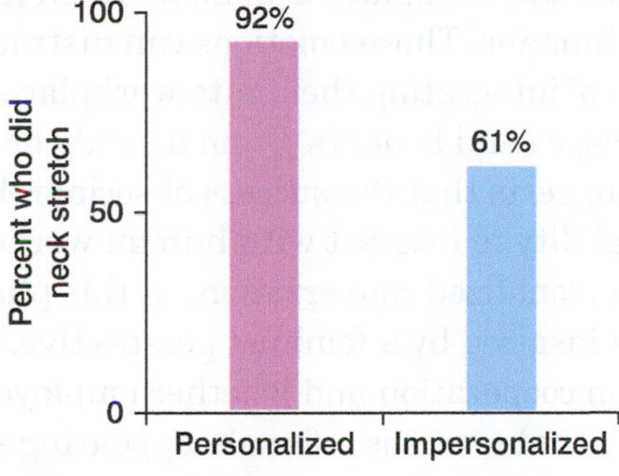

Source: Lee et al. (2012a)

Figure 1.6b: Percentage of Customers in Personalized and Impersonalized Conversation Groups Who Cooperated with Snackbot's Recommendation to Try a Mystery Snack

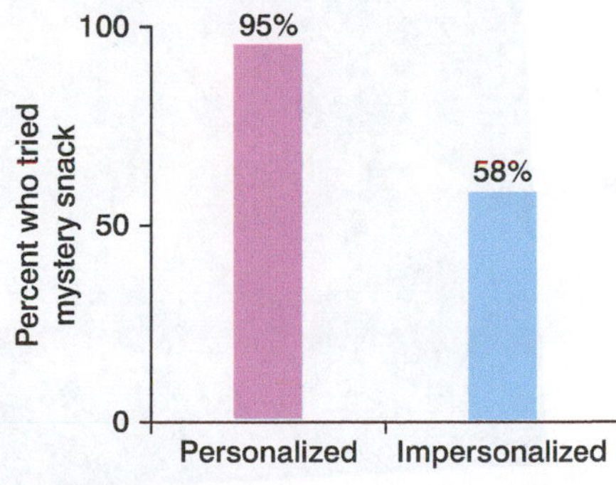

Source of data: Lee et al. (2012a)

Among other things, Lee analyzed the transcripts of conversations customers had with Snackbot, looking for patterns related to whether and when customers anthropomorphized the robot. She found that those in the personalized group were more likely to treat Snackbot like a person than those in the impersonalized group. Lee offered the following examples to illustrate. When Snackbot asked one customer assigned to the impersonalized conversation group to recommend locations to take tourists, that person answered, "Snackbot, let's not be ridiculous, can I take my snack?" In contrast, a customer assigned to the personalized group responded, "Let's see. You could visit the [exhibit name] on the first floor or the third floor. The second floor has a lot of cool other robotic stuff that you could check out or show people, I think they would like that."

Drawing Conclusions

For a research study to be significant, it must have implications that extend beyond the people or setting in the study. Lee maintained that her findings (not all of which are described in this module) have important implications for a future in which social robots will inevitably become part of the workplace. Her research confirmed that under the right conditions people can and do form collegial (even social) relationships with workplace robots.

We can also use the four perspectives to help clarify implications. From a functionalist's perspective, Lee's study speaks to anticipated (manifest) and unanticipated (latent) functions and dysfunctions. Robots can reduce human labor needed to do routine, repetitive tasks like snack delivery (a manifest function). People can become attached to robots (latent function). Emotions around social robots can fuel workplace conflicts and jealousies if people compete for the attention and favor of their robot "colleague" (latent dysfunction).

With regard to the conflict theory, Lee's findings suggest that one way to ease the transition of robots into the workplace is to make them likeable and even preferable to working with humans. Those emotions can distract people's attention away from the real purpose of integrating them into workplaces—to eliminate human labor and the costly wages and benefits. From a symbolic interactionist point of view, Lee's research suggests that the success of social robots in the workplace depends on a robot's ability to interact with human workers, most notably on their ability to engage in personalized conversation. In this phase of her research, Lee did not explore issues inspired by a feminist perspective, such as the robot's presumed gender effect on cooperation and whether employees or customers are more likely to comply with a male- versus a female-appearing social robot.

Josh Ellingson, Courtesy of Chris Caldeira

Lee's research also focused on human–robot interaction in the workplace and examined the process by which norms emerged to guide how customers treat robots. As such it fits especially well with the symbolic interactionist perspective. This is because Lee's research offers important insights about what factors contribute to successful human–robot interaction. The most important factor identified in Lee's research was the robot's ability to engage in personalized small talk.

Research studies conclude with a discussion about **generalizability**, the extent to which findings can be applied beyond the respondents studied. One might question the generalizability of Lee's study because she only studied one workplace setting, a floor in a computer science building on a college campus. In addition, that workplace setting involved computer science employees who may be predisposed favorably toward social robots. Finally, the customers she studied were not randomly chosen from all the employees who worked on that floor. Rather, the people she studied were self-selected. That is, they chose to sign up for a snack delivery service. We do not know if, and in what ways, the people who did not sign up were different from those who did. Perhaps they are less sociable, for example.

Lee acknowledged these potential shortcomings but she countered with the argument that the chosen setting might, in fact, prove more difficult to integrate a robot into because computer science types may be less likely to think of robots as social beings (to humanize them, for example). If that is in fact the case, then other audiences may prove even more susceptible to cooperating with, humanizing, and accepting robots into the work environment. Despite doubts about generalizability, Lee's research is still intriguing and offers valuable insights about human–robot interactions in workplace settings.

In evaluating the shortcomings of Lee's study, keep in mind that researchers cannot possibly study everyone. Instead, they study **samples**, or a subset of cases from a larger population of interest. Ideally, samples should be random, a situation in which every person or case has an equal chance of being selected. When a sample is randomly chosen, it increases the chances that it is representative of the population from which it is selected. We should note that there are compelling reasons to study nonrandom samples. For example, there had been no research about the process of integrating social robots into workplace settings. So Lee's research using a nonrandom sample fills a void.

Researchers can choose from a variety of other data-gathering strategies, including interviews, observation, secondary sources, case studies, and experimental design.

SURVEYS AND INTERVIEWS. Self-administered surveys and interviews are two ways in which sociologists gather data. A self-administered survey is a set of questions that respondents read and answer on their own. Respondents may be asked to write out answers to open-ended questions or to choose the best response from a list of potential responses (forced choice). The self-administered survey is one of the most common methods of data collection. Self-administered surveys have a number of advantages. No interviewers are needed to ask respondents questions, the surveys can be given to large numbers of people at one time, and respondents are less likely to feel pressure to give the socially acceptable response (especially when surveys are anonymous).

Some drawbacks of using self-administered surveys are that respondents can misunderstand the meaning of a question, skip over some questions, or just stop answering. In addition, self-administered questionnaires depend not only on respondents' decisions to fill them out and return them, but also on the quality of the survey questions asked, and a host of other considerations.

NKU Sociology, Missy Gish

In comparison to self-administered surveys, interviews are a more personal way to collect information. The interviewer asks questions and records the respondent's answers.

When respondents give answers during interviews, the interviewers must avoid pauses, expressions of surprise, inflections in their voice, or body language that reflect value judgments. Refraining from such conduct helps respondents feel comfortable and encourages them to give honest answers.

NKU Philosophy, Rudy Garns

Interviews can be structured or unstructured, or a combination of the two. In a structured interview, the wording and sequence of questions, and sometimes response choices, are set in advance and cannot be altered during the course of the interview. Recall that Lee surveyed participants to learn more about their background and interactions with Snackbot.

In contrast to the structured interview, an unstructured interview is flexible and open-ended. The question–answer sequence is spontaneous and resembles a conversation in that the interviewer allows respondents to take the conversation in an unplanned direction. The interviewer's role is to give focus to the interview, ask for further explanation or clarification, and probe and follow up on interesting ideas expressed by respondents.

No matter what type of interview is used, questions have to be clear and meaningful. Remember that writing good questions is much more difficult than it appears. Consider issues related to income. Researchers cannot simply ask, "What is your income?" but must specify if they mean hourly, weekly, monthly, or annual income; pretax or after-tax income; household or individual income; this year's or last year's income; or income from employment or other sources.

OBSERVATION. Observation involves watching, listening to, and recording human activity as it happens. This technique may sound easy, but the challenge lies in knowing how to observe and what is significant while still remaining open to other considerations. Good observation techniques are developed through practice and involve being alert, taking detailed notes, and making associations between observed behaviors. Observation is useful for (1) learning things that cannot be surveyed easily, and (2) experiencing the situation as those being observed experience it. One disadvantage is that observation research is time-consuming and specific to a particular setting. Observation can take two forms: nonparticipant and participant. Nonparticipant observation consists of detached watching and listening: The researcher only observes and does not become part of group life.

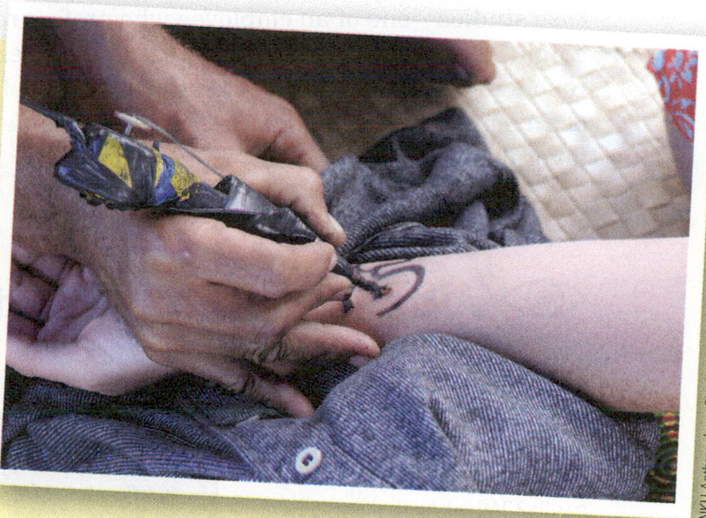
NKU Anthropology, Sharyn Jones

Researchers engage in participant observation when they join the group they are studying, assume a critical role in the life of a group, or participate in an experience that is critical to the group's identity. A researcher studying how the pain felt getting a tattoo plays a role in connecting people symbolically to some group (a gang or team) or to some valued idea (peace) might feel compelled to experience the pain associated with getting a tattoo.

In participant and nonparticipant observation, researchers must decide whether to hide their identity and purpose. One reason for choosing concealment is to avoid the **Hawthorne effect**, a phenomenon in which research subjects alter their behavior simply because they are being observed. This term originated from a series of worker productivity studies conducted in the 1920s and 1930s involving female employees of the Western Electric plant in Hawthorne, Illinois. Researchers found that no matter how they varied working conditions—bright versus dim lighting, long versus short breaks, piecework pay versus fixed salary—productivity increased. One explanation is that workers were responding positively to having been singled out for study (Roethlisberger and Dickson 1939).

SECONDARY SOURCES. Secondary sources or archival data has been collected for a purpose not related to the research study. This kind of data includes that gathered by census bureaus, research centers, and survey companies such as Gallup or Pew. Sociologists may use this already collected data to do their research. The advantages of secondary data are that it is often free or at least costs less to obtain than it would to collect. Governments and other large agencies have the resources to execute large-scale surveys of randomly chosen populations, something few researchers can do on their own.

Secondary data goes beyond surveys and includes biographies, photographs, letters, e-mails, websites, films, advertisements, graffiti, and so on. With this kind of data, sociologists often do what is called content analysis. That is, they identify themes, sometimes counting the number of times something occurs or specifying categories in which to place observations. A researcher who studies family photographs over time may look to see the extent to which pets are included or look for gender and age differences with regard to smiling.

CASE STUDIES. Case studies are objective accounts intended to educate readers about a person, group, or situation. Well-written case studies shed light through in-depth descriptions of an individual, an event, a group, or an institution. Case studies should tell a story with a beginning, a middle, and an ending. Researchers interested in social robotics in the workplace may choose to do four case studies—one of an employee who prefers robots to people; a second of an employee who is opposed to robots as colleagues; a third of an employee who saw robots in a favorable light at first but shifts feelings to unfavorable; and a fourth of an employee who begins with an unfavorable opinion of robots but changes to favorable.

What Do Sociologists See?

Sociologists see a potential research project. Why do sports teams' mascots seem to be predominantly male? What is the ratio of male to female mascots in high school, college, and professional sports? In what ways are schools with female mascots different from schools with male mascots?

Lance Cpl. Eric Quintanilla

Critical Thinking

Is there a topic you would like to do research on? Which method of data collection might you employ if you conducted research on this topic?

Key Terms

control variables
dependent variable
generalizability
Hawthorne effect
hypothesis
independent variable
operationalized
reliability
research design
research methods
samples
scientific method
validity
variable

Summary: Putting It All Together

Chapter 1

Sociology is the systematic and scientific study of society. To put it another way, sociologists work to understand how human interactions, relationships, and activities are organized with the aim of analyzing how that organization affects people's lives, thinking, and responses to others and to the world around them. Sociology offers a framework for analyzing the social forces that shape the way life around us is organized. This framework is especially relevant today, if only because the most pressing social issues we face are rooted in the ways in which human interactions, relationships, and activities have been organized. That framework includes the sociological imagination, a perspective that allows us to consider how outside forces shape our life stories or biographies and helps us to distinguish between troubles and issues.

Sociology emerged out of an effort to document and to explain a transformative social force, the Industrial Revolution. Those considered giants in the field—Comte, Marx, Durkheim, Weber, DuBois, and Addams—gave us conceptual tools for understanding how this transformative event is embodied in the division of labor, the means of production, solidarity, and the color line, all of which both connect and divide us from others in our community and beyond. Other conceptual tools allow us to think about the reasons and means we use to pursue valued goals and to achieve sympathetic understanding. In addition, sociology offers four major perspectives that guide any analysis. Those perspectives are functionalist, conflict, symbolic interaction, and feminist.

It is impossible to compile a list of the topics that sociologists study because almost any topic involving humans is a potential area of scrutiny. Sociology is distinguished not by the topics studied but by the four major perspectives it draws upon to frame any analysis. Each perspective presents us with guiding questions such as: (1) How does a part contribute to social order and stability? In what ways might a part disrupt order and stability? (2) Who benefits from a social arrangement and at whose expense? (3) How, when interacting, do people take account of others and then direct their conduct accordingly? (4) To what extent are gender inequalities reflected in the way human interactions, relationships, and activities are organized?

Sociologists adhere to the scientific method, a carefully planned research process with the goal of generating observations and data that can be verified by others. When sociologists do research, they decide on a research question, establish a research design, make observations, carefully collect and analyze data to test hypotheses, and draw conclusions.

Chris Caldeira

Chapter 2

Culture

Sociologists view culture as a key concept that captures the human capacity for devising ways to interact, to live together, and to negotiate the surrounding environment. Cultures do not exist in isolation; they "bump up against one another" and in the process "transform each other" (James Madison College 2014). The photograph that opens this chapter points to one way American and Korean cultures have "bumped" up against each other. The group shown on the flat screen is the Korean K-pop group Girls' Generation. Notice that the members are wearing blue jeans—an item of clothing that symbolizes "USA" in much the same way the American flag does (Timmerman 2012). But Girls' Generation members do not wear the jeans with the "come-as-I-am; I-am-an-individual"–like attitude that Americans tend to exude when they wear them (Timmerman 2012). Somehow the nine members of Girls' Generation have put a Korean touch on managing to present themselves as a cohesive unit with no one person dominating. In the meantime, K-pop has "bumped" up against American culture as college students in the U.S. fascinated with these Korean artists have pushed enrollments up in Korean language classes (Lee 2013).

Module 2.1 Culture

Objective

You will learn the meaning of culture and the challenges of defining a culture's boundaries.

Have you ever interacted with someone from another culture? What did you learn about that person's culture? Do you think that person learned something about your culture?

Tony Rotundo

Defining Culture

Wrestlers from Sweden (right) are known for technical precision; U.S. wrestlers (left) are known for being in extremely good shape and for their straightforward, aggressive, scoring-oriented style (Rotundo 2014). In international competitions, wrestlers consider the cultural influence their opponent brings to the sport. No doubt as the cultures "bump" up against each other, wrestlers see and perhaps incorporate another "way" into their own style.

In the broadest sense of the word, **culture** is the way of life of a people. To be more specific, culture includes the shared and human-created strategies for adapting and responding to one's surroundings, including people and other creatures that are part of those surroundings. The list of human-created strategies is endless: It includes the invention of physical objects such as cars to transport people from one place to another, values defining what is right and good, beliefs about how things in the world operate, a language with which to communicate, and rules guiding behavior.

A culture can be something as vast as U.S. culture or much smaller in scale, such as the culture of a family, a school, a community center, or a coffee shop. In our everyday use of the word *culture*, we often think of culture as something with clear boundaries and as something that explains differences among groups. You may be surprised to learn that sociologists face at least three challenges in defining a culture's boundaries:

- *Describing a culture*—Is it possible to describe something that encompasses the way of life of a people? How would you describe U.S. culture, Swedish or Korean culture? Or, on a smaller scale, how might you describe the culture of a high school or a local café?
- *Determining who belongs to a culture*—Does everyone who lives in the United States share a national culture? Does everyone who lives in Sweden or Korea share a national culture?
- *Identifying the distinguishing markers that set one culture apart from others*—Does a taste for pea soup and pancakes on Thursdays make someone a Swede? Does a taste for hamburgers make someone American? Or does a taste for rice make someone Korean?

Given these challenges, is culture a useful term? First, there is no question that cultural differences exist. For example, anyone who spends time in Sweden will note that people remove shoes before entering someone's home. Most people in the United States just walk into a home with shoes on. Likewise, anyone traveling to Saudi Arabia and then perhaps to Brazil will notice that women in the two countries dress quite differently when out in public. But once you think you have identified a clear marker, you can always find exceptions to the rule and see that it is difficult to find unique cultural markers that apply to everyone (Wallerstein 1990).

Culture as a Rough Blueprint

On some level culture is a blueprint that guides and, in some cases, even determines behavior. For the most part, people do not usually question the origin of the objects around them, the beliefs they hold, or the language they use to communicate and think about the world "any more than a baby analyzes the atmosphere before it begins to breathe it" (Sumner 1907, 76). Much of the time people think and behave as they do simply because it seems natural and they know of no other way.

Although culture is a blueprint of sorts, you will notice that people who are purported to share a culture are not replicas of one another. For example, Americans typically eat three meals per day—breakfast, lunch, and dinner associated with morning, noon, and evening. Still, all Americans do not eat at the same time or eat the same food. Of course, we can find people in the United States who skip breakfast or eat rice for breakfast, but they readily admit that in doing so they are not following what is considered typical in American culture. The fact that people are not cultural replicas makes it especially difficult to describe a culture and determine who belongs.

Cultural Universals and Particulars

Anthropologist George Murdock (1945) distinguished between cultural universals and particulars. **Cultural universals** are those things that all cultures have in common. Every culture has natural resources such as trees, plants, and rocks that people put to some use. In addition, every culture has developed responses to the challenges of being human and living with others. Those challenges include the need to interact with others, to be mentally stimulated, to satisfy hunger, and to face mortality. In every culture, people have established specific ways of meeting these universal challenges.

Cultural particulars include the *specific* practices that distinguish cultures from one another. For example, all people become hungry and all cultures have defined certain items and objects as edible. But the potential food sources defined as edible vary across cultures. That is, what is appealing to eat in one society may be considered repulsive or is simply unavailable in another.

Tony Rotundo

Tony Rotundo

This woman is eating a *kailuk*, as it is called in Laos. *Kailuk* is a duck embryo that is boiled alive and then eaten from the egg shell. Popular in Southeast Asian countries, *kailuk* is considered a snack food and a good source of protein.

All cultures provide formulas for expressing **social emotions**, feelings that we experience as we relate to other people, such as empathy, grief, love, guilt, jealousy, and embarrassment. In that sense, formulas for expressing these emotions are universal. Grief, for instance, is felt at the loss of a relationship; love reflects the strong attachment that one person feels for another person; jealousy can arise from fear of losing the affection of someone to another (Gordon 1981). People do not simply express a social emotion directly; they evaluate their true feelings and can modify the outward display of those feelings to fit with culturally established rules about how to express them. With regard to grief, some cultures embrace its open expression such that the bereaved are encouraged to wail loudly. Other cultures embrace self-control of grief to the point of encouraging their members to suppress any visible expression of grief (Galginaitis 2007).

Passing on Culture

The process by which people create and pass on culture suggests that it is a rough blueprint that people have the power to alter. Babies enter the world, and virtually everything they experience—being born, being bathed, being toilet trained, learning to talk, playing, and so on—involves others facilitating the experience. Those in a child's life at any one time—father, mother, grandparents, brothers, sisters, playmates, caregivers, and others—expose the child to their "versions" of culture. In this sense, people are carriers and transmitters of culture with a capacity to accept, modify, and reject the cultural blueprint to which they have been exposed. As a case in point, consider that Christmas is celebrated in the United States as if everyone in the country participates in the festivities. Businesses close on Christmas Eve and day; public schools give children Christmas week off; stores, houses, and streets are decorated as early as October; and Christmas-themed television commercials run for a month or more. Yet despite this exposure, many people in the United States reject this cultural option, as this reflection from one of my students illustrates:

> I grew up in a religion that did not celebrate Christmas or Easter even though we called ourselves Christian. . . . While growing up I never challenged this; I just did what I was told.

Yet this same student as an adult decided to celebrate Christmas and other holidays that her family rejected in her youth.

> Now that I am older and have a child . . . [t]his year was the first year I put up a Christmas tree. I didn't know how to decorate it and I didn't know any Christmas carols but I learned. I decided that I don't want my children growing up the way I did.

When sociologists study cultures, they do not get caught up in identifying distinct markers that set people of one culture apart from those of another (e.g., all Christians celebrate Christmas). Nor do they assume that physical cues (eye shape, hair texture, skin shade) qualify someone for membership in a culture. Instead, they are most interested in how culture shapes human behavior and in how people create, share, pass on, resist, change, and even abandon culture.

NKU Sociology, Missy Gish

MSgt Dawn Price

Children everywhere have relationships with animals. Children, by themselves, do not invent cultural uses for the animals around them; they learn what an animal means to their culture by observing how it is treated and listening to how others speak about it. Most American children are likely to encounter goats only if an adult takes them to a petting farm. They are likely to think a goat as having the name "Billy," as liking to eat tin cans, but not as a pet. Children from Uganda think of goats in much different ways: as a source of meat to be eaten at Christmas, Easter, weddings, and on other important occasions. They are likely aware that goat skins are used to make mats, clothing, carrying bags, knife handles, and covers for milk gourds (Okello 2013).

What Do Sociologists See?

It is a cultural universal that everywhere parents introduce their children to foods of their culture. The kinds of foods vary by culture. This photo reminds us that parents are carriers and transmitters of culture but that their children have some say in whether they will embrace or reject what parents present to them.

Cpl. Andrew Johnson

Critical Thinking

Do you think of yourself as belonging to a culture? Explain.

Key Terms

cultural particulars

cultural universals

culture

social emotions

Module 2.2

Material and Nonmaterial Culture

Objective

You will learn that culture consists of two components: material and nonmaterial.

Chris Caldeira

Chris Caldeira

What messages does each lawn mower convey about the larger culture of which it is a part?

The lawn mower on the left would elicit little, if any, attention in Cuba (the place where it was photographed) but would certainly elicit a reaction from the typical American. What does that lawn mower on the left say about Cuban culture? What does the lawn mower on the right say about U.S. culture? Both lawn mowers are part of what sociologists call material culture. But to understand why each mower is so different in appearance, we must look at the larger cultural context.

Material Culture

Material culture consists of all the natural and human-created objects to which people have assigned a name and attached meaning. Examples of material culture are endless and include smartphones, video games, clothing, tattoos, trees, diamonds, and much more. In order to grasp the social significance of material culture, sociologists strive to understand the larger context in which an object exists. They also work to identify the meanings people assign to that object and the ways it is used. From a sociological point of view, material objects are windows into a culture because they offer clues about how people relate to one another and about what is valued.

Chris Caldeira

The photo of the Cuban lawn mower that opens this module and this photo of a 50-year-old automobile under repair on the streets of Cuba suggest that Cubans throw away nothing. These photos prompt us to ask how Cuba became a nation of recyclers. In 1960, after Fidel Castro seized U.S. assets and subsequently declared Cuba a socialist country, the United States imposed an economic embargo and broke diplomatic relations with the country. Cuba formed close ties with the then Soviet Union, receiving $4 to $6 billion in foreign aid each year from its ally. When the Soviet Union collapsed in 1989, Cuba lost an important revenue source. This loss, in conjunction with more than 50 years of embargo, has made economic hardship a way of life for the Cuban people—a hardship that supported the creation of a culture where almost nothing is thrown away.

Nonmaterial Culture

In contrast to material culture, **nonmaterial culture** refers to the intangible human creations. Intangible means that these creations are not concrete objects that can be seen directly or touched with the hands. Nonmaterial culture includes values, beliefs, norms, and symbols.

VALUES. Values are general, shared conceptions of what is good, right, desirable, or important with regard to personal characteristics, ways of conducting the self, and other desired states of being. While it is impossible to make a complete list of shared values, examples of things valued include individual freedom, happiness, consumption, conservation, generosity, cleanliness, obedience, independence, and national security. Cultures are distinguished from one another not according to values that are unique to each culture, but rather according to which values are commonly invoked as reasons for taking some action (like going to war in the name of freedom or deciding to give a 6-year-old his own smartphone in the name of safety) and which values are most cherished and dominant (Rokeach 1973). One might argue, for example, that American culture values consumption over conservation and that Cuban culture values conservation over consumption.

BELIEFS. Beliefs are shared conceptions that people accept as true concerning how the world operates and the place of the individual in relationship to others. In contrast to values, beliefs are about what is or is not true or real. People hold a variety of beliefs that can be accurate or inaccurate. Beliefs may be descriptive (believing the earth is round), causal (believing that fluoride prevents cavities), or prescriptive (governments should not intervene in people's lives) (Rokeach 1973). Beliefs can be rooted in blind faith, lived experience, tradition, or science. With regard to beliefs about what makes for a happy life, many Americans hold the belief that wealth and consumption can bring happiness. In contrast, many Cubans hold the belief that happiness can be achieved by getting the most use out of what one has. The beliefs Cubans and Americans tend to hold about what actions result in happiness derive from the opportunities open to them and the kinds of behavior each society encourages.

For the most part, people in the United States are encouraged to throw away things without thinking of other uses to which they might be put. Americans are not encouraged, for example, to think about how they might use an empty toilet paper roll. People living in Cuba, on the other hand, are encouraged to think about such things—empty toilet paper rolls can function as hair curlers.

NORMS. A third type of nonmaterial culture is **norms**, written and unwritten expectations for behavior, thought, and appearance deemed appropriate to a particular social situation. Examples of written norms are rules that appear in college student handbooks (e.g., to be in good academic standing, maintain a 2.0 GPA), on signs (smoke-free area), and on garage doors of automobile repair centers (Honk Horn to Open). Unwritten norms exist for virtually every kind of situation: wash your hands before preparing food, raise your hand to indicate that you have something to say, and do not throw food away. Sometimes norms are formalized into laws. **Laws** are rules for behavior that are typically put in writing and enforced by agents holding jobs that exist to keep order (e.g., police, military, judges).

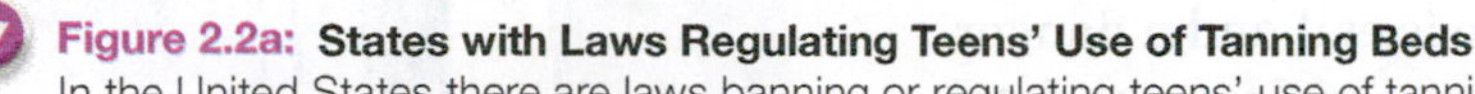

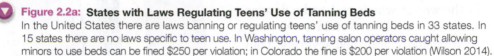

Figure 2.2a: States with Laws Regulating Teens' Use of Tanning Beds
In the United States there are laws banning or regulating teens' use of tanning beds in 33 states. In 15 states there are no laws specific to teen use. In Washington, tanning salon operators caught allowing minors to use beds can be fined $250 per violation; in Colorado the fine is $200 per violation (Wilson 2014).

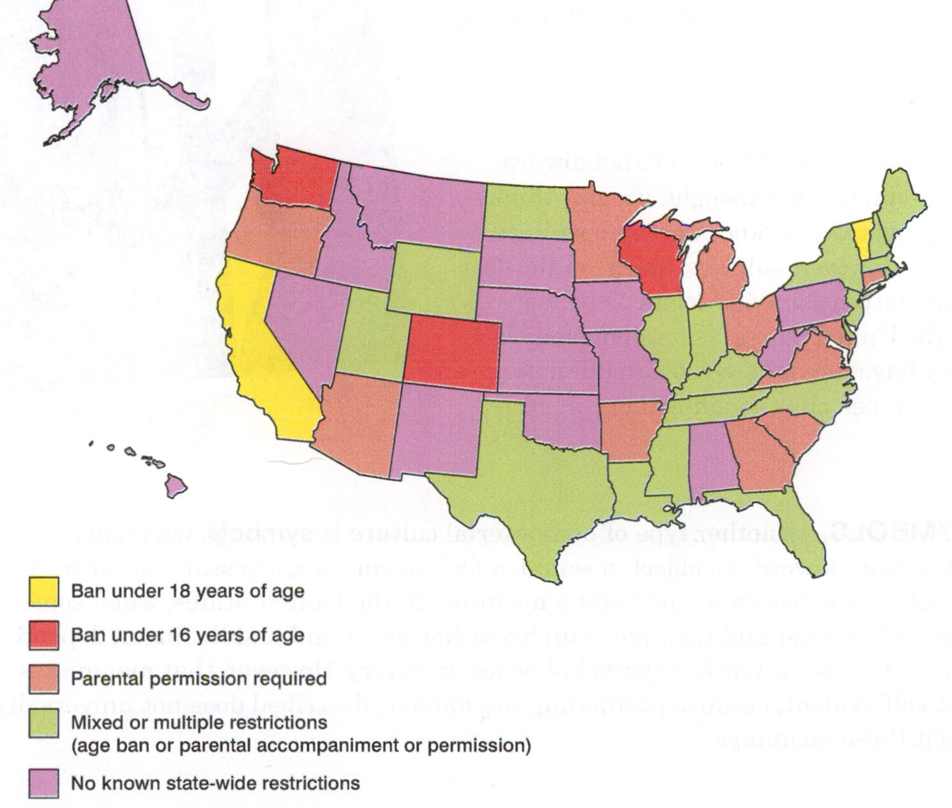

Source of data: National Conference of State Legislators (2014).

Laws specify penalties such as fines or jail time for violation. There are so many laws on the books that it would be impossible to list them. In 2014 alone, 40,000 new laws went into effect somewhere in the United States including the following: in Oregon mothers can now take home placentas after a baby is born; in Illinois animal rights activists are prohibited from using drones to watch hunters; and in California students are allowed to use bathrooms and try out for sports teams "consistent with their gender identity" (Johnson 2013).

Depending on the importance of the norm, punishment for violations can range from social rejection to the death penalty. With regard to importance, we can distinguish between folkways and mores. **Folkways** are norms that apply to the details of daily life: what time of day to eat, how to greet someone, and how to dress for a school event such as a prom. As sociologist William Graham Sumner noted, "Folkways give us discipline and support of routine and habit"; if we were forced constantly to make decisions about these details, "the burden would be unbearable" (1907, 92). Generally, we go about everyday life without asking "why?" until someone violates a folkway, at which point the violator is typically labeled as peculiar, strange, weird, and unconventional.

In contrast to folkways, **mores** are norms that mandate that a code of conduct be followed because adhering to that code is believed essential to a group's well-being. When someone violates a more, they are cast as someone in the wrong, who is evil and a danger to the society. People who violate a society's mores are usually punished severely: They may be banished, institutionalized, or killed. Mores are regarded as the only way.

NKU Sociology, Missy Gish

Folkways and mores are not always clear-cut. On first thought, we may think it is a folkway to "knock before opening a door." However, failure to knock on the door could be considered a violation of a more in the United States, as residents may feel they have the right to shoot to kill if they think a person is invading their property.

SYMBOLS. Another type of nonmaterial culture is **symbols**, which are anything—a word, an object, a sound, a feeling, an odor, a gesture, an idea—to which people assign a name and a meaning. In the United States, when someone makes a fist and then holds up his or her index and middle finger, depending on the context, it is a symbol of peace or victory. However, that meaning is not self-evident, because positioning the hand as described does not universally elicit these meanings.

Depending on place and time, a hand held in this way can convey victory or peace. As a victory sign it might be made in the context of a team win or after a defeat of another. But the gesture can have other meanings. When tourists take photos of children who live in Fiji, the children automatically gesture in the way pictured as if to say, "Hey, here we are!!!!"

NKU Anthropology, Sharyn Jones

In the broadest sense of the word, **language** is a symbol system that assigns meaning to particular sounds, gestures, characters, and specific combinations of letters. The complexity of human language is believed to set people apart from animals. Arguably, language is the most important symbol system people have created. When we learn the words of a language, we acquire a tool that enables us to establish and maintain relationships, convey information, and interpret experiences. Learning a language includes an expectation that we will communicate and organize our thoughts in a particular way (Whorf 1956, 212–214). For example, some languages are structured so that speakers have no choice but to address people using special age-acknowledging titles. For example, in Korea age relative to the speaker is acknowledged in every encounter, even when the other is a twin who is minutes younger or older. As another example, in the United States the word *my* is used to express ownership of persons or things over which the speaker does not have exclusive rights: *my* mother, *my* school, *my* school bus, *my* country. The use of *my* reflects an emphasis on the individual and not the group. In contrast, some languages such as Korean express possession as shared: *our* mother, *our* school, *our* school bus, *our* country.

Linguists Edward Sapir and Benjamin Whorf (1956) advanced the **linguistic relativity hypothesis**, also known as the Sapir–Whorf hypothesis, which states that "No two languages are ever sufficiently similar to be considered as representing the same social reality." The worlds of those who speak different languages "are distinct worlds, not merely the same world with different labels attached" (Sapir 1949, 162). Although languages channel thinking in distinct ways, do not assume that those speaking different languages cannot communicate. It may take some work, but it is possible to translate essential meanings from one language into another. For example, it is certainly possible to translate Korean words into English—a translator can emphasize *our* teacher, not *my* teacher—but lost in translation will be a Korean worldview that actually thinks in terms of *our* teacher. It is difficult for an English-speaking American who has no firsthand experience with Korean culture, which makes the group the central point of reference (rather than the individual), to completely grasp the meaning of "our."

Chris Caldeira

How might the meaning of *Venceremos!* be translated and conveyed to English speakers in the United States so that they understand its meaning in the same way people living in Cuba do? For Cubans, *Venceremos!* means "We shall overcome," and it is used in reference to the revolutionary socialist goal of achieving equality and dignity for all. Just as Americans view the word *freedom* as an ideal worth fighting for, Cubans see the word *Venceremos* as an ideal worth fighting for, even as the country adopts some capitalist principles. *Venceremos* has driven the Cuban revolution since 1960 to resist U.S. (and capitalist) influence and interference, which supporters believe undermines equality and dignity.

What Do Sociologists See?

When sociologists observe these young women in Vietnam taking out their umbrellas to shield their skin from the sun's rays, sociologists think about what a tan symbolizes to Americans considered white and how a tan likely symbolizes something different to those considered people of color. The photograph suggests that it is a norm to use an umbrella to protect skin from the sun. A tan is not viewed as something that makes Vietnamese feel younger or more attractive as it does for many Americans. Rather, a lighter complexion is a symbol of beauty.

Critical Thinking

Have you ever interacted with people you believed were from a culture different than your own? Using at least two concepts from this module (material culture, values, etc.), describe some features that distinguish the two cultures.

Key Terms

beliefs
folkways
language
laws
linguistic relativity hypothesis
material culture
mores
nonmaterial culture
norms
symbols
values

Module 2.3 Cultural Diversity

Objective

You will learn the meaning of cultural diversity and how to think about cultural variety.

Would you consider this elementary school class to be culturally diverse? If yes, why? If no, what does a diverse student body look like?

U.S. Air Force photo/Senior Airman Barry Loo

Sociologists use the term **cultural diversity** to capture the cultural variety that exists among people who share some physical or virtual space. That space may be as large as the planet or as small as a household. When the focus is on *cultural* diversity, sociologists examine the extent to which people in a particular setting vary with regard to physical appearance but also vary with regard to

1. material culture—the objects they possess and have access to and the meanings (positive or negative) assigned to those objects; and
2. nonmaterial culture—the values, beliefs, norms, symbolic meanings, and language guiding behavior and thinking.

Sociologists often look at the people who occupy a particular setting, such as an elementary school, and then try to establish the extent to which cultural diversity is present. In glancing at the photo of the class that opens this module, we might consider whether the people pictured reflect the racial and gender diversity of the United States. It is not that everyone classified as the same race and gender shares a specific culture; rather, they are likely to share similar cultural experiences related to being classified and perceived as a particular race or gender. That is, they experience, embrace, or struggle with norms, values, beliefs, and symbolic meanings that have come to be associated with their categories. For example, people believed to be a particular gender or race often face intense pressures to dress, live, and act in certain ways.

We do have to go beyond physical appearances to make a judgment about cultural diversity. Cultural capital is a useful concept for assessing the level of diversity that exists in a particular setting. In the most general sense, **cultural capital** includes all the material and nonmaterial resources a person possesses or has access to that are considered useful and desirable (or not) in a particular social setting. We can think of cultural capital as objectified, embodied, and institutionalized (Bourdieu 1986).

Objectified cultural capital consists of physical and material objects that a person owns outright or has direct access to. These objects have a monetary value tied to others' willingness to buy, sell, own, and hold on to them. These objects also have symbolic value because they convey meaning about the owner's, seller's, or buyer's status. Finally, objectified cultural capital includes the ability to understand, appreciate, and convey an object's meaning and value.

If we use the concept of objectified cultural capital to think about this homemade board game and its value in a specific setting, say American society, we might conclude that it would bring the owner little, if any, acclaim, nor would it be an object that most American children would covet (relative to electronic games or videos or the latest toy). In another setting, Cuba, this board game may be considered an innovative and clever way to recycle materials.

Chris Caldeira

Embodied cultural capital consists of everything that has been consciously and unconsciously internalized through the socialization process. The socialization process shapes a person's character and outlook. Embodied cultural capital includes the words and languages one hears, has acquired, and has used to communicate with others, think about the world, and present the self to others.

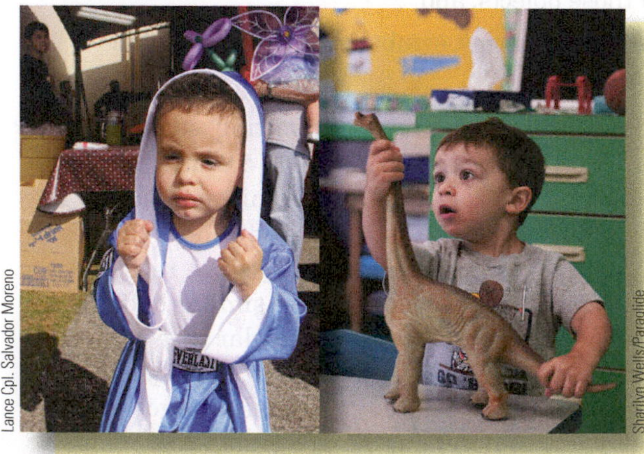
Lance Cpl. Salvador Moreno

Sharilyn Wells/Paraglide

Imagine that parents manage to instill in these toddlers an enduring interest in dinosaurs and in boxing. How might these respective interests shape each child's character and outlook? What words will each child learn as he pursues the cultivated interest? How might the respective experiences shape the way each child presents himself to others?

Institutionalized cultural capital consists of anything (material or nonmaterial) recognized by the larger society as important to success in a particular social setting. Examples include academic credentials for a job search, the ability to dress the part (to look like a doctor, professor, or carpenter), physical qualities such as straight white teeth noticed by clients/customers, a youthful appearance when entering a bar, and so on. Whatever the attribute or item—a professional degree, fashionable clothes, a nice smile—the institutionalized backing gives those who possess it advantages beyond the attribute or item itself.

NKU Sociology, Missy Gish

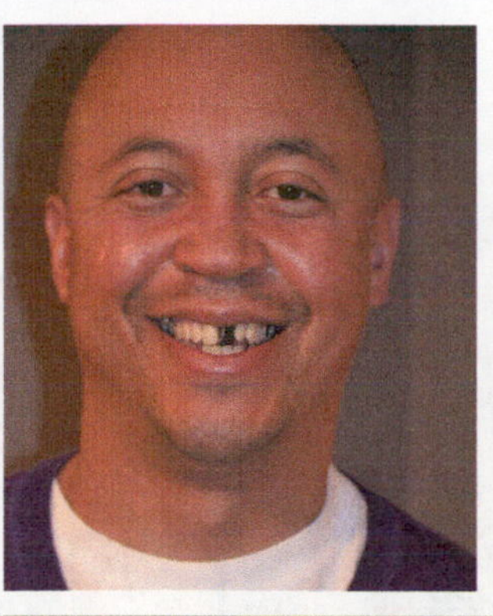
NKU Sociology, Missy Gish

In the United States (and elsewhere), teeth have important symbolic meanings that go beyond just the look of teeth per se. This is a photo of the same person with his front teeth intact and with a front tooth missing. Do assumptions about his intelligence, wealth, occupation, and marital status change depending on the presence or absence of a tooth? The perfect smile, institutionalized through commercials and dental hygiene products, has become a desired attribute.

The three concepts—objectified, embodied, and institutionalized cultural capital—help us think about what people bring with them to a setting. Given the endless variety of cultural experiences people can have, you see that establishing the degree to which a social setting is culturally diverse is not an easy task. Universities often highlight campus diversity by profiling their student body according to geographic location (in-state vs. out-of-state residents, international students), sex, race, ethnicity, and age. In addition, they highlight the number of student clubs and organizations to showcase the variety of cultural experiences available. Because universities present diversity as something that "enhances the quality of the living and learning environment," many offer diversity scholarships and leave it to the applicant to make a case for how "aspects of your identity, your life experiences, special skills or values equip you to make a positive contribution to help ensure that the [campus] is rich with diversity, yet inclusive of all its members" (Northern Kentucky University 2014).

If you were to try to inventory cultural diversity on a college campus (or in any setting, for that matter), it is likely that as you visited with various campus groups that attract students of color, veterans, women, disabled, or LBGTQ, you might encounter members who could not completely agree on how to portray their culture, nor could they say exactly what distinguishes them from other cultural groups on campus. Regardless, each group would likely remain "convinced of their singularity" and even react emotionally to any suggestion that they were not part of a distinct culture (Mahmood and Armstrong 1992). What, then, holds together people who present themselves as a culture? One answer is **cultural anchors**, some cultural component—material (a color, a mascot, a type of clothing, a book) or nonmaterial (a belief, value, norms, language)—that elicits broad consensus among members regarding its importance but also tolerates debate and dissent about its meaning (Ghaziani and Baldassarri 2011).

Chris Caldeira

Given the diversity that exists in the United States, is there even such a thing as American culture? We would be hard-pressed to describe THE culture that everyone in the United States shares. It is possible to identify a cultural anchor that unites most if not all Americans even in the face of bitter debate and dissent. One anchor is the flag, because it symbolizes a value considered important to all Americans—"freedom"—whether it be freedom to pursue happiness or to live a certain lifestyle free from interference. Notice that whenever there is protest in the United States, the protesting groups (in this case, LGBT [lesbian, gay, bisexual, and transgender] activists) wave the American flag to rally support. The concept of cultural anchor applies to subcultures in the United States such as LGBT activists, who come from all walks of life. Their cultural anchors include their gay pride flag and the belief that people should be free to choose whom they love.

Subcultures and Countercultures

When thinking about cultural variety, the concepts of *subcultures* and *countercultures* are especially useful. Every society has **subcultures** that share in certain parts of the mainstream culture but that possess cultural anchors—values, norms, beliefs, symbols, language, and/or material culture—not shared by those in the so-called mainstream. The anchors may be associated with a physical setting (church, community center, neighborhood) or some selected aspects of life, such as work, school, a type of recreation, marriage, fashion, or neighborhood.

Cowboy Churches are one example of a subculture. There are 862 such listings in the CowboyChurch.Net (2014) directory. The Cowboy Church movement began in the 1990s in the southwestern part of the United States, especially those areas that identify with the cultural anchor known as the cowboy. These churches are generally nondenominational and welcome all faiths and beliefs (Clark 2014; Hyslop 2011). Members come to church dressed in attire we associate with cowboys (hats, boots, jeans). Pastors may preach sitting on a horse. No collection plates are passed, although there may be a boot, hat, or wooden birdhouse to drop money in when leaving (Associated Press 2009).

Sociologists use the term **countercultures** in reference to subcultures that challenge, contradict, or outright reject the mainstream culture that surrounds them. Sociologist Milton Yinger maintains that members of countercultures feel strongly that the society as structured cannot bring them satisfaction; some believe that "they have been caught in very bad bargains, others that they are being exploited," and still others think the system is broken (1977, 834). Because countercultures emerge in response to an existing order, Yinger argues that "every society gets the countercultures it deserves." Countercultures deplore society's contradictions, present caricature-like descriptions of what are believed to be its weaknesses, and take action to revive neglected traditions (850). Countercultures "attack, strongly or weakly, violently or symbolically," the social order that they find so frustrating" (834). Yinger presents three broad, and at times overlapping, categories of countercultures:

- Communitarian utopians withdraw into a separate community where they can live with minimum interference from the larger society, which they view as evil, materialistic, wasteful, or self-centered. In the United States, the Old Order Amish constitute a communitarian counterculture in that they remain largely separate from the rest of the world, organizing their life so that they do not even draw power from electrical grids.
- Mystics search for "truth and for themselves" and in the process turn inward. "They do not so much attack society as disregard it, insofar as they can, and float above it in search of enlightenment" (838). Buddhist monks constitute such a counterculture because they make a point of rejecting the material trappings of capitalistic society. As monks, they are committed to simple living, modest dress, and a vegetarian diet—ways of living that run counter to the values of capitalist-driven societies.
- Radical activists preach, create, or demand a new order with new obligations to others. They stay engaged, hoping to change society and its values. Strategies to

bring about change can include violent and nonviolent protest; Gandhi and Martin Luther King Jr. considered radical activities in their day, and both famously used nonviolent protest to effect societal change.

*Library of Congress Prints and Photographs Division [LC-DIG-npcc-30454]

The Ku Klux Klan a radical activist subculture, gained membership and support at important transitional points in American history. Two points stand out in particular as times of great change to the existing social order: at the end of slavery (the late 1860s) and after World War II (1945), the time when the civil rights movement began building momentum. The support for the KKK grew at these times because supporters did not like the new direction the country was moving.

What Do Sociologists See?

Sociologists see rock climbers as a subculture whose members value risk, more accurately controlled risk (West and Allin 2010). To counter risk, climbing requires total concentration and engagement with the surroundings. Climbers cannot be afraid; they must be committed to reaching the next hold (The Rock Club 2014). Rock climbers also place high value on trust as they must build connections to other climbers who must be devoted to keeping each other safe during a climb.

Karl Weisel (USAG Wiesbaden)

Critical Thinking

Are you a member of a subculture or counterculture? Describe the cultural anchors important to the group.

Key Terms

countercultures
cultural anchors
cultural capital
cultural diversity
embodied cultural capital
institutionalized cultural capital
objectified cultural capital
subcultures

Module 2.4

Encountering Cultures

Objective

You will learn concepts for describing people's reactions to foreign cultures.

When you brush your teeth do you leave the water running while you brush? Do you ever think to turn the water off while brushing and then turn it back on to rinse the toothbrush before returning it to its holder?

If you lived in Germany all your life, you would very likely turn the water off while brushing your teeth. You would also be likely to take short showers and turn the shower water off while you lathered and scrubbed your hair and then turn the water back on to rinse.

Ethnocentrism

If you uncritically accept the idea that water should run the entire time one brushes one's teeth or showers, and you believe that it is absurd to conserve water in the ways mentioned above, then your point of view can be considered ethnocentric. **Ethnocentrism** is a point of view in which people use their home or other culture as the standard for judging the worth of another culture's ways. Sociologist Everett Hughes describes ethnocentric thought in this way: "One can think so exclusively in terms of his own social world that he simply has no set of concepts for comparing one social world with another. He can believe so deeply in the ways and the ideas of his own world that he has no point of reference for discussing those of other peoples, times, and places. Or he can be so engrossed in his own world that he lacks curiosity about any other; others simply do not concern him" (1984, 474).

Ethnocentrism puts one culture at the center of everything, and all other ways are "scaled and rated with reference to it" (Sumner 1907, 13). Thus, other cultures are seen as strange or inferior. Often the words people choose to evaluate cultural differences offer clues as to whether they are thinking ethnocentrically (Culbertson 2006).

Sgt. Steven King

Notice that this woman is writing in Arabic language which is written from right to left. If you conclude that she is writing backwards, then you have described the writing process in an ethnocentric way. From the point of view of the Arabic reader, writing English left to right could be considered backward.

Using one's home culture as the frame of reference to describe another culture's ways distorts perceptions. Many Americans are guilty of using their home culture to frame thinking about cultures that eat dog. Many Americans are appalled that some cultures eat dog because they believe that that dogs possess special qualities that other animals commonly eaten by Americans, such as pigs, cows, and chickens, do not.

SRA Diane S. Robinson

Keep in mind that people from cultures where dog is defined as a source of food don't eat *pet* dogs; rather, they eat a "special breed of large tan-colored dogs raised especially for canine cuisine" (Kang 1995, 267). In fact, those who eat dog would argue that Americans, including these men slicing meat from a roasted pig, are in no position to judge (Kang 1995). For some reason, many Americans protest that they could never eat Fido but seem to have no reservations about eating "Porky Pig" or "Babe the Pig."

The most extreme and destructive form of ethnocentrism is one in which people feel such revulsion toward another culture that they act to destroy it. Unfortunately, in human history there are many examples where one group acts to destroy another by banning the targeted culture's language or ethnic-sounding first and last names; destroying important symbols such as flags, places of worship, and museums; and even killing those who resist. In the United States between 1870 and 1920, Native American children were sent to boarding schools to ensure total immersion in white culture. This process separated them from their families and home cultures. Upon arrival children were issued uniforms, assigned names considered white, and given haircuts. They were punished for speaking their native languages.

Native Americans enrolled in boarding schools were also introduced to sports such as football to socialize them to American ideas about competition, winning, status, and teamwork. Taken in 1899, this photograph is of the Carlisle (PA) Indian Industrial School football team.

The concept ethnocentrism applies to those who regard their own culture as inferior to another culture. People who engage in this kind of thinking often idealize the other culture as a utopia. For example, an American might idealize Asian cultures as models of harmony, Swedish culture as a model of equality, and Native American cultures as models of environmental sustainability (Hannerz 1992).

Cultural Relativism

Cultural relativism is an antidote to ethnocentrism. **Cultural relativism** means two things: (1) that a foreign culture should *not* be judged by the standards of a home or some other culture, and (2) that a behavior or way of thinking must be examined in its cultural context—that is, in terms of that culture's values, norms, beliefs, environmental challenges, and history. For example, to understand the German propensity to conserve water when showering, we have to place that practice in the context of Germany's decades-long efforts to encourage environmentally conscious behavior. For one, the German government taxes household water consumption at a high rate. Likewise, to understand why people in some countries eat dog, we must consider that the practice is likely tied in large part to a shortage of grazing land to support large-scale cattle production.

Critics of cultural relativism maintain that this perspective encourages an anything-goes point of view, discourages critical assessment, and portrays all cultures as equal in value regardless of obviously cruel practices (Geertz 1984, 265). In response to this criticism, sociologists argue that there is no question that notions of rightness and wrongness vary across cultures, and if we look hard enough we can probably find a "culture in which just about any idea or behavior exists and can be made to seem right" (Redfield 1962, 451). A position of cultural relativism is not taken to condone or discredit a culture. Rather the primary aim is to understand a culture on its own terms. More than anything, cultural relativism is a point of view that acts as a check against an uncritical and overvalued acceptance of the home culture, thereby constricting thinking and narrowing sympathies (Geertz 1984).

When using their home culture as the standard, Americans often make fun of sumo wrestling and have a hard time appreciating it as a serious sport. An ethnocentric American might describe a sumo wrestling match as two "fat" guys trying to push each other out of the ring. Cultural relativism reminds us to place the sport in the context of Japanese culture and history. A serious study of the sport would reveal that it is "rich with tradition, pageantry, and elegance and filled with action, excitement, and heroes dedicated to an almost impossible standard of excellence down to the last detail" (JapanReference 2014).

Tony Rotundo

Culture Shock

Most people come to learn and accept the ways of their home culture as natural. Thus, when they encounter foreign cultures, they may experience **culture shock**, a mental and physical strain that people can experience as they adjust to the ways of a new culture. In particular, newcomers find that many of the behaviors and responses they learned in their home culture, and have come to take for granted, do not apply in the foreign setting. For example, newcomers may have to learn to squat (rather than sit) when using the toilet and learn to eat new foods. Or newcomers may experience "invisible" pressures they do not feel in their home culture, as described by this man who moved to the United States from a rural community in Mexico: "I lived in a laid-back, close knit community. The survival of the fittest attitude that exists in the United States made me feel that everyone was running from some sort of monster" (Anonymous [NKU student] 2012).

It is not that any one incident generates culture shock; rather, it is the cumulative effect of a series of such adjustments that can trigger an all-encompassing disorientation. The intensity of culture shock depends on several factors: (1) the extent to which the home and foreign cultures differ; (2) the level of preparation for living in a new culture; and (3) the circumstances—vacation, job transfer, or war—surrounding the encounter. Some cases of culture shock can be so intense and unsettling that people experience "obsessive concern with cleanliness, depression, compulsive eating and drinking, excessive sleeping, irritability, lack of self-confidence, fits of weeping, nausea" (Lamb 1987, 270).

A person does not have to live in a foreign culture to experience culture shock. People who move from West or East Coast cities to towns in the Midwest, and vice versa, can experience it. Children who attend private faith-based schools for eight years and then transfer to a public high school (or the reverse) can experience culture shock. When Hurricane Katrina devastated the Gulf Coast in 2005, 1 million people were forced from this region to live in scattered locations across the United States, resulting in the largest mass dislocation of people in U.S. history. Many from urban New Orleans found themselves living in very small towns and reported experiences of culture shock. As one example, 100 residents of New Orleans classified as black evacuated to Seguin, Texas, a town of 23,000 with 60 percent of residents considered Hispanic and 31 percent classified as white. They had nothing but the clothes on their backs and had to rely on the goodwill of strangers. Understandably they experienced cultural shock:

> Where were the city buses that they had relied on to get around in New Orleans? Where were all the black people? Where were folks sitting on porches, in a festive tangle of music and gossip? . . . Wrangler jeans and cowboy hats were among the donated items. . . . Heifers grazed nearby. . . . In the afternoon the breeze smelled like burned chicken from the nearby Tyson's poultry plant. (Hull 2005)

Reentry Shock

Do not assume that culture shock is limited to experiences with cultures outside of a home country. People can also experience **reentry shock**, or culture shock upon returning home after living in another culture (Koehler 1986).

Many people find it surprisingly difficult to readjust when they return home after spending a significant amount of time elsewhere. As with culture shock, returnees face a situation in which differences jump to the forefront. Imagine, for example, that you have lived for years in a culture such as a rural village in Vietnam where the pace of life was much slower and where parents are less worried about danger lurking around the corner that might harm their children.

Chris Caldeira

As with culture shock, the intensity of reentry shock depends on an array of factors, including the length of time someone has lived in the host culture and the extent to which the returnee has internalized the ways of the host culture. Symptoms of reentry shock mirror those of culture shock. They include panic attacks ("I thought I was going crazy"), glorification of and nostalgia for the foreign ways, a sense of isolation, and a feeling of being misunderstood. These comments by one American student returning from abroad illustrate: "Why do so many Americans buy such big cars? Why do Americans ask, 'How are you?' but really have no time to listen?" In fact, many people undergoing reentry shock feel anxious and guilty about being so critical. Many worry that family, friends, and other acquaintances will judge them as unpatriotic.

The experience of reentry shock points to the transforming effect of an encounter with another culture (Sobie 1986). That the returnees go through reentry shock means that they have experienced up close another way of life and that they have come to accept the host culture's norms, values, and beliefs. Consequently, when they come home, they see things in a new light and must come to terms with ways they once accepted without question.

What Do Sociologists See?

The boarding schools for Native Americans (1870–1920) were designed to totally immerse students in so-called white culture and to teach them to reject, or least forget their home cultures. Assuming this student with the umbrella learned her lessons, it is likely that she now holds an ethnocentric view toward her Pine Ridge culture, defining it as inferior to the cultural ways to which she has been exposed. Specifically, it appears that she has rejected the dress of her birth culture and that she avoids exposure to the sun because she views dark skin as inferior.

Library of Congress Prints and Photographs Division[LC-USZ62-100543]

Critical Thinking

Describe a time when you reacted to another culture's way of doing something or thinking with an ethnocentric response.

Key Terms

cultural relativism

culture shock

ethnocentrism

reentry shock

Module 2.5

Cultural Diffusion

Objective

You will learn that many of the items and ideas we take for granted in our daily lives originated in foreign settings.

Denim Day

Have you ever heard of Denim Day? While this day—April 24—is now observed in many places across the United States, the inspiration can be traced to Italy.

Hanging jeans containing words like "family" and "support" on chairs and walls is a ritual associated with Denim Day, a day set aside each year—April 24—to raise awareness about sexual assault and rape. On this day, supporters make a point of wearing jeans. The inspiration for Denim Day can be traced to 1988, the year an 18-year-old woman in Italy was raped by her 45-year-old driving instructor. The man was convicted of rape and sentenced. The Italian Supreme Court overturned his conviction on the grounds that the jeans worn by the 18-year-old were so tight that the attacker could not have removed them without her assistance. The assumption that the victim must have assisted made the act consensual. Females in the Italian legislature protested the decision by wearing jeans and waving signs reading "Jeans: An Alibi for Rape." Now on April 24 wearing jeans has become an international symbol of protest against blaming the victim. California became the first of the 50 states to recognize Denim Day in 1998, and now 20 states officially recognize Denim Day. In 2008 the Italian Supreme Court overturned that ruling so that "too tight denim jeans" can no longer be used as a defense (Bieltz 2012). Nevertheless, Denim Day is still observed.

Cultural Borrowing

The process by which an idea, invention, or way of behaving is borrowed from a foreign source and then adopted by the borrowing people is called **cultural diffusion**. The term *borrowed* is used in the broadest sense; it can mean to steal, imitate, purchase, copy, or be inspired by. The opportunity to borrow occurs whenever people interact or otherwise engage with people or ideas associated with another culture. Instances of cultural diffusion are endless. In fact, serious study would show that practically any idea or object has a multicultural history.

The Global Positioning System (GPS) is the most recent variation on a Chinese invention of the third century BC. That Chinese invention is the compass, an important navigational instrument that helped people know where they were on the earth. Countless cultures have borrowed the compass and made "upgrades" over the past 2,300 years or so.

Courtesy photo, U.S. Department of Defense

Selective Borrowing

People in one culture do not borrow ideas or inventions indiscriminately from another culture. Instead, borrowing is often selective. That is, even if people in one culture accept a foreign idea or invention, they are, nevertheless, choosy about which features of the item they adopt. Even the simplest invention is really a combination of complex elements, including various associations and ideas of how it should be used. Not surprisingly, people borrow the most concrete and most tangible elements and then shape the item to fit in with their larger culture (Linton 1936).

Sgt. Catherine Threat

The Afghan Olympic Committee "borrowed" the Lakers basketball team uniforms, but they did not borrow everything. Notice that all but one athlete's arms and legs are covered and most are wearing head and neck scarves.

The Diffusion Process

Keep in mind that cultural diffusion is a process that generates change in the borrowing society. The introduction of new items into a society such as fast foods, for example, changes people's eating habits in many ways, including decreasing the amount of time devoted to eating a meal, increasing the likelihood of eating away from home, and altering the types and amounts of food consumed.

Sociologists who study cultural diffusion are interested in the rate at which the borrowing people come to use or apply a new idea, behavior, or invention. After an innovation makes its debut, sociologists ask how quickly others in the culture acquire, learn about, and/or come to use or consume it. The answer depends on a variety of factors, including (1) the extent to which the borrowing causes people to change ways of thinking and behaving; too many changes often leads to resistance; (2) level of media—including social media—interest; the greater the interest, the faster and more far-reaching the diffusion; and (3) the social status of the first adopters. For example, first use by groups labeled as marginal or outsiders may cause those in the mainstream to reject the idea, behavior, or invention because it is viewed as disrespectful, dangerous, or subversive. Conversely, first use by groups considered elites may cause those with less status to reject it as lavish or out of out of touch.

The rise of the "Gangnam Style" song and accompanying dance by K-pop artist Psy illustrates cultural diffusion. Psy, once known only to South Korean audiences, became a global superstar with over 1.9 billion views on YouTube (2014). One reason the dance caught on was the moves were something most people could easily imitate.

Steven L. Shepard, Presidio of Monterey Public Affairs

Sociologist William F. Ogburn (1968) believes that one of the most urgent challenges facing people today is the need to adapt to new products and inventions in thoughtful and constructive ways. Any time something new comes on the scene there are always unanticipated and disruptive consequences. The mobile phone brought instant communication, but it also brought texting while driving. New medical technologies bring relief to some conditions, but then they are sometimes overused and misused. Orgburn uses the term **adaptive culture** in reference to the norms, values, and beliefs of the borrowing culture and the role each plays in the adjustment process. In this regard, one can argue that Americans adapted easily to the automobile because it saved people time and supported deeply rooted norms, values, and beliefs regarding individualism and personal freedom. On the other hand, the invention created many problems such as pollution, congestion, and so on.

What Do Sociologists See?

This girl is one of 180 schoolchildren in Djibouti City, Djibouti, who received school supplies and a T-shirt (with American flag) from soldiers of the U.S. Army. Sociologists see this as an act of cultural diffusion, but one that goes beyond the object itself. It is also an act of diffusing associated values, beliefs, and norms—"a rich density of meaning" related to the ways women should cover their bodies and present themselves (Jones 2011). It is unlikely that these girls will borrow the American way of dress without modification. Still, it is likely that the act of seeing this item of clothing and wearing it will introduce something new, however subtle.

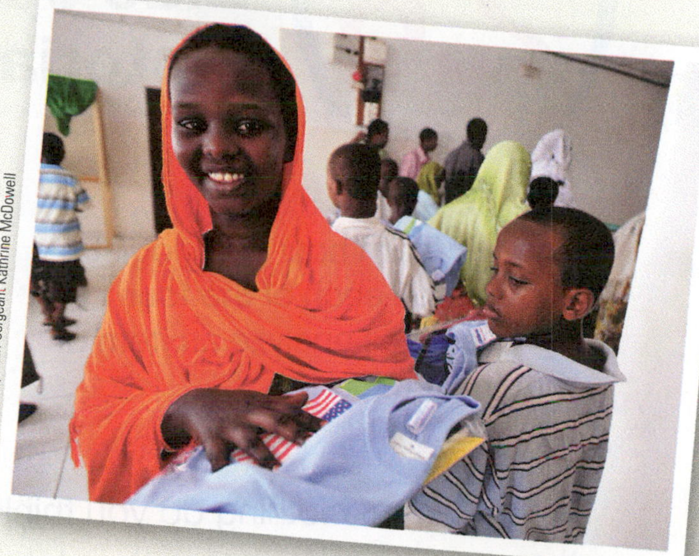

U.S. Air Force photo by Staff Sergeant Kathrine McDowell

Critical Thinking

Identify some object, idea, or way of behaving that you have "borrowed" from a "foreign" source. Explain its significance to your life.

Key Terms

adaptive culture

cultural diffusion

Module 2.6 Applying Theory: Blue Jeans as Material Culture

Objective

You will learn how sociologists inspired by each of the four perspectives analyze an item of material culture.

What item of clothing do you think is the most popular across the globe?

The photo shows mannequins wearing blue jeans, arguably the most popular item of clothing on the planet. While Americans and others tend to think of jeans as originating in the United States, the signature denim fabric is believed to have originated in France. A German immigrant to the United States (Levi Strauss) started a business in 1850 making copper-riveted denim pants for miners of the California Gold Rush. Since then jeans fashion has expanded to encompass many price ranges, varieties (low-rise, boot cut, straight leg, and skinny), and uses (for work, leisure, high fashion) (Banker 2011). At least 152 corporations sell jeans, including 7 For All Mankind, Black Peony, Gap, Lee, Levi-Strauss, and True Religion (Research and Markets 2014). What do sociologists gazing at this item of material culture think as they look?

A sociologist drawing on the functionalist perspective thinks about the contribution blue jeans make to order and stability. We can make the case that jeans qualify as an item of clothing that attaches the wearers to a global community and marks them as "fitting in." While jeans are a global product, they can also be counted as the "single most personal and intimate of outer garments," if only because they are the clothes of habit—the most worn item in our wardrobes (Miller and Woodward 2014). Billions of people around the world wear jeans and tens of millions are involved in their manufacture and sale. As one example, Levi's jeans are sold in 100 countries and manufactured in at least 20 Asian, Middle Eastern, and African countries (Levi-Strauss 2013).

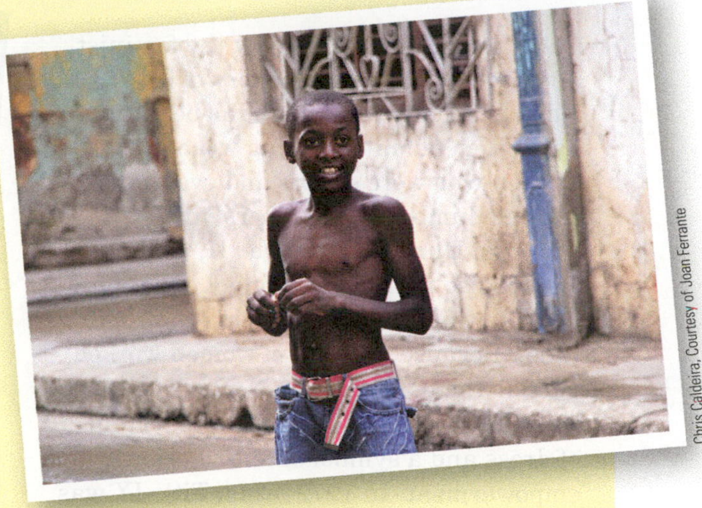
Chris Caldeira, Courtesy of Joan Ferrante

Sociologists thinking from a conflict theorist point of view see blue jeans as a global commodity. They think about the millions of workers who are part of the global-scale *commodity chain* that produces and delivers jeans to customers. The chain begins with those who grow and harvest cotton in Turkey or Azerbaijan, weave and dye fabric in Italy, cut cloth in India, sew patterns in Cambodia, wash/pack the jeans in Mexico, and distribute them from warehouses in Los Angeles or other cities (Holmquist 2008). The commodity chain includes those most and least advantaged by the process and everyone in between as measured by the income earned for their work, the labor conditions, health outcomes, and stress placed on their bodies (repetitive motions, breathing carcinogens, and so on). The conflict-oriented sociologist thinks about what price factories need to earn to guarantee safe, humane conditions. Factory owners in Bangladesh say 90 cents per pair, but contractors offer 75 cents or less, an amount that forces owners to cut corners (National Public Radio 2013).

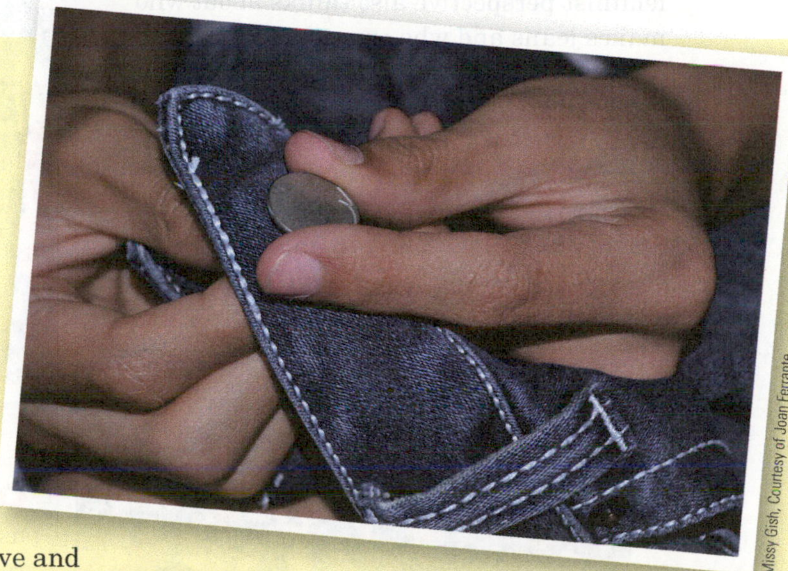
Missy Gish, Courtesy of Joan Ferrante

A sociologist inspired by the feminist perspective knows the history of blue jeans is a gendered one. Jeans began as an item of male clothing worn first by men who were sailors and eventually by male lumberjacks, miners, and cowboys. During World War II, denim was rationed so military personnel and others contributing to the war effort could wear them. Because millions of women worked outside the home in roles supportive of the war effort, jeans became a popular item of dress and a symbol of contribution and empowerment. In 1972, when Title IX was passed, it included a ruling that girls and women working and playing in taxpayer-supported environments could not be required to wear dresses. Jeans and other kinds of pants became part of women's wardrobes. The sociologist drawing on a feminist perspective also thinks about who makes jeans and who owns the factories and corporations that profit from the low-wage, repetitive, and dangerous labor of millions. No doubt females make up a disproportionate share of the bodies laboring in factories, and men dominate among the bodies that own and manage the process.

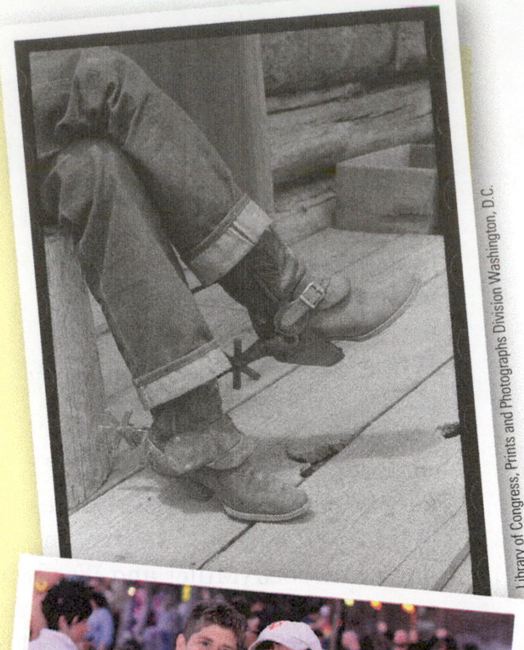
Library of Congress, Prints and Photographs Division Washington, D.C.

Chris Caldeira, Courtesy of Joan Ferrante

Chris Caldeira, Courtesy of Joan Ferrante

The sociologist thinking as a symbolic interactionist reflects on the countless local settings across the globe where people make and wear jeans. Their thoughts extend to the roles jeans play in shaping interactions and a person's sense of self. In particular, how does social context shape meanings assigned to the wearer or meanings a person assigns to the self in jeans (vs. the self in a dress). What messages are sent when a professor enters the classroom in jeans? Or a person wears jeans to church? How do blue jeans shape the identity of the these men who live in Mexico?

Summary: Putting It All Together

Chapter 2

Sociologists define culture as the way of life of a people. Culture includes the shared and human-created strategies for adapting and responding to the surrounding environment. On some level, culture is a blueprint that guides and, in some cases, even determines behavior. People of the same culture are not replicas of each other, because they possess the ability to accept, create, reject, revive, and change culture.

Culture consists of material and nonmaterial components. Material culture is the physical objects that people have invented or borrowed from other cultures. Sociologists seek to understand the ways people use these physical objects and the meanings they assign to them. Nonmaterial culture is the intangible aspects of culture, including beliefs, values, norms, and symbols.

People borrow ideas and objects from other cultures through a process known as cultural diffusion. Borrowing is usually selective in that people are choosy about which features of an item they adopt, and they reshape the item to fit the core values of their own cultures. Cultural diffusion is a process that generates change in the borrowing society. One key challenge facing people today is the need to adapt to new products and inventions in thoughtful and constructive ways.

Sociologists use the term *cultural diversity* to capture the cultural variety that exists among people who find themselves sharing some designated physical or virtual space. Cultural capital in all its forms (objectified, embodied, and institutionalized) is a useful concept for thinking about the diversity that exists in a particular setting (Bourdieu 1986). Sociologists use the concepts of subcultures and countercultures to further describe diversity. The difficult part of describing cultures is capturing what holds together people who believe they constitute a culture. One answer is cultural anchors, some cultural component—material (a color, a mascot, a sacred book) or nonmaterial (a belief, value, norms, language)—that elicits broad consensus among members regarding its importance but also allows debate and dissent about its meaning.

When we encounter different cultures, the natural tendency is to judge them using our home culture or some other culture as the standard. When people do this, they are engaging in ethnocentrism. Sometimes people find themselves in situations where they must leave their home culture and experience foreign ways of thinking and behaving. Culture shock is the mental and physical strain that people from one culture experience when they must reorient themselves to the ways of a new culture. People can also experience reentry shock upon returning home after living in and adapting to the ways of another culture. Cultural relativism is an antidote to ethnocentrism, in that it acts as a check against an uncritical and overvalued acceptance of the home culture that constricts thinking and narrows sympathies.

Chapter

3

Socialization

Chris Caldeira

Socialization is the link between society and the individual—a link so crucial that neither individual nor society could exist without it (Robertson 1988). To understand this link, think of society as analogous to a game with rules and expectations about how the game is played. For individuals to take part in the game, there must be mechanisms in place (socialization) to teach the rules and expectations. If there are no such mechanisms, the "game" ceases to exist. But the remarkable aspect is that people create the game and teach the rules and expectations to others. New "players" come on the scene, and they internalize some and alter other rules and expectations to meet changing circumstances.

Module

3.1 Nature and Nurture

Objective

You will learn about the significance of nature and nurture on physical and social development.

Do you consider yourself good at doing math? If yes, do you attribute the cause to natural ability, hard work, or some other factors?

NKU Sociology, Missy Gish

If you consider your math skills—whether good or poor—as natural or inborn, then you believe that nature is the responsible factor. If you attribute your math skills to effort or good coaching, then you believe that nurture is the responsible factor. You might find it interesting that in Taiwan, a country that is considered a top performer on international math tests, parents tend to attribute success in math to effort, interest, and practice. American parents, on the other hand, tend to attribute success to innate intelligence—math is something you are either good at or not (Eisenhart 2011).

Nature and Nurture

Nature is human genetic makeup, or biological inheritance. **Nurture** refers to the social environment, or to the interaction experiences that make up every person's life. Both nature and nurture are essential to socialization. Some scientists debate the relative importance of genetic makeup and social experiences, arguing that one is ultimately more important than the other. Such a debate is futile, because it is impossible to separate the influence of the two factors or to say that one is more critical. In fact, it is difficult to identify any human trait that can be explained by nature or nurture alone.

One might argue, for example, that height is genetically determined. But height depends, in part, on nutrition (nurture); children who are undernourished are considerably shorter than they might otherwise be. Likewise, one might argue that certain personality traits such as shyness are inherited. But no scientist has found a shyness gene. While it might seem that babies are born with personalities—that is, some babies seem to be born fussy, active, or serene—we must also consider that the babies' experiences are shaped by the mother's life, nutritional health, and emotional experiences while in the womb.

These quadruplets include three males and one female. Their lives are shaped by nurture when caretakers mark one baby with a flowered headband and give the boys a blue pacifier. Those objects signal others to treat each child according to their society's ideas about what boys and girls are and can be. Simply consider that the decision to place the baby girl in one of the middle positions suggests that she needs the protection of her brothers.

Trying to separate the effects of nature and nurture is like trying to determine whether the length or width of a picture frame is more important to the shape. The latest research suggests that nurture and nature are interwoven factors that collaborate to shape people's lives. To grasp this relationship, consider how language is learned. As part of our human genetic makeup (nature), we possess a cerebral cortex, which allows us to organize, remember, communicate, understand, and create. In the first months of life, all babies are biologically capable of babbling the essential sounds needed to speak any language. As children grow, however, this enormous language learning potential is reduced as the brain's flexibility is diminished. This is because the language or languages children hear spoken (nurture) organize the brain and its thought processes.

Research also tells us that babies cannot realize their biological potential (nature) unless they establish an emotional attachment with a caring adult (nurture). In other words, there must be at least one person who knows a baby well enough to understand his or her needs and feelings and who will act to satisfy them. Under such conditions, a bond of mutual expectation between caregiver and baby emerges. The bond gives the baby confidence that it can elicit predictable responses from caretakers: smiling causes the caretaker to smile; crying prompts the caretaker to soothe the child.

When there is a bond of mutual expectation, a child can expect caretakers to respond in predictable ways. This baby comes to learn and trust that when he looks at his father and smiles, his dad will respond in kind.

Ms. Jennifer M. Caprioli (Drum)

When researchers set up experimental situations in which parents failed to respond to their infants in expected ways, even for a few moments, they found that the babies suffered considerable tension and distress. Cases of children raised in situations of extreme isolation offer even more dramatic illustrations of the importance of caring adults in children's lives.

The Effect of Social Isolation

Sociologist Kingsley Davis (1940, 1947) did some of the earliest and most systematic work on the consequences of extreme isolation. His research shows how neglect and lack of social contact (that is, absence of nurture) can delay the development of human potential (that is, nature).

Davis documented and compared the separate yet similar lives of two girls: Anna and Isabelle. During the first six years of their lives, the girls received only minimal care. Both children lived in the United States in the 1940s. At that time, if mothers did not marry the fathers, babies were viewed and treated as illegitimate. Since Anna was considered illegitimate, she was forced into seclusion and shut off from her family and their daily activities. Isabelle, who was also considered illegitimate, was shut off in a dark room with her mother, who was deaf and could not articulate speech. Both girls were six years old when authorities intervened. At that time, they exhibited behavior comparable to that of six-month-olds. Anna "had no glimmering of speech, absolutely no ability to walk, no sense of gesture, not the least capacity to feed herself even when food was put in front of her, and no comprehension of cleanliness. She was so apathetic that it was hard to tell whether or not she could hear" (Davis 1947, 434). Anna was placed in a private home for mentally disabled children until she died four years later. At the time of her death, she behaved and thought at the level of a two-year-old.

While Isabelle had not developed speech, she did use gestures and croaks to communicate. Because of a lack of sunshine and a poor diet, she had developed rickets: "Her legs in particular were affected; they 'were so bowed that as she stood erect the soles of her shoes came nearly flat together, and she got about with a skittering gait'" (Davis 1947, 436). Isabelle also exhibited extreme fear of and hostility toward strangers. Her case shows how the "gene" for rickets is turned on by difficult social experiences.

Isabelle entered into a special needs program designed to help her master speech, reading, and other important skills. After two years, she achieved a level of thought and behavior normal for someone her age. Isabelle's success may be partly attributed to her establishing an important bond with her deaf-mute mother, who taught her how to communicate through gestures and croaks. Although the bond was formed under less than ideal circumstances, it gave Isabelle an advantage over Anna.

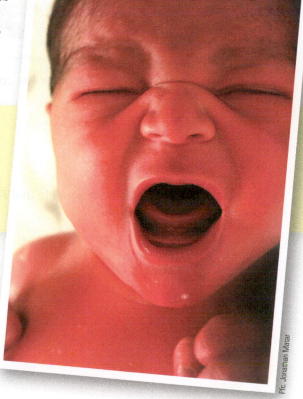

Imagine a situation where a baby was only fed and kept clean but that no one responded to the baby when it cried, held the baby when it was awake, or played with it. What would that mean to a child?

Pfc. Jonathan Matar

To answer this question, psychiatrist Rene Spitz (1951) studied 91 infants who were raised by their parents during their first three to four months of life but who were later placed in orphanages. When the infants were admitted to the orphanages, they were physically and emotionally normal. Orphanage staff provided adequate care for their bodily needs—good food, clothing, diaper changes, clean nurseries—but gave the children little personal attention. Because only one nurse was available for every 8–12 children, the children were starved emotionally. The emotional starvation caused by the lack of social contact resulted in such rapid physical and developmental deterioration that a significant number of the children died. Others became completely passive, lying on their backs in their cots. Many were unable to stand, walk, or talk (Spitz 1951). These cases and the cases of Anna and Isabelle teach us that children need close contact with and stimulation from others if they are to develop normally.

What Do Sociologists See?

NKU Anthropology, Sharyn Jones

Humans possess the capacity to move and position their fingers, arms, head, and legs in countless numbers of ways (nature). However, the ways in which these capacities are realized depend on the activities to which adults expose them and encourage them to master (nurture). Without the chance to learn some activity—to play a guitar, for example—a child cannot realize his or her potential to do so.

Critical Thinking

Give a specific example from your life that illustrates how what you do is a product of both *nature* and *nurture*.

Key Terms

nature

nurture

Module 3.2 Socialization

Objective

You will learn that socialization is an interactive process that prepares people to live with others in society.

Do you believe that infant babies should sleep alone in a crib or with a parent(s)? Why do you believe what you do?

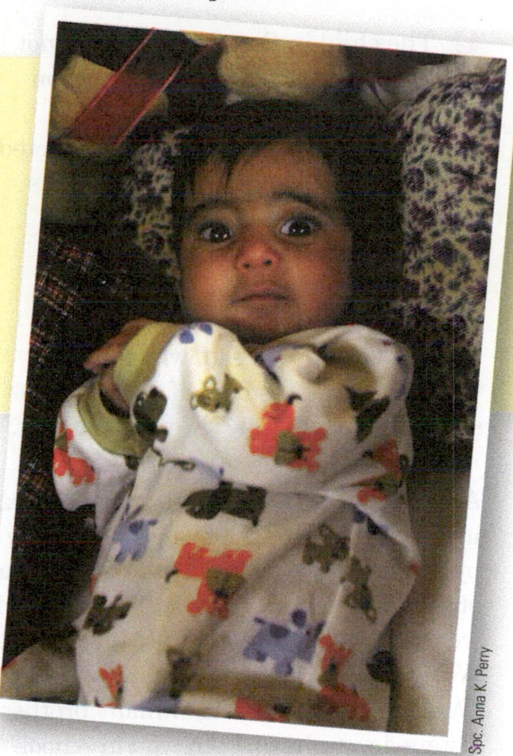
Spc. Anna K. Perry

Some parents believe that sleeping alone creates an independent child. Others argue that infants sleeping with the parent, especially the mother, is universal to all species and creates an emotionally secure child. Whatever your opinion, it is likely informed by the kind of child you believe each socialization experience will create.

Socialization is the lifelong process by which people learn the ways of the society in which they live. More specifically, it is the process by which humans

- acquire a sense of self or a social identity,
- develop their human capacities,
- learn the culture(s) of the society in which they live, and
- learn expectations for behavior.

Socialization is not a one-way process such that people simply absorb these lessons. It is a process by which people negotiate, resist, ignore, and even challenge those lessons.

Acquiring a Sense of Self

What does it mean to acquire a **sense of self**? From a sociological point of view, children acquire a sense of self when they can step outside the self and see it from another's point of view and also imagine the effects their appearance, words, and actions have on others. Having a sense of self also means that children have acquired a set of standards about how others expect them to behave and look in a given situation.

Researchers have devised an ingenious method for determining when a child has acquired a sense of self. A researcher puts a spot of blush on the child's nose and then places the child in front of a mirror. Presumably, a child who ignores the blush and does not try to remove it does so because he or she is not yet able to imaginatively step outside the self and evaluate the blush from another's point of view. Moreover, that child has not yet acquired a standard about how he or she is expected to look—in this case that a person does not wear red blush on their nose (Kagan 1989).

Lisa Southwick

Lisa Southwick

The child in the top photo does not seem bothered by the spot of blush on her nose. In fact, she does not even appear to know she is the baby reflected in the mirror. This is because she has not yet developed a sense of self. The child in the bottom photo appears to be concerned about the blush. Her facial expression tells us that she has acquired a sense of self, which means that she evaluates herself using a set of standards about how she ought to look.

Acquiring a sense of self also involves learning about the groups to which we belong and, by extension, do not belong. The importance of groups to identity is best illustrated when we think about how we get to know someone. We ask questions like "What is your name?" "Where do you live?" "How many brothers and sisters do you have?" "Are you in school?" "Do you play sports?" The answers to these questions reference group memberships in a family (last name), in a town, in a school, and so on. Consider how much time we spend teaching young children about the groups to which they belong, including working with them to know their family name, their age, their sex, their race, and their home country. Such ideas do not come easily. We have to go over them many times before children get it.

NKU Sociology, Missy Gish

The socialization process includes the process by which children come to think of themselves in terms of group memberships. The child on the right learns that because she wears pink she is a girl. The child on the left learns the meaning of the words ("little flirt") on his shirt and may come to identify as such.

Developing Human Capacities

Socialization is also a process by which human capacities are developed. Our parents transmit, by way of their genes, a biological heritage common to all humans but unique to each person. Our genetic heritage gives us, among other things, a capacity to speak or sign a language, an upright stance that frees arms and hands to carry and manipulate objects, and hands that can grasp objects. If these traits seem too obvious to mention, consider that, among other things, they allow humans to speak innumerable languages and to use their hands to create and use inventions. Babies cannot realize these biological potentials unless caregivers support and encourage their development. Likewise, babies born with physical impairments find ways to compensate if given the opportunity and support.

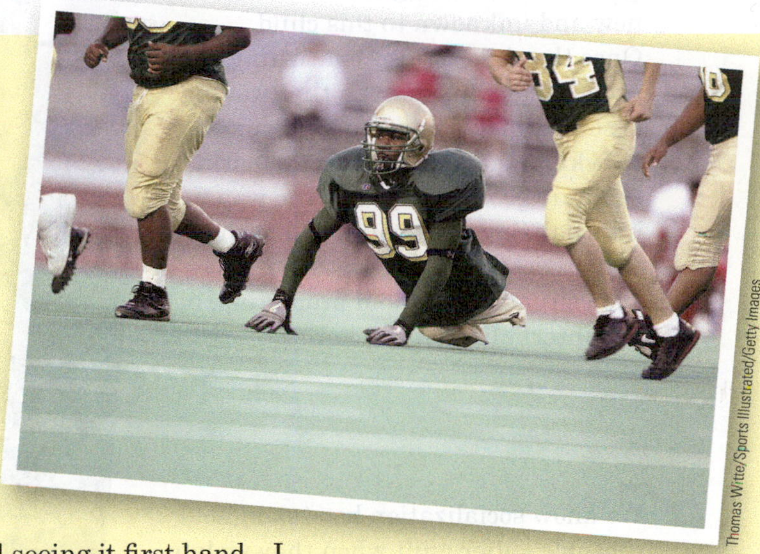

The story of Bobby Martin, a high school football player born without legs, captivated the sports world. One reporter described Martin's situation this way: "Never knowing a life with legs, Bobby from an early age just adapted to using his arms and the pendulum motion of his body for movement. And after perfecting this method of locomotion for 17 years—and seeing it first hand—I can tell you this kid can move with the best of them" (Witte 2005).

Learning Expectations for Behavior

By the time children are two years old, most are biologically ready to pay close attention to social expectations, or the "rules of life." They are bothered when things do not match their learned expectations: paint peeling from a table, broken toys, small holes in clothing, and persons in distress all raise troubling questions for children. To show this kind of concern, two-year-olds must first be exposed to information that leads them to expect people and objects in their world to be a certain way (Kagan 1989). They learn these expectations as they interact with others.

To understand how children learn the expectations of their culture, keep in mind that they learn by observing others and from the things others prompt them to do and say. In addition, as children learn a language, they learn names for things in the world and they acquire words that they use to express their thoughts and feelings. The clothes they wear and toys they play with convey messages about their culture and its values. In addition, routines—the repeated and predictable activities, such as eating at certain times of the day or taking baths, that make up day-to-day existence—teach children about what is important to the culture (Corsaro and Fingerson 2003).

Internalization

Socialization takes hold through **internalization**, a process by which people accept as binding learned ways of thinking, appearing, and behaving. To put it another way, internalization is a process by which an idea or a way of acting moves from something new and unknown to something thought and performed without question.

For now, waving is something new and unknown to this child. Once the child figures out, with the prompting of his mother, that waving is something you do to greet someone, to acknowledge that you see someone, or to say goodbye, he is on the way to internalizing the "wave." When the child reaches a point where he waves in such situations without prompting, he has internalized its meaning.

NKU Anthropology, Sharyn Jones

We know socialization has taken hold if people suffer guilt when they violate expectations (e.g., I saw my friend and I should have waved). We also know expectations have been internalized when a person conforms to some expectation even when no one is watching. In these instances, an inner voice urges conformity (Campbell 1964).

Internalization of cultural expectations is not automatic. Anyone who observes young children and their caretakers understands the effort caretakers expend to make living together "workable." As one example, consider how parents and children negotiate the experience of shopping. Parents let their children know through glances, words, and physical contact (e.g., firmly grabbing their hands) that they can't have everything they want. Children often respond by whining or throwing a tantrum. The two continue to negotiate an understanding about how they will shop together (Corsaro and Fingerson 2003). But their attempts to make shopping together workable must be placed in the larger social context. The U.S. government has relatively few laws limiting children's exposure to advertisements. The typical American child views between 25,000 and 40,000 television commercials per year. By contrast, Sweden, Norway, and Finland ban commercial sponsorship of television programming aimed at children 12 and under (Rabin 2008). Children in the United States are likely to have internalized the belief that happiness is achieved through having things.

What Do Sociologists See?

Children learn the ways of their culture through the things adults and others ask them to do. This child learns to expect medicine when he feels bad and to believe that the medicine will make him feel better. If taking medicine is the first response to physical discomfort, children will likely come to believe that the body cannot heal without some kind of medical intervention.

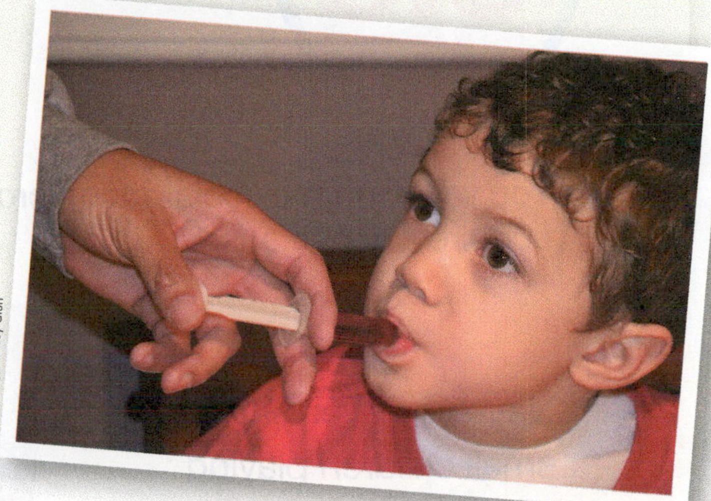

NKU Sociology, Missy Gish

Critical Thinking

Give an example of an expectation you learned but did not internalize.

Key Terms

internalization sense of self socialization

Module 3.3 The Social Self

Objective

You will learn the socialization processes by which children acquire a sense of self.

Have you ever watched small children playing at being someone significant in their lives, like Spiderman and Little Red Riding Hood?

Steve Roark (USAG Stuttgart Public Affairs Office)

Sociologists maintain that this kind of play fills a larger social purpose. It gives children practice at imaginatively stepping outside the self to take the role of the other. The ability to see the self from the point of view of another is known as role-taking. The ability to role-take is essential to acquiring a sense of self and for communicating and interacting with others.

Role-Taking

Role-taking means stepping into another person's shoes and imagining what things look like from that person's perceptual field, thereby allowing the role-taker to anticipate the person's behavior and respond accordingly (Coutu 1951). How do people learn to take the role of the other? George Herbert Mead (1934) maintained that we learn to do this through a three-stage interactive process. Those stages are (1) preparatory, (2) play, and (3) games. Each stage involves a progressively more sophisticated level of role-taking.

THE PREPARATORY STAGE (UNDER AGE 2). In the preparatory stage, children have not yet developed the cognitive ability to role-take. They mimic or imitate people in their environment but often do not know the meaning of what they are imitating. In this stage, children mimic what others around them are doing or repeat things they hear. In the process, young children learn to function symbolically; that is, they are in the beginning stage of learning to name things and learn that particular actions and words have meanings that arouse predictable responses from others.

Mr. Mark Weiman (AMC)

This little girl is in the preparatory stage, as the expression on her face suggests that she does not understand that she is a "princess," nor what it means to be a princess. At best she is mimicking behaviors she sees her mother do. At the same time, the little girl is coming to understand that dressing this way elicits positive responses from others around her.

THE PLAY STAGE (ABOUT AGES 2–6). Mead saw children's play as the mechanism by which they practice role-taking. **Play** is a voluntary, spontaneous activity with few or no formal rules. Play is not subject to constraints of time (e.g., 20-minute halves) or place (e.g., a regulation-size field). Children, in particular, play whenever and wherever the urge strikes. In play, children make the rules as they go; they are not imposed by rulebooks or referees. Children undertake play for entertainment or pleasure. These characteristics make play less socially complicated than organized games such as baseball.

In the play stage, children pretend to be **significant others**—people or characters such as cartoon characters, a parent, or the family pet—who are important in a child's life, in that they greatly influence the child's self-evaluation and way of behaving. When a little girl plays with a doll and pretends to be the doll's mother, she talks and acts toward the doll the same way her mother talks and acts toward her. By pretending to be a mother, the child gains a sense of the mother's expectations and perspective and learns to see herself as an object (through the doll) from her mother's point of view.

The children with helmets on are pretending to be fighter pilots and imagining the world from a pilot's point of view. In the process, the child pretends to think and feel as he imagines a fighter pilot does.

U.S. Air Force photo/Senior Airman Barry Loo

THE GAME STAGE (AGE 7 AND OLDER). In Mead's theory, the play stage is followed by the game stage. **Games** are structured, organized activities that involve more than one person. Games are characterized by a number of constraints, such as already established roles and rules and a purpose toward which all activity is directed. When people gather to play a game of baseball, for example, they do not have to decide the positions or the rules of the game. The rules are already in place. To be a successful pitcher, for example, one must understand not only how to play that position but how the position of pitcher relates to the other positions.

When young children first take part in organized games such as football, baseball, or basketball, their efforts are not coordinated. Everyone huddles around or runs toward the ball, mimicking the most basic elements of what they understand the game to be. Some forget to pay attention. The reason is that children have not acquired the mental capacity to see how each role fits with the positions that make up the game of football.

As they learn to play games, children also learn to (1) follow established rules, (2) imaginatively take the roles of all participants, and (3) see how their role fits in relation to an established system of expectations. In particular, children learn that what people in other positions do affects what they do in their position. Children learn that under some circumstances their position can take on added significance or assume lesser significance.

Through games, children learn to organize their behavior around the **generalized other**—a system of expected behaviors and meanings that transcend the people participating. An understanding of the generalized other is achieved by imaginatively relating the self to the many others playing the game. Through this imaginative process, the generalized other becomes incorporated into a person's sense of self (Cuzzort and King 2002). When children play games such as baseball, they practice fitting their behavior into an already established system of expectations. This ability is the key to living in society because most of the time, whether it is at school, work, or home, we are expected to learn and then fit into an already established system of roles and expectations.

Significant Symbols

We have learned that a sense of self involves the ability to role-take. In order to role-take, children must learn the meaning of **significant symbols**, gestures that ideally convey the same meaning to the persons transmitting and receiving them. Mead (1934) defined **gesture** as any action that requires people to interpret its meaning before responding. Language is a particularly important gesture because people interpret the meaning of words before they react. In addition to words, gestures also include nonverbal cues, such as tone of voice, inflection, facial expression, posture, and other body movements or positions that convey meaning.

Through gestures, people convey information about how they are feeling and others assign meaning—to shoulders hunched and face lowered into hands—and respond with gestures of comfort, perhaps a hand on the shoulder.

Ms. Rachel E Parks (IMCOM)

A sense of self emerges the moment children internalize **self-referent terms**, words and other symbols used to distinguish the self (with words like I, me, mine, first name, and last name) and to specify the statuses one holds in society (I am an athlete, doctor, child, and so on). Mead maintains that the self is always recognized in relationship to others. That is, one can be a student only in relationship to teachers and fellow students. Similarly, one can be an athlete only in relationship to other athletes, fans, a referee, and so on.

Mead described the self as having two parts—the I and the me. The **I** is the active and creative aspect of the self. It is the part of the self that carries out expectations for behavior in unique ways. It is the part of the self that can also reject the expectations guiding behavior and choose to act in unconventional, inappropriate, or unexpected ways. The **me** is the social self—the self that fits into an established system of roles and expectations.

Chris Caldeira

Chris Caldeira

As one example of how the "I" and the "me" shape the course of interaction, consider how hairdressers typically interact with customers. Their "me" knows not to stand too close to the customer, to stay on task, and to otherwise act in expected ways. But each hairdresser will have his or her own style—their "I"—of presenting themselves to customers and doing their job. In fact, every encounter each hairdresser has with customers, even repeat customers, is novel, "if only in the smallest ways" (Aboualafia 2012). The "I" is the part of the self that gives us a feeling of autonomy. The coexistence of the "I" and the "me" suggests that the self is dynamic and complex. In other words, a person is not a robot, programmed to behave and respond in completely expected ways.

Looking-Glass Self

Like Mead, sociologist Charles Horton Cooley (1961) assumed that the self is a product of interaction experiences. Cooley coined the term **looking-glass self** to describe the way in which a sense of self develops: specifically, people act as mirrors for one another. We see ourselves reflected in others' real or imagined reactions to our appearance and behaviors. We acquire a sense of self by being sensitive to the appraisals that we perceive others to have of us. As we interact, we imagine how we appear to others, we imagine a judgment of that appearance, and we develop a feeling about ourselves somewhere between pride and shame: "The thing that moves us to pride or shame is not the mere mechanical reflection of ourselves but . . . the imagined effect of this reflection upon another's mind" (Cooley 1961, 824).

Cooley went so far as to argue that "the solid facts of social life are the facts of the imagination." Because Cooley defined the looking-glass self as critical to self-assessment and awareness, he believed that people are affected deeply by what they imagine others' reactions to be, even if they perceive reactions incorrectly. The student who dominates class discussion may think that classmates are fascinated by their comments, when in fact their classmates are rolling their eyes. This student continues dominating discussions because he or she misinterprets others' assessment.

We have learned that self-awareness and self-identity emerge when people can role-take, understand meanings of significant symbols, and use self-referential terms. The question that still remains is how people develop the levels of cognitive sophistication to do these things. For the answer, we turn to the work of Swiss psychologist Jean Piaget.

Cognitive Development

Assessing the impact of Piaget on the study of cognitive development is "like assessing the impact of Shakespeare on English literature, or Aristotle in Philosophy—impossible" (Beilin, 1992 191). Piaget's ideas stemmed, in part, from his study of water snails, which spend their early life in calm waters. When transferred to tidal water, the size and shape of the snails' shells develop to help them cling to rocks and avoid being swept away (Satterly 1987). Building on this observation, Piaget arrived at the concept of active adaptation, a biologically based tendency to adjust to and resolve environmental challenges. Piaget believed that learning and reasoning are important cognitive tools that help children adapt to environmental challenges. These tools emerge according to a gradually unfolding genetic timetable in conjunction with direct experiences with people and objects.

Piaget's model of cognitive development includes four broad stages: (1) sensorimotor, (2) preoperational, (3) concrete operational, and (4) formal operational. According to Piaget's model, children cannot proceed from one stage to the next until they master the reasoning challenges of earlier stages. Piaget maintained that reasoning abilities cannot be hurried; a more sophisticated level of reasoning will not show itself until the brain is ready.

SENSORIMOTOR STAGE (BIRTH TO ABOUT AGE 2). In the sensorimotor stage, children are driven to learn how things work through a trial-and-error method of exploring that involves shaking and throwing things and putting things in their mouth.

NKU Sociology, Missy Gish

You might notice that very young children put into their mouths just about anything they pick up. This baby uses his mouth to explore things in the world around him.

The cognitive accomplishments of this first stage include an understanding of the self as separate from other persons and the realization that objects and persons exist even when they are moved out of sight. Before this notion takes hold, children act as if an object does not exist when they can no longer see it. At about eight months, children begin to actively look for an object that was once there.

PREOPERATIONAL STAGE (AGES 2 TO ABOUT 7). Children in the preoperational stage think anthropomorphically; that is, they assign human feelings to inanimate objects. They believe that objects such as the sun, the moon, nails, marbles, trees, and clouds have feelings and intentions (for example, dark clouds are angry; a nail that sinks to the bottom of a glass filled with water is tired). Children in the preoperational stage cannot grasp the fact that matter can change form but still remain the same in quantity. For example, they believe a 12-ounce cup that is tall and narrow holds more than a 12-ounce cup that is short and wide. In addition, children in this second stage cannot conceive how the world looks from another person's point of view, and they tend to center their attention on one detail and fail to process information that challenges that detail. That is, they believe that women have long hair, and that anyone with long hair qualifies as a woman even if that person also has a beard, something society attributes to a male.

CONCRETE OPERATIONAL STAGE (ABOUT AGES 7–11). By the concrete operational stage, children can take the role of the other. But they enter this stage having difficulty thinking abstractly. That is, they have trouble applying a specific situation to a general principle. But by the end of this third stage, the child can think abstractly. To illustrate: the child comes to truly understand the concept of brother apart from his own brother. So now a boy with a brother knows he is his brother's brother and that his dad and other men can be someone's brother, too.

FORMAL OPERATIONAL STAGE. In this stage (from the onset of adolescence forward), children learn to plan for the future, think through hypothetical situations, and entertain moral dilemmas. In this stage, children can conceptualize their life as being part of a much larger system. The world is not so black and white now, but has shades of gray.

Because children in the formal operational stage can think abstractly, they are able to imagine earth in relationship to the universe and, by extension, grasp the notion that they are one of more than 7 billion who live on the planet.

You are here

NASA

Piaget is criticized for, among other things, making conservative assessments of children's cognitive abilities at certain stages (Lourenço and Machado 1996). Some researchers, for example, have found that children at the age associated with earlier stages can do tasks associated with later stages. Piaget would respond by saying that he was not so much interested in the ages at which children move from one developmental task to another but the sequence through which they move (Lourenço and Machado 1996).

What Do Sociologists See?

A child playing doctor with his stuffed animal is seeing the world from viewpoints other than his own. Such play allows him to play two roles simultaneously and, in the process, to think about how a significant other in his life (a doctor) sees the patient and how the patient is supposed to behave relative to the doctor.

Pfc. Cory D. Polom

Critical Thinking

Name five self-referent terms that you use to describe yourself. Which one is most important to your sense of self?

Key Terms

games
generalized other
gesture
I
looking-glass self
me
play
role-taking
self-referent terms
significant other
significant symbols

Module 3.4 Primary and Secondary Agents of Socialization

Objective

You will learn how primary and secondary agents of socialization shape behavior, thinking, and social identities.

Kris Gonzalez, *Fort Jackson Leader*
Chris Caldeira

Think back to when you were a child. Do you remember a person or experience that shaped your identity in a way that carried into adulthood?

If you answered yes, then you have acknowledged a socializing agent in your life. **Agents of socialization** are significant people, groups, and institutions that shape our sense of self and social identity, help us realize our human capacities, and teach us to negotiate the world in which we live. It is impossible to list all the agents of socialization, but they can be divided into two categories: agents of primary and secondary socialization. **Agents of primary socialization** include family members and caretakers who prepare infants and children to live as family members and to go out into the larger society. It is parents and caregivers who expose infants and children to a language, ideas about what is considered right and wrong, and ways to express emotions and meet biological needs (eating, eliminating waste). **Agents of secondary socialization** include people other than family and caretakers who expose and teach people of all ages things they need to know to assume a particular role outside the home in the larger society, whether that role be associated with a school, sports, a club, a place of worship, a workplace, or any other.

In this module, we focus on three of the many agents of socialization—family, peer groups, and media. We begin with the family, most notably parents or guardians who are agents of primary socialization. Of course, parents/guardians make the decisions about when and whether to expose their children to agents of secondary socialization.

Family

The family is a primary agent of socialization because it gives individuals their deepest and earliest experiences with relationships and their first exposure to the rules of life. In addition, the family teaches its members about the world in which they live and ways to respond to it. During difficult times, the family can buffer its members against the ill effects of stressful events; alternatively, it can increase stress. Sociologists Amith Ben-David and Yoav Lavee (1992) offer a specific example. The two sociologists interviewed Israelis to learn how their families responded to missile attacks Iraq launched on Israel during the Persian Gulf War in 1991. During these attacks, families gathered in sealed rooms and put on gas masks. The researchers found that families varied in their responses to this life-threatening situation. Some respondents reported the interaction during that time as positive and supportive: "We laughed and we took pictures of each other with the gas masks on" or "We talked about different things, about the war, we told jokes, we heard the announcements on the radio" (39).

Other respondents reported little interaction among family members, though a feeling of togetherness prevailed: "I was quiet, immersed in my thoughts. We were all around the radio. . . . Nobody talked much" (40). Some respondents reported that interaction was tense: "We fought with the kids about putting on their masks, and also between us about whether the kids should put on their masks. There was much shouting and noise" (39). As this research illustrates, even under extremely stressful circumstances, such as war, the family can teach responses that increase or decrease that stress. Clearly, children in families that emphasize constructive responses to stressful events have an advantage over children whose parents respond in destructive ways.

U.S. Navy photo

The family can buffer its members against the effects of stressful events, or it can magnify the stress. This family is riding a boat to safety after their home and community were flooded. How do you think your family would react in such a situation?

While a family's income and wealth will surely shape how it responds to stressful events, there are many examples of families that, lacking even the most basic resources, still manage to respond constructively to stressful situations. Save the Children (2007) staff member Jerry Sternin offers one example. He was charged with a seemingly impossible assignment: to help save starving children in Vietnam. He drew inspiration from mothers he termed positive deviants—individuals "whose exceptional behaviors and practices enable them to get better results than their neighbors with the exact same resources."

Sternin identified those few Vietnamese children whose weight suggested they were well nourished and compared their situations with peers who were underweight and malnourished. He learned that the mothers of well-nourished children were behaving in ways that defied conventional wisdom. Among other things, these mothers were (1) using alternative food sources available to everyone. They were going to the rice paddies to harvest the tiniest shrimp and crabs, and they were picking sweet potato greens—considered low-class food—and mixing both food sources with rice; (2) feeding their children when the children had diarrhea, contrary to traditional practice; and (3) making sure their children ate, rather than "hoping children would take it upon themselves to eat." Save the Children successfully introduced these strategies to 2.2 million Vietnamese in 276 villages and to people in at least 20 other countries where malnutrition is widespread (Dorsey and Leon 2000).

Peer Groups

A **peer group** consists of people who are approximately the same age, participate in the same day-to-day activities, and share a similar overall social status in society. Examples of such status include middle school student, adolescent, teenager, or retired. Sociologists are especially interested in the process known as **peer pressure**, those instances in which people feel directly or indirectly pressured to engage in behavior that meets the approval and expectations of peers and/or to conform with what peers are doing. That pressure may be to smoke (or not smoke) cigarettes, to drink (or not drink) alcohol, and to engage (or not engage) in sexual activity.

It is through peer groups that children and others learn what it means to be male or female or to be classified into a racial/ethnic category. In the process, children are influenced by overall societal conceptions of race and gender but also create and integrate their own meanings.

NKU Sociology, Missy Gish

Sociologist Amira Proweller (1998) found that white middle-class students tended to perceive black counterparts as being more exotic and sensual in their styles, whereas blacks tended to perceive white middle-class styles as more accepted in school settings. Both blacks and whites adopted each other's styles but criticized each other when acting "out of race," labeling those blacks who did so as "acting white" and those whites who did so as "acting black."

Peer groups are important agents of socialization regarding gender as well. Among other things, peer groups negotiate what constitutes appropriate and inappropriate physical expression toward same- and other-sex persons. Close same-sex friendships are often labeled as gay, and the fear of being labeled as such can restrict how same-sex friends relate to one another. In addition, peers discuss and comment on physical changes accompanying puberty that revolve around girls wearing bras, menstruation, boys growing facial hair, girls removing (or failing to remove) body and facial hair, sexual development, sexual activity, rumors regarding romantic connections, and breakups. Praise, insults, teasing, rumors, and storytelling inform them about the meaning of being male, female, or something in between (Proweller 1998).

Mass Media

Another agent of secondary socialization is **mass media**, forms of communication designed to reach large audiences without direct face-to-face contact between those creating and conveying and those receiving messages. The tools of mass media—such as the printing press, television, radio, the Internet, and iPods—deliver content in the form of magazines, movies, commercials, songs, and video games to audiences and expose them to a variety of real and imaginary people, including sports figures, animated characters, politicians, actors, disc jockeys, and musicians.

Any exposure to a cartoon, video game, song lyrics, or other content delivered through the media presents an opportunity for socialization to occur, if only because it introduces viewers to possible ways to act, appear, and think. Sesame Street Muppets "Rosita" and "Elmo" are famous for socializing children about coping strategies to handle problems many children face.

Staff Sgt. Casey J. McGeorge (FORSCOM)

Because media is so pervasive, it is difficult to make any definitive statements about its effects as a whole, except perhaps to say that a very significant but unquantifiable portion of what we know about the world and the people in it is acquired through popular media sources. Since the 1950s (when televisions became widespread), there has been concern about the effects watching violence has on viewers, especially young children. A number of research studies have shown the following:

- The average child age six and under spends almost two hours per day in front of some type of screen media. Those between the ages of 8 and 18 spend 7 hours, 38 minutes in front of some form of media. Because youth are often engaged with two or more media at once (e.g., surfing the Internet and listening to music), exposure time increases to the equivalent of 11 hours per day (Kaiser Family Foundation 2010).

- If children were not watching, listening to, or playing with some media, it is likely they would be engaged in play or other physical activity, socializing with friends, reading, or doing homework (University of Michigan Health System 2010).
- A substantial proportion of children's programming includes violence, with the highest incidence found in animated programming, music videos, video games, and PG-13–rated movies (University of Michigan Health System 2010).
- Content analysis of media violence reveals that violence is often glamorized and is disproportionately aimed at women and racial minorities. In addition, it is something that goes unpunished, is often accompanied by humor, and fails to depict human suffering and loss as a consequence. Finally, violence is presented as an appropriate method for addressing problems and achieving goals (University of Michigan Health System 2010).
- A review of more than 2,000 research studies reveals a clear association between viewing violence and (1) aggressive behavior and thoughts, (2) a desensitization to violence, (3) nightmares, and (4) fear of being a victim (University of Michigan Health System 2010; Harvard Pediatric 2010).

DoD photo by Donna Miles

Some video games allow players to simulate aggressive and violent responses. These games embed players in virtual environments where they assume the role of the aggressor and are rewarded for injuring or killing others. Players are not simply watching violent scenes; they are strategizing and actively engaging in violent acts, albeit aimed at characters who are part of a virtual reality.

Clearly, viewing violence and other aggressive actions does not in itself lead viewers to engage in violent behavior such as bullying, abuse, homicide, or assault. Otherwise, everyone who viewed media violence would engage in such behavior. To understand the reasons people engage in violent and aggressive behavior, one must look beyond the media and consider other factors such as the viewer's gender, employment opportunities, age, and so on. Moreover, to understand the media's role in perpetuating violence and other aggression, one must consider a multitude of factors that mute or accentuate its effects. Those factors include the type and amount of violence being viewed; the age of the viewer; and the quality of a viewer's relationships with family, friends, and others. The fact that media violence is only one of many factors that contribute to violent and aggressive behavior does not mean it does not play an important role. Moreover, it is a factor worth studying because it is something to which exposure can be monitored (University of Michigan Health System 2010; Harvard Pediatric 2010).

What Do Sociologists See?

Drill sergeants are secondary agents of socialization. The drill sergeant's job is to socialize the recruits to perform under stress of battle and as a unit. The drill sergeant accomplishes this by relentlessly pressuring recruits to pay attention to details—a single hair on the face or a weapon held minutely out of position results in punishment to not just the offending soldier but to the entire unit.

Sgt. Virgil P. Richardson

Critical Thinking

Describe a secondary agent of socialization that shaped your sense of self and social identity in a significant way. Explain.

Key Terms

- agents of primary socialization
- agents of secondary socialization
- agents of socialization
- mass media
- peer group
- peer pressure

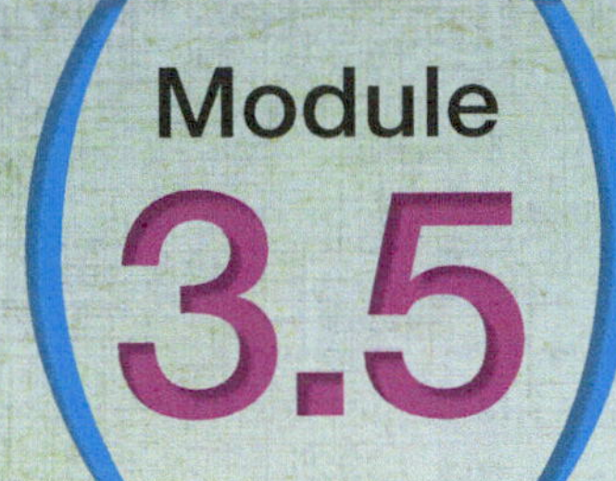

Module 3.5 Resocialization

Objective

You will learn that resocialization is an interactive process by which the affected parties reconstruct their identities and associated relationships.

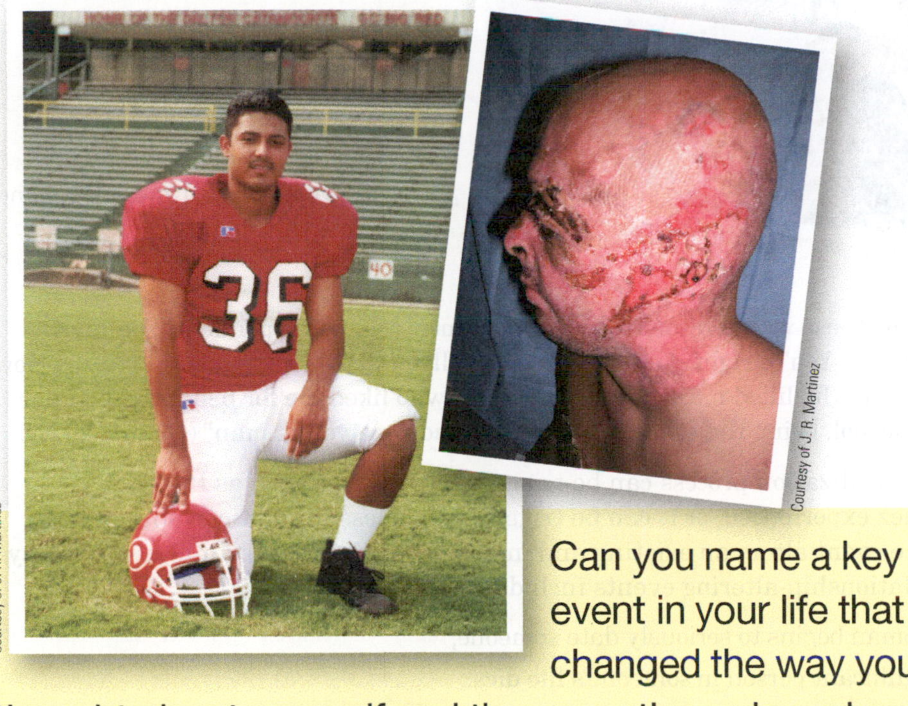

Courtesy of J. R. Martinez

Courtesy of J. R. Martinez

Can you name a key event in your life that changed the way you thought about yourself and the way others viewed and related to you?

For J. R. Martinez that event occurred in 2003 while he was serving in Iraq. The left tire of the Humvee he was driving hit a landmine. The resulting explosion left J. R. with severe burns over at least 40 percent of his body; he also suffered damage to internal organs from inhaling the hot air and smoke. Martinez spent three weeks in a coma and 34 months in recovery. He endured dozens of surgeries. When Martinez first looked at himself in the mirror, he remembers that he "just froze for a couple seconds" and thought that he would have been better off had he not survived (Martinez 2011). But Martinez did more than just survive; he went on to thrive as a soap opera star, a motivational speaker, and a contestant on *Dancing with the Stars*, which he won. In addition to the medical procedures, Martinez went through another process that sociologists call **resocialization**, an interactive process during which the affected party reconstructs his or her social identity.

The Process of Resocialization

Resocialization is a process by which the affected party renegotiates relationships with those who must also adjust to the changing person and circumstances. Significant others play an important role in the resocialization process because how they react is important to how the affected party reconstructs his or her identity and relationships (Daly 1992).

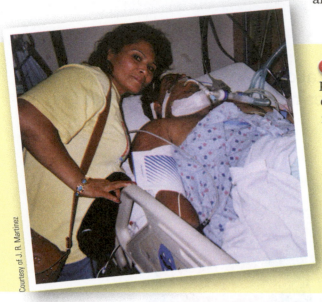

Courtesy of J. R. Martinez

For example, Martinez's mother, pictured here by her son's side after one of his 33 surgeries, played an important role when she told him that looks weren't everything and that "people are going to be in your life for who you are as a person and not what you look like." Martinez's mother told her son that when she was younger, "everyone told me I was pretty and gave me compliments. No one tells me that now."

Somehow these words resonated with Martinez, and he responded, "You know what, Mom? You're right. And now, I'm actually glad this happened to me. . . . Now I get to see who liked me as a person, versus who liked me for being the popular guy in school, being the athlete, being the handsome young man" (Collins 2011).

The resocialization process can be triggered by a crisis, such as the one Martinez experienced, or it can be triggered by a less dramatic event. Regardless, the event alters an existing and internalized identity. Examples of identity- and relationship-altering events include these situations:

- a woman begins to seriously date someone,
- a significant person in someone's life dies,
- a man who prides himself on being physically fit is diagnosed with a degenerative illness,
- someone retires from a job after a 40-year career,
- a student graduates from high school and enrolls in college,
- a person inherits a significant amount of money, and
- a financially independent person becomes unemployed.

In all these cases, the affected party must relinquish an existing identity and come to terms with a new one. Such identify transformations often involve a review of the life that once was and a protest against or celebration over what has changed, followed by a period of mourning (of varying intensity) over what has been lost. Martinez grieved for the handsome man he once was and his mother grieved over his near death and disfigurement. Then the affected parties must renegotiate their identities and relationships. "To do less is to remain mired in loss" (Fein 2011). For Martinez, that negotiation process was pushed further along when a nurse taking care of him asked if he would speak to a fellow burn patient who had become withdrawn upon seeing his body for the first time. After visiting with that patient for 45 minutes and gaining a positive response, Martinez realized that he could impact the lives of others simply by sharing experiences (Martinez 2011).

The Resocialization Experience

The resocialization experience varies depending on the extent to which the change is welcomed, the extent to which relationships are adversely affected, and a number of other factors including past experiences, beliefs about the possible self, and whether the new identity is considered normative.

Prior life experiences can shape the resocialization experience. For example, when people enter a new romantic relationship, the adjustment is affected by whether they have had previous romantic relationships and how significant others such as a best friend responded to those relationships in the past. Likewise, people's projections about what the new identity and changed relationships will mean for them affect the experience of resocialization. Will the new identity and relationships result in a "successful self, a rich self, the thin self, or a popular self? Conversely will it usher in a depressed, lonely, unemployed, or homeless self?" (Markus and Nurius 1986, 954). Martinez's view of his possible self included the belief that once people talked with him they would "not notice the scars anymore." They would see "a human being" with a "sense of humor" who likes to have a good time (Martinez 2004). Finally, the socialization experience is affected by whether the change is considered normative. That is, is the new identity expected or unexpected for someone of a given age, race, sex, social class, and so on? For the most part, in the United States we do not expect 85-year-olds to divorce, but are not too surprised if a 45-year-old divorces.

Much resocialization happens naturally and involves no formal training; people simply learn as they go. For example, people who marry; those who decide to live openly as gay, lesbian, or transgendered; and those who lose their job or make some other transition must incorporate the new status into their social identity and learn new ways to relate to others as they transition from single to married, from assumed heterosexual to another orientation, and from gainfully employed to unemployed. Sometimes circumstances are such that people voluntarily or involuntarily participate in programs created with the purpose of guiding or forcing them through a resocialization process.

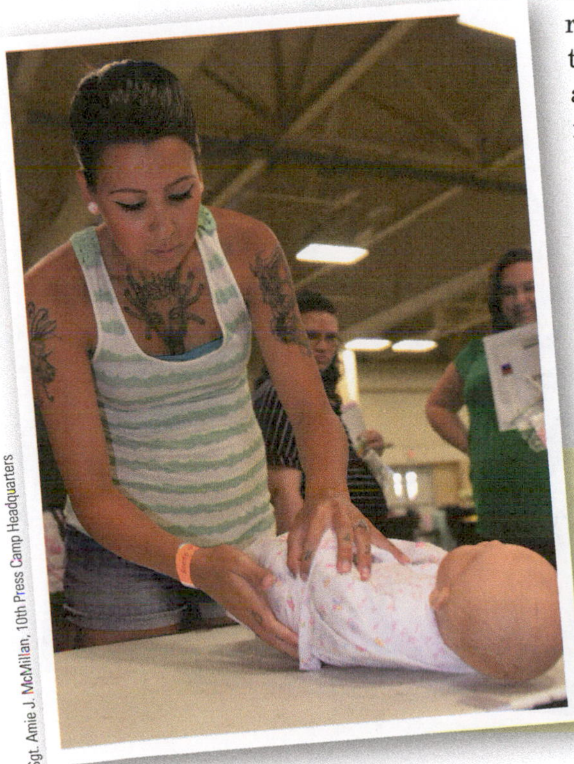

Sgt. Amie J. McMillan, 10th Press Camp Headquarters

This expecting mother participates in a support program designed to give mothers-to-be experiences and information that will help them during and after their pregnancy. Here she learns how to swaddle a baby in a receiving blanket.

Voluntary versus Imposed Resocialization

Resocialization can be voluntary or imposed. Voluntary resocialization occurs when people choose to participate in a process or program designed to remake them. Examples of voluntary resocialization are wide-ranging: the unemployed youth who enlists in the army to bring discipline to her life, the alcoholic who joins Alcoholics Anonymous to transform himself into a "recovering alcoholic," and the unemployed person who enrolls in college to begin training for an anticipated career.

Imposed resocialization occurs when people are forced into a program designed to train them, rehabilitate them, or correct some supposed deficiency. People drafted into the military, sentenced to prison, ordered by a court to attend parenting classes, or committed to mental institutions represent examples of those who undergo resocialization that is forced upon them.

In *Asylums*, sociologist Erving Goffman (1961) wrote about a particular type of setting called **total institutions** in which people are isolated from the rest of society to undergo systematic resocialization. Total institutions include homes for the blind, the elderly, the orphaned, and the indigent; mental hospitals; prisons; concentration camps; army barracks; boarding schools; and monasteries and convents. In total institutions, people surrender control of their lives, voluntarily or involuntarily. As "inmates," they carry out daily activities such as eating, sleeping, and recreation in the "immediate company of a large batch of others, all of whom are treated alike and required to do the same thing together" (6).

People who are confined to and/or join total institutions participate in activities (lining up, attending classes, making crafts, showering, running, attending group therapy sessions, praying) that fit in with institutionally planned goals. Those goals may be to care for the incapable, to keep inmates out of the community, or to teach people new roles (for example, to be a soldier, priest, or nun).

Goffman (1961) identified standard procedures that total institutions employ to resocialize inmates. Inmates-to-be arrive with a sense of self. Upon entering a total institution, that self undergoes **mortification**, a process by which the self is stripped of all its supports and "shaped and coded." In the case of those who join the military, staff shave "inmates'" heads, an act that symbolizes the transition from civilian to military life has begun.

Sgt. Vincent Fusco, West Point Directorate of Communications

Staff may make take life histories and photograph, weigh, fingerprint, and take personal possessions through which inmates presented themselves on the outside. Clothes and jewelry are taken and uniforms issued. Access to services such as a hairdresser or workout facilities are denied. Communication with those in the outside world is curtailed, and if restored, closely monitored. "Family, occupational, and educational career lines are chopped off" (17). Inmates have no say in how to spend the day; they follow a schedule of activities to which all adhere. Taken together, these policies enforce a clean break with the past and mark the beginning of a process that, if successful, will result in a new identity and way of thinking and behaving.

What Do Sociologists See?

Those who have lost limbs in accidents and war must go through a process of resocialization in which they adjust to their new identity, relearn how to do things they once did with limbs, and learn to relate to others in new ways. At this stage in his rehabilitation, this soldier can simulate the experience of horseback riding with his baby son.

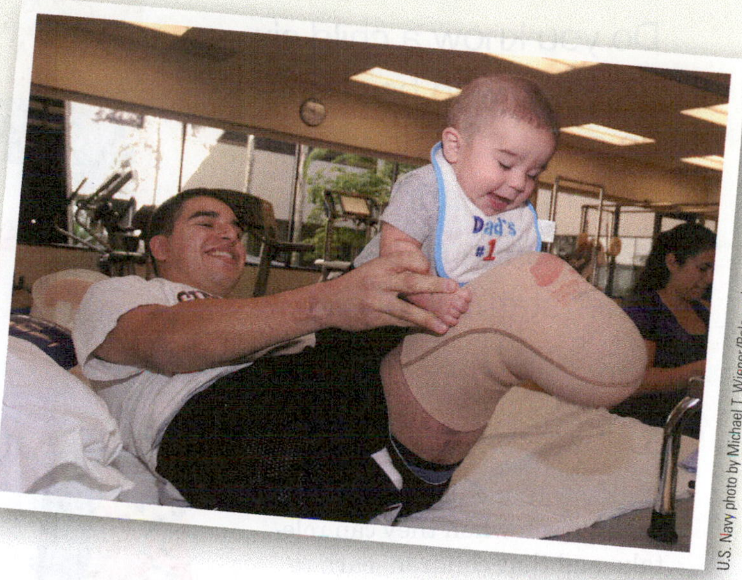

U.S. Navy photo by Michael T. Wiener/Released

Critical Thinking

Describe a time when your social identity changed voluntarily or involuntarily. Explain how your relationships changed (and adjusted or not) as a result.

Key Terms

mortification

resocialization

total institutions

Module 3.6 Applying Theory: Interactive Games as Agents of Socialization

Objective

You will learn how sociologists, inspired by each of the four perspectives, analyze interactive games as agents of socialization.

Do you know a child six or under who has used an app that helped him or her pretend to be a veterinarian or some other status in society?

NKU Sociology, Missy Gish

We know that children possess a sense of self when they can role-take and that children learn to role-take through imitation, play, and games. So what do we make of interactive technologies designed to teach children to imitate, play, and participate in games? Do these digitally delivered socialization tools cultivate role-taking skills? In this module we consider how sociologists, inspired by each of the four perspectives, answer these questions.

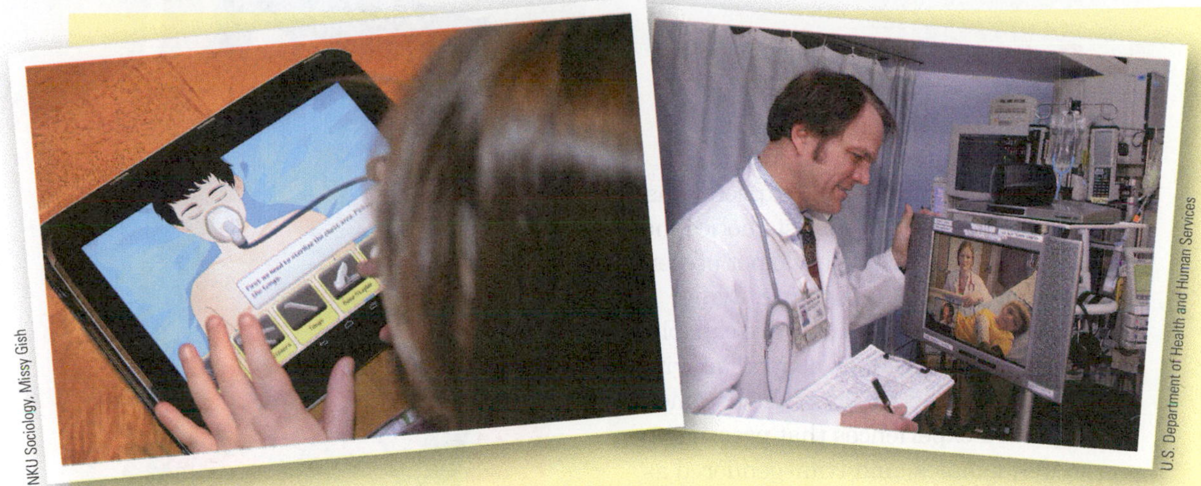

NKU Sociology, Missy Gish

U.S. Department of Health and Human Services

The photo on the left shows a child using interactive technology (also known as an app) to play at being a doctor—to simulate the things doctors do. The photo on the right shows a doctor engaging in telemedicine—real-time, interactive conversation between physician and patients who are in different locations mediated by digital technology. A sociologist inspired by the functionalist perspective sees this interactive technology as a socializing tool preparing children to succeed in their societies. Specifically, functionalists recognize interactive technologies' potential to prepare children to live in a knowledge economy. To be successful in that kind of society, people must know how to engage with technologies to communicate with and connect to others. Thus, playing virtually at being a doctor contributes to order and stability by preparing children to be doctors of the future who will likely practice medicine mediated by interactive technologies.

NKU Sociology, Missy Gish

Sociologists inspired by the conflict perspective question claims that interactive games are socializing tools that teach children specific skills. Rather, conflict theorists see them as tools that socialize children to become consumers. Ads are prominently displayed (e.g., "upgrade, only $2.99") so that children as young as six months cannot escape the reach of advertisers and marketers who personalize messages about what to buy—and what children need to do to get their parents to buy products. Ultimately, the interactive technologies benefit advertisers by creating legions of consumers who buy, not to satisfy real needs, but to satisfy artificially created desires.

A sociologist inspired by the feminist perspective observes ways interactive technologies operate as socializing tools that teach and reinforce gender-specific expectations and inequalities. Feminists notice that interactive technologies do not present activities—such as farming, flying, racing—that most people are capable of performing as gender-neutral activities. Rather, interactive activities become gendered experiences that present males, females, and transgendered as meeting or deviating from caricatures of masculinity and femininity.

NKU Sociology, Missy Gish

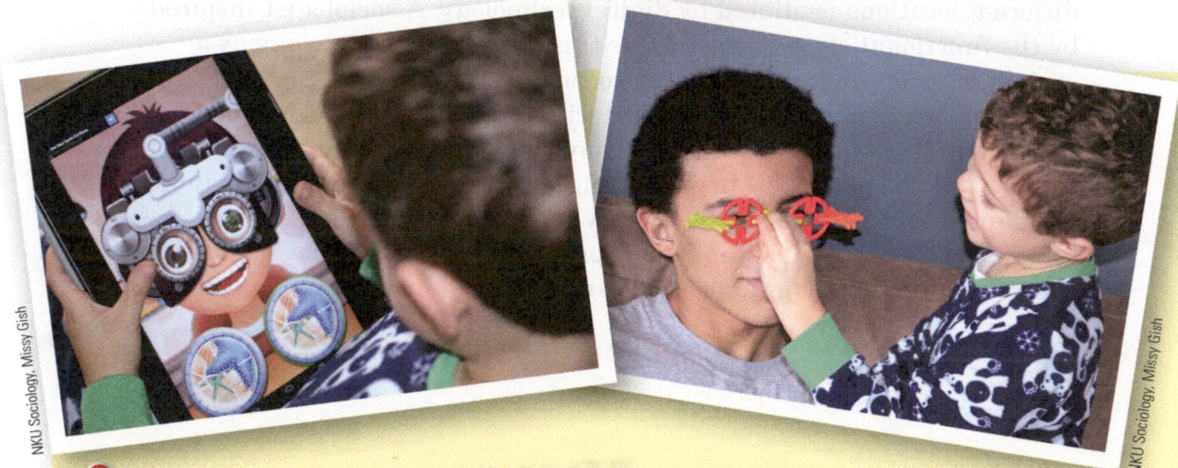
NKU Sociology, Missy Gish

NKU Sociology, Missy Gish

In the left-hand photo a child in the play stage pretends to be an eye doctor using interactive technology; in the right-hand photo that same child plays at being an eye doctor with his brother as the pretend patient. Sociologists thinking about the socializing effects of interactive technologies from a symbolic interactionist perspective would certainly point out that when children play with interactive games they are often alone, in a stationary position, and interacting with virtual characters, not "real humans" with all their idiosyncrasies and unpredictability. So while interactive technologies may be socializing children on some level, children do not have to work as hard to gain cooperation from virtual characters as from real-life humans. In this sense symbolic interactionists question whether interactive technologies are really socializing children to role-take, a skill necessary to fit into the larger society.

Critical Thinking

Which of the four theoretical perspectives best captures how you think about the interactive technologies and the development of the self? Explain.

Summary: Putting It All Together

Chapter 3

Socialization is the process by which people learn to live in society. We are socialized (and resocialized) by primary and secondary agents of socialization including the family, peer groups, and mass media. Socialization takes hold through internalization. Any discussion of socialization must take into account nature and nurture, both of which are essential to human and social development. We know social interaction is critical to social development from cases of children raised in extreme isolation, lacking meaningful social contact with others to realize their human potential.

Sociologists emphasize that a sense of self emerges out of interaction experiences. Children acquire a sense of self when they (1) take the role of the other; and (2) name, classify, and categorize the self relative to other social categories. Children learn to take the role of others through three stages: the preparatory, play, and game stages.

The concept of the looking-glass self sheds light on the role of the imagination in role-taking. The social experiences of the looking-glass self and the three stages of role-taking prepare the child to establish his or her place in a wider system of rules and expectations. Psychologist Jean Piaget's theory of cognitive development explains how children gain the conceptual abilities to role-take.

Socialization is a lifelong process because people learn new expectations that come with making the many transitions in life. All social transitions involve resocialization, an interactive process during which the affected party reconstructs identity. That process may be voluntary or involuntary, dramatic or subtle, and may involve a loss or gain in status and relationships with significant others that must be renegotiated. Significant others play an important role in the resocialization process because how they react is important to the transition process. Total institutions specialize in resocialization. They are settings in which people are isolated from the rest of society to undergo systematic resocialization.

Chapter 4

Social Structures

Chris Caldeira

NKU Sociology, Missy Gish

4.1 Institutions and Social Structure
4.2 Levels of Social Structure
4.3 Social Structure and Human Agency
4.4 Division of Labor and Social Networks
4.5 The Effects of Size
4.6 Formal Organizations
4.7 Rationalization and McDonaldization
4.8 Alienating and Empowering Social Structures
4.9 Applying Theory: The Social Structure of Nail Salons
Summary Putting It All Together

Look at the two signs for medical care. Both are part of an institution known as medicine. If you walk into each building, do you have expectations about how you will be treated? In which office do you expect the health care providers to touch your body more? In which office do you expect you will leave with a prescription? In which office do you expect the services rendered to be covered by health care insurance? In which of the two offices do you expect health care practitioners to focus on getting the mind, body, and spirit into balance? If you can answer these questions, then you already have some idea about how the institution of medicine is organized in the United States and the power of **social structures**, the largely invisible system that coordinates human behavior in broadly predictable ways. This chapter is about social structures, how they shape—even constrain—our relationships and experiences, and the role that human agency plays in creating, sustaining, and changing them.

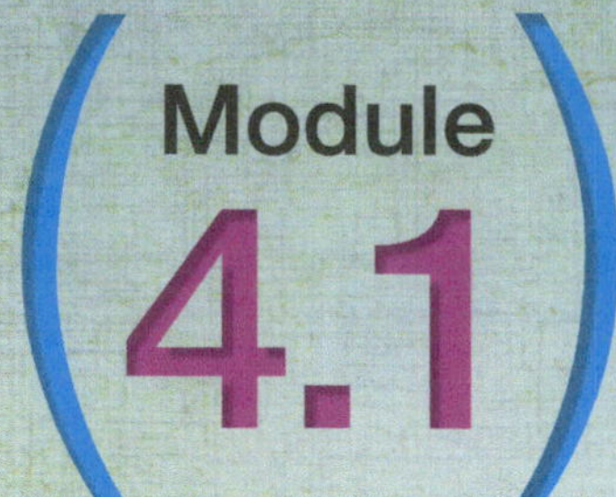

Institutions and Social Structure

Objective

You will learn how social institutions are structured to meet human needs.

NKU Sociology, Missy Gish

NKU Sociology, Missy Gish

Do you know someone, including yourself, who participated in a job fair or "found" parents?

In the most basic sense, sociologists see the two signs as advertisements to draw people into institutions. In the case of the drivers' job fair, those who attend are looking to become part of an institution known as the economy; in the case of those who answer the ad calling for foster parents, they give a child the opportunity to join the institution known as the family. These are just two of the core institutions sociologists study (see Table 4.1a).

Table 4.1a: Core Institutions

All human societies create institutions to meet universal social and biological needs; these needs include the need to care and be cared for, to secure shelter, to secure a livelihood, to feel meaning and purpose, to learn, and so on.

Institution*	Examples of Basic Needs Met
Family/Marriage	To socialize children; to care for members
Education	To transfer accumulated knowledge to new generations; to create new knowledge
Economy/Work	To produce and distribute goods and services; to provide ways to earn a livelihood
Religion	To provide a reason for being
Political/Government	To allocate power; to control resources; to make laws; to form a military
Medicine	To care for sick and physically vulnerable; to prevent illness and disease

*There is some debate within sociology over whether media is an institution in its own right or whether it is a tool of the core institutions such as the economy (e.g., corporations use media to promote and sell products; religions use the media to spread the word).

Institutions are not isolated or self-contained; they can overlap in their efforts to meet human needs. Consider that governments collect taxes to pay for education. Governments also make laws that affect family life—who can marry and whether new parents can take maternity/paternity leave with pay. The following broad characteristics apply to institutions.

Institutions have a history. Institutions have been around long enough that people have come to see them as key to achieving some goal—that is, to achieve a desired outcome one follows the "rules" or established ways of doing things. So, when we get sick, we go to a doctor to get a prescription. When we want to make a committed relationship officially, we marry. We believe that to be healthy, a child should be raised in a family. Over time, people come to accept these "rules" without question and fail to see that there might be other ways to meet needs.

Institutions continually change. Over time, ways of doing things become outdated and are replaced by new ways. When enough people stop or resist doing things in institutionally prescribed ways, institutions must adjust to survive. Change can be planned and orderly, forced, and/or chaotic. Forces from outside or within institutions can trigger change. Forces from outside changed the institution of the family when same-sex couples ("outsiders") demanded and gained the right to marry in some states. Forces from inside changed the institution of the family when larger proportion of mothers began working outside the home.

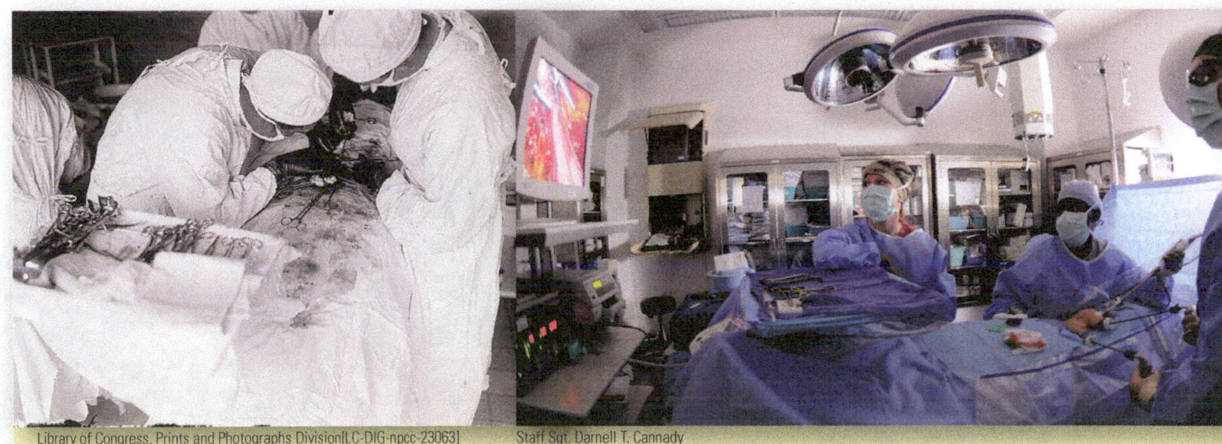

Library of Congress, Prints and Photographs Division[LC-DIG-npcc-23063] Staff Sgt. Darnell T. Cannady

Technology can also change institutions. Before the advent of tiny digital cameras and microchips, surgeons used their fingers to operate and control instruments (left). Today, surgeons may use new technologies (right) to see deep inside a patient's body and organs and direct instruments to do what fingers once did. Such technology means that surgeons now have the ability to operate on a patient from practically any location in the world. This change can potentially revolutionize the delivery of health care and the patient–physician relationship.

Institutions distribute income, wealth, and other resources in unequal ways. Thus, they play an important role in assigning people to advantaged and disadvantaged statuses, both within and outside the institution. Employers (who are part of the institution of work/economy) distribute different salaries, benefits, autonomy, and prestige to their employees depending on the positions they hold in the organization. Political institutions—which include federal, state, and local governments—set minimum wage laws, for example. And the institution of medicine offers higher salaries to some specialties such as anesthesiologists than to others such as primary care doctors.

Institutions promote ideologies that legitimate their existing structure. These legitimating ideologies are largely created and advanced by those occupying the most advantaged statuses or by those who benefit from institutionalized ways of doing things. The masses often accept these ideologies and resist efforts at change. For the longest time the government denied openly gay men and women the right to join the military, espousing the ideology that homosexuality was incompatible with military serve. That ideology has since been disproven.

Over time the needs met by one institution can be transferred—gradually or quickly—to other institutions. Before industrialization the family met needs that are now handled in large part by other institutions. At one time the family cared for children, taught their children the skills they needed to earn a livelihood, and operated as a self-sufficient economic unit, securing food and making clothes. Industrialization created new kinds of jobs with new kinds of skills, and in the process undermined the family as an economic unit. Needs the family once met were transferred in full or in part to the institutions of education, medicine, and the economy.

NKU Anthropology, Sharyn Jones

NASA/Sean Smith

In some societies that we call developing, the family is more likely to serve as an economic unit. It is common for children spend the day by their parents' side as they work. In such situations children observe and know what their parents actually do to earn their livelihood. In societies we call developed, the family does not function as an economic unit. Children may go to work with a parent one day out of the year. Outside of that one day they have no chance to observe their parents at work while they are in day care or school.

To participate in institutions, people must follow the rules and behave and interact in expected ways. When people do not behave as expected, they are asked to leave, are removed forcibly, or are denied benefits. This is because an institution's stability depends on people meeting expectations associated with the statuses they occupy and the roles they perform. When sociologists study an institution, they examine its social structure or the statuses, roles, groups, and organizations that make up that institution.

Statuses and Roles

Sociologists use the term **social status** in a very broad way to mean a human-created position that an individual occupies (Linton, 1936). Examples are endless but include female, teenager, doctor, patient, retiree, sister, homosexual, heterosexual, employer, employee, and unemployed. A social status has meaning only in relation to other social statuses. For instance, the status of a physician takes on quite different meanings depending on whether the physician is interacting with another physician, a patient, or a nurse. Thus, a physician's behavior varies depending on the social status of the person with whom he or she is interacting.

Note that some statuses are specific to an institution—doctor and patient are specific to medical institutions; teacher and student are specific to education; parent and child to family. But people also hold statuses apart from their institution-specific statuses—statuses related to race (e.g., Asian, white), gender (e.g., male, female, transgender), age, educational attainment, and social class. In thinking about statuses, sociologists distinguish between ascribed and achieved statuses.

Ascribed statuses are the result of chance in that people exert no effort to obtain them. By chance a person is born in a particular year and inherits certain physical characteristics. So birth order, race, sex, and age qualify as ascribed statuses. Other statuses, such as nurse's aide and college student, are **achieved statuses**; that is, they are acquired through some combination of personal choice, effort, and ability. The distinction between ascribed and achieved statuses is not clear-cut. One can always think of cases in which people take extreme measures to achieve a status typically thought of as ascribed; a person may undergo sex transformation surgery, lighten his or her skin to appear to be another race, or hire a plastic surgeon to create a younger appearance. Likewise, ascribed statuses can play a role in determining achieved statuses, as when women seemingly "choose" to enter a female-dominated career such as elementary school teacher, or when men "choose" to enter a male-dominated career such as car repair.

People usually occupy more than one social status. Sociologists use the term **status set** to capture all the statuses any one person assumes (see Figure 4.1a). Sometimes one status takes on such great importance that it overshadows all other statuses a person occupies. That is, it shapes every aspect of life and dominates social interactions. Such a status is known as a **master status**. Unemployed, retired, ex-convict, and HIV-infected can qualify as master statuses. The status of physician can be a master status as well, if everyone, no matter the setting (a party, church, fitness center), asks health-related questions or seeks health-related advice from the person occupying that status.

Figure 4.1a: Diagram of a Hypothetical Status Set
John was born with Down syndrome, and that status can be considered a master status because it shapes every aspect of his life, dominates his social interactions, and overshadows other statuses that he occupies—such as brother, video gamer, uncle, hospital employee.

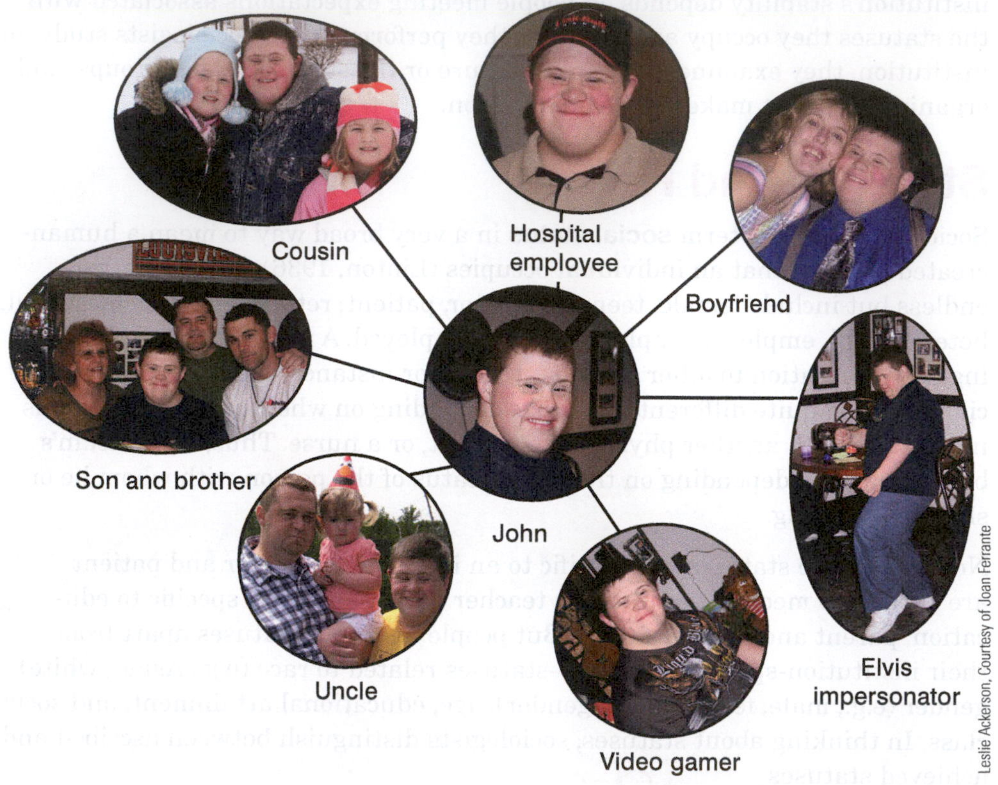

Sociologists use the term **role** to describe the behavior expected of a status in relation to another status—for example, the role of foster parent in relationship to foster child, or the role of physician in relationship to patient. Any given social status is associated with an array of roles, called a **role-set**, which consists of the various role relationships with which someone occupying a particular status is involved (Merton 1957a). The social status of son is associated with a role-set that can include a relationship with a mother and/or a father. The social status of school principal is associated with a role-set that includes relationships with students, parents, teachers, and other school staff members. The distinction between role and status is this: *people assume or are assigned to statuses and they enact or perform roles.*

Associated with each social status are **role expectations**, or norms about how a role should be enacted relative to other statuses. The role of physician in relation to a patient specifies that a physician has an obligation to establish a diagnosis, to not over treat, to respect patient privacy, to work to prevent disease, and to avoid sexual relations with patients (Hippocratic Oath 1943). Relative to a physician, patients have an obligation to answer questions honestly and to cooperate with a treatment plan. Quite often **role performances**, the actual behavior of the person occupying a role, do not meet role expectations—as when some physicians knowingly perform unnecessary surgery or when some patients fail to comply with treatment plans. The concept of role performance reminds us that people carry out their roles in unique ways. Still, there is predictability in our interactions with others, because if people deviate too far from the expected range of behaviors, negative sanctions (penalties)—ranging in severity from a frown to imprisonment and even death—may be applied (Merton 1957a). If we consider that people hold multiple statuses and that each status is enacted through its corresponding role-set, we can identify at least two potential sources of strain that lead to role performances that deviate from role expectations: role conflict and role strain.

Role conflict is a predicament in which the roles associated with two or more distinct statuses that a person holds conflict in some way. For example, people who occupy the statuses of college student and full-time employee often experience role conflict when professors expect students to attend class and keep up with coursework and employers expect employees to be available to work hours that leave little time for schoolwork. The person holding the statuses of both student and employee must find ways to address this conflict, such as working fewer hours, quitting the job, skipping class, or studying less.

Role strain is a predicament in which there are contradictory or conflicting role expectations associated with a single status. For example, doctors have an obligation to do no harm to their patients. At the same time, doctors who work for a health care corporation may feel unspoken pressure to meet quotas related to patient load or number of medical procedures recommended. In response, some physicians may recommend unnecessary medical procedures. Role expectations are embedded within a larger cultural context.

Sgt. Jose A. Torres Jr.

NKU Philosophy, Rudy Garns

Many countries in the world have a Tomb of the Unknown Soldier where soldiers are expected to stand guard 24/7 no matter the weather. But the role expectations are culturally influenced. In the United States the guard (left) moves 21 steps up and down a mat in front of the tomb, then stops for 21 seconds and then shifts direction (the 21 symbolizing a 21-gun salute). At each turn the guard must place the weapon on the shoulder nearer to visitors to show he represents a line of protection to the tomb. The guards at Greece's Tomb of the Unknown Soldier march in a highly deliberate manner and at times remain totally motionless (role expectations that hold even in rare instances when the tomb has been attacked).

Groups

Like statuses and roles, groups are an important component of social structure. A **group** consists of two or more people interacting in largely predictable ways who share expectations about their purpose for being. Group members hold statuses and enact roles that relate to the group's purpose. Groups can be classified as primary or secondary. **Primary groups** are characterized by face-to-face contact and/or by strong emotional ties among members who feel an allegiance to one another. Examples of primary groups can include teams, couples, siblings, small military units, cliques, and peer groups.

Secondary groups consist of two or more people who interact for a very specific purpose. Secondary group relationships are confined to a particular setting, and the involved parties relate to each other in terms of specific roles. People join secondary groups as a means to achieve some agreed-upon end, whether it be as fans of a sports team coming together to cheer their team on; students "joining" a college with the hope of one day becoming a college graduate; and cancer survivors setting a day aside to fund-raise. Secondary groups can be small to extremely large in size. They include a work unit, a college classroom, a parent–teacher association, and a church. Examples of larger secondary groups include the employees of a larger corporation or fans that support a particular sports team. Certainly, some people who are part of secondary groups form close relationships with each other and can constitute a primary group.

NKU Philosophy, Rudy Garns

Staff Sgt. Bernhard Lashleyleidner, 8th Army Public Affairs

A group of students on a month-long study abroad research trip (left) constitute a primary group. The students eat, work, and learn together; they travel together in a van and they spend their free time with one another. Runners at a marathon constitute a secondary group because their relationships with other runners, which can be as many as 47,000, are confined to a specific setting such as the New York or Boston Marathons.

What Do Sociologists See?

Sociologists see two statuses—those of drill instructor and military recruit. The recruits are enacting their role in relationship to the drill instructor. To be successful, recruits undergoing basic training must meet a number of role expectations including allowing drill instructors to invade their personal space, looking straight ahead while experiencing their wrath, responding "yes ma'am," and following commands.

Sgt. Jose Nava

Critical Thinking

Identify an institution of which you are a part and write about your experiences using concepts in the module (e.g., status, role expectations, and so on).

Key Terms

achieved statuses
ascribed statuses
group
master status
primary groups
role
role conflict
role expectations
role performances
role-set
role strain
secondary groups
social status
status set

Module

4.2 Levels of Social Structure

Objective

You will learn that there are three broad levels of social structure.

How do you analyze the social structure of an institution such as the economy that includes so many employees, businesses, regulatory agencies, consumers, and much more?

Imagine trying to study the economy of the United States, valued at about $16 trillion dollars. Or imagine trying to study the political structure of the United States that consists of a federal government and 90,106 state and local governments, and then add 4,500 political action committees and 1,310 super-PACs and 40,000 registered federal and state lobbyists to the mix (U.S. Census Bureau 2012a). Now imagine trying to figure out the interworkings and interrelationships among people who are part of political and economic institutions. Given the size and complexity of institutions, when sociologists study them they analyze only some parts of their social structure (see Module 4.1). What they choose to study may be part of one of three levels: the micro, meso, and macro.

Three Levels of Social Structure

If we place the focus on how an individual's identity, thought, behavior, and interactions are shaped by social structure—more specifically by the statuses they occupy or the roles they enact—we are involved with a micro-level analysis. At the micro level, sociologists are also interested in small groups that form within an institution. Examples include two status groups such as mother–daughter, doctor–patient, teacher–student, and other small groups such as a campus sorority/fraternity, a team, a book club, a classroom, a police unit, and so on.

When sociologists focus on middle-sized social structures within institutions such as neighborhoods, prisons, corporations, or college campuses, they are doing mesosociology. The meso level also includes any analysis of how these structures reproduce within their boundaries divisions of the larger society related to income, race, ethnicity, gender, occupation, and other differently valued statuses. Meso level structures can also defy those divisions.

An analysis of how the U.S. military depends on and integrates noncitizens to fill its ranks qualifies as a meso analysis. The photo shows 170 individuals from 35 different countries taking the oath to become U.S. citizens, something they are eligible to do after one year of service in the military.

U.S. Navy photo by Photographer's Mate 2nd Class Charles E. Alvarado

Finally, if the focus is on social structures that are very large and complex, either because of the sheer numbers of people involved or because a social structure's reach is national, regional, or global in scale, then we are doing macro-level analysis. Macro-level analysis looks at how social institutions such as medicine are structured to affect people's access to health care or how the economy is structured to create jobs that pay $10.00 or less per hour. A macro analysis also looks at how governments around the world structure opportunities.

Figure 4.2a: Duration of Paid Maternity Leave by Country

Governments are institutions that establish policies and enact laws that shape opportunities. A macro-level structural analysis of the duration of paid maternity leave around the world shows some clear patterns. The United States is one of three national governments in the world that do not mandate maternity leave at the federal level. Some governments mandate women be paid for less than 14 weeks and others mandate 52 weeks or more. Sweden grants the longest paid leave of 450 days, or about 15 months per child. Could it be that the lower the birth rate in the country, the longer the maternity leave as governments use leave as an incentive to women to have children? Or could it be that the greater the percentage of female population employed in the paid labor market, the longer the maternity leave?

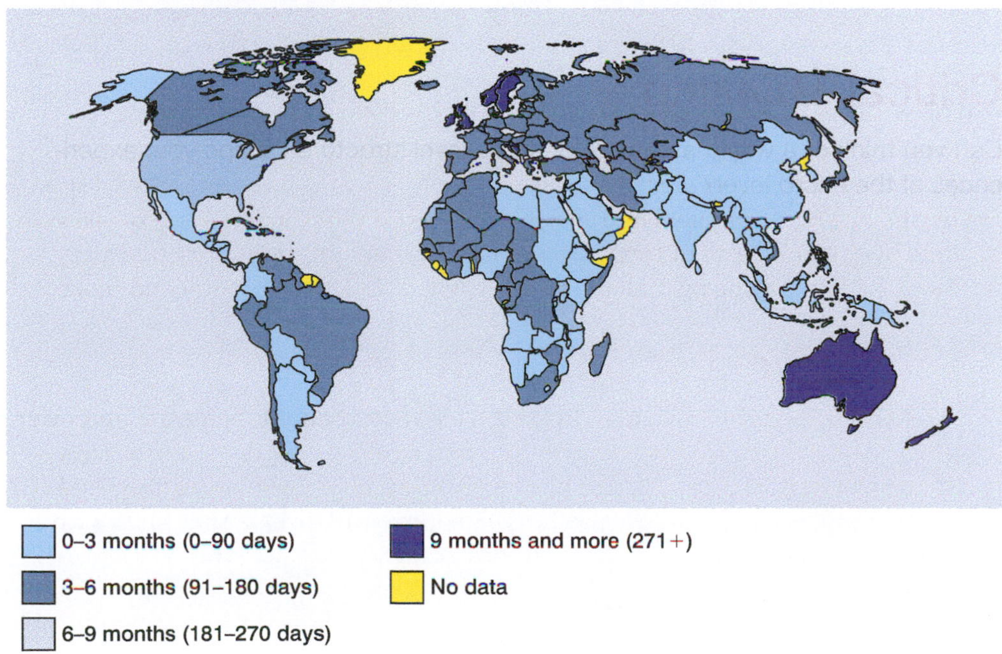

Source of data: International Labour Organization (2010).

Sociologists are also interested in interrelationships between and among the three levels. For example, to what extent are micro-level social structures (say a family) shaped by the macro (the global economy) and meso structures (a corporation that offers flex time)? To what extent are meso structures (corporations) shaped by macro-level structures (global economy)? How do activities at the micro level shape meso and macro structures? How do macro-level dynamics trickle down to affect the meso and micro levels?

What Do Sociologists See?

When sociologists see national currency, they think about more than money; they think about the clues the images on the currency offer about the interrelationship between a country's government and economy (two major institutions). The Fiji two-dollar note places children front and center and suggests that the government values its youth. The images in the background of the Korobas Mountains and of the 30,000-seat ANZ Stadium (located in Suva, the capital of Fiji) offer hints that these sites are important to national identity and to the economy as tourist attractions.

NKU Anthropology. Sharyn Jones

Critical Thinking

Can you think of a way that macro- or meso-level structures shape your experiences at the micro level?

Module 4.3 Social Structure and Human Agency

Objective

You will learn about the power of social structures to constrain behavior but also about the power of human agency to change social structures.

Look at the passengers' facial expressions in this photo. Why do you think most if not all the passengers appear to be paying no attention to those around them?

Rachel Ellison

Most people can walk onto any subway train, take their position among the other passengers, and behave as a passenger even if they have never been on a subway. This is because a largely invisible social structure broadly shapes and constrains passengers' behavior in predictable ways. To gain some idea of social structure's behavior-shaping and -constraining power, consider what would happen if someone entered a subway train but did not behave as expected. That person may be stared at, asked to leave, or reported to the police. Social structures (1) prompt people to assume a social identity, (2) shape the content and course of interactions, and (3) present people with varying opportunities and constraints.

Like any social setting that people step into, a subway train has a **social structure**, a largely invisible system that broadly shapes and constrains human activity in noticeable and predictable ways. If you think about it, knowing expectations for behavior in an endless variety of settings is key to living our lives with others. Our first challenge as babies and toddlers was to learn the basic expectations for living with our family and for negotiating the settings into which our family took us. Those expectations were often framed in terms of don't—don't play in the toilet, don't hit your sibling, don't run in the house, don't scream in the restaurant, and so on. Throughout our lifespan we find ourselves moving in and out of social settings and faced with the challenge of learning related expectations for behavior. To put it another way, we have to learn to fit our behavior into already established systems with recognizable patterns of behavior and interactions that give predictability and order to social settings.

Rachel Ellison

This photo of a subway train with passengers on board is a snapshot of subway behavior—that is, there is a predictable pattern to the way riders are behaving. Anyone who rides a subway or even watches movies containing subway scenes notices that this is typically the way passengers behave.

The seemingly scripted behavior of passengers on a subway raises a question about the relative influence of social structure and agency on human thought and behavior. Here we turn to sociologist Émile Durkheim's concept of social facts to describe the constraining power of social facts (his term for social structure) to limit choice and opportunity.

The Constraining Power of Social Facts

Social facts are collectively imposed ways of thinking, feeling, or behaving that have "the remarkable property of existing outside the consciousness of the individual" (Durkheim 1901, 51). That is, social facts are usually not the creations of the people experiencing them.

To illustrate the power of social facts to constrain behavior, can you guess whether this child stepping out of a swimming pool is a boy or a girl? Does it matter? This child happens to be a boy. However, if this were a girl, do you think she should cover her chest? If you said "no," you understand the power of social facts. If you think it is okay for a girl to go topless, you likely recognize that others will challenge such a view.

Chris Caldeira

From the time we are born, the people around us seek to impose upon us ways of thinking, feeling, behaving, and expressing ourselves that we had no hand in creating. In their earliest years of life, caretakers oblige children to "eat, drink and sleep at regular hours . . . to be mindful of others, and to respect customs and conventions" (Durkheim 1901, 55).

The expression on this boy's face speaks to the pressures to which children are subjected—to wear their hair a certain way, to speak a particular language, or to play with certain toys. According to Durkheim, society "seeks to shape a child in its own image" (1901, 54). Parents, teachers, and other caretakers act as the intermediaries, insisting on ways of being that even they did not invent.

Lisa Southwick

Over time what was once imposed becomes habit. According to Durkheim (1901), habit disguises the pressures once applied to turn an imposed behavior into something people do without thinking, but it does not erase that pressure. As Durkheim put it: "air does not cease to have weight," even though we cannot feel that weight (53).

NKU Sociology, Missy Gish

The effort required to learn and then to make expected behavior feel like a habit can be equated with learning to drive and follow the rules of the road. "The first time you drive a car, you have to think about every move" (Brooks 2011). But after weeks or months of practice, driving becomes automatic. No longer does the new driver have to think about how much pressure to apply to the brakes in given situations—the decision to brake hard or lightly happens without thought. It is amazing that we come to "know" without thinking just how much pressure to place on the brakes in any situation and that we apply the right amount of pressure without thinking.

Just like learning to drive a car, when we enter a new social setting, we have to think about our every move. Over time and with practice we learn expectations for behavior, and what was once strange and unnatural becomes a part of us. When a social fact becomes habit, its power "is not felt or felt hardly at all." Only when people resist do they come to know and experience its power (Durkheim 1901, 55). At that point people around us enforce social facts in a variety of ways—casting looks of approval or disapproval, sending people to prison, offering money, giving detention, and so on.

Human Agency

Human agency is the capacity of individuals to act autonomously—to resist the constraining pressure of social structure. To put it another way, agency is about the individual's ability to question the way things are done and to "choose" to do it another way, or at least to try to do it another way. Keep in mind that decisions or the wish to break with established expectations and ways of doing depend on how much a person believes in the "binding power of the past, the malleability of the future," and his or her confidence in their ability to effect change (Emirbayer and Mische 1998).

There is no question that some social structures offer greater opportunities for people to exercise agency than others. That is, certain settings—a prison, a church service, an oppressive work environment—require, even demand, that individuals behave in expected ways. Other settings challenge people to exercise human agency—to reject old ways of doing things and invent rules and expectations.

Chris Caldeira

This man exercised human agency when he challenged the social fact that people should wear clothes in public settings. It is likely easier, however, to exercise human agency and walk around nude if a city has no laws on the books prohibiting nudity. Still, even in cities with no laws prohibiting nudity, the unspoken expectation is that people should wear clothes. In such settings human agency is still exercised but with no legal penalty.

Tech Sgt. Anthony Iusi

Of course, human agency is only one factor driving change. Just because someone works, even puts their life on the line, to change social structures defined as oppressive does not mean that such actions will lead to change. Those who have the most to lose if things change can demonize those exercising human agency. If agents of change are successfully demonized, the status quo is supported.

What Do Sociologists See?

Sociologists see a Buddhist monk exercising human agency as he sits outside the U.S. embassy in Afghanistan protesting military action. The fact that he is alone suggests that he is going against widespread support for military action.

Ph2 Anthony J. Pugliani

Critical Thinking Question

Can you think of a time when you exercised human agency by challenging, even defying expectations for behavior?

Key Terms

human agency

social facts

social structure

Module 4.4

Division of Labor and Social Networks

Objective

You will learn about how the division of labor and social networks are social structures that connect people to one another and to society.

How many pairs of hands does it take to make a pair of athletic shoes?

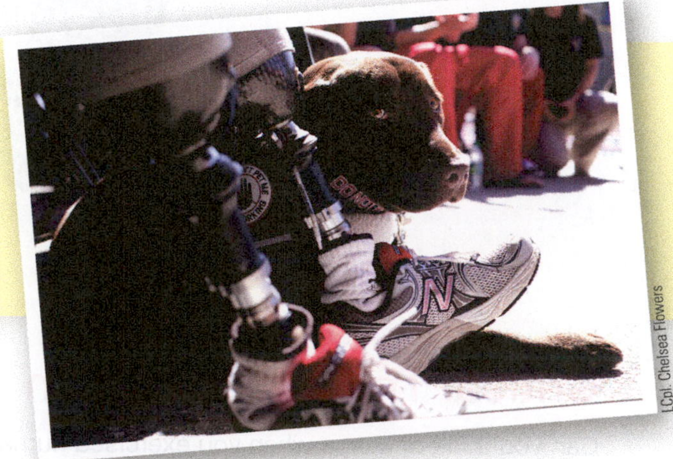
LCpl. Chelsea Flowers

In her book *Factory Girls: From Village to City in a Changing China*, Leslie T. Chang describes the incredibly involved process of making a pair of athletic shoes:

> It takes two hundred pairs of hands. . . . Everything begins with a person called a cutter, who stamps a sheet of mesh fabric into curvy irregular pieces, like a child's picture puzzle. Stitchers are next. They sew the pieces together into the shoe upper, attaching other things—a plastic logo, shoelace eyelets—as they go. After that sole-workers use infrared ovens to heat pieces of the sole and glue them together. Assemblers—typically men, as the work requires great strength—stretch the upper over a plastic mold, or last, shaped like a human foot. They lace the upper tightly, brush glue on the sole, and press the upper and sole together. A machine applies 90 pounds of pressure to each shoe. Finishers remove the lasts, check each shoe for flaws and place matched pairs into cardboard boxes. The boxes are put in crates, ten shoes to a crate and shipped to the world. (2009, 98)

Although the headquarters of the major athletic footwear corporations are located in places like Oregon, Massachusetts, and Germany, the shoes are most likely made in Asian countries, most notably China, Indonesia, and Vietnam. Keep in mind that the 200 pairs of hands represent only some of the hands involved in getting the shoes to consumers. Other hands navigate the freighters carrying the shoes from China to international ports, where dockworkers unload them. Then truck drivers transport them to stores located around the world. And we cannot forget the labor (pairs of hands) that creates the commercials that promise superior athletic prowess when worn.

Division of Labor

In *The Division of Labor in Society* (1933), sociologist Émile Durkheim wrote about how the Industrial Revolution not only transformed the ways in which workers related to each other, but also transformed human interaction. The **division of labor** refers to work that is broken down into specialized tasks, each performed by a different set of workers trained to do that task. When labor is broken down into specialized tasks, people relate to one another in terms of the specialized roles they perform. As a result, most day-to-day interactions inside and outside the workplace are largely impersonal and instrumental (that is, we interact with people for a specific reason, not to get to know them). We do not need to know personally the people with whom we work or who make the products we wear and use. In spite of this anonymity, the largely invisible ties that bind people to one another and to their society are very strong, because few individuals possess the knowledge, skills, and materials to be self-sufficient. Consequently, people need one another to survive.

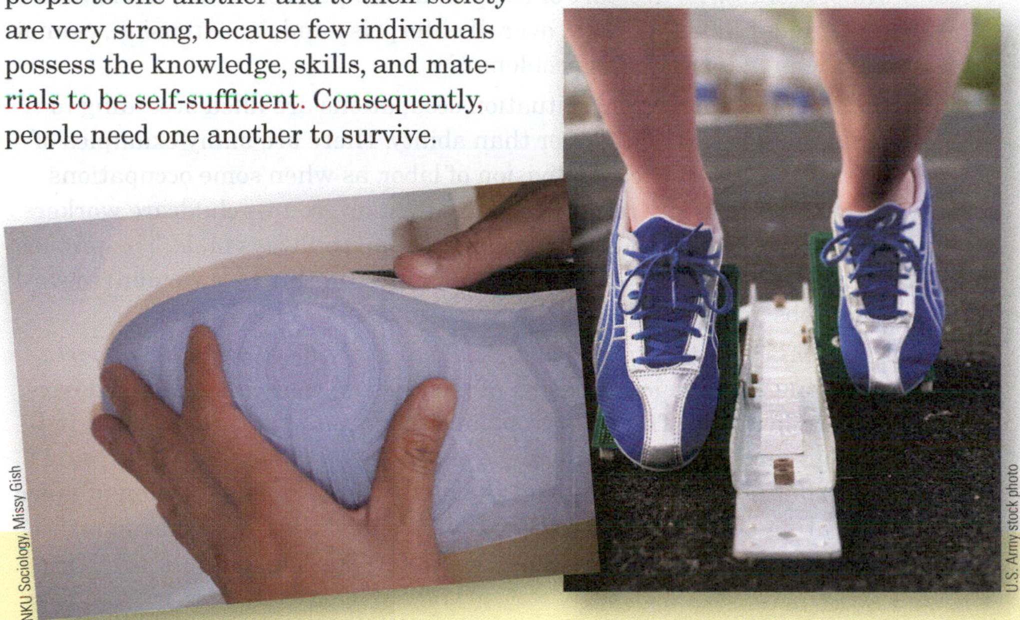

NKU Sociology, Missy Gish

U.S. Army stock photo

Many Americans do not make a conscious connection between the athletic shoes they wear and those who labor to make them. The chances are quite high that workers whose hands make the shoes are laboring in Asian factories. People who wear the shoes do not personally know the workers who produce the products they depend on to run a race or engage in other activities.

Disruptions to the Division of Labor

Durkheim hypothesized that societies become more vulnerable as the division of labor becomes more complex and interconnected. Durkheim was particularly concerned with the kinds of events that disturb the division of labor. Such disturbances affect people's ability to connect to each other and their community through their labor. As one example, consider how life changes for those who have been laid off from work. After losing a job, the daily routine shifts from interacting with colleagues at work to the search for work. Moreover, the newly unemployed person loses the structure that comes with a job. Now, instead of working, the person may watch play video games or search the Internet for job openings. Relationships with family members may become strained if a partner or children must work more hours or give up activities to compensate for the lost job.

Durkheim identified five key events that disrupt the division of labor and its ability to bind people to each other and their communities:

1. *Industrial and commercial crises resulting in plant closings and massive layoffs.* These crises can be triggered by technological revolutions that eliminate human labor or by corporate decisions to outsource work to locations where wages are lower.
2. *Worker strikes in which workers stop working.* The goal of strikes is to force employers to meet a demand, raise pay, increase benefits, or improve working conditions. When employers resist such demands, they search for replacement workers or shut down operations. The striking workers suffer lost wages.
3. *Job specialization creating a situation in which workers labor in isolation.* When people work in isolation, they are unaware of how their work relates to the work of others who are part of the production process. Thus, workers lack a complete understanding of the overall enterprise and, by extension, cannot see ways the process might be problematic.
4. *Forced division of labor.* In this situation occupations are filled according to nationality, age, race, or sex, rather than ability. There are many examples of what Durkheim calls the forced division of labor, as when some occupations are formally or informally reserved for a particular group—childcare workers are almost always female; airplane pilots are almost always male. The problem is that the division of labor channels some groups (in this case women) toward low-paying jobs and other groups (men) toward higher-paying jobs.
5. *Workers' talents and abilities do not match up with the existing job opportunities.* A strong and efficient division of labor is one that cultivates and uses all its available talent. The division of labor does not operate efficiently if, in times of high unemployment, there are job openings in some sectors (like IT or nursing) but not enough qualified people to fill them. The division of labor is not operating efficiently when work that the unskilled can do is nonexistent, irregular, intermittent, or subject to high turnover.

Since 1907, but especially since the 1970s, Detroit autoworkers have faced a series of ongoing industrial and commercial crises that have shut down auto factories and moved work elsewhere. These crises severed ties many workers had to the division of labor and to a source of income. As a result, on some streets of Detroit and surrounding suburbs, more houses and businesses are abandoned than are occupied because people no longer have the money to maintain the lifestyle they attained when employed.

Social Networks

Imagine you are looking for a job. Do you know someone who might help you find one? If you do, then you are using your social network to gain employment. If no one in your social network can connect you to employment opportunities, you are at a disadvantage in the job market. A **social network** is a web of social relationships linking people to one another.

Figure 4.4a: Examples of Social Networks
Think about Maddy's social network, as depicted in this image. You can see some of the people who are part of her social network. You can also see how, through her dad (Greg), Maddy is also connected to Bill's network. Likewise, Maddy is connected to the networks of all those who are part of her network. Here only Bill's network is shown.

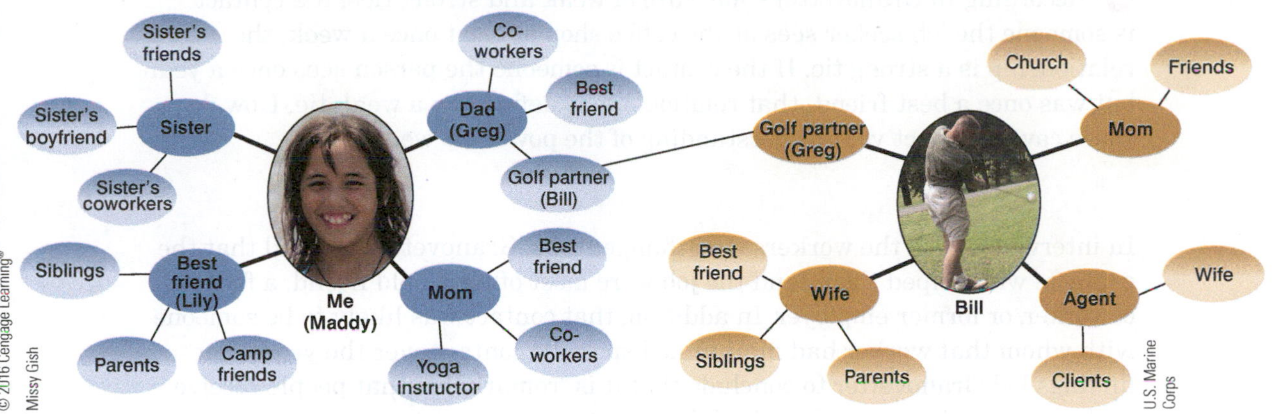

In looking at Maddy's network, we might ask which ties are most likely to connect her to a job if she were looking. The work of sociologist Mark Granovetter (1973) provides some insights. Granovetter, who wrote one of the most cited articles in sociology, "The Strength of Weak Ties," studied 100 professional, technical, and managerial workers living in a Boston suburb who had recently changed jobs. Granovetter hoped to learn the role social networks played in securing their new jobs. He learned that 54 of these workers found their new jobs through a personal contact, someone in their social network. To measure the strength of workers' ties to their named contacts, Granovetter asked how often they saw that contact. He found that 27.8 percent saw their contact rarely (once a year or less), 56 percent saw their contact occasionally (more than once a year but less than twice a week), and 16.7 percent saw their contact often (at least twice a week). This led him to conclude that "the skew is clearly in the weak end of the continuum" (1371).

According to Granovetter's measure of weak and strong ties, if a contact is someone the job seeker sees at the coffee shop at least once a week, the relationship is a strong tie. If the contact is someone the person sees once a year but was once a best friend, that relationship is defined as a weak tie. How do these caveats affect your understanding of the power of "weak" ties?

In interviews with the workers who changed jobs, Granovetter learned that the contacts who helped them land the job were most often an old friend, a former coworker, or former employer. In addition, that contact was likely to be someone with whom that worker had maintained sporadic contact over the years. These findings led Granovetter to conclude that it is "remarkable that people receive such crucial information from individuals whose very existence they have [mostly] forgotten" (1372). In his interviews, Granovetter asked workers how their contacts knew about the job opening. He found that

- 45.0 percent said that their contact knew the prospective employer,
- 39.1 percent said their contact was a prospective employer, and
- 12.2 percent said that their contact knew someone who knew the potential employer.

Granovetter also learned, to his surprise, that the information paths to new employment opportunities involved only one to three linked contacts. He hypothesized that if the paths had involved more linked contacts, larger numbers of people "might have found out about any given job, and no particular tie would have been crucial" (1372). Granovetter's findings do not discount the importance of strong ties and individual initiative in finding a job. But his research does speak to the importance of cultivating weak ties within social networks when searching for employment.

Electronic-Supported Social Networks

Sociologists, of course, are interested in how digital technologies affect the size and character of social networks and people's ability to connect to others through them. Even before digital technologies such as the Internet, e-mail, and cell phones, sociologists realized that it is not easy to map a person's social networks. Digital technologies are not the first innovations to expand the size and reach of social networks. The telephone, airplane, train, and car expanded social networks to the point that it is impossible to manage and sustain connections with everyone we have come to know or hope to know.

In what ways do digital technologies further enhance the ability of people to connect with others? They

- allow people in different time zones and on different schedules to communicate at their convenience,
- increase the speed of communication,
- expand the number of people with whom one communicates to theoretically include anyone with access to the Internet, and
- offer people who live geographically far apart a convenient and inexpensive tool with which to remain in touch.

Weak ties that may have faded without these technologies are more likely to remain strong, and a person's strong ties are maintained.

Digital technologies do not just expand the reach of social networks; they also enhance local connections. Preliminary research suggests that most of the e-mails and text messages we send each day are to those who live nearby—to significant others, work colleagues, and friends—all of whom use the technology to check in, send reminders, or arrange a face-to-face meeting (Wellman and Hampton 1999).

What Do Sociologists See?

When sociologists see the accessible symbol, they think of the Americans with Disabilities Act of 1990, and how that Act opened opportunities to connect with those inside buildings. This act guaranteed people with disabilities the right to access public spaces and facilities, widening their social networks and allowing them to gain access to goods, lodging, transportation, education, and dining, among other things.

Chris Caldeira

Critical Thinking

Think about jobs you have held over your lifetime. How did you learn about the job opening? If you have never worked, then think about how a friend or family member learned about the job he or she holds.

Key Terms

division of labor

social network

Module

4.5 The Effects of Size

Objective

You will learn some ways in which group size affects social dynamics.

Think about a time when you were you having a conversation with one person, and a third person joined in. How was your conversation affected?

Lisa Southwick

A third party joining a two-person conversation already in progress will change the dynamics between those two people. The third person might direct his or her attention toward one of the two, and the other may feel left out or decide to leave. Sociologist Georg Simmel (1950) argued that one of the most important criteria for understanding group dynamics is size. Size affects the relationships among members and shapes the character of interactions.

Dyads, Triads, and Beyond

The smallest group is a **dyad**, which consists of two people. Sociologists are interested in the reasons dyads form, whether as a consequence of birth (mother–son), by choice (two friends), or because of social necessity (doctor–patient, teacher–student, author–editor). In assessing a dyad, sociologists ask: How much does each party know about the other? Keep in mind that no person can know everything about another. But we can say that in **comprehensive dyads** the involved parties have more than a superficial knowledge of each other's personality and life; they know each other in a variety of ways. In **segmentalized dyads** the parties know much less of the other's personality and personal life; and what they do know is confined to a specific situation, such as the classroom, a hair salon, or other specialized setting (Becker and Useem 1942).

Sociologists consider this husband–wife dyad (top) comprehensive, as it is very likely that after five decades of marriage each knows a lot about the other as a person. This street vendor–tourist dyad is segmentalized because the involved parties interact out of social necessity—one party interacts out of a need to purchase a jacket, the other interacts out of a need to make a living. It is also likely that the two will never see each other again after this transaction is complete.

Rachel Ellison

NKU Philosophy, Rudy Garns

If a third member is added to a dyad—whether it be a stepfather to a mother–son dyad, a nurse to a doctor–patient dyad, a new friend to an existing pair of friends—the result is a **triad**, or a three-person group. The addition of a third person significantly alters the pattern of interaction between the two people. Now, with a third person added to the mix, two members can form an alliance against the third. Obviously, as more members are added—a fourth, a fifth, and so on—the possible patterns of interaction and alliance increase.

Lisa Southwick

At some point—perhaps around seven members—the group breaks down into subgroups because it is impossible to focus everyone's attention on the group per se. That is, it is difficult to have a single, focused conversation in which everyone is listening, taking turns talking, or focusing on the task at hand. Unless someone steps forward and directs communication among the members, the group breaks down into dyads and triads, with each smaller group carrying on its own conversations (Becker and Useem 1942).

Oligarchy

Political analyst Robert Michels (1962) was interested in very large groups involving thousands of people. He believed that large organizations inevitably tend to become oligarchical. **Oligarchy** is rule by the few, or the concentration of decision-making power in the hands of a few persons who hold the top positions in a hierarchy. Michels maintained that one of the most bizarre features of any advanced industrial society is that life-and-death choices are made by a handful of people, usually men, who cannot possibly consider all who will be affected.

Organizations become oligarchical because democratic participation is virtually inconceivable in large organizations. Size alone makes it impossible to discuss matters and settle controversies in a timely and orderly fashion. For example, Walmart is the largest employer on the Global Fortune 500 list, with 1.9 million employees and millions of stockholders. Obviously, such a large number of employees cannot engage in direct discussion.

The greater an organization's size, the less likely it is that members can comprehend its workings. As a result, leaders may push employees to advance organizational goals, the full consequences of which no one may be able to know or understand. There is also the danger that key decision makers may become so preoccupied with preserving their own leadership or with the bottom line that they do not consider the greater good or the full implications of their choices.

We can apply the concept of oligarchy to the Great Recession of 2008. Some blame this economic crisis on a small group of powerful executives running financial institutions that were "too big to fail," for creating and abusing unregulated financial products for personal and institutional gain. Others point to complex financial instruments that few, if any, people understood until it was too late. Regardless of the real source of the crisis, a few people at the top of some of the largest organizations in the world ultimately decided to use these products to build profits, with disastrous consequences for national and global economies (Gross 2009).

What Do Sociologists See?

Sociologists do not just see three people together. They think about how each woman's relationship with each one of the other two women is shaped by the presence of the third. Sociologists ask, which two women feel closest and tend to form alliances against the third?

NKU Sociology, Missy Gish

Critical Thinking

Think of a time when a third person was added to a dyad to which you belonged. Describe how that third person changed the interactional dynamics between you and the other person.

Key Terms

comprehensive dyads
dyad
oligarchy
segmentalized dyads
triad

Module

4.6 Formal Organizations

Objective

You will learn that formal organizations coordinate human activity with the aim of achieving some valued goal.

Have you ever participated in an organization-sponsored event where you did something for a good cause, say breast cancer awareness?

Mark Iacampo, U.S. Army Garrison Hohenfels Public Affairs

One organization considered a global leader in breast cancer awareness and the search for a cure is Susan G. Komen. The organization started out as Nancy G. Brinker's promise to her dying sister (Susan) to "do everything in her power to end breast cancer forever." Thirty years later that organization has invested $2 billion in research and programs with the goal of increasing awareness and ending breast cancer deaths. The nonprofit has affiliated programs in more than 50 countries (Susan G. Komen 2014).

If you have been part of such an effort, you have experienced the coordinating power of formal organizations to achieve a goal. **Formal organizations** are coordinating mechanisms that bring together people, resources, and technology and then direct human activity toward achieving a specific outcome. That outcome may be to increase awareness about breast cancer, to maintain order in a community (as does a police department), to challenge an established order (as does People for the Ethical Treatment of Animals), or to provide a credentialing service (a university). Formal organizations are a taken-for-granted part of our lives. If you were born in a hospital, attended a school, acquired a driver's license from a state agency, secured a loan from a bank, worked for a corporation, received care at a hospital, or purchased a product at a store, you have been involved with a formal organization (Aldrich and Marsden 1988).

Formal organizations can be voluntary, coercive, or utilitarian, depending on the reason that people participate (Etzioni 1975). **Voluntary organizations** draw in people who give time, talent, or money to address a community need. Voluntary organizations include food pantries, political parties, religious organizations, and fraternities and sororities.

There are millions of voluntary organizations worldwide to which people give time, talent, and money. One example is a homeless coalition that provides winter shelter for those without homes.

NKU Sociology, Missy Gish

Coercive organizations draw in people who have no choice but to participate. Examples include public schools that students are required by law to attend, the military when there is a draft, and treatment facilities to which people have been ordered to report because a judge or other authority deems them medically or socially unfit (Spreitzer 1971). Prisons also qualify as coercive organizations.

Utilitarian organizations draw in those seeking to achieve some desired goal in exchange for money. That goal may be to earn an income as an employee, to acquire a skill by enrolling in a special program, or to purchase a desired product at a department store.

From a sociological perspective, formal organizations, whatever their type, have a life that extends beyond the personnel and clients/customers. Indeed, formal organizations prevail even as the people die, quit, retire, or get fired. One concept that sociologists use to think about the ways in which organizations coordinate resources and structure human activity is bureaucracy.

Bureaucracy

Sociologist Max Weber (1925) defined a **bureaucracy**, in theory, as a completely rational organization—one that uses the most efficient means to achieve a valued goal, whether that goal is feeding people (McDonald's), recruiting soldiers (military), counting people (Census Bureau), collecting taxes (IRS), drilling for oil (ExxonMobil), or providing a service (hospitals).There are at least six means by which a bureaucracy efficiently coordinates human activity to meet organizational goals.

1. *A clear-cut division of labor exists.* Each office or position is assigned a specific task geared toward accomplishing the organizational goals. Google, for example, employs 20,000 people worldwide in occupations including office managers, technicians, laborers, chemists, engineers, and so on. All employees—software engineers, product managers, web developers, ad quality raters, and facilities managers—work toward achieving Google's mission, which is "to organize the world's information and make it universally accessible and useful" (Google 2014).
2. *Authority is hierarchical.* Most bureaucracies publish organizational charts depicting how authority and responsibility are distributed among the positions that make up the organization. Within bureaucracies there may be vice presidents, regional managers, managers, assistant managers, and entry-level employees, all of whom report to someone in the chain of command.

Chris Caldeira

This office building evokes images of a bureaucracy. We can imagine employees, each with a job description, assigned to an office, cubicle, or work space, working to achieve organizational goals. The building suggests a hierarchy, with people holding the most power and authority in the organization occupying offices on the top floors.

3. *Written rules specify the way positions relate to each other and describe the way an organization should operate.* Administrative decisions, rules, procedures, and job descriptions are recorded in operations and training manuals, handbooks, or bylaws.
4. *Positions are filled according to objective criteria.* Criteria include academic degrees, seniority, merit points, or test results, but not emotional considerations, such as family ties or friendship.
5. *Authority belongs to the position.* It does not belong to the particular person who fills a position. This characteristic implies that while on the job, managers, for example, have authority over subordinates because they hold a higher position. Managers cannot demand that subordinates do tasks unrelated to work, such as washing a manager's car or babysitting.

6. *Organizational personnel treat clients or customers as cases.* That is, "without hatred or passion, and hence without affection or enthusiasm" (Weber 1947, 340). To put it another way, no client receives special treatment. This objective approach is believed to be necessary because emotion and special circumstances can interfere with the efficient delivery of goods and services. Many organizations require employees to greet every customer with standard lines such as, "Thank you for shopping at Kmart. May I help you?"

The "take a number" system is designed to ensure that employees treat clients/customers as "cases." This system eliminates emotion and bias in determining who is next. Taking a number sends the message that no person is special or can demand to be served before another.

Taken together, these six characteristics describe a bureaucracy as an **ideal type**—a deliberate simplification or caricature in that it exaggerates essential traits. Ideal does not mean desirable; an ideal is simply a standard against which real cases can be compared. Anyone involved with a bureaucracy realizes that actual behavior departs from this ideal. Still, the six ideal traits of bureaucracy can be used to evaluate the extent to which any bureaucracy follows or departs from these traits. Such an evaluation may reveal problems or strengths in an organization's structure. In this regard, sociologists distinguish between formal and informal dimensions of organizations.

Formal and Informal Dimensions

The **formal dimension** is the official, by-the-book way an organization should operate. The formal dimension is known through an organization's job descriptions and through its written rules, guidelines, and policies. The **informal dimension** encompasses any aspect of the organization's operations that departs from the way it is officially supposed to operate. The informal dimension includes employee-generated responses that violate official policies and regulations. A manager who demands that employees work off the clock, employees who give their friends free food and soft drinks, and servers who spit in a rude customer's drinks are all displaying behavior that departs from official policies.

Lisa Southwick

Lisa Southwick

These photos speak to informal and formal dimensions of organizations. When employees sleep on the job or engage in activities unrelated to work, such as browsing the Internet while on the clock, they are behaving in ways that violate official organizational policies. When checkout clerks at Home Depot wear a smock proclaiming "I Put Customers First," they are complying with official policy.

Performance Measures

Many bureaucracies have performance measures in place, quantitative indicators of how well their employees or clients are performing with reference to some valued goal. Managers often evaluate employees using statistics related to sales, customer satisfaction, and production quotas. Such measures can be useful management tools for two reasons: They are considered to be objective, and they permit comparison across individuals, over time, or across departments. Management can tie pay increases and promotions to objective measures of good performance and can use objective measures of poor performance to justify actions such as firings, pay reductions, or demotions. But performance measures have shortcomings, one of which is that they can encourage employees to concentrate on meeting performance goals and to ignore problems generated by their drive to score well.

Kari Hawkins, USAG Redstone

Does your employer use mystery shoppers as a tool for evaluating your performance? Mystery shoppers purchase products and ask questions of employees with the goal of assessing the shopping experience and whether employees follow company protocol. From an employer's point of view, the possibility that a customer might be a mystery shopper incentivizes employees to be at the top of their game. This method of evaluating performance has its drawbacks as employees may follow protocol so closely that they don't treat customers as people with unique needs.

What Do Sociologists See?

When sociologists see the words "billions and billions served" on the McDonald's sign, they think about the organizational structure that makes this possible. The structure is that of a bureaucracy, employing the most efficient means to coordinate the actions of 1.8 million employees in 119 countries to fill 25 billion food orders each year.

NKU Sociology, Missy Gish

Critical Thinking

Name two organizations that play important roles in your life. Explain. Classify each as voluntary, coercive, or utilitarian.

Key Terms

bureaucracy
coercive organizations
formal dimension
formal organizations
ideal type
informal dimension
utilitarian organizations
voluntary organizations

Module 4.7 Rationalization and McDonaldization

Objective

You will learn how rationalization and McDonaldization shapes the social structures of organizations.

National Science Foundation

DARPA

What motivates CEOs and other decision makers to invest in social robots?

One answer is that decision makers see robots as innovations that will do work that is hard, dangerous, and requires patience but without complaining, getting tired, or needing salaries and benefits. In this module we consider what motivates people to make decisions and take action. Sociologist Max Weber identified four sources of motivation: tradition, emotion, value-rational, or instrumental rational action. As we will see, these motivations do not emerge out of the blue but are cultivated by the social structure guiding decision making and action.

U.S. Army photo by Tim Hipps

Thought and action driven by tradition are guided by a respect for the ways things were done in the past. The performers' actions shown here are motivated by *tradition*. These performers were part of the opening ceremonies of the 2008 Summer Olympic Games held in China. Over two thousand performers played the most ancient of Chinese percussion instruments, the *fou*. The performers practiced endless hours to create the sound of spring thunder and to move in unison, even stitching needles into shirt collars pointed at their necks to keep neck and head held high (LaFraniere 2009).

Cheryl Harrison

When thought and action are driven by *emotion*, they are accompanied by a physical sensation, such as feelings of love, hate, or fear. The women who made these quilts and the men who are loading them into a truck are motivated by emotion. The quilts are part of a support effort known as Operation Wounded Warrior. The quilts are to be sent to Warriors in Transition as a way of saying "Thank you for what you do for us."

SEVEN SOCIAL SINS
Politics without Principles
Wealth without Work
Pleasure without Conscience
Knowledge without Character
Commerce without Morality
Science without Humanity
Worship without Sacrifice
Young India, 22-10-1925

NKU Anthropology, Sharyn Jones

People who keep the seven social sins in mind when they make decisions and take action are guided by *value-rational action*. Value-rational action is behavior guided by a code of conduct—a "right" way to do something. For people motivated by value-rational action, adhering to that code is more important than achieving a goal.

Instrumental Rational Action

Instrumental rational action is result-oriented behavior that emphasizes the most efficient methods or means for achieving some valued goal, without regard for any adverse consequences those methods may have. As such it rejects tradition, emotion, and codes of conduct as considerations and places emphasis on the most cost-effective and time-saving way to achieve a goal—whether that goal is producing eggs, satisfying hunger, or recovering from an illness. Instrumental rational action, for example, drives the treatment of animals raised on factory farms. Obviously, the more chickens a factory farm can house and the faster it can raise them to egg-laying maturity, the more eggs it can produce and sell.

Jeff Vanuga

Corporations that process pigs are driven by instrumental rational action. Farmers contract with corporations to raise pigs that the corporation has provided. The pigs are confined in crowded and controlled environments, never having the opportunity to move around freely outdoors.

Weber maintained that the rise of instrumental rational action as a dominant way of organizing human activity is a product of industrialization. Through a process he labeled rationalization, instrumental rational action came to replace tradition, emotion, and codes of conduct as guides to thought and behavior.

Tony Rotundo

This photo of three little pigs in a barn represents the way many of us would imagine or would like to believe pigs are raised. This image suggests that the farmer who raises these pigs is driven by emotion or an obligation to care for them. In reality, most pigs are raised on corporate farms and live in crowded conditions with little room to move or even turn around. Instrumental rational action, with its emphasis on efficiency and means-to-ends thinking, leaves no room for emotion.

Rationalization, then, is a process in which thought and action organized around emotion, superstition, respect for mysterious forces, or tradition are replaced by instrumental rational action. Keep in mind that Weber used the term to refer to the way daily life has come to be organized so that people are forced to use the "efficient" structures that are already in place to meet their needs (Freund 1968).

Sociologist George Ritzer (1993) describes an organizational trend guided by instrumental rational action in which people use the most cost-efficient and quickest means to achieve some valued end. He called this trend the McDonaldization of society.

McDonaldization of Society

Ritzer defines **McDonaldization of society** as the process by which the principles of the fast-food industry have been applied to other sectors of American society and the world. Those principles are (1) efficiency, (2) calculability, (3) predictability, and (4) control.

EFFICIENCY. **Efficiency** involves using methods that will achieve a desired end in the shortest amount of time. Many organizations advertise products and services touting claims that buying them will help consumers move most quickly from one state of being to another—from hungry to full, from fat to thin, from uneducated to educated, or from sleep-deprived to rested. In some cases, an organization puts a system in place where customers serve themselves. In the name of saving time, customers scan, bag, and pay for purchases; clear their table after eating in a quick-service restaurant; and check themselves into the airport without expecting to be compensated for their labor.

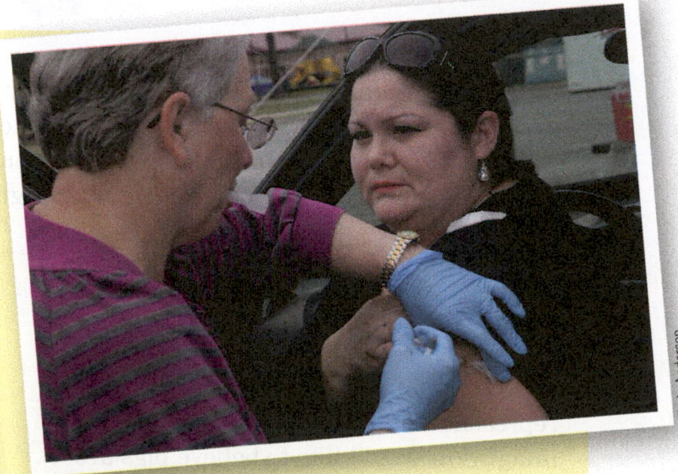

Mindy Anderson

Drive-through flu vaccination centers have applied the efficiency principle of McDonaldization to health care services. According to one nurse who administered the shots, "The drive-through concept is so popular because of its ease and convenience. For many people it is very difficult to get themselves and/or children out of the car and into a doctor's office, or other inside location. It is very convenient for those who just don't want to get out their car" (*University of Kentucky News* 2003).

CALCULABILITY. The second principle of McDonaldization is **calculability**, which emphasizes numerical indicators as a way customers can judge the results of a product or service or the speed with which it is delivered (e.g., delivery within 30 minutes, lose 10 pounds in 10 days, earn a college degree in 24 months, limit menstrual periods to four times a year, or obtain eyeglasses in an hour). These are clear measures that customers can draw upon to assess results. Ritzer argues that the emphasis on quantity promotes the idea that size matters, more is better, and something obtained quickly is superior to a product that takes time to deliver.

PREDICTABILITY. The principle of **predictability** is the expectation that a service or product will be the same no matter where in the world or when (time of year, time of day) it is purchased. With regard to food products, this kind of predictability requires that the product be genetically modified. If the consumer expects strawberries year-round, the produce must be able to withstand shipment from one continent to another. The fruit is genetically modified so that the berry stays firm and does not decay before it reaches its destination (Barrionuevo 2007).

How do apples get their predictable uniform and unblemished appearance? Notice the "cosmetically challenged" apples are not predictable in size and color like the apples shown in the bottom photo. Among other things, the apples in the bottom photo have been subjected to an aggressive regimen of chemical sprays to maintain the smooth skin, coated with wax for shine, and perhaps stored in a controlled environment for 10 months or more. Apples like those on the bottom are considered to be among the most pesticide-contaminated fruit (Lloyd 2011).

Chris Caldeira

Scott Bauer

CONTROL. The fourth principle of McDonaldization, **control**, involves replacing employee labor with "smart" technologies and/or requiring, even demanding, that employees and customers behave in a certain way. From an employer's point of view, humans are a source of uncertainty and unpredictability. The quality and consistency of people's work, for example, is affected by any number of factors, including how they feel, if they are paying attention, the personal problems they face, and their relationship with the boss. Consider how customers dialing 411 to obtain a phone number are greeted, not by a human voice, but by a computerized voice asking for the city and state and then asking for the name of the person or business they wish to contact. Customers cannot proceed unless they give clear and precise answers to the questions. When a computer-generated voice asks for the city and state, the customer cannot use free-flowing speech expressing uncertainty about a city's name. It is easy to see how smart technology controls the interaction and saves labor costs.

Assessing McDonaldization

Clearly, there are advantages to McDonaldization—people can check bank balances, take college classes, and pay bills at the hours of their choosing. Though McDonaldization has facilitated amazing feats, the model has drawbacks. In the quest to deliver products and services quickly and efficiently, an organization can create dehumanizing structures that can lower the quality of life. Ultimately, we must ask whether so-called rational processes create irrationalities

(Weber 1994; Ritzer 2008). Sociologists use the term **iron cage of irrationality** to label the process by which supposedly rational systems produce irrationalities. As one example, consider that the pharmaceutical industry creates medicines to "cure" or alleviate just about every social, physical, and psychological problem people face. It is irrational, however, that this industry seeks to fix the conditions that make us human by offering medications that eliminate monthly menstruation, minimize wrinkles, and alter moods (even normal sadness and grief).

What Do Sociologists See?

Sociologists do not just see a chicken with no feathers; rather, they see a bird that would be of great interest to the corporations that process billions of chickens each year for consumption. Feathers must be removed from these billions killed for food (Satya 2014). The forces of rationalization support the development of chickens so they do not grow feathers and push away any considerations of emotion, tradition, or a code of conduct.

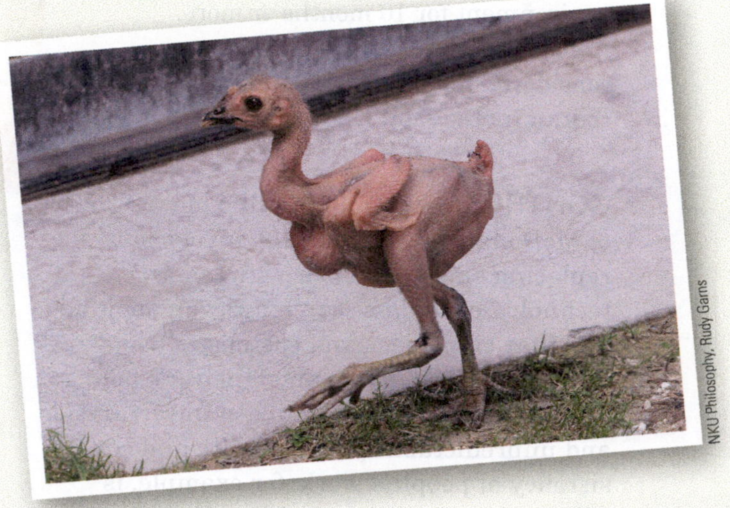
NKU Philosophy, Rudy Garns

Critical Thinking

Describe a social setting that supports decision making and action guided by one of the four motivating forces—tradition, emotion, value-rational, or instrumental rational.

Key Terms

calculability

control

efficiency

instrumental rational action

iron cage of irrationality

McDonaldization of society

predictability

rationalization

Module 4.8 Alienating and Empowering Social Structures

Objective

You will learn about social structures that promote alienating and empowering workplaces.

If you are employed, how satisfied are you in your current job?

Rachel Ellison

In his book *Three Signs of a Miserable Job*, Patrick Lencioni (2007) names three signs of dissatisfaction:

1. *anonymity*—my boss has little interest in me as a person with a unique life, interests, and aspirations;
2. *irrelevance*—my work makes no real difference in people's lives; and
3. *cluelessness about impact*—there is no way of knowing whether I make an impact or contribution through my work.

Karl Marx wrote about such issues more than 150 years ago. Marx (1844) wrote: "labor is *alien* to workers when they do not affirm themselves through labor." The alienated worker "does not feel happy, but rather unhappy; he does not grow physically or mentally, but rather tortures his body and ruins his mind."

Alienation

Human control over nature increased with mechanization and the growth of bureaucracies. Both innovations allowed people, with the help of machines, to extract raw materials from the earth quickly and more efficiently. Mechanization also reduced the amount of human labor needed to complete a task and increased the speed by which necessities such as food, clothing, and shelter could be produced and distributed. Karl Marx believed that increased control over nature is accompanied by **alienation**, a state of being in which humans lose control over the social world they have created and are dominated by the forces of their inventions. The phrase "social world they have created" refers to the way people relate to and interact with each other.

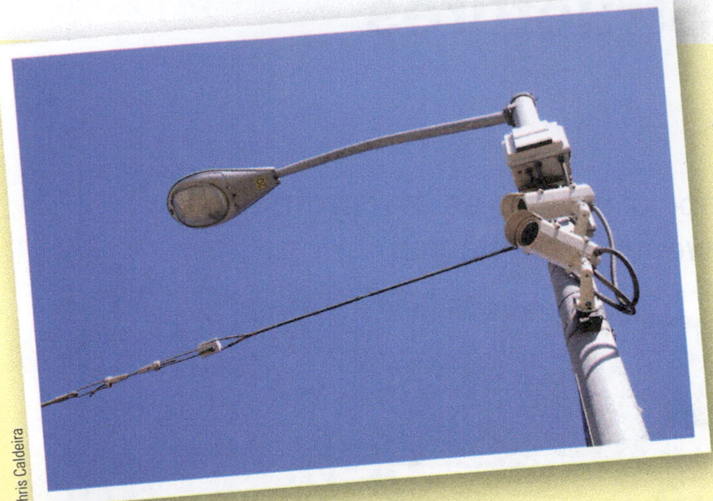
Chris Caldeira

Surveillance via cameras, drones, mobile phones, and computers has given governments, corporations, and individuals the power and ability to watch and monitor others. As a result, people have lost control over their private lives, and many may never know when someone is watching them or for what purpose. In addition, it is hard to know how the behavior caught on surveillance will be interpreted by those watching or reviewing tapes.

Although Marx discussed alienation in general, he wrote more specifically about alienation in the workplace. Marx maintained that workers are alienated on four levels: (1) from the process of production, (2) from the product, (3) from the family and the community of fellow workers, and (4) from the self.

ALIENATION FROM THE PROCESS AND PRODUCT. Workers are alienated from the *process* when they produce not for themselves or for people they know but rather for an abstract, impersonal market. In addition, workers are alienated when they do not own the tools they use to produce things and when what they produce has no individual character and sentimental value to either the worker or the consumer (Marx 1888).

Workers are alienated from the *product* when their roles are rote and limited and when their employers treat them like replaceable machine parts. Marx believed that most jobs do not allow people the chance to be active, creative, and social (Young 1975).

ALIENATION FROM FAMILY AND COMMUNITY OF FELLOW WORKERS. Workers are alienated from the *family* because households and workplaces are separate spheres. Specifically, the workplace makes few, if any, accommodations to family life unless forced to do so. Workers can lose touch with their families when they work shifts late at night, early in the morning, or on weekends, keeping them from participating in family life. They also lose connection to their families when they must relocate to find employment. An estimated 25 percent of the Philippines' labor force works outside the country in 190 countries, leaving children to be cared for by relatives (Migration Policy Institute 2007). The United States and Western European countries, for example, recruit 85 percent of all nurses trained in the Philippines. The exodus of nurses is so great that the World Health Organization has expressed concerns about its effect on the Philippines' system of medical care (Coonan 2008).

Workers are also alienated from the *community of fellow workers* because they compete for jobs, business, advancement, and salary. Migrant workers, especially the undocumented, are often viewed as taking the jobs of people in the host country. As workers compete for jobs, they fail to consider how they might unite as a force to better control their working conditions.

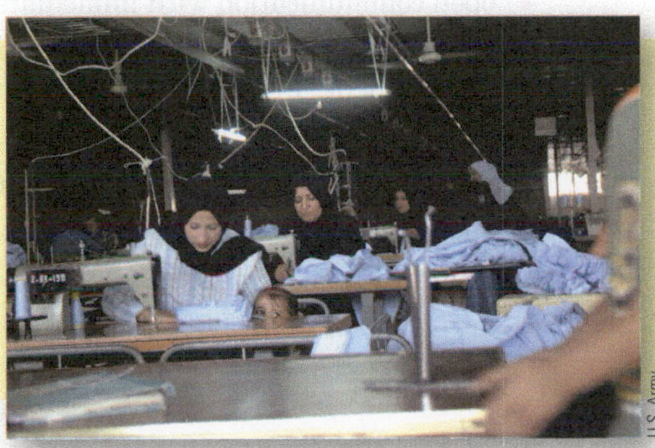

Wherever they work in the world, textile workers do not earn living wages to support a family. In the United States, the average wage is $1,661 per month; in Kuwait it is $477 per month; in Thailand it is $106 per month (International Labour Organization 2010).

U.S. Army

ALIENATION FROM THE SELF. Finally, workers are alienated from the *self*, or from the human need to realize one's unique talents and creative impulses. When Karl Marx developed his ideas about alienation in the late 1800s, he was describing alienation from self as it relates to industrial society. More recently, sociologist Robin Leidner has described the alienation from self that can occur in service industries when management standardizes virtually every aspect of the service provider–customer relationship, so that neither party feels authentic, autonomous, or sincere: "Employers may try to specify exactly how workers look, exactly what they say, their demeanors, their gestures, even their thoughts" (1993, 8). The means available for standardizing interactions include giving employees scripts to follow, requiring workers to follow detailed dress codes, and issuing specific rules and guidelines for dealing with customers and with coworkers. Employers may use surveillance cameras or computer software to monitor worker performance and enforce compliance.

Lorin T. Smith/Northwest

Computer technology is an alienating tool if management uses it strictly as a means for monitoring things such as how long customer service workers take to respond to people on hold, the number of calls handled per hour, or the time that lapses between calls.

The Best Work Environments

There are many work environments designed to reduce alienation among employees. Each year *Fortune* magazine lists the 100 best companies to work for. There are a variety of characteristics that earn a company a place on this list, including:

- ethical and collaborative values and principles held by the top leadership, including the president;
- open communication with immediate supervisors;
- opportunities for personal growth through challenging work, training, career development, and prospects for advancement;
- programs that help employees manage workplace stresses;
- policies that bring balance to work and personal life;
- quality relationships with colleagues;
- chances to give something back to the community and society; and
- fairness as it relates to pay and benefits (*Times Online* 2008).

What Do Sociologists See?

Gerald L. Nino

Sociologists see law enforcement officers guarding workers without documents in a holding cell. Workers without documents are alienated from family and their community of fellow workers. Because of limited employment opportunities in their home country, they must leave their families to find work in the United States. Their undocumented status alienates them from workers in the host country, many of whom see them as taking their jobs or lowering wages.

Critical Thinking

Think about the work you or someone close to you does for a living. Is the workplace empowering, alienating, or some combination of the two?

Key Term

alienation

Module 4.9 Applying Theory: The Social Structure of Nail Salons

Objective

You will learn how sociologists using each of the four perspectives analyze Vietnamese-owned nail salons.

How do you think the Vietnamese came to dominate the nail industry in the United States?

Chris Caldeira

According to nail industry data, 43 percent of all nail salons in the United States are Vietnamese-owned and 45 percent of all salon staff are Vietnamese (Bates 2012). This is impressive in light of the fact that people of Vietnamese ancestry make up less than 1 percent (0.4 percent) of the U.S. population (U.S. Census Bureau 2013b). Vietnamese-owned salons transformed the nail industry in at least two ways: They moved manicure and pedicure services out of neighborhood hair salons into nail salons offering quick service, no appointment, assembly-line-like service. The business model is often referred to as "McNail" (Eckstein and Nguyen 2011). The "McNail" model made manicures and pedicures affordable, expanding regular clientele beyond the wealthy to include diverse lower- and middle-class patrons of all racial and ethnic categories (Schlesinger 2011). The demand for pedicures increased as sandals and open-toed shoes became fashionable (a fashion trend supported by McNails). In this module we will consider how each of the four perspectives facilitates analysis of the social structure underlying Vietnamese-dominated nail salons.

Chris Caldeira

A sociologist using the functionalist perspective to analyze the structure of Vietnamese-dominated nail salons emphasizes that immigration to the United States (and elsewhere) has always functioned to meet labor demands and create new businesses to stimulate the economy. Beginning in 1978 (after the end of the Vietnam War), Vietnamese refugees found a niche—providing a personal care service that, before the Vietnamese came on the scene, women did only on special occasions or, if done routinely, involved high-income clientele. The Vietnamese refugees and immigrants gravitated toward jobs as manicurists because they spoke little or no English and because other Vietnamese had established salons or vocational training programs in California. The demand for affordable manicures—even when the economy is not doing well—has "sustained a steady stream of Vietnamese nail technicians" who usually get their start in a California salon with some moving to other locations in the United States to set up their own salons (Eckstein and Nguyen 2011; Bates 2012).

Department of Defense

This photo shows refugees from Vietnam in a crowded boat fleeing their country as a U.S. aircraft carrier approaches to rescue them. The number who attempted the risky journey is estimated at 1.5 million (History Learning Site 2014). Sociologists inspired by the conflict perspective emphasize that the flow of refugees from Vietnam to nail salons in the United States must be placed in the context of the Vietnam War, a war between communist North Vietnam and noncommunist South Vietnam in which the United States fought on the side of South Vietnam until 1973. When the United States withdrew its support, South Vietnam eventually fell to the communists. By 1978 and 1979, the number of refugees leaving Vietnam was so large that it assumed the status of a humanitarian crisis. To ease this strain, the United States agreed to admit 823,000 refugees, some of whom found work in nail salons and established a path for others to follow. Thus, a conflict of epic proportions was behind the structure of "formal and informal ethnic networks that fueled their growing monopolization of jobs in the sector" (Eckstein and Nguyen 2011).

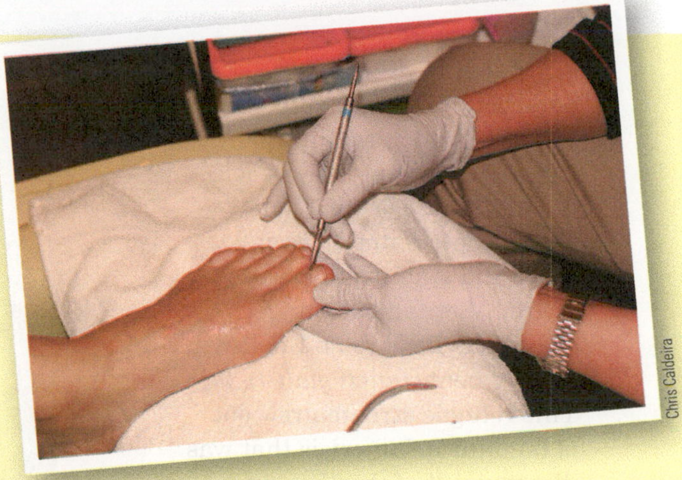
Chris Caldeira

Sociologists who study the Vietnamese dominance of nail salons from a symbolic interactionist point of view focus on the structure guiding interaction between clients and salon workers as well as symbolic meanings surrounding nails. Among other things, symbolic interactionists would notice that it is very common to hear nail salon workers speak to each other in Vietnamese as they work on clients' hands and feet. It is also very common to see workers giving manicures and pedicures while clients are fiddling with an iPad or smartphone. Symbolic interactionists would surely note that in Vietnam and other Asian cultures feet are perceived as the dirtiest part of the body. But Vietnamese, when offered a chance to earn a living taking care of feet, looked past this belief to dominate the industry (Eckstein and Nguyen 2011).

Chris Caldeira

Sociologists looking at Vietnamese nail salons from a feminist perspective would immediately notice the gendered structure of nail salons. Nationally only 4 percent of nail technicians are men. Among Vietnamese employed in the industry, 28 percent are men and 62 percent females. One in five Vietnamese women in the labor market are employed as manicurists (Eckstein and Nguyen 2011). Clearly there is much for feminists to admire about a refugee population that established an economic niche in the American economy. On the other hand, feminists see the McNail industry as adding yet another fashion imperative on women—to have manicured finger- and toenails. It would also not escape their attention that the salon workers are predominantly women and serving predominantly female customers. Feminist sociologists are interested in the interaction dynamics—especially interactions that involve prejudice and stereotyping such as the account presented in a popular YouTube video featuring comedian Anjelah Johnson that has over 36 million hits (Johnson 2014).

Together all four perspectives give a more complete understanding of an industry dominated by people of Vietnamese origin. Nail salons contributed to social order and stability because they offered Vietnamese refugees who spoke little to no English employment (*functionalist*). This refugee stream grew out of humanitarian crisis that was direct result of the Vietnam War, a conflict to control power in the region (*conflict*). A social structure emerged that allowed Vietnamese- speaking manicurists to interact with English-speaking clientele (*symbolic interaction*). The industry was dominated by female manicurists charging low enough prices that the nail salons drew female customers from diverse class backgrounds. In the end the nail industry grew to a size and level of influence that it introduced another fashion demand that many women feel compelled to meet–manicured nails on hands, but especially feet (*feminist*).

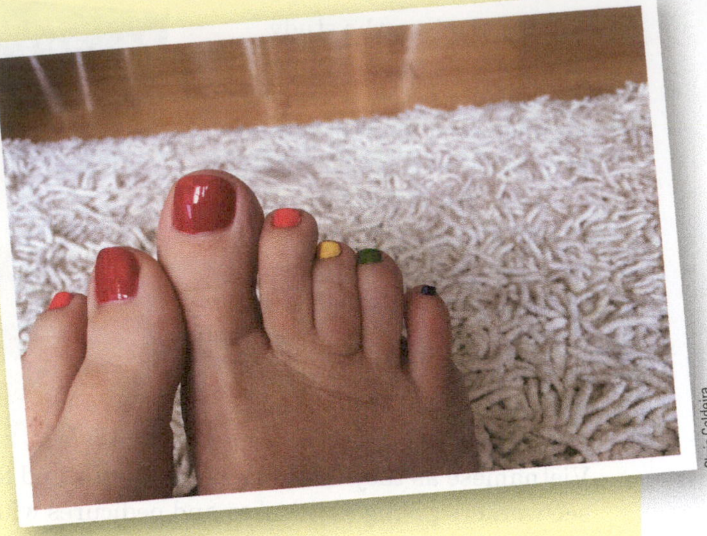
Chris Caldeira

Summary: Putting It All Together

Chapter 4

A social structure is a largely invisible system that coordinates human activities in broadly predictable ways. The concept of social structure is key to understanding how institutions meet human needs. The core institutions are the family, education, economy/work, religion, government, and medicine. When sociologists study an institution, they examine its social structure. Important components of social structure include statuses, roles, and groups. Since institutions are large and complex, when sociologists study them they focus on some specific setting or dynamic underlying social structure. What they choose to study may be part of one of three levels: micro, meso, or macro. Whatever level sociologists focus on, one thing is clear: social structures shape people's sense of themselves and their relationships and opportunities to connect with others.

The concept of social facts is useful for thinking about the constraining power of social structures. While social structures act to constrain thought and behavior, sociologists do not discount the power of human agency, or the capacity of individuals to act autonomously. In other words, sociologists recognize that people have the power to question the way things are structured and take action to resist, make change, and create new social structures.

The division of labor is a social structure that connects people to one another and to their society. Durkheim was particularly concerned with the kinds of events that disrupt the division of labor or that break down people's ability to connect with others and their society through their labor. These events include economic crisis, strikes, and job specialization. In the most basic sense, such disruptions affect people's ability to secure a livelihood.

The social networks (face-to-face and digital) to which people belong are also important social structures that connect people to others and to their society. Sociologists employ a variety of other concepts—including dyads, triads, oligarchy, and strong and weak ties—to capture the structural dynamics underlying social interactions and relationships. The concept of formal organization emphasizes how these social structures can operate as coordinating mechanisms, bringing together people, resources, and technology to achieve specific outcomes. The concept of bureaucracy describes specific mechanisms employed, including a clear-cut division of labor, an authority structure that is hierarchical, and written rules governing operations. Of course, people embedded in bureaucracies depart from official ways of doing things. To capture these dynamics, sociologists use the concepts of formal and informal dimensions of organizations. The concepts rationalization and McDonaldization alert us to a process by which organizations channel human behavior in the most efficient ways. Alienation in the workplace describes the downside of efficiency.

Chapter 5 The Social Construction of Reality

Aleena Ferrante

The process by which people make sense of the world is known as the social construction of reality. Imagine you went on a service learning trip to Zanzibar, Tanzania (as the two college students pictured did), and found yourself constructing a school using recycled plastic water bottles filled with sand. Would the meanings you assign to plastic water bottles change? When sociologists study the social construction of reality, they focus on how people go about assigning meaning to what is going on around them, including to objects such as plastic bottles and ways to use them. Sociologists also focus on the knowledge people draw upon to create a reality upon which they act. For the most part, people do not consider alternative realities until they encounter something that challenges an existing reality. So, for example, there is little about the way plastic water bottles are used in the United States that would lead one to think that they could be put to good use after they have been emptied of water to build a house or a school.

Module 5.1 Definition of the Situation

Objective

You will learn that people do not approach a situation with fresh eyes.

What thoughts run through your mind about these plastic water bottles? Do you wonder why they are filled with sand? Did you know a plastic water bottle filled with properly compacted stone-free sand inside, secured in place with cement mix, is 20 times stronger than a brick (Hubpages 2014)?

Aleena Ferrante

Definition of the Situation

W. I. Thomas (1923) points out that the human nervous system carries "memories or records of past experiences." When we see something, like an empty plastic water bottle, we do not see it as something new and fresh. Rather, what we see and how we define it are shaped by past experiences. Those experiences begin accumulating as soon as babies are born, at which point the caretakers and others in babies' lives use words and gestures to define situations for them. Caretakers usually talk to babies about what is happening, defining the situation for them. Examples include: "Let me see what time it is—oh, it's 3:00, time to pick your big brother up from school!" "Grandpa's coming over; hurry up, we have to take a bath!" or "Time to eat breakfast—let's have some cereal." Gradually, by observing definitions conveyed by caretakers, playmates, teachers, and others, the child learns the codes of society (e.g., a school day ends at 3:00; cereal is a breakfast food).

Thomas (1923) maintains that before people respond or take action there is a fleeting moment during which they deliberate about the meaning of the situation. That is, they attach a meaning, informed by the context and past experiences, to the situation. So after the typical American sees an empty plastic bottle on the ground, they deliberate for a fleeting moment. Seeing no value beyond its function as a one-time-use container, they may pick it up and throw it away or let it lie.

What if you grow up in an environment where you see plastic bottles only as things thrown away or objects to be recycled but that have no further associations? Perhaps that is why an estimated 70 percent of plastic bottles in the United States are thrown away and never recycled because most Americans cannot see their value beyond holding water to drink (J. Johnson 2013).

Sgt. 1st Class Michael Hagburg, 116th Public Affairs Detachment

Assigning Meaning

In *The Social Construction of Reality,* Peter L. Berger and Thomas Luckmann offer a "sociological analysis of the reality of everyday life" by emphasizing the knowledge people draw on to create the reality upon which they act (1966, 19). The two sociologists were interested in understanding the factors that determine how the world appears to people; it does not matter if the meanings assigned are "real." In this regard Berger and Luckmann made a number of generalizations about how people construct reality, seven of which are described below.

1. *For the most part, just about everything going on around us has been named and assigned meaning before we were born or otherwise arrived on the scene.* That is, our world is already ordered with names, meanings, and associations. We live in a geographic location that has a name and a reputation, and we use tools with names like "can opener" and "computer." We are part of a web of human relationships in which people are named for what they do—teacher, student, employee, mother, skate boarder, criminal, and so on. It is not easy to challenge reality, if only because the language we use to describe and think about it reinforces that reality.

Most Americans can see these are "grasshoppers on a stick." Our language—which includes not just names but associated meanings—tells us grasshoppers are most certainly not food. But seeing grasshoppers on a stick for sale as a food item reminds us that we likely never think to question the meaning of a grasshopper beyond that of a category of insect.

Tony Rotundo

2. *The reality of everyday life can be divided into a continuum of zones*. At one end of the continuum are the zones closest to us: the zones in which we actually live our physical lives—home, work, school, neighborhood—and experience directly. The zones closest to us include those whom we know and keep in touch with, including digitally. At the other end of the continuum are the zones we consider remote or farthest from our direct experience. People give greatest attention to the zones in which their lives are immersed because their interest is heightened by "what they are doing, have done or plan to do in it" (1966, 22). Typically, people are less, or not at all, interested in the zones that they define as remote from their lives.

Do you know where Zanzibar is? Does it interest you that Zanzibar includes a series of islands (known for spices and thus referred to as the Spice Islands)? Zanzibar is considered to be part of the East African country of Tanzania. If you studied abroad there, as this student did, and got to know the people, Tanzania would no longer be a remote zone of your life but one close to your heart.

Aleena Ferrante

3. *Events in everyday life can also be divided into a continuum of sectors with the most routine and the most unusual as endpoints*. The **routine** includes the usual ways of thinking and doing things. The unusual includes any unanticipated disruption that challenges the "reality" we have come to take for granted or challenges our understanding of what is usual. Until something disrupts our routine, there is no need to question existing assumptions. When the routine is disrupted, people work to keep their beliefs about the reality they know intact.

4. *We draw upon typificatory schemes to organize the world and our relationships.* **Typificatory schemes** are mental frameworks that prompt broad and simplistic generalizations to social statuses and situations. For example, we apply a typificatory scheme to the social categories "elderly" and "celebrity." Likewise we apply typificatory schemes to the social categories brothers. Whether we know it or not, when we learn that we are about to meet two brothers, we anticipate meeting two people who should physically resemble one another, be the offspring of the same parents, share memories of growing up, and be of the same race. These thoughts seem obvious only until we meet brothers who possess characteristics that violate one or more of these anticipations.

Generally in the United States, most people assume that brothers will appear to be the same race. Thus upon learning that the two boys pictured are brothers, many people would immediately question whether they are half brothers or whether one is adopted. They might label the pair as an exception to the rule. Seeing these brothers as an exception to the rule has the effect of preserving the rule that "brothers are of the same race."

NKU Sociology, Missy Gish

Typifications are not just applied to those who live in the present. We also apply typifications to those in generations that precede and that will succeed us. We relate to our predecessors through typifications that are often projections of imagined reality; that is, we may describe our ancestors as immigrants who worked hard to realize the American dream.

DARPA

Typifications made about future generations are projections of a current generation's hopes and worries. Some project a future where, when people engage in physical challenges, they would wear lightweight bionic suits with technology that augments the wearers' muscles, allowing them to perform at superhuman levels.

5. *Our place in history informs our daily reality*. The calendar places us in the context of history (born on a particular day, graduated from high school in a particular year, joined the army at a particular age). Consider how those who wake from a coma are compelled to locate themselves in time and "re-enter the reality of everyday life" by asking what day it is, and what year. Others help by filling them in on key events that occurred. Imagine what you would say to someone who fell into a coma and awoke ten years later. What events would you have to recount to catch that person up with the current state of affairs?

6. *People attribute cause to one of two types of factors: (1) dispositional or (2) situational.* **Dispositional factors** are things that people are believed to control, including personal qualities related to motivation, interest, mood, and effort. **Situational factors** are things believed to be outside a person's control—such as the weather, bad luck, and another's incompetence. Usually, people stress situational factors in explaining their own failures ("I failed the exam because the teacher is terrible") and dispositional factors in explaining their own successes ("I passed the exam because I studied hard"). With regard to others' failures, people tend to emphasize dispositional factors ("She failed the exam because she parties too much"). With regard to others' successes, however, people tend to stress situational factors ("She passed the test because it was easy").

7. *When people "define situations as real they are real in their consequences"* (Thomas and Thomas 1928, 572). Perhaps the most dramatic illustration of this is the self-fulfilling prophecy. Sociologist Robert K. Merton (1957b) defined a **self-fulfilling prophecy** as a false definition of a situation that is assumed to be accurate. People behave, however, as if that false definition is true. In the end, the misguided behavior produces responses that confirm the false definition. Only when that definition is questioned and a new definition is introduced will the situation correct itself. The dynamics underlying the self-fulfilling prophecy have the consequence of making reality conform to the false belief. The actors who hold the false belief "fail to understand how their own belief has helped to construct that reality because their belief is eventually validated" (Bearman and Hedström 2009).

What Do Sociologists See?

Sociologists see a mother making a "welcome home" sign as her daughter looks on. W. I. Thomas notes that the human nervous system stores and carries memories or records of past experiences. We draw upon these ever-accumulating past experiences to give meaning to situations in which we find ourselves. This little girl will carry with her the experience of "welcoming" someone home and apply it to future situations in which loved ones are returning into her life.

Cpl. J. R. Stence

Critical Thinking

Use one or more concepts in this module to write about a definition of reality that you hold. Explain that reality.

Key Terms

dispositional factors
self-fulfilling prophecy
typificatory schemes
routine
situational factors

Module 5.2 Dramaturgical Model

Objective

You will learn an approach for thinking about how people manage interactions and the presentation of self.

What about this encounter is like theatre? Can you draw an analogy to scripts, actors, and costumes?

Tony Rotundo

Life as Theater

Sociologist Erving Goffman's writings revolve around the assumption that "life is a dramatically acted thing" (1959, 72). He offered the dramaturgical model for analyzing social encounters. **Dramaturgical sociology** studies social interactions, emphasizing the ways in which those involved work (much like actors on stage) to present and manage a shared understanding of reality (Kivisto and Pittman 2007).

Mindy Campbell, U.S. Army Garrison Kaiserslautern

Goffman uses the analogy of the theatre to describe the work of impression management. In the theatre, actors use scripted dialogue, costumes, gestures, and settings to convey a particular reality to the audience. Social encounters are like staged performances, and the people involved are like actors in costume performing roles "on stage." Like actors, they must work to convince their "audience" that they are who they appear to be or who they say they are. Like actors, people depend on other "cast members" to support the scene.

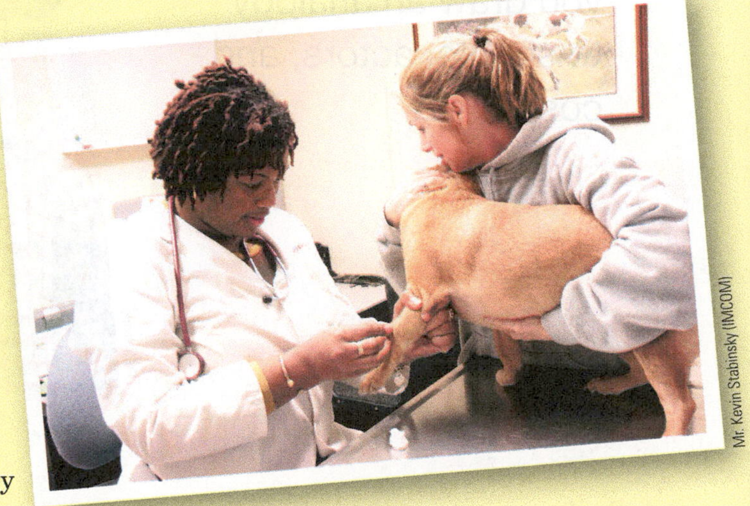

Mr. Kevin Stabinsky (IMCOM)

In social encounters, as on a stage, people manage the setting (stage), their dress (costumes), and their words and gestures (script) to correspond to the impression they are trying to make. This process is called impression management. This photograph shows two people, one performing the role of veterinarian and the other that of pet owner. From a dramaturgical point of view they are both in "costume," on a stage (an examining room) with appropriate props, and following a script we can all recognize. If one "actor" doesn't "play along," the encounter collapses. As one example, if the veterinarian comments, "I think it is ridiculous that you feel so emotional about your pet; it is just a dog," the scene would collapse because a basic assumption of what veterinarians should think and say has been violated.

Goffman maintains that the self "is a product of a scene that comes off." To put it another way, the self is a performed character, rising out of performances (1959, 252–253). The self cannot be thought of as a lone actor; rather, it depends on a supporting cast and a "stage." For example, a firefighter exists only in relation to a social setting that includes a fire station, a burning building, other firefighters, and victims. Goffman (1959) argues that any analysis of interaction cannot focus on the individual per se (Kivisto and Pittman 2007); it must focus on the entire cast, the script, and the props.

Managing Impressions

When managing impressions, some people behave in completely calculated ways with the goal of evoking a particular response (Goffman 1959). An example of such calculation involves someone posting a 10-year-old photograph as an up-to-date likeness of himself on a social networking website. Likewise, thieves posing as utility workers, who knock on doors and request permission to enter under false pretenses, are thoroughly calculating in their attempt to gain entrance (the desired response) so they can rob the occupants.

Impression management is not always self-serving; sometimes people have to talk or behave in a particular way because the status they occupy requires them to do so. Coaches work to hide any doubts about whether they think their team can win, and instead express great optimism in team meetings. They engage in impression management because that is what coaches are expected to do. At other times people are unaware that they are engaged in impression management because they are simply behaving in ways they regard as natural. Women engage in impression management when they seek to meet gender ideals by putting on makeup, dyeing their hair, or shaving their legs. Likewise, men engage in impression management when they seek to meet gender ideals by hiding their emotions in stressful situations so that no one questions their masculinity.

Goffman (1959) judges the success of impression management by whether an audience "plays along with the performance." If the audience plays along, the actor has successfully projected a desired definition of the situation or has at least cultivated an unspoken agreement with the audience that they will uphold their end of the bargain.

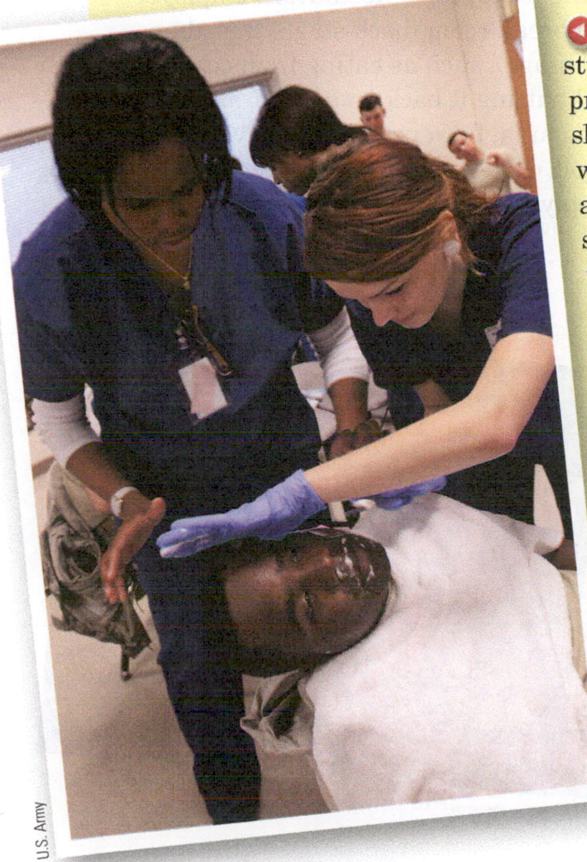
U.S. Army

Before they work with "real" patients, students in nursing assistant certification programs practice things such as shaving patients' faces or changing beds with patients in them on mannequins and then on volunteer patients. Here students practice shaving the face of a volunteer patient. Live volunteers allow students to get immediate feedback on how they are doing. This feedback gives them experience and confidence, increasing the likelihood that nursing assistants can induce real patients to, as Goffman would say, "play along" and allow them to carry out their roles as nursing assistants.

Goffman (1959) recognized that there is a dark side of impression management that occurs when people manipulate their audience in deliberately deceitful and hurtful ways. But impression management can also be constructive. If people said whatever they wanted and behaved entirely as they pleased, social order would break down. According to Goffman, in most social interactions people weigh the costs of losing their audience against the costs of losing their integrity. If keeping the audience seems more important, impression management is viewed as necessary; if being completely honest and upfront seems more important, we may take the risk of losing our audience.

Front and Back Stage

Goffman used a variety of concepts to elaborate on the process by which impressions are managed—including the idea of front and back stage. Just as the theater has a front stage and a back stage, so too does everyday life. The **front stage** is the area visible to the audience, where people feel compelled to present themselves in expected ways. Thus, when people step into an established social role, such as a teacher in relation to students or as a doctor in relation to patients, they step onto a front stage such as a classroom or an examining room (Goffman 1959).

The **back stage** is the area out of the audience's sight, where individuals let their guard down and do things that would be inappropriate or unexpected in a front-stage setting. Because back-stage behavior frequently contradicts front-stage behavior, we take great care to conceal it from the audience. In the back stage, a "person can relax, drop his front, forgo his speaking lines, and step out of character" (Goffman 1959, 112). The division of front stage and back stage can be found in nearly every social setting. Often that division is separated by a door or sign signaling that no one or only certain people can enter the back stage without permission or knocking.

Goffman uses a restaurant as an example of a social setting that has clear boundaries between the back stage and the front stage. Restaurant employees may do things in the kitchen, pantry, and break room (back stage) that they would never do in the dining areas (front stage), such as eating from customers' plates, dropping food on the floor and putting it back on a plate, and yelling at one another. Once they enter the dining area, however, such behavior stops. Of course, a restaurant is only one example of the many settings in which the concepts of front stage and back stage apply.

Most of time customers do not get to see how food in made. That typically takes place in the back stage beyond the gaze of patrons. Some restaurants deliberately showcase employees preparing food, or certain signature food items. This noodle maker must always be aware that someone could be watching her, even from afar, making noodles. She must perform her job just as an actress performs for an audience.

What Do Sociologists See?

Sociologists see a campsite with single-person tents. To sociologists the tents function as more than a shelter, but as a back stage for campers where they can relax out of sight and take a break from the stresses and energy expended related to managing behavior and conversation in face-to-face interactions to meet audience expectations.

NKU Anthropology, Sharyn Jones

Critical Thinking

Describe a part of your life that has a front stage or back stage. The part may relate to school, work, home, leisure, or something else.

Key Terms

back stage

dramaturgical sociology

front stage

Module 5.3

Emotional Labor and Emotion Work

Objective

You will learn about the dynamics of managing emotions in the workplace.

Do you hold a job in which you are expected to hide either negative or excessively positive emotions when relating to customers, clients, or the public?

If your job requires you to serve customers, clients, or the public, your employer is asking that you engage in what sociologists call **emotional labor**, work that requires employees to display and suppress specific emotions and/or manage customer/client emotions.

Firefighters must manage not only their emotions but also the emotions of people they rescue and of bystanders who may fear for the safety of loved ones the firefighters are trying to rescue.

Dramaturgical Theory

Dramaturgical theory, which is most associated with the work of sociologist Erving Goffman (1959), views social interaction as if it were theater and people as if they were actors giving performances before an audience in a particular setting. In social situations, as on a stage, people manage the setting, their dress, their words, and their gestures so that they correspond to an impression they are trying to make.

Sociologist Arlie Hochschild (2003a) extended the work of Erving Goffman by presenting actors as not only managing their outer impressions but also working at managing their inner feelings. Hochschild recognized that emotions—whether sadness, anxiety, anger, boredom, or nervousness—are more than simply reflex-like responses; they are shaped by expectations about how one should feel in a particular situation. Thus, people are sad at funerals, not simply because someone has died, but because that is how people are supposed to act (Appelrouth and Edles 2007). Likewise, people suppress feelings of envy at award ceremonies when not chosen for an award as they congratulate a competitor. They suppress this emotion because losers are supposed to be gracious.

Emotion Work

Hochschild argues that people do **emotion work**—that is, they consciously work at managing their feelings by evoking an expected emotional state or suppressing an inappropriate one. That work may involve presenting an outward display of an emotion that does not match inner feelings or convincing themselves to actually feel what others expect them to feel. For Hochschild, emotion work involves *effort*—with emphasis placed on trying to feel a certain way and not the actual outcome of that effort, "which may or may not be successful" (241). Such statements as "I psyched myself up," "I squashed my anger down," "I tried hard not to feel disappointed," "I made myself have a good time," "I tried to feel grateful," and "I let myself finally feel sad" represent examples of emotion work (Hochschild 2003a, 241). Hochschild acknowledges that there are often discrepancies between "what one does feel and what one wants to feel," which are further complicated by what one thinks one should feel (242). Even if people fail at managing an emotion, their efforts are still influenced by social expectations.

Emotional labor is a requirement for those jobs in which employees:

1. must engage with the public,
2. are directed to produce a specific emotional state in clients/customers, whether it be feelings of satisfaction with a service delivered or a reasonable outlook (such as police officers who calm excited citizens),
3. are required to follow scripts and are penalized for deviating from them (such as clerks who must greet customers with "Welcome to . . .").

Service jobs—customer service representatives, massage therapists, food servers, sales clerks, funeral directors, cashiers, doctors, nurses, social workers, teachers—involve emotional labor. People in such jobs must manage their emotions in ways that suggest they are out to please, and that includes suppressing any aggressive or negative emotions felt toward customers, especially difficult ones.

Bonnie Heater

In her research on flight attendants, Hochschild found their job required them to manufacture expressions of warmth, caring, and cheerfulness and suppress expressions of anger or boredom. Hochschild believes that the emotional labor associated with service work can be an alienating experience in much the same way as the repetitive physical labor we associate with the assembly line. In the case of factory work, employers control workers' bodies and movements; in the case of service work, employers manage workers' emotional states and reactions toward customers.

Hochschild writes that "those who perform emotional labor in the course of giving service are like those who perform physical labor in the course of making things—both are subject to the demands of mass production. But when the product is a smile, a mood, a feeling, or a relationship, it comes to belong more to the organization and less to the self" (2003b, 198).

U.S. Air Force photo/Staff Sgt. Stephenie Wade

Hochschild argues that emotional labor is not performed equally across race, class, gender, and age. Consider that flight attendants are predominantly female. There are an estimated 26 male attendants for every 100 females in the business. Hochschild makes the case that women especially face greater pressure than their male counterparts to manage their emotions and to present themselves in nurturing or sexually inviting ways.

Why is this the case? First, there is the deeply rooted cultural idea that women are practiced at managing emotions—that they "have the capacity to premeditate a sigh, an outburst of tears, or a flight of joy. In general, women are thought to manage expression and feelings, not only better than, but more often than men" (248). Second, when women fail to manage emotions in expected ways, they are considered less "feminine." Finally, women feel pressure to manage their emotions because when they lose control of their emotions they are labeled as "unstable" or "too emotional" to do the job.

While it is true that men face cultural pressures to manage their emotions in so-called masculine ways, Hochschild found that these cultural pressures specific to women help explain why female flight attendants are more likely to handle the babies, deal with the children, and comfort older passengers. Male flight attendants (when on board) are less likely to engage with passengers in these ways because they are not expected to. As a result, they are more likely to enforce rules about where to stow oversized luggage and to monitor seat belt usage. These cultural pressures may also explain why women dominate teaching and medical occupations (e.g., nurses, nurse's aides, home health care workers).

What Do Sociologists See?

U.S. Air Force photo by Master Sgt. Edward Holzapfel

Starbuck's and other coffee house employees are expected to present themselves as interested in their clients, often by making an effort to remember the first names of customers who come in regularly. When customers purchase coffee, they are buying more than coffee; they are buying an experience of getting coffee—that includes friendly encounters with staff who appear to really like them. Leaving the transaction feeling good is an important ingredient in anchoring customers to products and creating brand loyalty Blount, Heather. 2013. "Starbucks Introduces Cross-Channel, Multi-Brand Loyalty Program." Sunbelt FS (March 21) http://www.sunbeltfoodservice.com/2013/03/21/starbucks-introduces-cross-channel-multi-brand-loyalty-program/

Critical Thinking

Give an example of emotional labor from your own work life or from your observations of someone whose job it is to serve the public.

Key Terms

emotional labor

emotion work

Module 5.4 Ethnomethodology

Objective

You will learn an observational, investigative method of studying how people construct social order.

How would you react if you saw a mother lighting a cigarette while holding a toddler?

NKU Sociology, Missy Gish

Most of the time our encounters with others can be considered as simply ordinary and routine. That is, we have little reason to question what is going on. Think about the number of times we see a parent carrying a small child and take for granted that parents want to protect their children from harm. We think to question that assumption only when something is out of place, as when we see parents smoking in front of their children. It is the taken-for-granted quality of most social encounters that makes the commonplace "impervious to deeper analysis" (Cuzzort and King 2002, 307). The commonplace lulls us into complacency.

Ethnomethodology

Sociologist Harold Garfinkel coined the term *ethnomethodology* more than 50 years ago. **Ethnomethodology** is an investigative and observational approach that focuses on how people make sense of everyday social activities and experiences. Ethnomethodologists assume that people work at making social encounters meaningful. Social encounters are meaningful when both parties share essentially the same understanding of the situation. Ethnomethodologists seek to penetrate a reality that those caught up in it cannot even begin to question (Cuzzort and King 2002). Garfinkel maintains that the only way we can possibly get at the structure and character of this social order is by disrupting expectations. Once expectations are disrupted, ethnomethodologists observe how people react and/or take action to restore normalcy.

Disrupting Social Order

Disrupting the social order is perhaps the best-known investigative technique ethnomethodologists employ. By disrupting expectations, and then observing how the participating parties handle that disruption, ethnomethodologists gain insights into the work people do to maintain the social order. To reveal such efforts, ethnomethodologists ask, "What can be done to make trouble . . . to produce and sustain bewilderment, consternation, and confusion; to produce the social structured effects of anxiety, shame, guilt, indignation; to produce disorganized interaction?" (Garfinkel 1967, 37–38). As one example, Garfinkel asked his students to engage an acquaintance in conversation and insist that that person clarify everything he or she says by asking "what do you mean?" or "would you please explain?"

> **Acquaintance:** (waving hand cheerily) How are you?
>
> **Student of Garfinkel:** How am I in regard to what? My health, my finances, my school work, my peace of mind, my . . . ?
>
> **Acquaintance:** (red in face and suddenly out of control) Look I was just trying to be polite. Frankly I don't give a damn how you are! (1967, 44)

Look at the facial expressions and decide the meaning each conveys. Now think about making one of these facial expressions in a situation where it is not an expected response. For example, imagine your good friend asks you to meet for dinner and a movie and you make one of the faces shown. That is the kind of thing that Garfinkel asks his students to do.

Sgt. Mark Fayloga

Garfinkel also instructed his students to engage someone in conversation and to pretend that the chosen person was deliberately misleading them in some way. In one case a student chose to question her husband about his reasons for working the evening before and about whether he was really at a poker game several days earlier, as he claimed.

Garfinkel noted that in carrying out these assignments only two of his 35 students chose to engage strangers. "Most admitted being afraid to carry this assignment out with a stranger thinking things could get out of hand." So virtually all of his students selected "friends, roommates, and family members" as subjects (1967, 51). Even then, students reported imaginatively rehearsing for the kinds of repercussions they might face and making plans for how to respond.

Other kinds of order-disrupting assignments Garfinkel devised included asking his students to:

- initiate a game of tic-tac-toe with someone. After the chosen opponent made the first move, the student should erase the mark and move it to another square;
- engage a person in an ordinary conversation. Sometime later in the conversation, the student should show the device recording the conversation and announce, "See what I have."

Garfinkel noted that his students found it very difficult to disrupt routine expectations. Depending on the specific experiment, students expressed a range of emotions, including anxiety, distrust, hostility, anger, frustration, and isolation. Garfinkel related his students' reluctance to disrupt expectations of **trust**, the taken-for-granted assumption that in a given social encounter others share the same expectations and definitions of the situation and that they will act to meet those expectations. When one or more of the involved parties, including the student, are put in a position in which they are forced to violate those expectations and/or to mistrust another, the relationship becomes problematic, deteriorates, and eventually collapses.

What Do Sociologists See?

Sociologists know that in the right setting, this way of displaying chickens—with feet intact and seemingly reaching toward the sky—would not raise an eyebrow. In the context of food markets in the United States, displaying chickens in this way would likely "produce and sustain bewilderment, consternation, and confusion" and even make "trouble." Why? There is little about the way chickens are presented in the United States that reminds shopper that the meat was once living. This is because chicken is usually sold with feet severed from their bodies and packaged by parts (8 breasts to a package) wrapped in plastic.

Chris Caldeira

Critical Thinking

Imagine you were a student in Garfinkel's class. How would you feel about the kinds of assignments he required? Which of Garfinkel's assignments described in this module would you feel most comfortable carrying out? Least comfortable? Why?

Key Terms

ethnomethodology   trust

Module 5.5 Constructing Identities

Objective

You will learn that the groups to which we belong, aspire to belong, and do not belong play important roles in constructing our sense of self and the way we see what is going on in the world.

Think about the groups to which you belong. Which one of those groups would have the most fun eating and drinking while they pedaled around town, as this group is doing?

Chris Caldeira

The company, PedalTavern, bills the experience of pedaling around town while drinking with 12 friends or coworkers as the perfect team bonding and building activity. How do you think others watching the PedalTavern as it rolls down the street might view the group? How would this group experience affect your sense of self?

In what ways do the groups to which you belong affect your sense of self and others' views of you? What about the groups to which you do not belong, would never aspire to join, or are barred from joining? What does nonmembership in some groups say about who you are and how others see you? Group membership or nonmembership affects a person's sense of self on three levels—a cognitive level (awareness of being or not being a member), an affective level (feelings of belonging to, being rejected by, or rejecting a group), and an evaluative level (perceived value/importance of membership or nonmembership) (Christian et al. 2012). Simply think how the sense of self is affected by the high school or college a person attends. Every school has a reputation. Did you go to a high school with a high dropout rate, a high graduation rate, a high teen pregnancy rate, a school considered elite or troubled. When you tell people the name of your school, what do they say? The answers to these questions speak to the power of reputation in shaping a sense of self. Now think about the groups you associated with in, say, high school.

Tony Rotundo, Courtesy of Joan Ferrante

Stephen Oertwig, U.S. Army, Pacific Public Affairs

At your high school, how were students in the band viewed relative to football players? The answer to this question speaks to the power of groups to shape identity.

In this module we examine the role reference groups and ingroups and outgroups play in shaping identity and assessments of self and others.

Reference Groups

Sociologists define a **reference group** as any group whose standards people take into account when evaluating something about themselves or others, whether it be personal achievements, aspirations, or their current place in life. In this regard the family is an important reference group, as are classmates, teammates, and coworkers. A person does not have to belong to a reference group to be influenced by its standards. A reference group can also be a group to which a person belonged in the past or hopes to belong to. There are three broad types of reference groups: normative, comparison, and audience (Kemper 1968).

Normative reference groups provide people with norms that they draw upon or consider when evaluating a behavior, appearance, or a course of action. The mark of a normative reference group's influence is that the person simply takes its norms into consideration; in the end the person may follow those norms or pursue a contradictory course of action (Kemper 1968). Regardless, the person takes the group's norms into account.

Comparison reference groups provide people with a frame of reference for:

- judging the fairness of a situation in which they find themselves (i.e., "Everyone at work earns $15 per hour while I only earn $10 per hour. What's up with that?");
- rationalizing or justifying their actions or way of thinking (i.e., "All taxpayers cheat the government; why should I be the exception?" "Everyone I work with takes home office supplies; it's a way to make up for the low salaries we earn."); and
- assessing the adequacy of their performance relative to others, as when a student earns a 68 on a chemistry test and feels terrible until she assesses that grade in light of her peers' test scores. When she learns that the class average was a 48 and the highest grade was a 70, she feels pretty good. With the curve, a 68 is an A.

Audience reference groups consist of those who are watching, listening, or otherwise paying attention to someone seeking to influence or be noticed by that audience. In addressing an audience, people consider what they believe that audience wants or needs and then adjust the message accordingly (Kemper 1968). Candidates running for elected office often tailor their message to the audience they are addressing. Imagine a politician scheduled to speak before an audience of college students versus retired adults. The politician considers issues important to each audience group, then tailors the message accordingly. When speaking to college students, the politician is likely to mention efforts to increase access to college loans and grants. When speaking to retired adults, the candidate is sure to mention efforts to safeguard Social Security benefits.

Ingroups–Outgroups

From the moment babies are born, and perhaps in utero, they are exposed to information about the groups to which they belong and do not belong. Gradually, as babies mature into toddlers and beyond, they learn who they are and how people see them—as a boy or girl, a particular race, a member of a family, a good reader, a soccer player, a musician, a princess, and so on. In the process of learning who they are and how others see them, children also learn who they are not.

A group distinguishes itself by the symbolic and physical boundaries its members establish to set it apart from nonmembers. Examples of physical boundaries may be gated communities, special buildings, or other distinct locations. Symbolic boundaries include membership cards, colors, or dress codes such as a uniform. A group also distinguishes itself by establishing criteria for membership—for example, members must have a certain kind of last name, be of a certain race, profess a particular interest (in the environment, in hunting), or have "achieved" some status (a certain level of education, GPA, or talent).

Daily we encounter signs and symbols that designate spaces reserved for people who aspire to, identify with, or belong to certain groups. The boots sign and the image of a bicyclist establish boundaries that distinguish between who belongs in the space and who will likely feel uncomfortable or out of place if they venture into a space. Obviously someone driving a car or truck does not belong in a bike lane. Likewise, someone seeking to perform as a rapper, a jazz musician, or a classically trained musician will likely avoid buying country boots for their wardrobe.

Sociologists use the terms *ingroup* and *outgroup* in reference to a type of intergroup dynamics. An **ingroup** is the group that a person belongs to, identifies with, admires, and/or feels loyalty to. An **outgroup** is any group to which a person does not belong. Obviously, one person's ingroup is another person's outgroup. Ingroup formation is built on established boundaries and membership criteria. Ingroup members think of themselves as "us" in relation to some "them" (Brewer 1999).

"Us" versus "them" dynamics create a sense of oneness that is especially evident when an ingroup and an outgroup compete for some valued outcome—in this case winning a tug-of-war game. But the tug-of-war outcome symbolizes more than a victory—a victory over an outgroup.

Ingroups and outgroups are everywhere. This gated community is advertising for an ingroup who can afford to buy an executive home. The emphasis on "gated" sends the message that walls, fences, and/or guards control outgroup access to the community.

Depending on the circumstances, ingroup and outgroup dynamics may be characterized as indifferent, cooperative, competitive, or even violent. When sociologists study ingroup and outgroup dynamics, they ask: Under what circumstances does the presence of an outgroup unify an ingroup and create an us-versus-them dynamic? Three such circumstances are described.

1. *An ingroup assumes a position of superiority.* **Moral superiority** is the belief that an ingroup's way of being represents the only way. In fact, there is no room for negotiation and there is no tolerance for another way (Brewer 1999). Ingroup members convey feelings of superiority when they refuse to

interact with anyone who belongs to an outgroup, or they enact laws banning outgroup members from using facilities considered the domain of an ingroup.

2. *An ingroup perceives an outgroup as a threat.* In this situation an ingroup believes (rightly or wrongly) that an outgroup threatens its way of life. The ingroup holds real or imagined fears that the outgroup is "invading" its space or taking steps to increase its political power or control over some scarce and valued resource such as jobs.

Real or imagined belief fuels fear and hostility. The motel owner is using an implicit ingroup–outgroup dynamic of American- versus foreign-owned to advertise his motel to potential customers. The suggestion is that "American-owned" is in danger of disappearing. In the United States, an estimated 50 percent of all motels are owned by people of Indian (as in India) origin. Of course, that owner may still be a U.S. citizen or even American-born (NPR 2012).

Chris Caldeira

3. *Ingroup–outgroup tensions may be evoked for political gain.* Those with political ambitions may deliberately evoke ingroup–outgroup tensions as a strategy for mobilizing support for some political purpose (Brewer 1999). Thus, some candidates running for elective office may, as a strategy for rallying support for their campaigns, declare an outgroup such as undocumented workers or gays a threat to the American way of life.

What Do Sociologists See?

Sociologists see a sign designating a space for those who consider themselves part of an ingroup known as smokers. As the number of public spaces—colleges, bars, restaurants, apartment complexes—that welcome, or even tolerate, smokers decreases, smokers are increasingly designated as outgroups in those spaces. Some businesses are creating spaces where smokers can feel safe in environments where nonsmokers are the outgroup.

Chris Caldeira

Critical Thinking

Give a specific example of some ingroup–outgroup dynamic that has shaped your sense of self.

Key Terms

audience reference groups

comparison reference groups

ingroup

moral superiority

normative reference groups

outgroup

reference group

Module 5.6 Applying Theory: Language and Reality Construction

Objective

You will learn how language shapes the way we see ourselves and the world around us.

What do the symbols or words carved into this rock mean? Why did someone take the time to do this? What insights do the symbols hold about the world and the place of the individual in it?

NKU Anthropology. Sharyn Jones

The symbols carved into this rock are read as Om Mani Päme Hum, the well-known Buddhist mantra. When the words *Om Mani Päme Hum* are thoughtfully spoken or chanted, they inspire one to practice and strive for perfection in achieving generosity over ego (om), pure ethics over jealousy (ma), tolerance and patience over passion and desire (pä), renunciation over greed (me), and wisdom over aggression (hum). Now imagine that you live in a society where the words or symbols carved into this rock are heard or written everywhere. In Dharamsala, India, where this photo was taken, and where the Dalai Lama resides, these words constitute a sacred formula to guide thought and behavior. *Om Mani Päme Hum* is written on rocks, paper, signs, and other surfaces; and it is tattooed on bodies and chanted.

This mantra speaks to power of language—the words we learn, hear spoken, see signed or written move us to action, shape a view of the world, help us feel connected to others, allow us to preserve and pass on meaning, and constrain and liberate. Language is not static; it is always changing. The language we speak shapes reality while at the same time it responds to new realities. While we could debate the extent to which language is constraining or empowering and the extent to which language helps us capture "reality" or mask it, there is no question that we can find examples to support either view.

Sociologists inspired by a functionalist perspective are attuned to the ways in which language contributes to order and stability. On the most basic level, when we learn a language, we learn the ways of our culture and a vocabulary that allows us to communicate with others who speak the language. Some words are so powerful that they act as cultural anchors uniting everyone, even in the face of internal debate or protest. If we think of Americans as a group, we can point to the word "freedom" (freedom of religion, freedom of speech) as a cultural anchor because that word is believed to express THE value upon which the country was founded (in spite of evidence to the contrary) and is believed to represent an ideal worth dying for.

John D. Banusiewicz

NKU Anthropology, Sharyn Jones

Sociologists inspired by the conflict perspective alert us to the ways in which language can be used to camouflage social arrangements that give advantage to some groups at the expense of others. Instances where language is used to sell products such as water offer examples. FIJI, a Los Angeles–headquartered corporation that sells its products in 50 countries, uses words to suggest that its bottled water from the remote island of Fiji is pristine, as it is protected by the earth. The language on the bill board asks us to believe that Fiji somehow benefits from this commercial activity. In reality, Fijian water is sold on the world market even as 57 percent of its communities lack access to clean drinking water (Raz 2010).

Official White House photo by Pete Souza

A sociologist is informed by the symbolic interaction perspective when the focus is on the role of language in interaction and presenting the self. When Tiger Woods first came on the scene as a child prodigy golfer, he announced himself to be a "Cablinasian"—a mixture of Caucasian, African, American Indian, and Asian descent. While this self-referent term he made up did not stick, he used language in a unique way. President Obama told the nation he had a white mother but he is still referred to as our "first black president." These two example show that people are active agents when they apply, invent, and challenge meaning.

U.S. Embassy
NASA

Sociologists informed by a feminist perspective are attuned to gendered language or uses of language that act to maintain and perpetuate inequalities and stereotypes and act to insert considerations of gender into situations where it need not be a factor. Feminists do not accept the argument that words like *mankind, policeman*, and *freshman* are inclusive of all genders. Instead, they prefer and advocate for gender-neutral terms like people, police officers, and first-year students. Sociologists drawing on a feminist perspective also point to the associative qualities of language that make a certain gender the default, such as assuming that nurses are women and doctors are men.

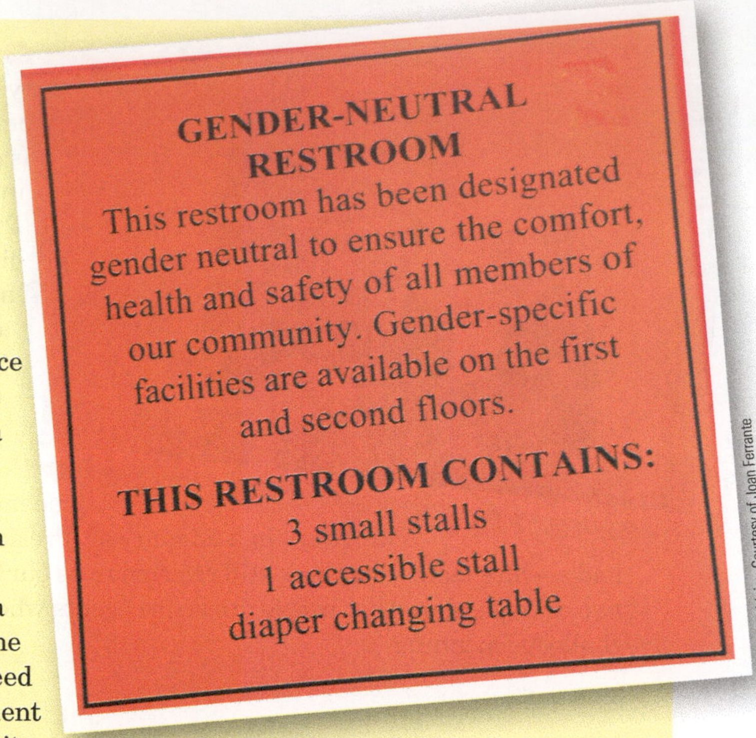

Taken together the four perspectives give a more complete understanding of the power of language to express ideas that go beyond the words themselves. Study the words on this sign. What meanings do the words convey? From a *functionalist* perspective the words announce the emergence of a new social order—an order that creates a space for people of any gender. The existence of a gender neutral space means a son can now accompany a frail elderly mother to the bathroom and a mother can accompany her nine year old son without fear or need to explain. The new arrangement contributes to order and stability as it solves a longstanding question of what to do when someone needs help going to the bathroom and the available caregiver is another gender. From a *conflct* perspective the words suggest a disadvantaged group is challenging a social arrangement that had previously ignored the needs of the transgendered, the frail elderly, the very young and disabled. When there is no gender-neutral bathroom option these groups must find a way to make do. Conflict theorists believe these words are behind a social movement that challenges an ideology that defines it as dangerous for people of different genders to share a public bathroom. From a *symbolic interactionist* point of view when people choose a gender-neutral restroom they are free to present themselves as the gender they identify. In gender neutral settings those who must accompany another gender to the bathroom need not set the stage by waiting for a bathroom to empty, yelling to "warn" patrons before going in or rushing to get out before someone comes in. From a *feminist* perspective, a seemingly intrenchable gender-barrier that directed people into one of two spaces is dissolving. Because this a new phenemenon, the gender-neutral restoom currently serves as a "social experiment" where people can feel safe interacting with other genders and begin the work of establishing norms for sharing a once gender-restrictive public space–norms that the society will eventually adopt.

Chris Caldeira, Courtesy of Joan Ferrante

Summary: Putting It All Together

Chapter 5

The process by which people make sense of the world is known as the social construction of reality. When people observe something, they assign a meaning, informed by past experiences. That meaning becomes the basis for a response. We identified a number of points about how people construct reality, including the following: (1) Just about everything going on around us has been named and assigned meaning before we arrived on the scene. (2) We draw upon typificatory schemes to organize the world and our relationships. (3) Our place in history informs our daily reality. (4) People attribute cause to one of two types of factors: dispositional or situational. (5) When people "define situations as real they are real in their consequences" (Thomas and Thomas 1928, 572).

Dramaturgical sociology, inspired by Erving Goffman, views social interaction as if it were theater and people as if they were actors giving performances before an audience in a particular setting. In social situations, as on a stage, people manage the setting, their dress, their words, and their gestures so that they correspond to an impression they are trying to make. This process is called impression management. Sociologist Arlie Hochschild extended Goffman's analysis by presenting actors as not only managing their outer impressions but also working at managing their inner feelings. Hochschild argues that people do emotion work—that is, they consciously work at managing their feelings by evoking an expected emotional state or suppressing an inappropriate one. When emotion work is a requirement of a job, a person engages in emotional labor.

Ethnomethodology is an investigative and observational approach that focuses on how people make sense of everyday social activities and experiences. Ethnomethodologists seek to penetrate a reality that those caught up in it cannot even begin to question. Ethnomethodologists maintain that the only way we can possibly get at the structure and character of this social order is by disrupting expectations. Once expectations are disrupted, ethnomethodologists observe how people react and/or take action to restore normalcy.

The groups to which we belong, aspire to belong, and do not belong play important roles in constructing our sense of self and the way we see what is going on in the world. A reference group is defined as any group whose standards people take into account when evaluating something about themselves or others, whether it be personal achievements, aspirations, or their current place in life.

Sociologists use the terms *ingroup* and *outgroup* in reference to a type of intergroup dynamics. An ingroup is the group that a person belongs to, identifies with, admires, and/or feels loyalty to. An outgroup is any group to which a person does not belong. Obviously, one person's ingroup is another person's outgroup. Ingroup members think of themselves as "us" in relation to some "them." "Us" versus "them" dynamics create a sense of oneness that is especially evident when an ingroup and an outgroup compete for some valued outcome or any "victory" over an outgroup.

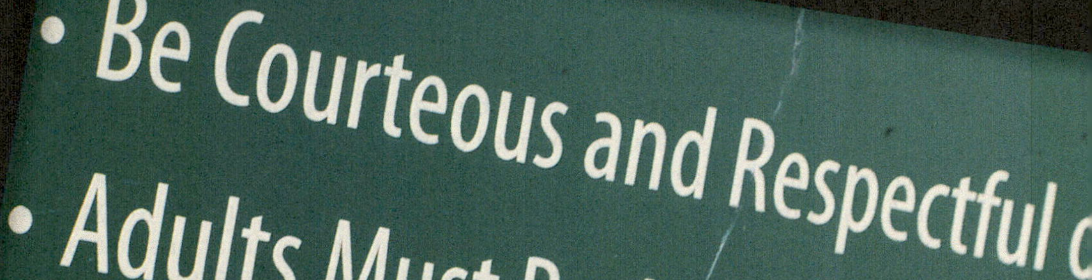
• Be Courteous and Respectful
• Adults Must Be Accompanied
• No Dogs
• Smoking Prohibited
• Feeding of Birds or Animals
• No Alcohol or Glass Containers
• No Riding of Skateboards,
or Rollerblades

Chris Caldeira

Chapter 6

Deviance

Sociologists are interested in the process by which behavior gets defined as deviant; that process is usually not a simple one because any behavior or appearance can be defined as deviant under the right circumstances. For this reason sociologists focus on the context, which includes the social audience with the power to define and punish what it considers to be deviant.

Module 6.1

Defining Deviance

Objective

You will learn how sociologists define deviance and how almost any behavior or appearance can be defined as deviant depending on context.

Would you label this girl as deviant? No matter your answer—yes, no, don't know, don't care—what criteria did you use to arrive at that assessment?

In the context of the school this student attends, she is considered deviant. As background, a group of students at her school gave symbolic meaning to these bracelets beyond that of a colorful fashion statement. These students named them "sex bracelets" and decided that each color stood for a specific kind of sexual act. The rules were such that if another student managed to break a band, then the wearer has to perform the sexual act the color represented. Once the school administrators discovered the meaning, they banned wearing them on school grounds. Any student caught wearing them faced punishment including detention and suspension.

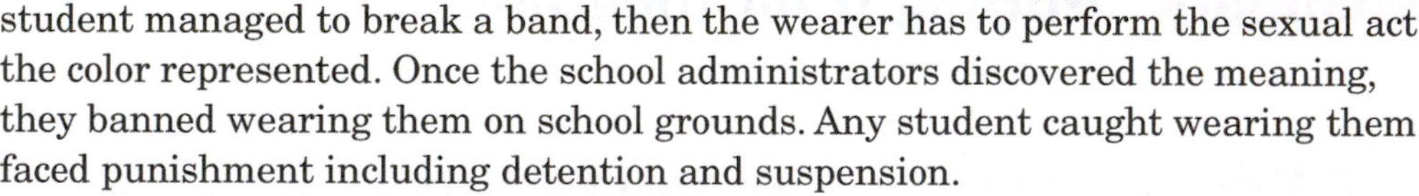

Defining Deviance

From a sociological point of view, **deviance** is any behavior or physical appearance that is socially challenged and/or condemned because it departs from the norms and expectations of some group. In the case of the sex bracelets described above, administrators challenged the students wearing them and took action to end the practice on school grounds. Sociologists are most interested in the process by which some group decides something is deviant and then takes action to punish it. Placing primary interest on understanding the process does not mean sociologists take an "anything goes" approach to behavior; rather, it means they seek to capture the complex dynamics that underly definitions of deviance and actions taken to address it.

For example, who would label taking an empty, uncontested seat at the front of the bus as deviant—something one could be arrested for? But this act was considered deviant in 1956 when Rosa Parks sat down in such a seat. Today we clearly see her act as challenging discrimination, not an act of deviance in the popular sense of the word. Sociologists believe that understanding the process by which something becomes deviant is critical because just knowing the act, thought, or appearance by itself does not enable us to know it is deviant. We must also know the context and something about the audience that has the power to define an act as deviant.

Rachel Ellison

As the definition suggests, deviance involves violating norms, rules, or shared expectations for the way people should behave, feel, and appear in the presence of a particular group or in a particular social situation. Norms vary across groups and exist for virtually every kind of situation, including how many times a day to eat, how to greet a friend, what to wear to school, how to handle the American flag, and when to use a gun.

The consequences people face when breaking a norm vary according to whether an audience (1) knows it exists, (2) accepts it as just, (3) enforces it uniformly, (4) thinks it important to follow, and (5) backs it up with the force of law (Gibbs 1965). Consider norms guiding speed limits, which are backed by the force of law. Most drivers know the speed limits from observing posted signs but, depending on the setting, do not find it important to follow them to the letter of the law. In fact, most drivers exceed posted limits by 10 or 15 miles per hour without fear of getting caught by police. Even when they are caught, police officers do not always cite speeders for the exact number of miles they were driving over the limit, and sometimes officers let them go with a warning.

Norms vary by group. Some people, depending on the groups with which they identify, celebrate wearing sex bracelets as a sign of solidarity and support for a group; others treat wearing them as a sign of some character flaw. Some norms exist for seemingly valid reasons—for example, to prevent harm to self and/or others. But there are just as many examples of norms for which there seems to be no valid reason; that is, they aim at controlling behavior that is for all practical purposes harmless, such as a certain appearance or way of dressing (Erikson 1966).

The Sociological Perspective

From a sociological point of view, what makes something deviant is the presence of some social audience that regards a behavior or appearance as deviant and takes some kind of action to discourage and punish it. Deviance is not inherent to a specific behavior. Marrying a first cousin, for example, is not in itself deviant; if it were, that action would be deviant everywhere in the world. In fact, from a global perspective, marrying first cousins is a rather common practice. Deviance is something that is conferred. For sociologists the critical factors are not the behavior itself or the individual who violates norms. The critical factors are an audience's response and whether authorities take notice and punish behavior deemed as deviant. The sociological contribution to understanding deviance, then, lies in studying the larger context. In this regard sociologists note that almost any behavior or appearance can qualify as deviant under the "right" circumstances. How is it that anything can be defined as deviant? Émile Durkheim offered an intriguing answer to that question.

Durkheim (1901) argued that while ideas about what is deviant vary, deviance is present in all societies. He defined deviance as those acts that offend collective sentiments. The fact that there are some acts that offend always and everywhere led him to conclude that there is no such thing as a society without deviance. According to Durkheim, deviance will be present even in a "community of saints" (100). Even in a seemingly perfect society, acts that most persons would view as insignificant or minor may offend, create a sense of scandal, or be treated as crimes. To explain this, Durkheim (1901) drew an analogy to those who consider themselves "perfect and upright" but judge their smallest failings with a severity that others reserve for the most serious offenses. The same holds for groups that present themselves as "exemplary." They too can be severe in their judgment toward members (or outsiders) who commit seemingly minor offences. Thus, what makes an act or appearance deviant, even criminal, is not so much the act itself, but rather the fact that the group has defined it as something dangerous or threatening to its well-being.

U.S. Air Force Tech. Sgt. Jacob N. Bailey

The military represents a setting in which behaviors most people would, at worst, view as minor offenses, are treated as crimes. A speck of facial hair, a chin strap slightly off kilter, or a gun not held at a perfect 90-degree angle takes on critical significance for new military recruits in basic training.

Durkheim argued that the ritual of identifying and exposing a wrongdoing, determining a punishment, and/or carrying it out is an emotional experience that binds together the members of a group and establishes a sense of order and community. Durkheim maintained that a group that went too long without noticing deviance or doing something about it would lose its identity as a group.

Who Defines What Is Deviant?

In answering the question "Who defines what is deviant?" sociologists focus on the ways in which specific groups (such as undocumented workers), behaviors (such as using tanning beds), conditions (such as teenage pregnancy, infertility, or pollution), or artifacts (such as song lyrics, guns, art, or tattoos) become defined as problems. In particular, sociologists examine claims makers and claims-making activities.

Claims makers are those who articulate and promote claims and who tend to gain in some way if the targeted audience accepts their claims as true. Anyone can articulate claims, but not everyone with a claim is heard. Examples of claims makers include parents, children, government officials, advertisers, scientists, and professors. Claims-making activities are actions taken to draw attention to a claim—actions such as "demanding services, filling out forms, lodging complaints, filing lawsuits, calling press conferences, writing letters of protest" (Spector and Kitsuse 1977, 79). Studying claims-making activities can help us understand why smoking in public places came to be largely banned in the United States. Before 1980, people could freely smoke in public. In fact, professors and students smoked during class without eliciting a raised eyebrow. Today, such behavior would be unthinkable.

Negative health effects of smoking were claimed as early as 1900, as evidenced by this poster showing a skeleton rising from the smoke of a burning cigarette. However, other claims makers argued successfully for decades that because celebrities and even doctors smoke, it must be safe.

Library of Congress Prints and Photographs Division [LC-USZ62-89928]

Library of Congress Prints and Photographs Division [LC-USZC4-5780]

The success of a claims-making campaign depends on a number of factors: claims makers' status; personality; access to the media; available resources; and skill at fund-raising, promotion, and organization. When sociologists study the process by which a group or behavior is defined as deviant, they focus on who makes claims, whose claims are heard, and how audiences respond to them (Best 1989). Sociologists also pay attention to any labels that claims makers apply because labels tend to evoke a specific cause, consequence, and/or solution to a problem (Best 1989). For example, labeling an addiction—whether it be to gambling, credit card use, prescription drugs, or alcohol—as a medical problem is to locate the cause in the biological workings of the body or mind and to suggest that the solution rests with a drug, a vaccine, or surgery. Labeling an addiction as a personal failing, on the other hand, is to locate the cause in the character of the person, such as an inability to delay gratification or a lack of discipline.

What Do Sociologists See?

As city ordinances and other laws increasingly prohibit smoking in all workplaces and public areas, new norms must be established about where smoking is permitted. This sign represents one step introducing smokers to these norms or rules guiding them about where and when it is appropriate to light up without fear of being fined.

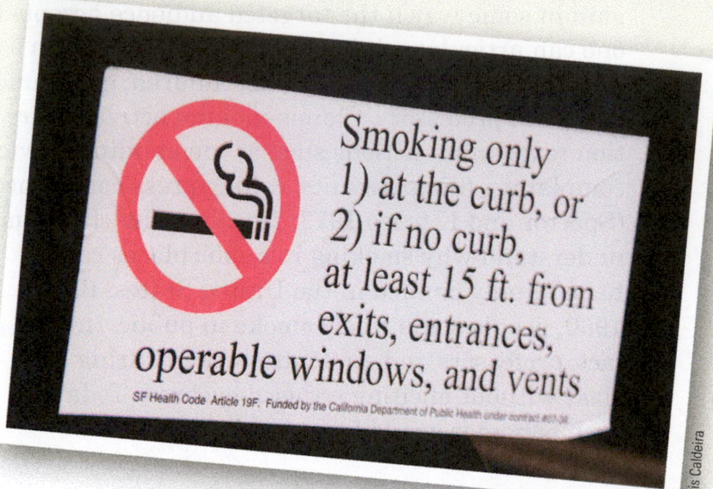

Chris Caldeira

Critical Thinking

Describe something that was once considered deviant but that is no longer the case (or something that is now deviant but was not in the past).

Key Terms

claims makers

deviance

Module

6.2 Mechanisms of Social Control

Objective

You will learn the mechanisms groups use to elicit conformity and punish deviance.

What is a referee's purpose? Can you think of an activity outside of the sports world that has the equivalent of a referee?

Tony Rotundo

Most sports have referees, someone in a position of authority to watch athletes (and coaches and fans) to make sure they abide by the rules. If you stop and think about it, most human activities have the equivalent of referees. "Referees" can include parents who correct their children's behavior, children who censor their parents' behavior, teachers who watch students as they take exams, police officers who arrest people they believe have committed crimes. "Referees" also include bystanders who stare in a disapproving manner when they observe someone doing something they consider odd.

Society's referees employ what sociologists call **mechanisms of social control**, strategies people use to encourage, often force, others to comply with social norms. For example, a referee may charge athletes with a foul when they violate rules, or, if the violation is deemed serious enough, even eject an athlete from the game. Police officers, school principals, and others act as referees when they issue tickets, put students in after school detention, and arrest or expel suspected offenders. Ideally, from society's point of view, it is easier to control people when they want to conform or when they conform out of habit. When conformity cannot be achieved voluntarily, mechanisms of social control are employed. Methods of social control include positive and negative sanctions, censorship, surveillance, authority, and group pressure.

Sanctions

Sanctions are reactions of approval or disapproval to behavior that conforms to or departs from group norms. Sanctions can be formal or informal and positive or negative. **Formal sanctions** are reactions backed by laws, rules, or policies specifying the conditions under which people should be rewarded or punished for specific behaviors. By contrast, **informal sanctions** are spontaneous, unofficial expressions of approval not backed by the force of law or official policy. **Positive sanctions** are expressions of approval for compliance. In contrast, **negative sanctions** are expressions of disapproval. Police officers employ negative formal sanctions when they arrest people suspected of breaking the law. Teens employ informal sanctions when they share images via cell phones with the intent of inviting people to praise or ridicule a particular person's behavior.

From a sociological point of view, this restaurant is announcing that women who breast-feed their babies will not experience negative informal sanctions (staring, whispers, verbal demands to stop) from patrons and the staff. In addition, the sign is essentially inviting in a social audience that approves of and supports public breast-feeding.

Jeremiah Evans, Courtesy of Joan Ferrante

Censorship and Surveillance

Censorship is an action taken to prevent information believed to be sensitive, unsuitable, or threatening from reaching some audience, whether that audience be children, voters, employees, prisoners, or others. Censorship relies on censors—people whose job it is to remove or block access to information deemed problematic in movies, books, e-mail, on TV, the Internet, and other media. Reporters Without Borders, which ranks countries according to the level of Internet freedom, named the following governments as engaging in the most extensive Internet censorship: Saudi Arabia, Burma, China, North Korea, Cuba, Iran, Syria, Tunisia, Turkmenistan, and Vietnam (*USA Today* 2014b).

Surveillance, another mechanism of social control, involves monitoring movements, conversations, and associations with the intent of catching people in the act of doing something wrong. Surveillance activities include tapping phones; intercepting letters, e-mail, and documents; videotaping/recording; and electronic monitoring.

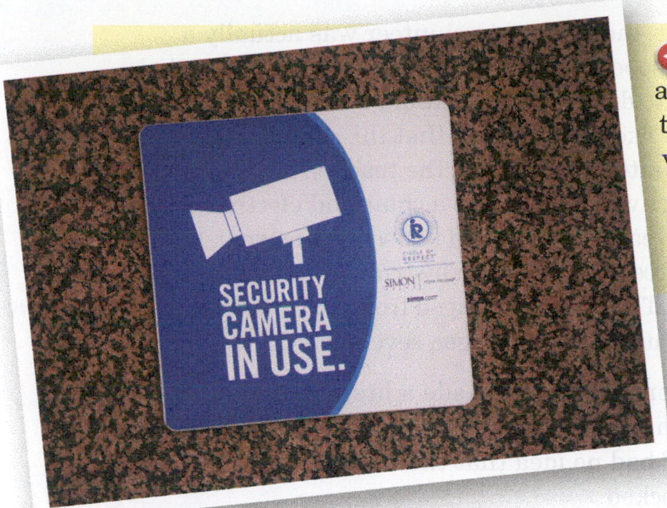

With the widespread availability of portable technologies that can record voices and capture images, there is no situation in which a person is potentially free from the possibility of surveillance.

Sociologist Kingsley Dennis (2008) describes the rise of new forms of personal surveillance with which people are able to monitor others through portable devices such as mobile phones. Dennis cites 1992 as the year this technology came of age. In that year a bystander videotaped Los Angeles police officers brutally beating Rodney King, a black man who led police on a 100-mile-per-hour car chase and, after being stopped, appeared to disobey police commands to lie down. The tape showed that officers used excessive force. The acquittal of three participating officers triggered five days of protest, which turned violent, in South Central Los Angeles, resulting in dozens of deaths, thousands of injuries, and $900 million in property damage.

Dennis (2008) points out that in the hands of responsible parties, personal recording devices can be liberating tools, allowing the public to record wrongdoings. In the wrong hands, such devices can also be used to harass, stalk, intrude, and otherwise ruin lives. Dennis offers the following example to illustrate virtual vigilantism. In South Korea a girl became known as the "Dog Shit Girl" after a bystander used his mobile phone to photograph a girl who failed to pick up after her dog when it defecated on a subway train floor. The bystander posted photographs online to mobilize the Internet-based community to humiliate her as punishment. The photo became a globally shared story. The girl eventually issued an apology over the Internet saying, "I know I was wrong, but you guys are so harsh. I regret it . . . if you keep putting me down on the Internet I will sue all the people and at the worst I will commit suicide" (350).

Obedience to Authority

In his now-classic study, social psychologist Stanley Milgram (1974) studied the commands of recognized authority figures as a mechanism of social control. Milgram wanted to learn why some people obey an authority's command to behave in ways that conflict with their conscience. Milgram designed an experiment to see how far people would go before they would refuse to conform. He placed an ad in a local paper asking for volunteers. When participants arrived at the study site—a university—they were greeted by a man in a laboratory coat who explained to them and another apparent volunteer that the study's purpose was to find out whether the use of punishment improves a person's ability to learn. Unknown to each participant, the apparent volunteer was actually a confederate—someone working in cooperation with the investigator. The participant and the confederate drew lots to determine who would be the teacher and who would be the learner. The draw was fixed, however, so that the confederate was always the learner and the real volunteer was always the teacher. In Milgram's experiments, the confederate-learner was strapped to a chair and electrodes were placed on his wrists. The volunteer-teacher, who could not see the learner, but could hear him moan and scream, was placed in front of an instrument panel containing a line of shock-generating switches. The switches ranged from 15 to 450 volts and were labeled accordingly, from "slight shock" to "danger, severe shock."

The researcher explained that when the learner made a first mistake, a 15-volt shock would be administered. With each subsequent mistake, the teacher should increase the voltage. The "teacher" had no idea that the learner was a confederate and was not actually being shocked.

NKU Sciology, Missy Gish

At one point during the experiment, the learner would even say that his heart was bothering him and would go silent. If a volunteer-teacher expressed concern, the researcher firmly said to continue administering shocks. In each case, as the strength of the shock increased, the volunteer could hear the learner express greater discomfort, even yelling out in distress. Although many volunteers protested, a substantial number (65 percent) obeyed and continued on, "no matter how painful the shocks seemed to be, and no matter how much the victim pleaded to be let out" (Milgram 1987, 567). These results are especially significant because the volunteers received no penalty if they refused to administer the shocks.

In Milgram's experiment, obedience was founded simply on the firm command of a person with an assumed status (university-based researcher) and supporting symbols (white lab coat) that gave minimal authority over the participant. The authority was rated as minimal because the research participants were under no obligation to be involved in the experiment.

Milgram's study offers insights as to how events such as the death of 6 million Jews during the Holocaust, the death of 1.7 million Cambodians under Pol Pot, and, more recently, prisoner abuse by U.S. military at Abu Ghraib prison in Iraq could have taken place. These kinds of events require the cooperation of many people.

U.S. Navy

This willingness to obey authorities raises important questions. Specifically, how is "behavior that is unthinkable" to an individual when acting on his or her own "executed without hesitation when carried out under orders" (1974, xi).

As a case in point, Staff Sergeant Ivan L. Frederick II was caught up in the 2003 Abu Ghraib prison scandal. Abu Ghraib, one of the world's most notorious prisons under Saddam Hussein, was converted to a U.S. military prison after Hussein was removed from power. Frederick faced charges of cruelty toward prisoners, maltreatment, assault, and indecent acts. He was sentenced to eight years in prison. Letters and e-mail messages that Frederick wrote to family members asserted that he was simply carrying out orders (Hersh 2004).

In January 2004, Frederick wrote, "I questioned some of the things that I saw . . . such things as leaving inmates in their cell with no clothes or in female underpants, handcuffing them to the door of their cell—and the answer I got was, 'This is how military intelligence [MI] wants it done.'"

Group Pressure

Social psychologist Irving Janis (1972) coined the term **group think**, a phenomenon that occurs when a group under great pressure to take action achieves the illusion of consensus by putting pressure on its members to suppress expression of doubt and ignore the moral consequences of their actions. Group think is most likely to occur when members are from similar backgrounds, do not seek outside opinions, and unquestioningly believe in the rightness of their cause. The concept of group think grew out of Janis's research on the group dynamics underlying the making of foreign policies with disastrous consequences and comparing them with the group dynamics underlying the making of foreign policies with successful outcomes.

National Archives and Records Administration

One disastrous decision Janis investigated was a 1950 U.S. military order requiring soldiers stationed in South Korea to cross into North Korea. This order came after the North Korean army had crossed into and marched across South Korea, almost taking it, but Americans and South Korean soldiers pushed the North Korean army out. At that point the American military decided to invade North Korea. That decision had the effect of prolonging the Korean War another three years. In the end, 2.5 million civilians and 2 million soldiers on both sides died, North and South Korea were reduced to rubble, and neither side gained territory.

Janis and other researchers have applied group think theory to group dynamics underlying other historical events with disastrous results, including the

Watergate break-in, the Iran hostage crisis, the Kent State massacre, and the space shuttle *Challenger* explosion (Esser 1998). There have also been attempts to simulate group think in laboratory settings. Figure 6.2a shows some of the antecedents, elements, and symptoms of group think (Esser 1998).

Master Sgt. Gerold Gamble

Group think dynamics can also be applied to any group under pressure to perform at a high level or maintain a tradition of excellence, as was the case with the elite Florida A&M Marching Band. In 2011, one of the band's drum majors died after being the focus of a hazing ritual in which he was repeatedly punched by a small group of band members. The antecedents of group think (see Figure 6.2a) applied to this elite marching band, whose members held celebrity status on campus. That elite reputation and celebrity status gave some members the illusion of invulnerability and suppressed any objective assessment of risk associated with such actions (Alvarez and Brown 2011).

Figure 6.2a: **Group Think : Its Antecedents, Elements, and Symptoms**

Antecedents to Group Think

- Leaders are directive in style
- Members share similar social backgrounds and beliefs
- Group is isolated from outside sources of information and analysis
- Group under pressure to take action

Group Think

- An illusion of invulnerability accompanied by excessive optimism and risk taking
- Guiding assumptions are not questioned
- Certainty about the group's purpose and morality
- Labeling the opposition as weak, evil, biased, vindictive, disloyal, or uninformed
- Internal pressure on members to not raise questions
- An illusion that members are unanimous; interpreting silence to mean agreement
- Dissenting information cannot surface

Symptoms of Group Think

- Alternative strategies are not considered
- Objectives are not clear
- No assessment of potential risks associated with decisions
- Weak effort at gathering information
- No thought given to a contingency plan if things go wrong

What Do Sociologists See?

Sociologists see a man in a white lab coat, a symbol of expertise and authority. In matters of science the lab coat signifies that the wearer has knowledge and credentials over those without lab coats who challenge scientific innovations, such as a farmer who warns of environmental or human costs of genetically modified seeds.

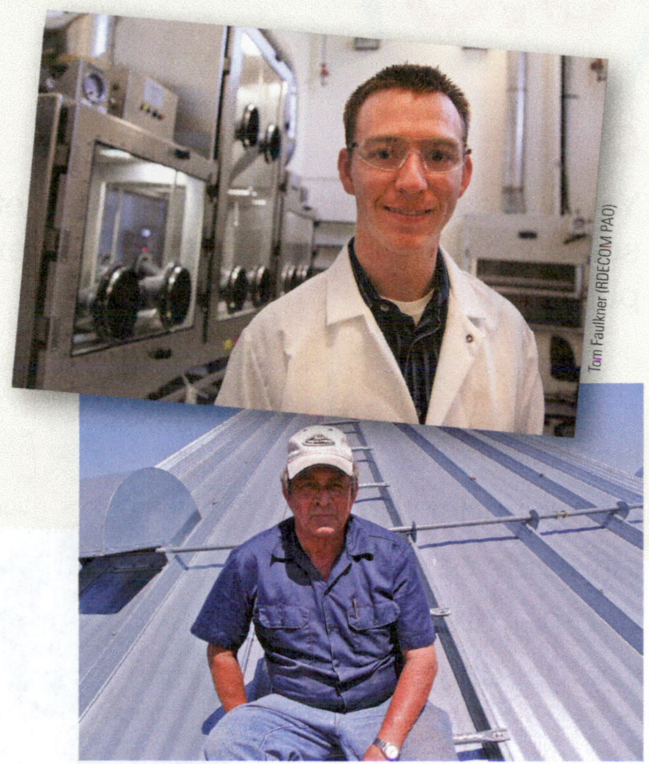

Tom Faulkner (RDECOM PAO)

Chris Caldeira

Critical Thinking

Can you describe a time in your life when you wanted to do something and decided against it because you feared the sanctions that someone important to you might apply? Explain.

Key Terms

censorship

formal sanctions

group think

informal sanctions

mechanisms of social control

negative sanctions

positive sanctions

sanctions

surveillance

Module 6.3 Labeling Theory

Objective

You will learn that labeling theorists define a deviant as someone who is noticed as violating expectations and punished.

Have you ever driven under the influence of alcohol or other drugs? Did you get caught?

Ms. Brittany Carlson (IMCOM)

When I asked my students to answer these questions anonymously, about 40 percent said they have engaged in drugged driving. Less than one percent said they were caught by police. Those caught went to drivers' school or had their license suspended. The small percentage caught raises questions about the nature of deviance, but especially about just who gets caught. Such questions are at the heart of labeling theory.

Labeling Theory

In *Outsiders: Studies in the Sociology of Deviance*, sociologist Howard Becker states the central thesis of labeling theory: "All social groups make rules and attempt, at some times and under some circumstances, to enforce them. When a rule is enforced, the person who is supposed to have broken it may be seen as a special kind of person, one who cannot be trusted to live by the rules agreed on by the group." That person is "regarded as an outsider" (1963, 1).

As Becker's statement suggests, labeling theorists are guided by two assumptions: (1) rules are socially constructed; that is, people make rules; and (2) rules are not enforced uniformly or consistently. Labeling theorists maintain that whether an act is deviant depends on whether people notice it and, if they do notice, whether they label it as a violation of a rule and then proceed to apply sanctions. In other words, simply violating a rule does not automatically make someone deviant. From a sociological point of view, a rule breaker is not deviant unless someone notices the violation and decides to take corrective action.

Categories of Deviants

Labeling theorists point out that for every rule a social group creates, four categories of people exist: conformists, pure deviants, secret deviants, and the falsely accused. The category to which someone belongs depends on a combination of two factors: whether a rule has been violated and whether sanctions are applied (see Table 6.3a).

Table 6.3a: Typology of Deviance Applied to Drugged Driving

The table summarizes the four categories of people as they relate to a particular social norm or law. Think of some offending behavior, such as drugged driving. Either a person engages in that behavior or not. But for the behavior to be defined as deviant, someone has to notice and apply sanctions. If no one in a position of authority notices or takes action to punish him or her, the person is a secret deviant.

	Noticed/Sanctions Applied	Not Noticed/Sanctions Not Applied
Engaged in offending behavior (drugged driving)	Pure deviant (Engaged in drugged driving; authority noticed and applied sanctions)	Secret deviant (Engaged in drugged driving; no authority noticed or applied sanctions)
Did not engage in offending behavior (drugged driving)	Falsely accused (Did not engage in drugged driving but is noticed and charged)	Conformist (Did not engage in drugged driving and no authority suspected such behavior)

From Table 6.3a we can see that **conformists** are people who have not engaged in offending behavior and are treated accordingly. **Pure deviants** are people who have engaged in offending behavior and are caught, punished, and labeled as outsiders. **Secret deviants** are those who have engaged in the offending behavior but no one notices or, if it is noticed, no one applies sanctions. Becker maintains that we cannot really know how many secret deviants exist, but he is convinced that the number is sizable, many more than "we are apt to think" (1963, 20).

Sociologist Kai Erikson (1966) identified the situation under which people are likely to be falsely accused of a crime—when the well-being of a country or a group is threatened. The threat can take the form of an economic crisis (such as a depression or recession), a health crisis (such as AIDS or MERS), or a national security crisis (such as war). Whenever a catastrophe occurs, it is common to blame someone for it. Identifying the threat gives an illusion of control. In such crises, the person blamed is likely to be someone who is at best indirectly responsible, someone in the wrong place at the wrong time, or someone who is viewed as different. This defining activity can take the form of a witch hunt.

Witch hunts are campaigns to identify, investigate, and correct behavior that has been defined as dangerous to the larger society. In actuality, a witch hunt rarely accomplishes these goals because the real cause of a problem is often complex, extending far beyond the behavior of a targeted category. Often, people who are identified as the cause of a problem are simply being used to make the problem appear as if it is being managed. Although the FBI did not keep statistics on the ethnicity or religious affiliation of the people questioned about the September 11, 2001, attacks, it is believed that most people interrogated were or appeared to be Muslim or Middle Eastern. Many federal agents, haunted by the attacks, acted on "information from tipsters with questionable backgrounds and motives, touching off needless scares and upending the lives of innocent suspects" (Moss 2003, A1).

What Do Sociologists See?

Sociologists use labeling theory to frame the arrest of this man. He has been labeled as deviant because someone with enough power noticed his behavior and decided to take corrective action. He may be a pure deviant or falsely accused; the fact that he is handcuffed tells us he has been labeled as deviant.

U.S. Army photo by Staff Sgt. Sean A. Foley/Released

Critical Thinking

Can you think of situations in your life when you assumed the status of falsely accused, secret deviant, and/or pure deviant? Explain.

Key Terms

conformists

pure deviants

secret deviants

witch hunt

Module 6.4 Stigma

Objective

You will learn that stigmas are deeply discrediting attributes.

Are you surprised to learn that people with a higher body mass index are often viewed as being undisciplined, lazy, lacking in will power, and without ambition? Such judgments are most likely made by people who have no knowledge of them as persons.

Chris Caldeira

Sociologist Erving Goffman defined a **stigma** as an attribute that is deeply discrediting. Goffman identified three broad categories of stigmas:

1. stigmas of the body or physical conditions that some audience defines as an imperfection, a deformity, or a disability;
2. stigmas related to behavior or behaviors some audience considers deviant, such as an addiction, sexual orientation, or a criminal record; and
3. stigmas that an audience has defined as racial, ethnic, religious, or national.

Stigmas are discrediting because they break the claims of all other attributes a person possesses. To illustrate, Goffman refers to a letter written by a 16-year-old girl born without a nose. In that letter she writes, "Although I am a good student, have a good figure, and am a good dancer, no one I meet seems able to get past the fact that I have no nose." The absence of a nose is considered so significant that those she interacts with fail to notice her other attributes.

Goffman was particularly interested in social encounters known as **mixed contacts**—interactions between stigmatized persons and so-called normals. Note that Goffman did not use the term *normal* to mean "well-adjusted" or "healthy." Instead, he used it to refer to those people who possess no stigma. Note that whether something is labeled as normal or stigmatized depends on the social context.

Goffman wrote that mixed contacts occur when the stigmatized and normals find themselves in the same social setting—whether they be "in a conversation-like encounter" or simply part of some "unfocused gathering" (Goffman 1963, 12). According to Goffman, when normals and the stigmatized interact, the stigma dominates the interaction.

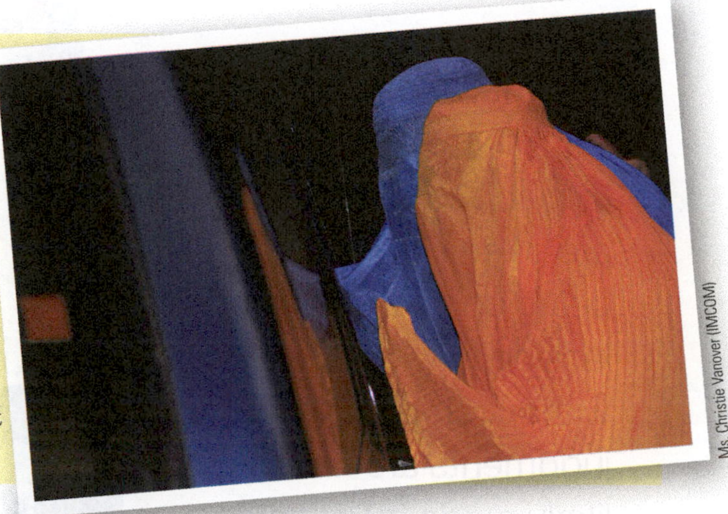

The two Afghan women approaching a U.S. military vehicle with soldiers inside qualifies as a mixed contact. In this interaction, we might question which party—the soldiers or the Afghan women—qualifies as the stigmatized and which qualifies as the normals? The answer depends on which point of view is being considered.

Ms. Christie Vanover (IMCOM)

Patterns of Mixed Contact

A stigma can come to dominate interactions in many ways. First, the very anticipation of contact can cause normals and stigmatized to avoid one another. One reason they avoid contact is to escape anticipated discomfort, rejection, disapproval, and suspicion.

The response of avoidance is related to a second pattern that often characterizes mixed contacts: upon meeting, both normals and stigmatized are unsure how the other views them or will act toward them. For the stigmatized, the source of the uncertainty is not that everyone they meet views or treats them negatively; rather, the stigmatized consider the chances high that they might encounter someone who will.

A third characteristic of mixed contacts is that normals often define any accomplishment of the stigmatized—even minor achievements—"as signs of remarkable and noteworthy capacities" (Goffman 1963, 14) or as evidence that they have met someone who is an exception to the rule.

In a fourth pattern, normals tend to attribute any major or even minor failing on the part of the stigmatized—being late for a meeting, cashing a bad check, or tripping on a step—to the stigma.

NKU Sociology, Missy Gish

Drivers of all ages get into car accidents, but factors like age, race, gender, and social class affect how people judge the parties involved. If after we learn, for example, that an elderly person or a teen was the driver causing the accident, we jump to the conclusion that age "explains" the crash without learning the details, we have fallen victim to the dynamics Goffman describes in his theory of stigma.

A fifth pattern common to mixed contacts is that the stigmatized are likely to experience invasion of privacy, especially when normals stare.

Chris Caldeira

A person considered elderly who walks slowly and who suffers from a degenerative condition is made self-conscious when all eyes watch as he makes his way down the street. The staring qualifies as an invasion of privacy.

Responses to Stigmatization

Goffman describes five ways that the stigmatized respond to normals who cannot see beyond the discrediting attribute. First, the stigmatized may respond by making an attempt to erase the source of stigma. This response includes eliminating the visible markers that act as barriers to success and belonging, whether it be getting cosmetic surgery to alter the shape of the nose, eyes, or lips; coloring or to cover grays; or wearing tinted contacts so eyes appear the "desired" color.

A second way the stigmatized respond involves devoting a great deal of time and effort to overcoming stereotypes or to appearing as if they are in full control of everything around them. The stigmatized may try to be perfect—to always be in a good mood, to be extra friendly, to outperform everyone else, or to master an activity not expected of them.

Soldiers who lose limbs while fighting wars may choose to train for and compete in what are called Warrior Games. The games are designed to showcase the abilities of those who have been stigmatized as disabled and counter stereotypes that they lead severely limited lives.

MC1 R. Jason Brunson

As a third response to being stigmatized, people may use their subordinate status for secondary gains, including personal profit, or as "an excuse for ill success that has come [their] way for other reasons" (Goffman 1963, 10). The stigmatized may also view discrimination as a blessing in disguise, especially for its ability to build character or for what it teaches about life and humanity. Finally, a stigmatized person can condemn all the normals and view them as incapable of forming a friendship or caring for someone with a stigma.

What Do Sociologists See?

When sociologists see two males supervising preschool-age children, they point out that only about 1 percent of preschool teachers are male. In this context, being male is a stigmatized status because many are often suspicious of males who choose a career in child care or early childhood education. Such suspicions encourage men to avoid working with young children.

Cpl. Claudio Martinez

Critical Thinking

Do you have a stigma—a discrediting attribute that breaks the claims of other qualities you possess? Explain.

Key Terms

mixed contacts

stigma

Module 6.5 Sociological Perspective on Crime

Objective

You will learn that sociologists focus not just on those who are known to commit crimes, but on the process by which behaviors are defined as crimes and the contexts that offer opportunities to commit crime.

How does this image shape ideas about what crimes are?

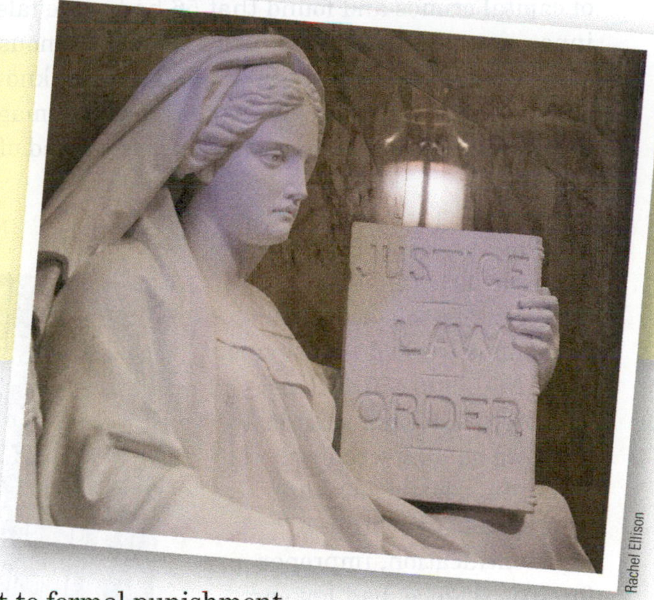

Rachel Ellison

Crime is behavior that violates a law. People in positions of power make laws with the goal of protecting some "desired" social order. Those caught breaking laws are subject to formal punishment such as fines, imprisonment, community service, or the death penalty. Behaviors defined as crimes reflect the values and interests of lawmakers and other dominant groups (Henry 2009).

Sgt. 1st Class Howard Reed (Japan)

While we like to think justice, law, and order are synonymous, there are countless examples that suggest otherwise. Simply consider that Mrs. Rosa Parks was taken into police custody, booked, fingerprinted, and briefly incarcerated for refusing to obey a bus driver who ordered her to move to the back of the bus (National Archives 2014). Today when we reenact this historical event, we celebrate Rosa Parks's "crime" of challenging racial segregation laws.

In addition to the fact that laws are not always just, they are also enforced unevenly. A U.S. Department of Justice (2013) survey of crime victims in the United States suggests that a large percentage of lawbreakers are never caught. That survey found that just 44 percent of violent victimizations and 54 percent of serious violent victimizations were reported to police. The percentage of property victimizations reported is 34 percent. In other words, 56 percent of violent victimizations, 45 percent of serious victimizations, and 64 percent of property crimes are not reported. The unreported assume statuses as secret deviants. (Of course, simply reporting a crime does not mean that the perpetrator will be caught.)

To complicate matters, we can never know the number of people who are **falsely accused**—those who have not committed crimes but are treated as if they have. The ranks of the falsely accused include victims of eyewitness errors and police cover-ups and innocent suspects who make false confessions under the pressure of interrogation. For their book *In Spite of Innocence*, sociologist Michael L. Radelet and colleagues (1994) reviewed more than 400 cases of innocent people convicted of capital crimes and found that 56 had made false confessions. Apparently, some innocent suspects admitted guilt, even to heinous crimes, to escape the stress of interrogation (Jerome 1995). While no one knows how often false accusations occur, it probably occurs more often than we imagine. Moreover, the taint of guilt lingers even after the falsely accused is cleared of all charges.

The Innocence Project, a legal initiative that works to free those falsely accused of crime, has worked to overturn more than 300 convictions. The organization is currently working on 300 active cases. Wrongful convictions can be traced to eyewitness misidentification, improper forensic science, bad lawyers, and government misconduct (Innocence Project 2014).

Federal Bureau of Investigation

Who Goes to Prison?

We know then that prisons are filled with those who got caught or who are falsely accused. Worldwide, an estimated 7.6 million people are in prisons and jails. The incarceration rate varies by country, with the United States having the highest rate at 730 prisoners per 100,000 population. After the United States are the Republic of Georgia (536) and the Russian Federation (525) (International Centre for Prison Studies 2012).

It is important to remember that a high incarceration rate is not necessarily a measure of a criminal justice system's effectiveness. It could be that countries with the highest rates are the ones with the most serious crime problems, are the most effective at bringing to justice those who commit crimes, or can pay the cost of housing a large prison population (Walmsley 2002). Regardless, it is still important to understand why the United States with 4.6 percent of the world's population holds 29 percent of all incarcerated persons around the world (International Centre for Prison Studies 2012).

Going to prison is a multistage process. In studying this process, one thing is clear: only a small portion of those who commit crimes are in prison. Why? First, someone must know about the crime and then decide to report it (or self-report it). As we have learned, only about 40–50 percent of known crimes are reported to police. Among those reported, only 44 percent lead to an arrest (the equivalent of 16 percent of all crimes committed). Of those arrested, some smaller portion are prosecuted, but not everyone prosecuted is convicted. And not everyone convicted is imprisoned (see Figure 6.5a).

Figure 6.5a: The Road to Prision

This flowchart shows the process by which people are arrested and then make their way through the criminal justice system. The prison population includes those who could not avoid one or more of the following: (1) detection, (2) someone reporting their offense, (3) being arrested, (4) arraignment, (5) prison or jail.

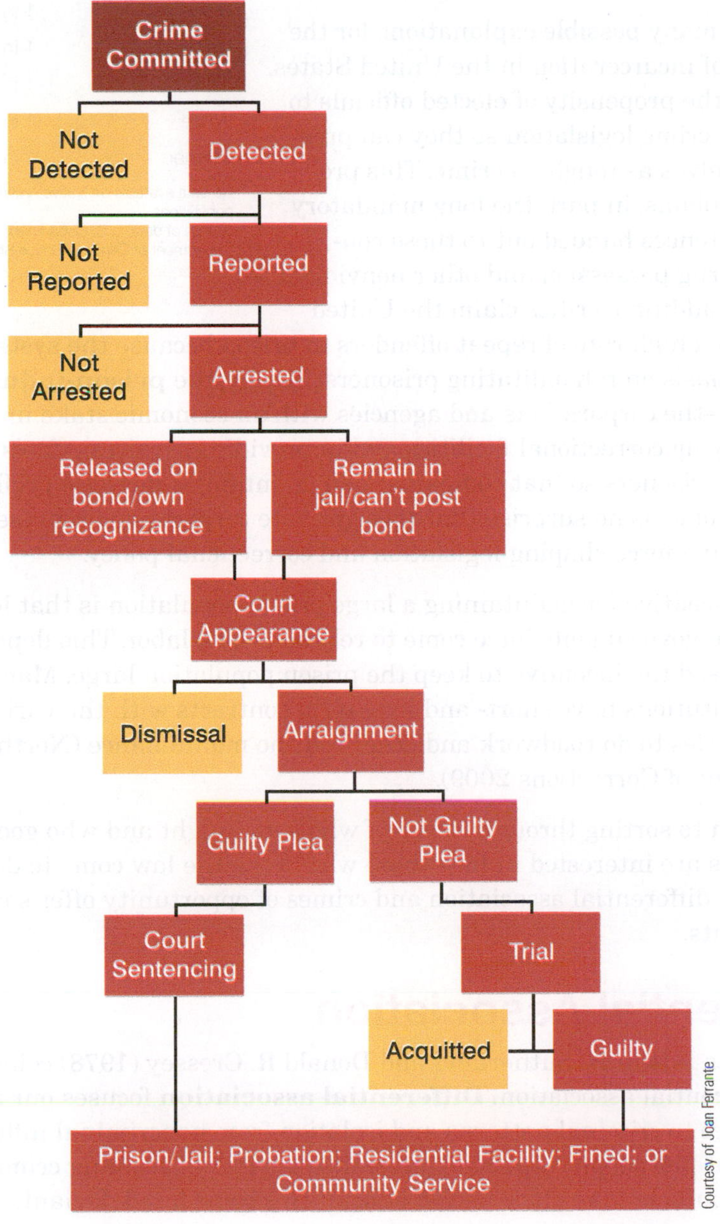

Courtesy of Joan Ferrante

According to the U.S. Department of Justice (U.S. Department of Criminal Justice Statistics 2010), 1 in 15 people in the United States will end up in prison at some point in their lifetime. But that rate is particularly high for men classified as black and Hispanic. Over the course of a lifetime, an estimated 32 percent of all black males and 17 percent of all Hispanic males will spend some time in prison. For white males that percentage is 5.9 percent. If we ask who is under some form of correctional control—on probation, under home incarceration, in jail, in a residential facility, performing community service—we see that at any one time 1 in every 32 adults (or about 6.1 million) is involved in some way with the criminal justice system.

Table 6.5a: Sex and Race Characteristics of People Involved with U.S. Correctional System (in prison, jail, on probation or parole), 2010

Characteristics	Number Involved in U.S. Correctional System
Blacks	1 in every 11
Men	1 in every 18
Hispanics	1 in every 27
All	1 in every 32
White	1 in every 45
Women	1 in every 89

No data available on Native Americans or Asian populations.
Source of data: U.S. Department of Justice (U.S. Department of Criminal Justice Statistics 2010)

There are many possible explanations for the high rate of incarceration in the United States, including the propensity of elected officials to pass strict crime legislation so they can present themselves as tough on crime. This propensity explains, in part, the long mandatory prison sentences handed out to those convicted of drug possession and other nonviolent crimes. In addition, critics claim the United States has a high rate of repeat offenders in prison because the system places little emphasis on rehabilitating prisoners. Finally, the **prison-industrial complex**—the corporations and agencies with an economic stake in building and supplying correctional facilities and in providing services—fuels an ongoing "need" for prisoners so that companies can maintain or increase profit margins. It should come as no surprise that these private corporations represent a significant lobbying force shaping legislation and correctional policy.

Another incentive for maintaining a large prison population is that local, state, and federal governments have come to rely on prison labor. This dependency has increased the incentive to keep the prison population large. Many correctional institutions have short- and long-term contracts with the various government agencies to do roadwork and other routine maintenance (North Carolina Department of Corrections 2009).

In addition to sorting through issues of who gets caught and who goes to prison, sociologists are interested in how those who break the law come to do so. The theories of differential association and crimes of opportunity offer some important insights.

Differential Association

Sociologists Edwin H. Sutherland and Donald R. Cressey (1978) coined the term differential association. **Differential association** focuses our attention on exposure to criminal patterns and isolation from noncriminal influences as factors that put people, especially juveniles, at risk of becoming criminals. These criminal contacts take place within subcultures considered deviant.

When you were a teenager, did your parents or some other authority figure ever tell you that they did not like the people you were hanging out with because they thought you might do something that would get you arrested? If so, they were guided by assumptions underlying differential association theory.

NKU Sciology, Missy Gish

When thinking about who commits crime, Sutherland and Cressey put the focus on subcultures considered deviant or groups that are part of the larger society but whose members share norms and values that encourage violation of that larger society's laws. People learn criminal behavior from closely interacting with those who engage in and approve of law-breaking activities. It is important to keep in mind, however, that contact with deviant subcultures does not by itself make people criminals. Rather, there is some unspecified tipping point of exposure to criminal influences that offsets any exposure to law-abiding influences (Sutherland and Cressey 1978).

If we accept the premise that criminal behavior is learned, then criminals constitute a special type of conformist in that they are simply following the norms of the subculture with which they associate. The theory of differential association does not explain how people make initial contact with a deviant subculture or the exact mechanisms by which people learn criminal behavior, except that the individual learns the deviant subculture's rules the same way any behavior is learned.

Sociologist Terry Williams's (1989) research shows how teenagers can make contact with a deviant subculture. He studied a group of teenagers, some as young as 14, who sold cocaine in the Washington Heights section of New York City. Major drug suppliers recruited teenagers because, as minors, they could not be sent to prison if caught. Williams argues that the teenagers he studied were susceptible to recruitment for two reasons: the teens saw little chance of finding high-paying jobs, and they perceived drug dealing as a way to earn enough money to escape their disadvantaged circumstances. Williams's findings suggest that once teenagers become involved in drug networks, they learn the skills to perform their jobs the same way everyone learns to do a job. Indeed, success in an illegal pursuit is measured in much the same way that success is measured in mainstream jobs: pleasing the boss, meeting goals, and getting along with associates.

Crimes of Opportunity

Williams's research suggests that criminal behavior is not simply the result of differential association with criminal ways. There are other factors at work, including what he terms **illegitimate opportunity structures**, social settings and arrangements that offer people the opportunity to commit specific types of crime. In the case of Williams's study, drug suppliers recruited 14-year-olds because as minors they would not go to prison. The larger society offers minors an "opportunity" structure that allows them to engage in criminal activity without risking the full penalties of the law. Opportunities to commit crimes are also shaped by the environments in which people live or work (Merton 1997). For someone to embezzle money, another person has to have entrusted the would-be embezzler with a large sum of money; the act of entrusting money has to occur before the embezzler can set money aside for some unintended purpose.

Illegitimate opportunity structures figure into the types of crimes people commit. Working as a pharmacist offers opportunities to steal prescription drugs to feed an addiction or to sell on the street. Working as a firefighter offers an opportunity to steal jewelry or other valuable items after gaining access to a house.

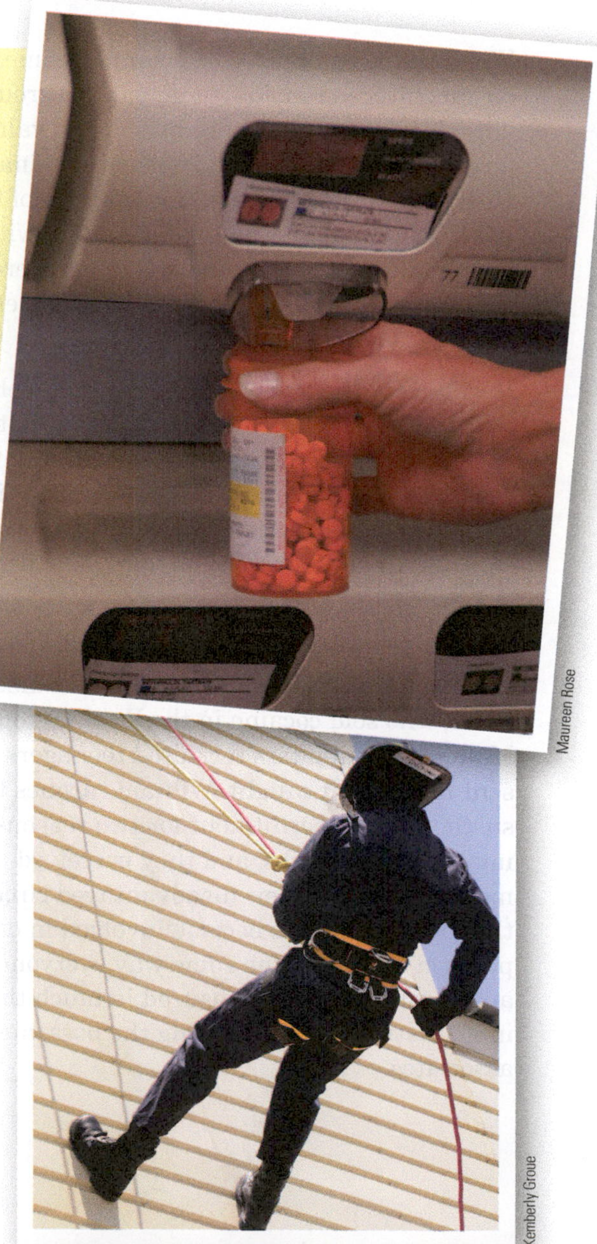
Maureen Rose
Kemberly Groue

The concept of an illegitimate opportunity structure undermines the belief that the uneducated and members of minority groups are more prone to criminal behavior than are those in other groups. In fact, crime exists in all social strata, but the type of crime, the opportunities to commit crime, the extent to which the laws are enforced, access to legal aid, and the power to sidestep laws vary by social strata (Chambliss 1974).

White-collar and corporate criminals also benefit from differential opportunity. **White-collar crime** consists of "crimes committed by persons of respectability and high social status in the course of their occupations" (Sutherland and Cressey 1978, 44). **Corporate crime** is committed by a corporation in the way that it does business as it competes with other companies for market share and profits. In the case of white-collar crime, offenders are part of the system: They occupy positions in the organization that permit them to carry out illegal activities discreetly. In the case of corporate crime, everyone in the organization contributes to illegal activities simply by doing their jobs. Even activities considered legal can have adverse consequences on people and the environment. Both white-collar and corporate crimes are often aimed at impersonal—and often vaguely defined—entities; they may involve evading taxes, polluting the environment, putting competitors out of business, and so on. They are "seldom directed against a particular person who can go to the police and report an offense" (National Council for Crime Prevention in Sweden 1985, 13).

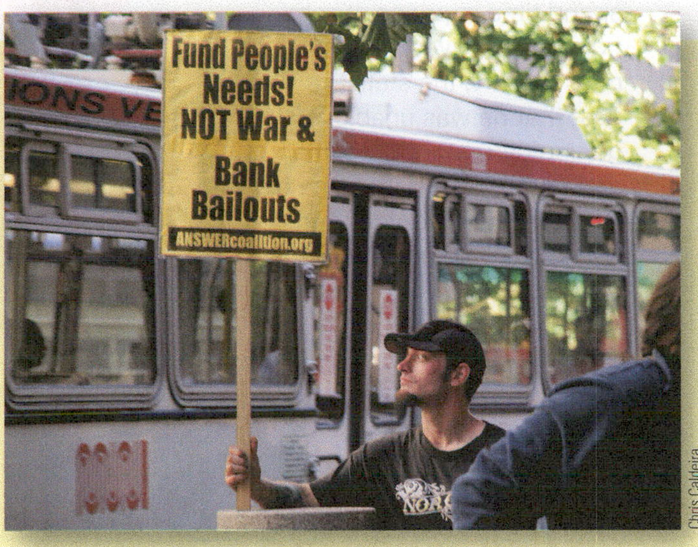

The Occupy Wall Street and other Occupy movements targeted Wall Street and financial institutions for committing "corporate crimes" that triggered the Great Recession of 2008. Wall Street and financial institutions engaged in risky lending practices backed by Moody's Investors Service and Standard & Poor's triple-A ratings of subprime mortgage securities. Despite the rating, these investments proved to be worthless when millions defaulted on loans. To compound matters, many holders of mortgage securities took out insurance policies that would pay off in the event of mass default. This backup protection had the effect of leading banks to make risky loans to those who could not afford them.

White-collar and corporate crimes—such as engaging in risky lending practices, manufacturing and marketing unsafe products, unlawful disposal of hazardous waste, tax evasion, and money laundering—are usually handled not by the police but by regulatory agencies (such as the Environmental Protection Agency, the Federal Bureau of Investigation, and the Food and Drug Administration), which have minimal staff to monitor compliance. Escaping punishment is easier for white-collar and corporate criminals than for other criminals.

What Do Sociologists See?

This man is among the 16 percent of all people who commit crimes who are eventually arrested. He is also among a relatively small percentage of people who commit crimes who eventually end up in prison. This man's chances of going to prison are increased by the fact that he is classified as black. Approximately 1 in 3 black males spend time in prison at some point in their lives. Assuming this man is guilty, the statistics suggest that he was unable to avoid getting caught for his crime, and once in the system, he did not have the resources and connections to avoid a prison sentence.

LCpl. Alfredo V. Ferrer

Critical Thinking

Give an example of an illegitimate opportunity structure that could provide you with the means to commit a crime.

Key Terms

corporate crime

crime

differential association

falsely accused

illegitimate opportunity structures

prison-industrial complex

white-collar crime

Module 6.6 Structural Strain Theory

Objective

You will learn how structural strain generates deviant responses.

What is the American dream? How do you know you have achieved it?

Chris Caldeira

The American dream is grounded in the belief that everyone has the opportunity to achieve material prosperity regardless of social position in life. What does it mean to live in a society in which people are told that they can be anything they want if they work hard enough? As we will see, sociologist Robert K. Merton's (1938, 1957b) theory of structural strain shows how belief in the American dream is connected to deviant behavior.

The Structure of Strain

Robert K. Merton's theory of structural strain takes two elements of social structure into account:

1. the *goals* a society defines as valuable (such as economic success, upward mobility, home ownership); and
2. the culturally legitimate *means* to achieve those valued goals (such as go to college, work hard), including the actual number of legitimate opportunities available to achieve valued goals (such as the number of jobs paying over $100,000).

Structural strain is a situation in which there is an imbalance between culturally valued goals and the legitimate means to obtain them. An imbalance exists when:

1. opportunities for reaching the valued goals are limited or closed off to a significant portion of the population; that is, there are not enough opportunities to satisfy demand;

2. people are unsure whether following the legitimate means will lead to success; or
3. the sole emphasis is on achieving valued goals by any means necessary.

Merton (1938) argues that structural strain induces a state of cultural chaos, or **anomie**. Under such conditions people are susceptible to abandoning the legitimate means to achieve culturally valued goals and are even susceptible to abandoning those goals.

Responses to Structural Strain

Merton identified five ways that people respond to structural strain. The responses involve some combination of acceptance and rejection of the valued goals and means.

Table 6.6a: Typology of Responses to Structural Strain
The table summarizes the five responses to structural strain. Which response involves acceptance of culturally valued goals and the rejection of legitimate means to achieve them? Which responses involve rejecting (or abandoning) culturally valued goals?

Mode of Adaptation	Goals	Means
Conformity	+	+
Innovation	+	–
Ritualism	–	+
Retreatism	–	–
Rebellion	+/–	+/–

+ Acceptance/achievement of valued goals or means
– Rejection of/failure to achieve valued goals or means
Source: Adapted from Merton (1957b, "A Typology of Modes of Individual Adaptations," 140)

Conformity is the acceptance of cultural goals and the pursuit of those goals through legitimate means. The category includes those who play by the book. They use legitimate means such as earning a college education and working hard to achieve success.

Innovation is the acceptance of cultural goals but the rejection of legitimate means to achieve them. For the innovator, success is equated with winning the game rather than with playing by the rules; that is, the innovator seeks to achieve valued goals of financial success but uses means considered illegitimate, such as evading taxes, selling drugs, identity theft, or embezzlement.

In 2013 the IRS initiated about 1,500 tax identity investigations involving a thief stealing a victim's Social Security number and using it to file a false tax return claiming a refund. In one case a woman stole the identity of more than 400 individuals (many of whom were deceased) and collected $835,883. She pled guilty and was sentenced to 144 months in prison. Under Merton's typology, this woman qualifies as an innovator (IRS 2014).

IRS

People who engage in innovation as a response to structural strain accept the culturally valued goals, but reject the culturally valued means. Some who engage in innovation confront the system using protests aimed to change the means to achieve culturally valued goals. Examples include the thousands of California college students who protested the means—high tuition and debt—by which many achieve the culturally valued goal of a college degree. The National Guard was called in to assist local police in preventing riots and to remove students blocking entrances to administration and other buildings.

Tech Sgt Dave Deremer

Ritualism involves the rejection of the cultural goals but a rigid adherence to the legitimate means society has in place to achieve them. This response is the opposite of innovation: the game is played according to the rules despite defeat. Merton (1938) maintains that this response can be a reaction to the status anxiety that accompanies the ceaseless competitive struggle to stay on top or to get ahead. For ritualists, the culturally valued goal, even though valued, is defined as being beyond their reach. This response applies to those who work full-time jobs that do not pay a living wage. They work hard but do not achieve financial success.

Retreatism involves the rejection of both culturally valued goals and the legitimate means of achieving them. Retreatism is the response of those who have internalized the culturally valued goals, but the legitimate means promising success have failed them. According to Merton, retreatists face a mental conflict in that it is against their moral principles to use illegitimate means, yet the legitimate means no longer apply.

Frustrated and handicapped because they believed they played by the rules yet the system failed them, retreatists find that they cannot cope and decide to drop out. Examples include prominent high-status professionals who thought they did "everything they were supposed to do in life" but whose investments have been wiped out (Kotbi 2009).

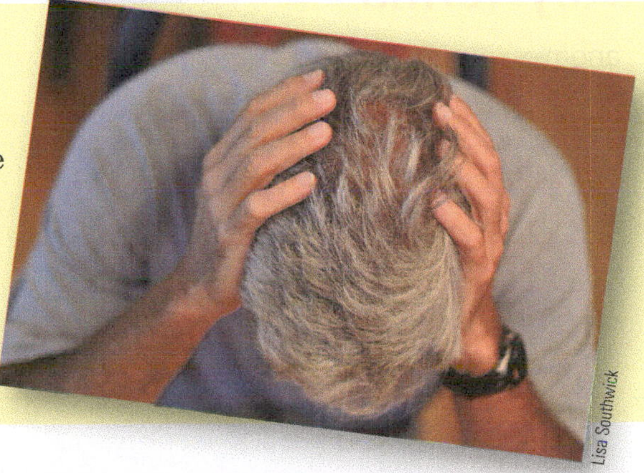
Lisa Southwick

Rebellion also involves the rejection of both the valued goals and the legitimate means of attaining them. Whereas retreatists simply give up, rebels seek a new set of goals and means of obtaining them. If enough people rebel, a great potential for revolution exists.

What Do Sociologists See?

No matter what you think of it, the Affordable Care Act is a response to structural strain between culturally valued goal of "free market system of health care" and the legitimate means to achieve it (e.g., securing insurance coverage or paying out of pocket). Under the "free market" system, many could not afford insurance or were denied coverage because of preexisting conditions. While free market principles still operate (as insurers compete for clients), there are now rules in place to protect those with preexisting conditions. President Obama and those who voted for the law rebelled against some elements of the free market system and rewrote the rules specifying who can legitimately access health insurance.

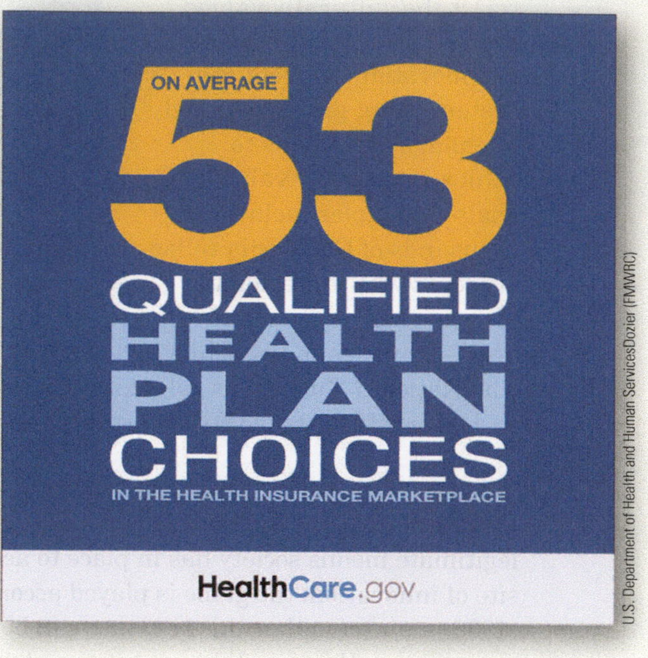

ON AVERAGE
53
QUALIFIED
HEALTH
PLAN
CHOICES
IN THE HEALTH INSURANCE MARKETPLACE
HealthCare.gov

U.S. Department of Health and Human ServicesDozier (FMWRC)

Critical Thinking

In the context of achieving the American dream, do you see yourself as a conformist, innovator, ritualist, retreatist, or rebel? Explain.

Key Terms

anomie
conformity
innovation
rebellion
retreatism
ritualism
structural strain

Module 6.7 Surveillance Society

Objective

You will learn how being watched—including the possibility that someone is watching—shapes behavior.

During the course of your day, how often does it occur to you that your activities are being monitored?

National Archives and Records Administration

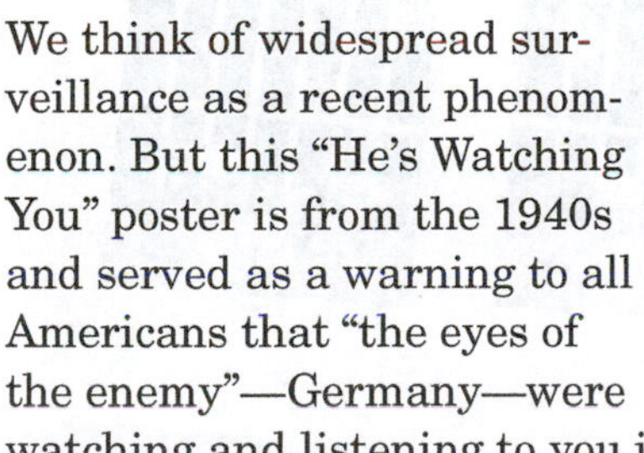

We think of widespread surveillance as a recent phenomenon. But this "He's Watching You" poster is from the 1940s and served as a warning to all Americans that "the eyes of the enemy"—Germany—were watching and listening to you in your work place. At that time the U.S. government was concerned that careless talk could pass critical information to enemy saboteurs and spies.

Most of us know that we are being watched at some point over the course of a typical day, if only by a store or ATM surveillance camera. What does it mean to know that behavior is being monitored at any time? How does that possibility shape behavior? These are the questions that interested French philosopher and social theorist Michel Foucault (1977). Foucault sought to identify the turning points that make the society we live in today fundamentally different in structure from the society that preceded it. In this regard Foucault identified a historical shift or turning point in the way society punishes people from what he called a culture of spectacle to a carceral culture.

A **culture of spectacle** is a social arrangement in which punishment for crimes—torture, disfigurement, dismemberment, and execution—is delivered in public settings for all to see.

Library of Congress Prints and Photographs Division [LC-USZ62-64582]

This very public way of punishing began to change in Europe and in the United States as part of the prison reform movements (1775–1889). This era ushered in what Foucault called a **carceral culture**, a social arrangement under which the society largely abandons physical and public punishment and replaces it with surveillance as the method of controlling people's activities and thoughts. Foucault attributes this change not to a rise in humanitarian concern but to a transformation in the technologies available to control others.

To understand the magnitude of this shift, consider that before what we call the prison reform movement in the United States and Europe, the death penalty was applied to any number of crimes, including murder, denying the existence of God, and homosexuality. Severe physical punishments were issued to those who committed less serious crimes. Prisons existed, but they were used to hold those awaiting trial, debtors, and sometimes even eyewitnesses to crime. Everyone, regardless of age, race, or sex, was crowded together in holding pens (Johnston 2009). From Foucault's point of view, prison reforms—and by extension reforms related to punishment—were connected to the need to establish discipline and order, not to achieve some moral objective.

Crowded prison quarters facilitated the spread of disease and increased opportunities for prisoners to conspire against the guards and plan an escape. This kind of holding pen also made it difficult to monitor relationships among prisoners.

Library of Congress Prints and Photographs Division[LC-USZ62-137901]

The Panopticon

The prison reform movement coincides with the period of history in which "a whole set of techniques and institutions emerged for measuring, supervising and correcting" those considered abnormal, including criminals (Johnston 2009, 401). The **panopticon** was designed by British philosopher Jeremy Bentham in 1785. *Pan* means a complete view and *optic* means seeing. The design represented his effort to create the most efficient and rational prison—the perfect prison. Foucault (1977) used it as a metaphor for the mentality driving 19th-century society.

The architectural plan of the panopticon included this signature feature: a guard tower positioned in the center of the facility inside a circular gallery of cells. The front of each cell was barred and the guard standing in the tower could see into each cell. The side and back walls of each cell were solid so that the prisoners could not see or interact with each other. This architectural design allowed just one guard to watch the inmates housed in hundreds of cells. Since the inmates could not see the guard in the central tower, they were never sure when or if they were being watched. The threat of surveillance pushed inmates to discipline themselves (Foucault 1977).

Tabitha Adams, Courtesy of Joan Ferrante

The panopticon is a metaphor for what Foucault calls the **disciplinary society**, a social arrangement that normalizes surveillance, making it expected and routine. The disciplinary society that Foucault wrote about in the mid-1970s has been further expanded by new technologies. Those technologies can monitor offenders, keep an eye on frail elderly people in their homes, follow teens as they drive, supervise workers, track Internet use, and survey public spaces (Felluga 2009).

What Do Sociologists See?

Sociologists see a diagram showing vision-enhancing contact lenses "allowing a wearer to view virtual and augmented reality images without the need for bulky apparatus. . . . [D]igital images are projected onto tiny full-color displays that are very near the eye . . . allowing users to focus simultaneously on objects that are close up and far away" (DARPA 2012). The technology adds to the pervasiveness of the disciplinary society—a society where surveillance is built into the very structure of society so that fear of being watched shapes virtually every aspect of our personal and social lives.

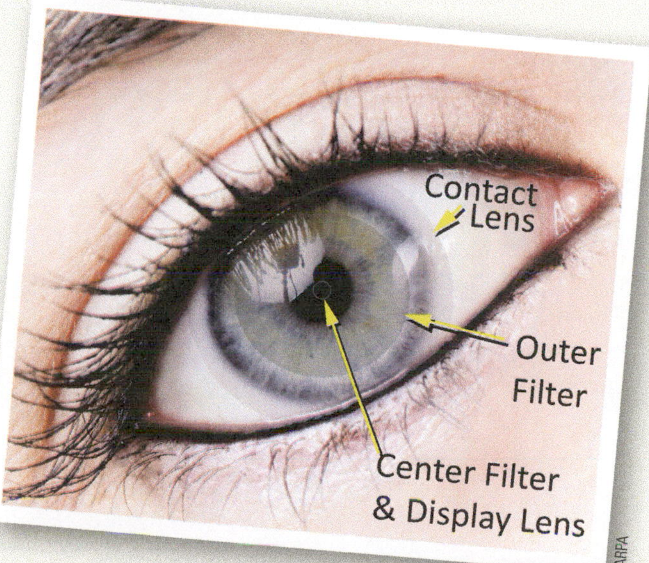

DARPA

Critical Thinking

Have you ever censored or disciplined yourself because of the possibility that someone was watching?

Key Terms

carceral culture

disciplinary society

culture of spectacle

panopticon

Module 6.8 Applying Theory: Laws

Objective

You will learn that people who violate laws are not always "criminals."

Have you ever considered raising chickens in your backyard so you could have access to fresh eggs?

NKU Anthropology, Sharyn Jones

Before you make having access to fresh eggs part of your lifestyle, it is important to check to see if there are any ordinances banning the raising of chickens in backyards and, if not, check to see if you need to get a permit, to ask permission of neighbors, limit the number of chickens, or provide certain kind of shelter (Smith 2012). The fact that there are laws controlling ownership of chickens suggests that crime is not a matter of good versus evil.

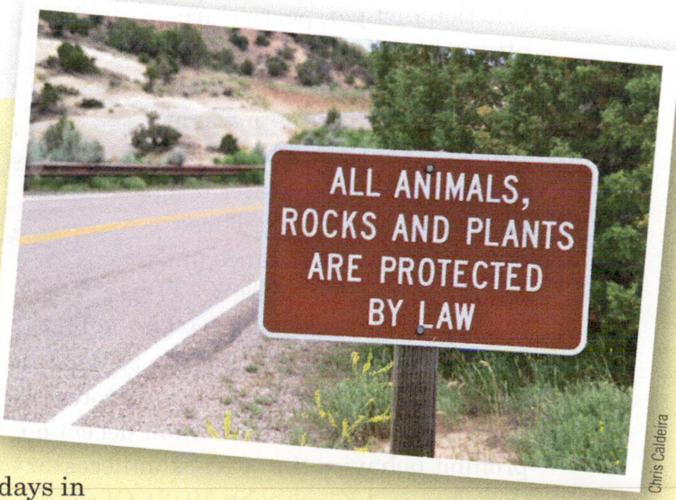
ALL ANIMALS,
ROCKS AND PLANTS
ARE PROTECTED
BY LAW
Chris Caldeira

It is impossible to list all the federal, state and local ordinances that are on the books. Regardless we can say that laws specify penalties including fines, time in prison or jail and hours of community service. In Nevada and other states there are laws that protect "all animals, rocks and plants" and violation of those laws can result in fines up to $10,000, 180 days in jail, and seizure and forfeiture of property used to violate the laws.

NKU Sociology, Jibril McCaster

In the course of your day, think about all the laws you see posted that just relate to cars. This sign reminds drivers that, unless there are traffic-control lights and signals telling pedestrians when to walk or stay put, drivers must yield to pedestrians crossing a roadway. When these crosswalks are on college campuses and students are walking to class, you might wonder why it is that seemingly endless strings of students are allowed to cross streets at will and hold up lines of cars coming onto campus.

NKU Sociology, Jibril McCaster

Have you ever wondered why some parking lots on campus are designated for faculty and staff and others for students? Has the thought ever crossed your mind that faculty and staff have the parking spaces that are closest to campus buildings and students have spots that are further away?

Chris Caldeira

Sociologists inspired by a functionalist perspective are attuned to the ways in which laws (and ordinances) contribute to order and stability. This sign warns residents that an ordinance is in place restricting water usage to 110 gallons per day. Those who violate the ordinance must pay $100, and if overuse does not cease after three fines, water will be turned off (Stinson Beach County Water District 2014). From a functionalist perspective, laws are in place to protect society. In fact, all societies appear to have laws in place that prohibit deception, prohibit acts of unjustified violence, prohibit behavior that causes unnecessary suffering, and restrict sexual activity (Kroeber and Kluckhohn 1952). Of course, groups have different ideas about what constitutes deception, unjustified violence, and so on. The point is that from a functionalist point of view laws exist to ensure order and stability.

Sociologists inspired by the conflict perspective focus on laws that protect the interests of some advantaged group at another group's expense. Tax laws and immigration laws represent legal areas where the advantaged benefit from laws that reduce taxes paid on capital gains or make immigration laws that give preference to immigrants from some countries (Canada) but not others (Cuba).

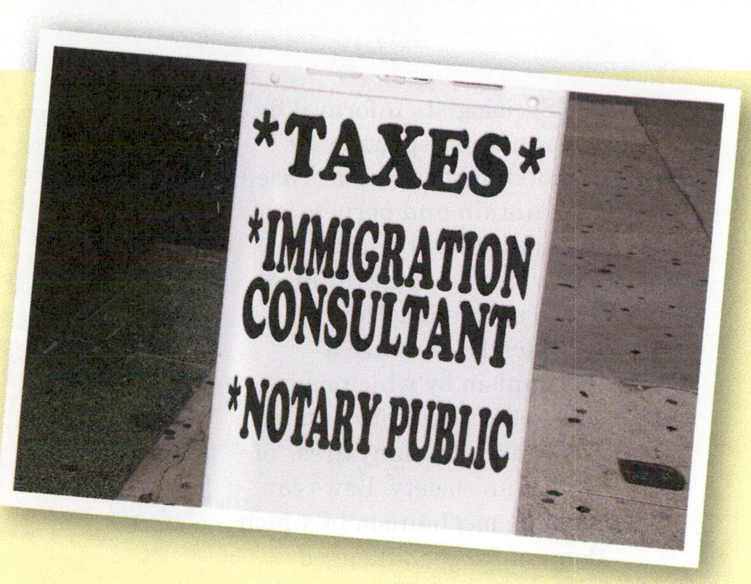

Chris Caldeira

A sociologist is informed by the symbolic interactionist perspective when focusing on the role of laws in shaping interaction and presenting the self. Consider that some local governments have passed ordinances requiring dog owners to have their pets on leashes. How do such laws shape interactions between owners and their dogs and between owners and people passing by? What message do laws send when owners can walk with dogs unleashed? Symbolic interactionists are also interested in acts of resisting laws and in actions taken to change laws or influence the way they are written, such as this sign announcing support for keeping an off-leash system in place.

Chris Caldeira

Chris Caldeira

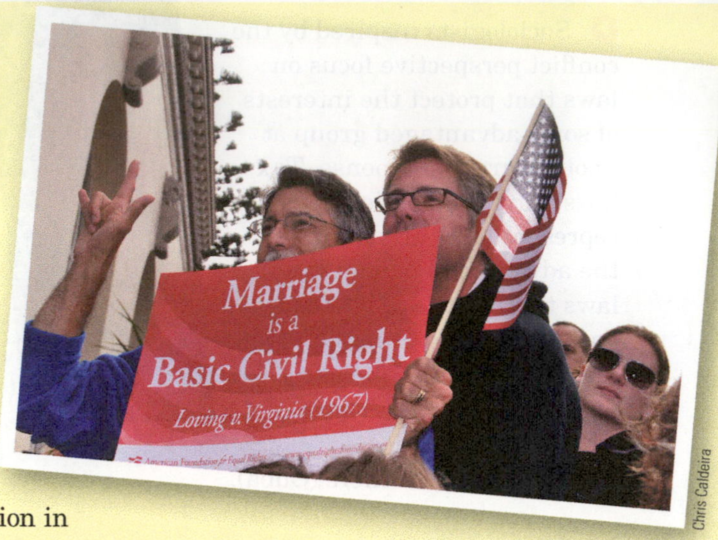

Sociologists informed by a feminist perspective are attuned to how laws are used to maintain and perpetuate inequalities or to mandate behavior and opportunities based on gender. More specifically, laws are a mechanism by which males and heterosexuality assume dominance in many areas of American society. Laws can also the mechanism by which dominance is challenged and corrected. The court decision in *Loving v. Virginia* struck down laws outlawing interracial marriage. Challenges to the Defense of Marriage Act have opened the door for same-sex couples to secure the rights of heterosexual couples.

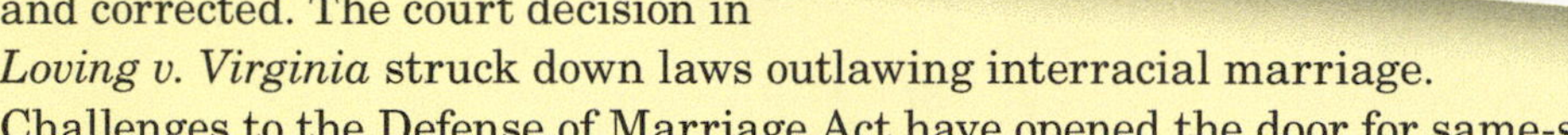

Summary: Putting It All Together

Chapter 6

Deviance is any behavior or physical appearance that is socially challenged and/or condemned because it departs from the norms and expectations of a group. The critical factor in defining deviance is not the behavior per se, but whether someone takes notice and, if so, whether the offender is "punished."

Ideally, from society's point of view, people should want to conform. When conformity cannot be achieved voluntarily, other mechanisms of social control are employed to enforce norms. Methods of social control include positive and negative sanctions, censorship, surveillance, authority, and group pressure.

The sociological contribution to understanding deviance lies not with studying deviant individuals per se, but with studying the context under which something is deemed deviant. To understand the context, sociologists employ labeling theory, structural strain, and stigma.

A crime is a subcategory of deviance that involves breaking a law. When sociologists study crimes, they ask: Who avoids detection? Who gets caught? Who goes to prison? From a global perspective, the United States sends the largest proportion of residents to prisons. Still, its prison population includes only a small portion of those people who have actually committed crimes. Prisons house those who did not have the resources, connections, or luck to avoid being detected, noticed, arrested, arraigned, convicted, and sentenced to prison. Possible explanations for the high rate of imprisonment include the propensity of elected officials to pass tough crime legislation, laws that mandate long prison sentences for drug possession and other nonviolent crimes, the prison-industrial complex, and the fact that many communities have come to depend on prisoners who do work at a low cost.

In addition to examining who gets caught, sociologists explore structural opportunities to commit crimes. The concept of differential association focuses our attention on how exposure to influences considered criminal presents people with opportunities to learn criminal behavior. Differential opportunity reminds us that just about any environment contains within it opportunities to commit crimes. Corporate and white-collar crimes are obvious examples of crimes supported by larger social structures that support such actions.

Chapter 7 Social Inequalities

Chris Caldeira

Have you seen headline news proclaiming the death of the middle class? Evidence that the middle class is, at the very least, declining comes from survey data showing a record 40 percent of Americans now identify as lower or lower middle class, up from 25 percent in 2008. Some populations are more likely than others to identify as such. Among those with some college, 47 percent now identify as lower or lower middle class (up from 24 percent in 2008) compared to 20 percent of college graduates who do so (up from 12 percent in 2008). Today 50 percent of young adults say they see themselves as lower to lower middle class (up from 25 percent) compared to 31 percent of adults age 65 and older (up from 21 percent in 2008). These perceptions reflect declines in median household income, which has fallen from $55,627 to $51,017 (in 2012 dollars) since 2008 (Kochhar and Morin 2014). In contrast, the top 1 and 5 percent of households have dramatically increased their economic standing relative to others (Hill 2013). In this chapter we explore the larger social forces behind the increasing shares of income and wealth being concentrated into the hands of a relatively small number of people and households.

Module 7.1

Assigning Social Worth

Objective

You will learn about the processes by which people are categorized and ranked on a scale of social worth and rewarded accordingly.

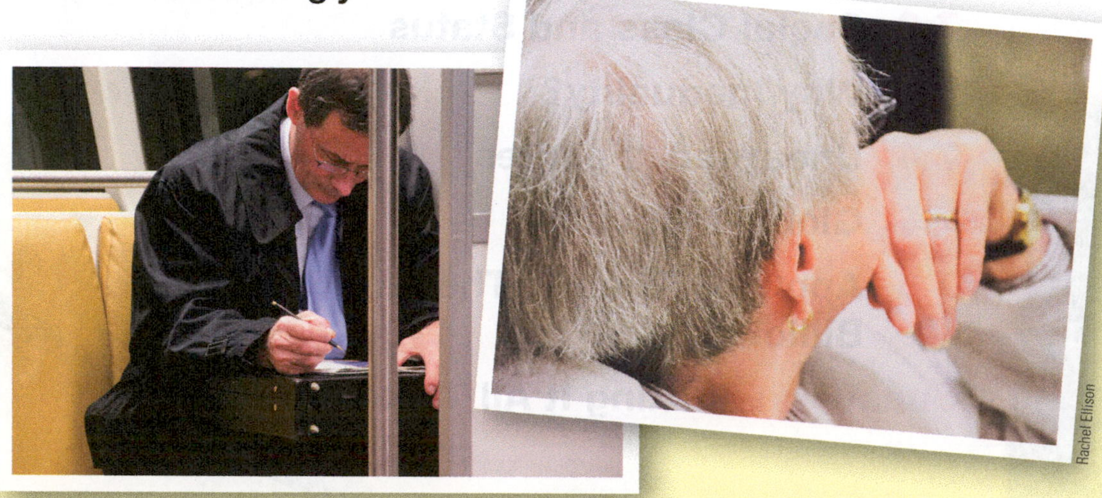
Rachel Ellison

Rachel Ellison

Imagine that you had to assign a dollar value between $250,000 and $8.6 million to each life pictured. Could you do it?

While this question might seem offensive, this is what administrators of a Victim's Compensation Fund did—they assigned a monetary value to each of the 2,800 people who died as a result of the September 11, 2001, attacks on the United States. By what criteria do you think the administrators awarded money? Those criteria speak to the issues of social stratification.

Social Stratification

When sociologists study inequality, they give attention to a society's system of **social stratification**, the systematic process of categorizing and ranking people on a scale of social worth such that one's ranking affects **life chances** in unequal ways. Sociologists define life chances as the probability that a person's life will follow a certain path and turn out a certain way. Life chances apply to virtually every aspect of life—the chances that someone will survive the first year after birth, complete high school, see a dentist twice a year, work while going to school, travel abroad, major in elementary education, own 50 or more pairs of shoes, and live a long life.

Every society in the world has a stratification system. Almost any criterion—eye color, hair texture, age, sexual preference, marital status, weight, occupation, income, test scores—can and has been used to rank people from most valued to least valued.

Which color eyes do you think is more valued, blue or brown? How do you know this? One way we learn which eye color is more "valuable" than another is through advertisements. A disproportionate percentage of models and celebrities have blue eyes (or light-colored eyes) rather than brown.

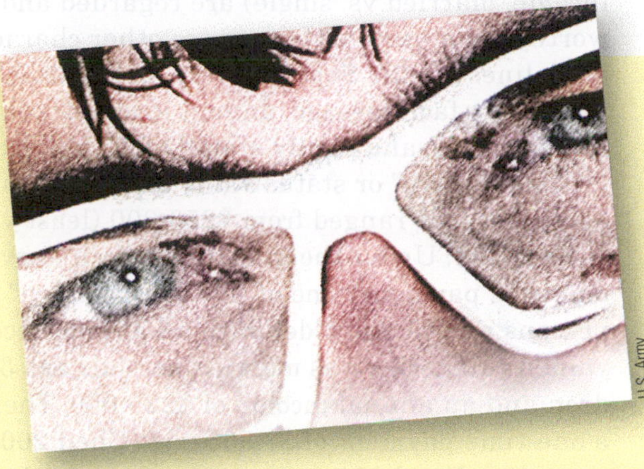
U.S. Army

People's status in society can be ascribed or achieved. **Ascribed statuses** are social positions assigned on the basis of attributes people possess through no fault of their own; these attributes may be inherited (such as skin shade, sex, or hair color), develop as a result of time (such as height or wrinkles), or be otherwise possessed through no personal effort (inherited wealth). **Achieved statuses** are attained through some combination of choice, effort, and ability. That is, people must act in some way to acquire an achieved status. Achieved statuses include earned wealth, income, occupation, and educational attainment.

The various achieved and ascribed statuses hold **social prestige**, a level of respect or admiration for a status apart from any person who happens to occupy it.

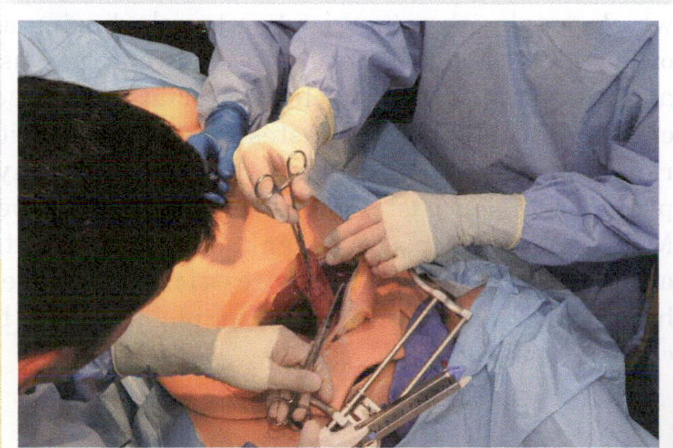
Sgt. Mark Cloutier

Rachel Ellison

The achieved statuses of a surgeon and a concession worker bring different levels of social prestige, not connected to any person, but connected to each position. The level of prestige can be enhanced by the organizational setting in which the surgeon operates (Johns Hopkins, a world-class hospital) or the stadium at which the concession worker sells snow cones (the Staples Center).

Sociologists are especially interested in the status value assigned to ascribed and achieved characteristics. **Status value** is the social worth of a status relative to others such that people who possess one characteristic (white skin vs. brown skin, blond hair vs. dark hair, high income vs. low income, married vs. single) are regarded and treated as more valuable or worthy than people who possess other characteristics. The compensation guidelines for the September 11, 2001, attacks on the United States show how many factors—age, annual income, occupation, marital status, potential earnings, and family status (with children versus childless)—affected victims' "worth" or status value and the money awarded to survivors. The actual awards ranged from $250,000 (least valued life) to $8.6 million (most valued life). Under the guidelines, their survivors were to be awarded a one-time payment. One of the "least valued" categories was single, childless persons age 65 and older with an annual income of $10,000. One of the more valued categories was married persons age 30 and younger with two children and an annual income of $225,000. Their survivors were to be awarded a one-time payment of $3,805,087 (Chen 2004; September 11 Victim Compensation Fund 2001). Some survivors of the most highly paid executives resisted settlement on the grounds that their loved ones were "worth" more than the $8.6 million cap.

Stratification systems fall somewhere on a continuum between two extremes: a **caste system**, in which people are ranked according to ascribed statuses, and a **class system**, in which people are ranked on the basis of their achievements related to merit, talent, ability, or past performance.

Caste Systems

Sociologists use the term *caste* to refer to any form of stratification in which people are categorized and ranked using ascribed characteristics over which they have no control and that they usually cannot change. In a caste system there is a clear association between caste rank and life chances. People in lower castes are seen and portrayed as innately inferior in intelligence, morality, ambition, and many other traits. Conversely, people in higher castes consider themselves to be superior in such traits. Moreover, a person's caste rank is treated as if it is absolute and unalterable. Finally, there are heavy restrictions on interactions between people in higher and lower castes. For example, marriage between people of different castes is forbidden.

Library of Congress Prints and Photographs Division [LC-USF34-052508-D]

Between 1876 and 1965 racial segregation laws supported a caste system of stratification. People considered white and nonwhite used separate, but so-called equal, public facilities including public schools, restrooms, theaters, restaurants, hospitals, and drinking fountains.

Class Systems

While class systems contain inequality, that inequality is in theory based on differences in talent, ability, and past performance, not on ascribed characteristics such as skin shade or sex. In true class systems, people assume that they can achieve a desired education, income, or other outcome through personal effort. Furthermore, people can raise their class position during their own lifetime, and their children's class position can differ from (and ideally be higher than) their own. Movement from one social class to another is termed **social mobility**.

Staff Sgt. Patrick N. Moes

American Idol winner (season 4) Carrie Underwood represents the ideals of a class system. She experienced *inter*generational upward (vertical) mobility as her father worked at a sawmill and her mother taught at an elementary school. She also experienced *intra*generational upward mobility. Before winning *Idol*, Underwood worked at odd jobs and was three credit hours from completing a college degree (which she eventually completed in 2007). As winner of *American Idol*, Underwood received a million-dollar recording contract, a car, and use of a private plane (Associated Press 2006). According to *Forbes* magazine, Underwood earned $35 million in 2013 (Harrison 2013).

Many Americans believe the United States has a class system and that it is possible to move from rags to riches through individual merit and hard work alone. However, as the middle class declines in size relative to upper and lower classes, many are questioning whether merit and hard work are enough to achieve success and point to other factors such as those listed in Table 7.1a.

Table 7.1a: Selected List of External Factors Affecting Probability of Mobility, Upward and Downward

There is no question that personal qualities and many external factors shape the probability a person will experience success or failure. Of course, external factors cultivate personal qualities, and personal qualities have a bearing on the kinds of external factors that lead to success. Study the list of factors. Which three factors have played the largest roles in your "successes" and "failures"?

Personal Qualities
Work ethic (strong to poor)
Attitude (good vs. bad)
Level of integrity
Ability/inability to delay gratification
Guiding principles (integrity/honesty vs. lack of integrity/dishonesty)
External Forces
Inherited socioeconomic position
Inheritance (wealth passed on)
Can count on parental rescue/family safety net in difficult times
Social networks (who you know)
Location (residence) in the world (e.g., rural, urban, suburban, country)
Unearned job loss resulting from automation, downsizing, outsourcing, business closings, or disability from repetitive and strenuous labor
Discrimination (positive or negative) based on . . . age, race, sexual orientation, gender, disability, weight, and perceived attractiveness
Money earned from profit on labor of others (for example, from stocks)

Social Mobility in the United States

When thinking about social mobility, sociologists consider the factors that predict mobility—both upward and downward. Specifically, sociologists seek to identify the factors that predict intragenerational mobility—upward or downward mobility over the course of a person's lifetime—and those that predict intergenerational mobility—children's upward or downward mobility relative to their parents' status. Many Americans believe that they live in a country where it is possible to move from rags to riches. But how many people born into poor households become wealthy as adults? Data to answer this question are hard to come by, if only because it is difficult to measure status in society. This is because status is based on more than income—it is also based on occupation, education, wealth, and much more. Still, most researchers use increases or decreases in income large enough to place someone in a lower or higher income bracket as measures of upward or downward mobility.

The most current data on intergenerational mobility comes from the Panel Study of Income Dynamics, a longitudinal survey that began in 1968 and continues to the present and that follows a representative sample of 5,000 households with 18,000 members and their descendants. Researchers have used this data to compare the family income of native-born adults who were in their late 30s and early 40s with that of their parents when they were in that same age range. The study found that the probability of moving from the lowest to the highest or from the highest to the lowest income categories is actually quite low (see Figure 7.1a).

The mobility study also showed that there is considerable mobility in the United States, as 61 percent of children from the top 20 percent of household incomes dropped to a lower household income category. Among those children who grew

Figure 7.1a: Probability of Children Staying in and Moving Out of the Top 20 and Bottom 20 Percent of Household Incomes As Adults

Probability of . . .

staying in lowest income category = 42%
moving from bottom 20 percent to top 20 percent = 9%
moving up from bottom 20 percent to a higher income category = 58%
moving up one income category from the lowest 20 percent = 25%

Probability of . . .

staying in highest income category = 39%
moving from top 20 percent to bottom 20 percent = 6%
moving down from top 20 percent to a lower income category = 61%
moving down one income category from the highest 20 percent = 26%

Source of data: Isaacs, 2011

up in households with the lowest income category (bottom 20 percent), 58 percent as adults lived in households in a higher income category. Almost 9 percent of those who lived in the lowest income households as children went on to live in a household in the top 20 percent as adults. Taken together these probabilities show that parents' income is a significant factor in predicting intergenerational mobility. If parents' income had *no* bearing on mobility, the probability a child would be in any one of the five income categories would be 20 percent. That is, 20 percent of those adult children raised in the most advantaged households would be in each of the five income categories.

It should come as no surprise that children from households in the top 20 percent received more education and were healthier as adults, factors that surely contributed to helping them maintain their economic status. Children classified as white are advantaged over children classified as black with regard to financial mobility. Specifically, 63 percent of black children who grow up in lowest-income households (bottom 20 percent) retain that status as adults, compared to 32.3 percent of the poorest whites. Black children from the lowest-income households have a 3.6 percent probability of moving into the highest 20 percent income category as adults, compared to 14.2 percent for poor white children.

While it is clear from the data that there is a relationship between parents' and their children's economic status as adults, it is also clear that parents do not simply pass on their economic advantages or disadvantages to their children. If that were the case, 100 percent of children from each income group would live in households of the same income group as their parents.

What Do Sociologists See?

Sociologists think about the prestige associated with the occupation of firefighter and note that it is clearly derived from their life-saving role and expected bravery in the face of danger. That level of prestige is complicated when we learn that a firefighter is a woman. Many immediately question whether she is qualified physically to do the job. Of course, news stories remind us that women can do this work, as evidenced by a female firefighter who ran up 1,316 stairs in Seattle's Columbia Tower for a fundraising event. From her point of view this was something routine for any firefighter, not extraordinary (Grindeland 2014).

Critical Thinking

Assign a monetary value to your life. Explain your rationale. Do not try to deflect this question by saying it's impossible to put a price tag on anyone's life. Rather, think about who in your life would be affected if you were to die and your value to that person.

Key Terms

achieved statuses
ascribed statuses
caste system
class system
life chances
social mobility
social prestige
social stratification
status value

Module 7.2

Social Class and Status

Objective

You will learn about two theoretical traditions in sociology that give meaning and significance to the concept of social class.

Imagine that the only way you could make calls was by using a pay phone. What would it convey about your class position or social status?

Chris Caldeira

The phones we use are **status symbols**, visible markers of economic position and social rank. Sociologists use the term **social class** to designate a person's overall economic and social status in a system of social stratification. A person's social class is difficult to determine because it depends on many factors, including occupation, sources of income, marketable abilities, access to consumer goods and services, and group and organizational memberships. In our exploration of the concept of class, we begin with the writings of Karl Marx and Max Weber, who represent the "two most important traditions of class analysis in sociological theory" (Wright 2004, 1).

Karl Marx and Social Class

In *The Communist Manifesto*, co-written with Friedrich Engels and published in 1848, Marx observed that the rise of factories and mechanization created a fundamental class divide between those who owned the means of production (the bourgeoisie) and the largely propertyless workers (the proletariat) who sell their labor to the bourgeoisie. For Marx, then, the key variable in determining social class is source of income.

In *Das Kapital*, Marx names three classes: wage laborers, capitalists, and landowners. Each class is comprised of people whose incomes "flow from the same common sources" (p. 1032, 1887). For wage laborers, the source is wages; for capitalists, the source is profit from the labor of others; for landowners, the source is ground rent. In the *Class Struggles in France 1848–1850*, Marx (1865) named a fourth class, the **finance aristocracy**, who lived in obvious luxury among masses of starving, low-paid, and unemployed workers. The finance aristocracy includes bankers and stockholders seemingly detached from the world of work. Marx described the finance aristocracy's source of income as "created from nothing —without labor and without creating a product or service to sell in exchange for wealth." The finance aristocracy speculates and, "while speculation has this power of inventiveness, it is at the same time also a gamble and a search for the 'easy life'; as such it is the art of getting rich without work." According to Marx, the financial aristocracy gains income "without giving anything . . . in exchange; it is the cancer of production, the plague of society and of states" (Bologna 2008; Marx 1856; Proudhon 1847).

Rachel Ellison

When we break down sources of income for those at the very top of the income scale—the top 1 percent—33 percent derives from salary and 66 percent from nonlabor sources, most notably investment income or capital gains, which are taxed at a rate of 15 percent in the United States (Frank 2013).

Max Weber and Social Class

Karl Marx clearly states that a person's social class is based on sources of income. Max Weber (1947) defined a social class as being composed of those who hold similar life chances, determined not just by income but also by marketable abilities (work experience and qualifications), access to consumer goods and services, and ability to generate investment income.

Weber called those in the very top social class the **"positively privileged" property class**; they monopolize the purchase of the highest-priced consumer goods, have access to the most socially advantageous kinds of education, occupy the highest-paying positions, and live on income from property and other investments.

Weber labeled those in very lowest social class the **"negatively privileged" property class**. This class is completely lacking in skills, property, or employment and depends on seasonal or sporadic employment.

Chris Caldeira

Between the top and the bottom of this social-status ladder is a series of rungs. Weber (1948) argued that a "uniform class situation prevails only among the negatively privileged property class." We cannot describe a lifestyle common to each of the other classes because the status groups to which people belong complicate matters. People with similar income and wealth can belong to very different status groups. Weber defines a **status group** as consisting of people held together by a lifestyle "expected of all those who wish to belong to the circle" and by the level of social esteem accorded to those circles (Weber 1948, 187). People in the same social class are also accorded different levels of social esteem depending on the status groups to which they belong, a particular talent, their race, gender, and so on.

Status groups are held together by a lifestyle that is expected of all who belong. What lifestyle characteristics might bind together the people who live in the housing complex pictured?

Mr. Justin Matthew Ward (USACE)

Power adds another dimension to any analysis of social class position. Weber defined **power** as the probability that someone can exercise his or her will in the face of resistance. Power derives in part from **political parties**, organizations established to secure and maintain the level of power needed to influence others. Parties are organized to represent the interests of their members. The means of obtaining power include exerting force, engaging in nonviolent protest, securing votes, bribery, investing money, and running media campaigns. The list of political parties is endless. Some examples include NOW (National Organization for Women), the NRA (National Rifle Association), and AARP (American Association of Retired Persons). Weber also recognized that people derive power from positions they hold (e.g., CEO, president) that give them the authority to rule and influence others.

Distribution of Wealth in the United States

Wealth refers to the combined value of a person's income and other material assets such as stocks, real estate, and savings minus debt. In the United States, the share of net wealth held by the top 10 percent of households is 75 percent (the equivalent of $43.6 trillion); the top 1 percent control 34.5 percent of all wealth. The bottom 50 percent of households hold just 1.1 percent of net wealth (Levine 2012a). If we consider the average household wealth by race/ethnicity, the median net worth for households classified as white is $110,729 versus $69,500 for Asian, $7,240 for Hispanics, and $4,995 for black households (Luhby 2012).

Researchers Michael Norton and Dan Ariely (2011) asked a sample of 5,522 U.S. respondents a series of questions about how wealth is and should be distributed. Before answering questions, respondents were given a definition of wealth—"net worth or the total value of everything someone owns minus any debt that he or she owes. A person's net worth includes his or her bank account savings plus the value of other things such as property, stocks, bonds, art collections, etc., minus the value of things like loans or mortgages."

Respondents were then shown three hypothetical distributions of wealth and asked to choose under which one they would like to live (see Figure 7.2a). In hypothetical distribution A, the richest 20 percent of households have 36 percent of all wealth; the poorest 20 percent have 11 percent. In hypothetical distribution B, the wealthiest 20 percent of households have 84 percent of all wealth, and the bottom 20 percent have 0.1 percent. In hypothetical distribution C, the wealth is divided evenly among all five household groups. Very few respondents chose hypothetical distribution B, and virtually no one chose C. Almost everyone in the study chose distribution A.

Figure 7.2a: Three Hypothetical Distributions of Wealth

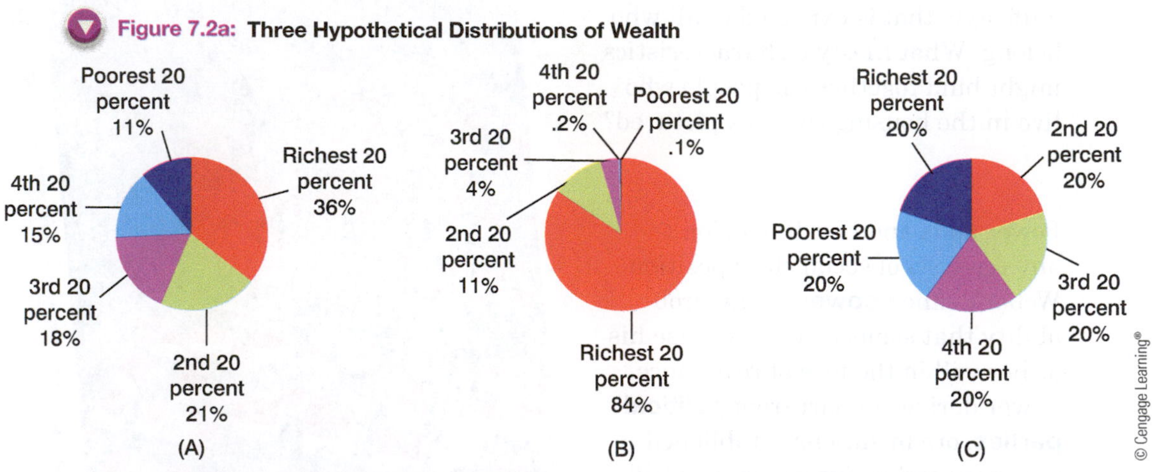

Distribution of Income in the United States

Income refers to the money a person earns, usually on an annual basis through salary or wages. Within the United States, the average after-tax income of the richest 20 percent is $198,300, or 11.2 times that of the poorest 20 percent, who average $17,700. That is, for every $1,000 of taxed income earned by the poorest one-fifth, the top one-fifth earns $11,200. When we compare the after-tax income of the top 1 percent with that of the bottom 20 percent, the inequality is even greater. That 1 percent's after-tax income is $1.32 million, or 74.5 times greater than the bottom 20 percent (see Table 7.2a).

 **Table 7.2a: Distribution of Household Pre-Tax Money Income by Quintile, 2011**

In 2011, the last year from which we have data, there were a total of 121,084,000 U.S. households with income. If we divide them into five groups or quintiles, there are 24,217,000 households in the lowest 20 percent income category as well as in each of the other four. The table also includes data on the 6,054,000 households in the top 5 percent.

	Quintiles					
	Lowest	Second	Third	Fourth	Highest	Top 5%
Number of households	24,217,000	24,217,000	24,217,000	24,217,000	24,217,000	6,054,000
Range of income within each quintile	$20,262 or less	$20,263 to $38,520	$38,521 to $62,434	$62,435 to $101,582	$101,583 or more	$186,000 or more
Share of household income (%)	3.2	8.4	14.3	23.0	51.5	22.3

Source: U.S. Census Bureau, 2011, Annual Social and Economic Supplement to the Current Population Survey

As shown in Table 7.2a, the top 5 percent earned 22.3 percent of total pre-tax income; the top 20 percent of households earned 51.5 of that total. As you can see, the lowest 20 percent earned 3.2 percent of all household income. Of course, the data in the table only offer a broad understanding of inequality as some households vary by size (Levine 2012a).

What Do Sociologists See?

Sociologists see an advertisement for diamonds, a status symbol associated with wealth, celebrity, royalty, and love. The size and cut of diamonds allow people looking at a ring to roughly estimate the ring's cost and to make a guess about the wearer's economic status or social class.

NKU Sociology, Boni Li

Critical Thinking

What is your social class? In answering this question use the language of Marx and Weber.

Key Terms

finance aristocracy
income
negatively privileged property class
political parties
positively privileged property class
power
social class
status group
status symbols
wealth

Module 7.3 Why Inequality?

Objective

You will learn four sociological perspectives on social inequality.

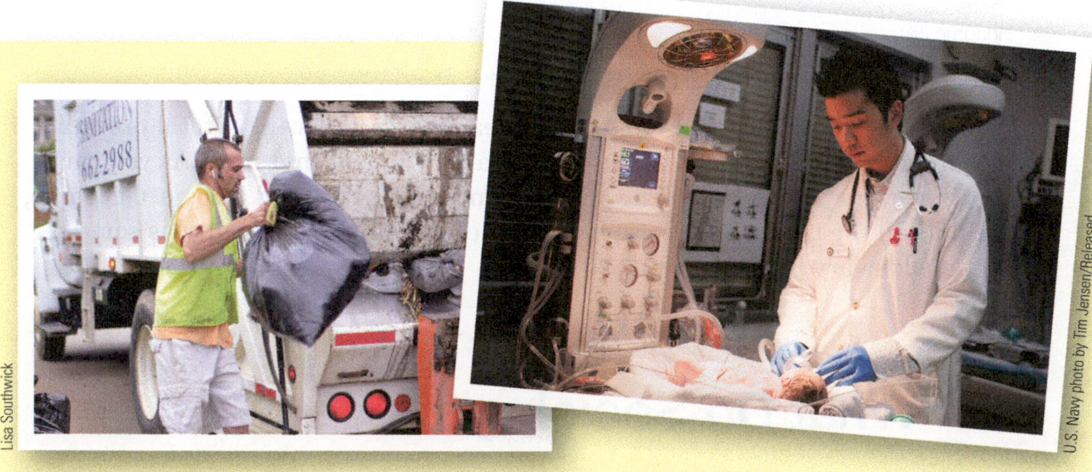

Lisa Southwick

U.S. Navy photo by Tim Jensen/Released

Which occupation is more important to society—physician or sanitation worker?

Social inequality is the unequal distribution of income, wealth, and other valued resources. Sociologists draw upon four perspectives to explain inequality and to understand how it is manifested in daily life. Those perspectives are functionalism, conflict, symbolic interaction, and feminism.

Functionalist Perspective

Sociologists Kingsley Davis and Wilbert Moore (1945) argue that social inequality is the device by which societies ensure that the best-qualified people fill those occupations considered to be the most functionally important. From a functionalist perspective, that is the reason sanitation workers in the United States earn an average of $34,150 per year and physicians earn an average of $190,060 per year. The $155,910 difference in average salary represents the greater functional importance of the physician relative to the sanitation worker (U.S. Bureau of Labor Statistics 2013a, 2013b).

Davis and Moore argue that society need not offer extra incentives to entice people to work as garbage collectors because that job requires few skills and little training. Society does have to offer extra incentives to entice the most talented people to undergo the long, arduous training to become skilled physicians.

Davis and Moore concede that the stratification system's ability to attract the most talented and qualified people is weakened when

- capable people are overlooked,
- some categories of people are deemed ineligible to apply, and
- factors other than qualifications are involved in filling positions.

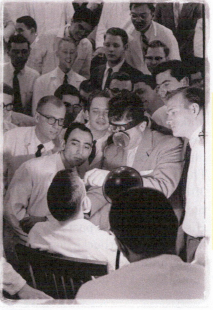

This 1958 photo of all-male medical students at George Washington University Medical School shows that, for the most part, capable women and nonwhites were denied access to medical schools at that time. As a result, few women and minorities applied to medical school and all but a few faced rejection.

Library of Congress Prints and Photographs Division [LC-DIG-ppmsca-03097]

According to Davis and Moore's theory, these kinds of systematic discrimination weaken a stratification system's ability to attract the most talented and qualified people. Davis and Moore argue, however, that society eventually corrects the problem, as evidenced by the fact that medical schools eventually admitted people from categories once denied access.

Conflict Perspective

Melvin M. Tumin (1953) and Richard L. Simpson (1956) challenge the functionalist assumption that social inequality is a necessary device societies use to attract the best-qualified people for functionally important occupations. These sociologists point out that some positions command large salaries even though their functional importance is questionable. Consider the salaries of Division I college basketball coaches; the median salary is around $1.1 million. The pay of the 68 coaches whose teams qualified for the NCAA tournament in March 2014 ranged from a high of $9.6 million to a low of $115,000 (*USA Today* 2014a). We might argue that Division I basketball coaches deserve large, in some cases enormous, salaries because they generate income for the university and surrounding communities (Tyler 2011). But is the coach's functional importance more essential than that of the players, most of whom are only compensated with tuition, housing, and meals? Should we assume that coaches would refuse to coach if salaries were lower? Moreover, a close look at the data shows that only about 10 percent of Division I schools generate enough revenue to cover expenses (Wilson et al. 2011).

In further critiquing the functionalist perspective on inequality, Tumin and Simpson ask why some employees earn a lower salary than others for doing the same job, just because they are of a certain race, age, sex, or national origin. After all, the employees are performing the same job, so functional importance is not the issue. For example, why do women working full time as registered nurses in the United States earn a median weekly wage of $1,085 while their male counterparts earn $1,189? In fact, in only 2 of 300 occupational categories do the median weekly earnings for females exceed those of their male counterparts: counselors ($855 vs. $833) and health practitioner support technologists and technicians ($621 vs. $599) (U.S. Bureau of Labor Statistics 2013b).

In addition to the issue of pay equity, we must raise the question of comparable worth. Assuming comparable worth, why should full-time workers at a child day care center, who are overwhelmingly female, earn a median weekly salary of $21,310 while auto technicians who are overwhelmingly male earn $39,460 per year (U.S. Bureau of Labor Statistics 2013b)?

Conflict theorists also ask if dramatic differences in pay are really necessary to make sure that someone takes the job of CEO over, say, the job of a factory worker. Probably not. In the United States, the median base salary of the CEO of the largest corporations (those with $1 billion or more in revenues) is $15.1 million. This median base salary (excluding bonuses, stock options, and other benefits) is 301 times the median household income ($50,054). Consider that the compensation package for the CEO of Starbucks for fiscal year 2012 was $41.99 million. His labor over the past five years was rewarded with $200.9 million. In addition, the CEO of Starbucks holds $1.1 billion in company stock (*Forbes* 2013). Notwithstanding this CEO's skills, conflict theorists would argue that the financial success of Starbucks can be largely attributed to the fact that the 95,000 baristas who work for the company earn between $7.50 and $10.00 an hour.

NKU Sociology, Missy Gish

When you drink a hot or cold beverage from Starbucks or another quick-service establishment, do you ever think about the wages the people who served you make relative to the CEO or others at the highest levels of management? If we assume that Starbucks' CEO puts in a 60-hour workweek, he earns the equivalent of $13,458 an hour, or an hourly wage 1,583 times that of the barista. Note that Starbucks does offer baristas who work at least 20 hours a week, a health benefits package, and 62 percent of them enroll (CNN Money 2013; Starbucks 2013).

Finally, Tumin (1953) and Simpson (1956) further criticize the functionalist position by arguing that specialization and interdependence make every occupational category necessary. Thus, to judge that physicians are functionally more important than sanitation workers is to not consider the historical importance of sanitation relative to medicine. Contrary to popular belief, advances in medical technology had little influence on death rates until the turn of the 20th century—well after improvements in nutrition and sanitation had caused dramatic decreases in deaths due to infectious diseases.

Symbolic Interactionist Perspective

When symbolic interactionists study social inequality, they seek to understand the experience of social inequality; specifically, they seek to understand how social inequality shapes interactions and experiences. In the tradition of symbolic interaction, journalist Barbara Ehrenreich (2001) studied inequality in everyday life as it is experienced by those working jobs that pay $8.00 or less per hour. Ehrenreich, a "white woman with unaccented English" and a professional writer with a Ph.D. in biology, decided to visit a world that many others, as many as 30 percent of the workforce, "inhabit full-time, often for most of their lives" (6). Her aim was just to see if she could "match income to expenses, as the truly poor attempt to do every day" (6). In the process, Ehrenreich worked as a "waitress, a cleaning person, a nursing home aid, or a retail clerk" (9). Ehrenreich learned that "low-wage workers are no more homogeneous in personality or ability than people who write for a living, and no less funny or bright. Anyone in the educated classes who thinks otherwise ought to broaden their circle of friends" (8).

Ehrenreich's on-the-job observations reveal the many ways inequality is experienced. Ehrenreich tells of a colleague who becomes "frantic about a painfully impacted wisdom tooth and keeps making calls from our houses (we are cleaning) to try and locate a source of free dental care" (80). She tells of a colleague who would like to change jobs, but the act of changing jobs means "a week or possibly more without a paycheck" (136); and then there is the colleague making $7.00 per hour at K-Mart thinking about trying for a $9.00-per-hour job at a plastics factory (79).

Among the low-wage workers to whom Ehrenreich's observations apply are the hundreds of thousands of workers who assemble sandwiches and other quick-service foods.

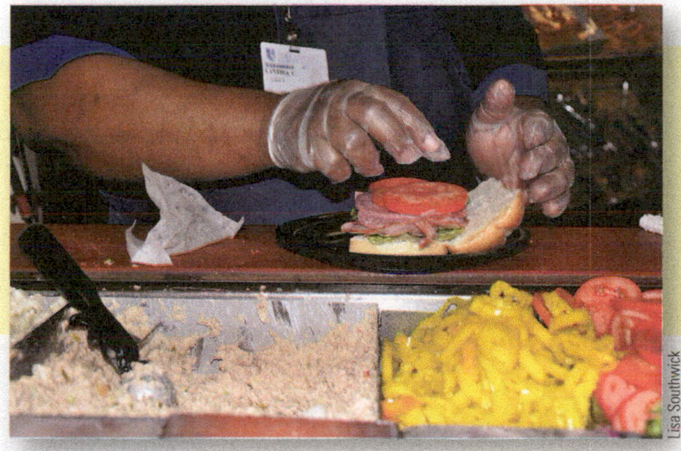
Lisa Southwick

Feminist Perspective

Feminists look at how the various categories people occupy affect their chances of experiencing inequality, whether it takes the form of rejection, ridicule, marginalization, prejudice, discrimination, or low wages for hard work. In this regard, the concept of intersectionality is a particularly useful concept. **Intersectionality** focuses our attention on the interconnections among the various categories people occupy, including race, class, gender, sexual orientation, religion, ethnicity, age (generation), nationality, and disability status. The sociologist most associated with this concept is Patricia Hill Collins. Collins argues that the various categories people occupy are interlocking and when taken together they "cultivate profound differences in our personal biography" and experiences with others as we move through the world (2000a, 460).

Chris Caldeira

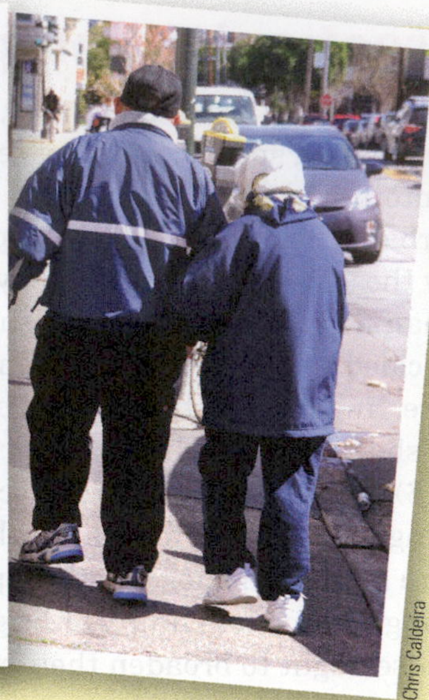
Chris Caldeira

To understand Collins' ideas about intersectionality, study the people in each of the two photographs. Notice that at a glance we "see" people as a certain race, gender, sexual orientation, age, weight, AND class. We do not just see people as a particular race or a gender, for example. We see them as a race AND gender together. Can you see how considering all the categories a person occupies complicates the meaning of any one category a person holds?

The overall effects of each of our multiple categories on how people see us and treat us cannot simply be added together to obtain some grand effect—their effects can only be understood as a gestalt, that is, as something more than and different from the sum of the parts.

Collins maintains that people derive varying amounts of penalty and privilege from positions they hold in the social structures of which they are part. **Penalties** include constraints on a person's opportunities and choices, as well as the price paid for departing from expectations associated with the categories we occupy. That price may be rejection, ridicule, or even death. A **privilege** is a special, often unearned, advantage or opportunity. We know some categories of people—whites, but especially white males—enjoy unearned privileges.

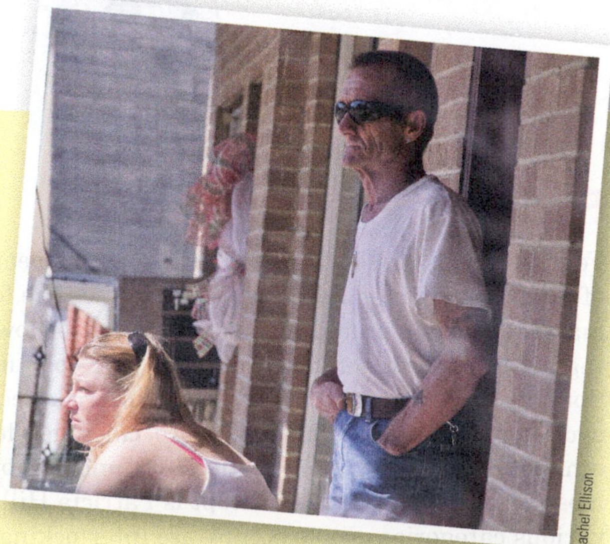
Rachel Ellison

But it is simplistic to think that any "man" or "white" or any "white man" is advantaged. While statistics show that these race and gender categories as a whole are advantaged over other gender and race categories, we can find many contexts in which "whites" and "men" hold disadvantaged statuses. Some of the poorest counties in the United States consist of populations considered entirely white. In such settings, how can we talk of white as an advantaged status?

Depending on the setting, social structures empower and privilege some categories of people while disempowering other categories. We can gain some insights by asking these questions:

- Can you think of times that you have felt penalized by one or more categories to which you belong or have been assigned?
- Can you think of times you felt empowered by one or more of those categories?
- Have you ever resisted being labeled as belonging to a category?
- Have you ever taken pride in a category to which you belong or felt superior to someone in another category?
- How are each of the various categories represented in the media? Do those representations make you proud or defensive?

What Do Sociologists See?

NkU Sociology, Boni Li

Sociologists see an advertisement that is clearly in a setting with a Chinese population, as evidenced by the language under the product name. Yet the model in the advertisement represents an ideal of beauty that is associated with another racial category. This advertisement privileges the category "white" as the ideal of beauty marginalizing those in other racial category, in this case Asian.

Critical Thinking

Use the concept of intersectionality to write about the ascribed characteristics by which others know/perceive you.

Key Terms

intersectionality

penalties

privilege

social inequalities

Module 7.4

Unearned "Failures"

Objective

You will learn that "failures" in life are often unearned.

What does it mean for a rural community when a school closes because the community cannot generate enough tax revenue to keep it open?

Rachel Ellison

If you lived in a community where the school closed, would you move to a new community with better schools or send your child to the closest school which is underfunded and a 30-minute drive away? What if you could not afford to move? Generally speaking, when a community cannot generate enough tax revenue to keep its schools open it is because the community has suffered an economic blow such as a plant closing. Those whose jobs disappeared, even as they worked hard each day, became unemployed through no fault of their own. In the sections that follow we consider how unearned failure can be traced to economic restructuring, the forces of creative destruction, turbulent unpredictability and economic dependence on low-wage labor.

Economic Restructuring

What happens to people when a factory or business closes? How does this loss affect the surrounding communities where the employees worked and where businesses once thrived serving them? The automobile industry and its many restructurings is a case in point. Detroit, once known as the "motor city," was the capital of the automobile industry. The Detroit plants began to close in 1957, laying off 130,000 autoworkers by 1967, as car companies restructured their operations, relocating plants to the suburbs and automating production facilities (Sugrue 2007). A second wave of restructuring and layoffs began in the 1970s and continues through today, as Detroit automakers have steadily moved operations to overseas locations, automated production lines, and downsized their operations in the United States. Recently GM closed 16 of 47 operating plants, laid off 23,000 production and 10,000 white-collar workers, and shut down 50 percent of its 6,200 dealerships (Goldstein 2009).

This kind of restructuring and downsizing was on the mind of sociologist William Julius Wilson when he wrote *The Truly Disadvantaged in the United States*. Wilson (1987) describes how structural changes in the U.S. economy going back to the 1970s helped create what he termed the "ghetto poor," now referred to as the or **urban underclass**—diverse groups of families and individuals living in the inner city who are "outside the mainstream of the American occupational system and consequently represent the very bottom of the economic hierarchy" (Wilson 1983, 80). Those economic transformations include:

- the restructuring of the American economy from manufacturing based to service and information based;
- the rise of a labor surplus marked by the entry of women and the large "baby boom" segment of the population into the labor market;
- a massive exodus of jobs from the cities to the suburbs;
- the transfer of manufacturing jobs out of the United States; and
- the transfer of customer service and knowledge jobs out of the United States over the past decade.

These job losses, along with an out-migration of stably employed working-class and middle-class families attracted by new housing opportunities in the suburbs, profoundly affected the daily life of people left behind in the inner cities. The single most significant consequence of these economic transformations was the disruption to "the networks of occupational contacts that are so crucial in moving individuals into and up job chains." Inner-city residents now knew fewer "parents, friends, and acquaintances" who were stably employed and could serve as bridges to employment opportunities "by telling them about a possible opening" and coaching them about how to apply for jobs and retain them (Wacquant 1989, 515–516).

Rachel Ellison

A restaurant in a rural town closes down because it loses its customer base after a factory closes. When it closes, 20 employees are also added to the list of casualties. The dynamics Wilson describes apply to any community—urban, rural, suburban—made insecure by what is sometimes called the creative destruction or turbulent unpredictability underlying industrial and postindustrial capitalism (Muller 2013; Smith 2010).

Creative Destruction and Turbulent Unpredictability

Karl Marx and Friedrich Engels wrote about the creative destruction and turbulent unpredictability that characterized capitalism. In fact, their 1848 description is eerie in the way it captures the "restless, anxious, and competitive world of today's global economy" (Lewis 1998, A17). Marx and Engels (1848) described capitalism as a system that destroys as it creates new products and ways to produce and distribute them. What sets capitalism apart from the economic systems that preceded it is its "constant revolutionizing of production, the uninterrupted disturbance of all social conditions, and everlasting uncertainty and agitation." As capitalism destroys the old to make way for the new, it introduces life and death changes to nations, states, cities, workers, and families (how will we survive? how to make a living now?). This turbulence and unpredictability manifests itself in job losses that take the form of firings, layoffs, downsizings, restructurings, and business closings which fuel the size of unemployment lines and the ranks of the chronically unemployed and underemployed. These job losses are often accompanied by the emergence of new kinds of occupations and career paths. This perpetual insecurity has become the "new normal" and has expanded beyond agricultural and manufacturing sectors to include "all levels of the occupation hierarchy" (Smith 2010).

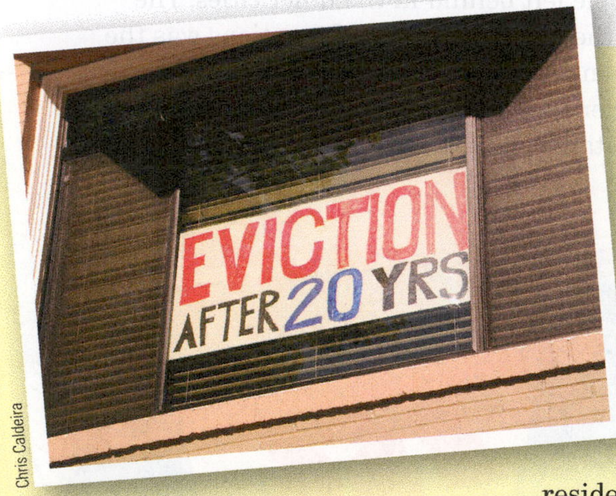

Chris Caldeira

Economic turbulence and unpredictability is reflected in housing markets when rental and mortgage prices rise and fall dramatically. When new high-tech startups, for example, take off and thrive, they lure high-wage talent to the area looking for housing and able to pay high rent and housing costs. The newcomers inadvertently push out long-time residents of 20 years or more who can no longer afford the apartments and houses that they once lived in.

The destructive side of capitalism inevitably prompts a response from government and other agencies to reduce risk for investors, to bail out companies considered too big to fail, to assist the unemployed/underemployed (food banks, food stamps, free and reduced lunch, secondhand stores), retrain workers, and offer compensation (disability insurance, unemployment insurance).

The latest job-related insecurities have generated a range of programs to help people figure out where they are going next and to enhance their employability in an economy that places high value on scientific and technical knowledge; information-gathering, processing, and analyzing skills; mental dexterity; and social and emotional intelligence.

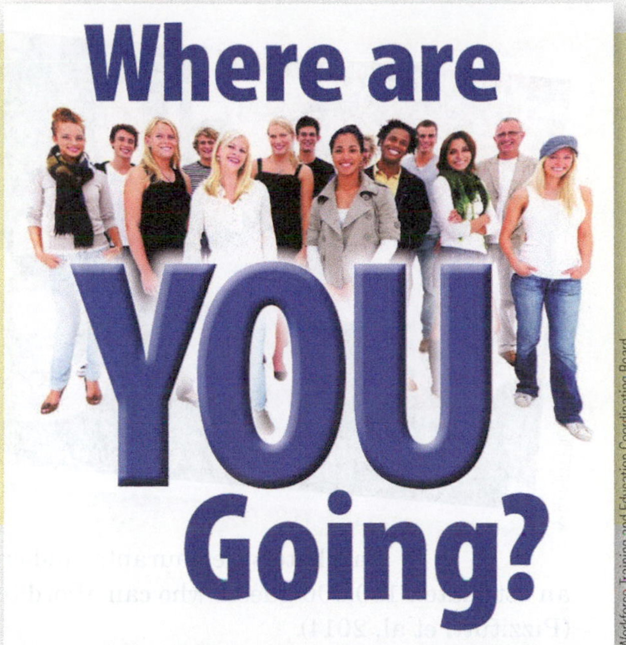

Workforce Training and Education Coordinating Board

Structural "Need" for Poverty-Wage Labor

According to the U.S. Bureau of Labor Statistics (2012), there are 4.8 million full-time workers and 6.5 million part-time workers whose income falls below the poverty threshold because they earn less than $360 a week or $16,640 a year. The working poor constitute 7.2 percent of the labor force. In addition to the working poor, there are 35.7 million poor who do not work, mostly because they are too young, too old, or too impaired. In the now-classic essay "The Functions of Poverty," sociologist Herbert Gans (1972) asked, "Why does poverty exist?" He listed about 15 ways the economy "depends" on poverty-level wages; several of those ways are described below.

1. *People who do the unskilled, dangerous, temporary, dead-end, undignified, menial work of society earn low wages*. As a case in point, someone must assist the 8.1 million people in the United States who are clients of the 4,800 adult day centers, 12,200 home health agencies, 3,700 hospices, 15,700 nursing homes, and 22,200 assisted living and residential care communities—to "help them in and out of baths, make beds, and take residents to and from the toilet" (Charney 2010; Harris-Kojetin et al. 2013). The aides who do the lifting have the highest rate of occupational injuries, especially to the back. They earn an average salary of $11.54 per hour (U.S. Bureau of Labor Statistics 2013b).
2. *The United States economy depends on cheap labor from around the world and within its borders.* Obviously, the lower the wage, the lower the associated labor costs to the hospitals, hotels, restaurants, factories, and farms that draw from the pool of laborers forced or "willing" to work at minimum wage or below. Without low-cost labor, "fruits and vegetables would rot in the fields. Toddlers in Manhattan would be without nannies. Towels at hotels . . . would go unlaundered . . . bedpans and lunch trays at nursing homes would go uncollected" (Murphy 2004).

NKU Philosophy, Rudy Garns

Low-wage labor is the lynchpin of restaurant, hotel, fast food, tourism, and other industries. The Galápagos, a popular tourist destination that attracts tourists from the United States and elsewhere, depends on the labor of Ecuadorians and others from South American countries who migrate to the islands to work in the boat- and land-based tourism sectors. They staff hotels, restaurants, and cruise ships that each year serve an estimated 180,000 guests who can afford to take vacations and time off work (Pizzitutti et al. 2014).

3. *Affluent people contract out and pay low wages for many time-consuming activities*, such as housecleaning, yard work, and child care. On a global scale, millions of poor women work outside their home countries as maids in middle- and upper-class homes.
4. *The poor often volunteer for over-the-counter and prescription drug tests*. Most new drugs, ranging from vaccines to allergy medicines, must eventually be tried on healthy human subjects to determine their potential side effects (such as rashes, headaches, vomiting, constipation, or drowsiness) and appropriate dosages. Money motivates people to volunteer as subjects for these clinical trials. Because payment is relatively low, however, the tests attract a disproportionate share of low-income, unemployed, or underemployed people (Morrow 1996).
5. *Many businesses, governmental agencies, and nonprofit organizations exist to serve those in poverty*. The employees of these organizations draw salaries for performing such work. And the products and services the poor buy with food stamps and medical cards pay corporations and providers.
6. *Those counted as poor use goods and services that would otherwise be discarded*. The poor purchase day-old bread, used cars, and secondhand clothes. The poor also play a role in extending the life of the estimated 307 million pieces of electronic equipment—such as cell phones, computer hard drives, and computer and TV monitors—that are discarded each year in the United States. The Environmental Protection Agency (EPA) estimates that 50 to 80 percent is exported to countries like Ghana, China, Indonesia, and Pakistan (Frontline 2009). There, in some of the worst recycling facilities, people pick through e-waste, salvaging what they can by hand.

Rachel Ellison

If people did not donate clothes that they no can longer wear (or want to wear) to charitable organizations such as Goodwill or St. Vincent de Paul, the estimated 28 billion tons of clothes that end up in landfills would be even greater.

Gans (1972) outlined the functions of poverty to show how a part of society that everyone agrees is problematic and should be eliminated remains intact: It contributes to the supposed stability of the overall system. Based on this reasoning, the economic system as we know it would be seriously strained if we completely eliminated poverty; industries, consumers, and occupational groups that benefit from poverty would be forced to adjust.

Fueling Economic Growth through Debt

Since the 1970s, credit has helped drive the U.S. and global economies lending people money to spend beyond their means. Over this time many people acquired unmanageable levels of debt, creating a division in society between the debt-free and the indebted. Simply put, *debt* is money owed to another party. Debt is one way to fuel economic growth because credit puts money in the hands of consumers to purchase goods and services. Some of the most common sources of borrowed money are credit cards, payday loans, and other financing arrangements (two years same as cash, equity loans). Although debt temporarily frees borrowers from their financial constraints, it can severely constrain their life chances if it becomes unmanageable.
Often the borrowers least able to afford credit and to pay off credit card debt each month are subjected to the highest interest rates.

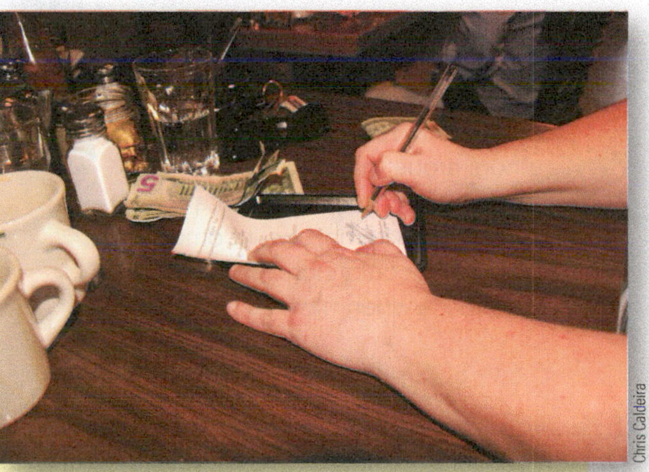

Chris Caldeira

Have you ever needed some cash, so when you were out with friends for dinner and drinks you put everyone's purchases on your credit card and collected their share of the bill in cash to put in your pocket?

Americans who are late making credit card payments pay an estimated $15 billion in penalty fees a year. One in every five credit card holders carries over debt each month and pays interest rates of 20 percent or more (Baker 2009). While people have a responsibility not to use credit cards to live beyond their means, Baker points out that credit-granting businesses "hook us with the promise of low rates while keeping the right to raise those rates at any time for any reason," even on past purchases made when the interest rate was at a certain level.

Rachel Ellison

Payday loans represent a lending practice that can trap its users in a cycle of debt. Payday loan companies offer credit in the form of cash advances to be repaid when borrowers receive their next paycheck (usually one to two weeks later). Borrowers who fail to repay the loan in full at the designated time can renew the loan for an added fee (plus interest). A Pew Charitable Trusts (2013) research study estimated that 12 million Americans take out payday loans each year. The average borrower takes out a payday loan of $375 and remains in debt for five months, paying $520 in finance charges.

What Do Sociologists See?

Sociologists see a function of poverty. One study titled "The Effect of Bottle Laws on Income" identified the recycling function of poverty when it found that people in households with an annual income under $10,000 who collect cans and bottles for recycling add an additional $340 to their income as they contribute to recycling efforts. The authors argue that if states mandated a refund of, say, 5 cents on recyclable items, low-income households could increase earning further (Ashenmiller 2011).

Tom Zaniello

Critical Thinking

Look over the list of factors that help to explain unearned failures and successes. Think of a "success" or "failure" in your life and identify three factors that were most influential in your achievement or failing.

Key Term

urban underclass

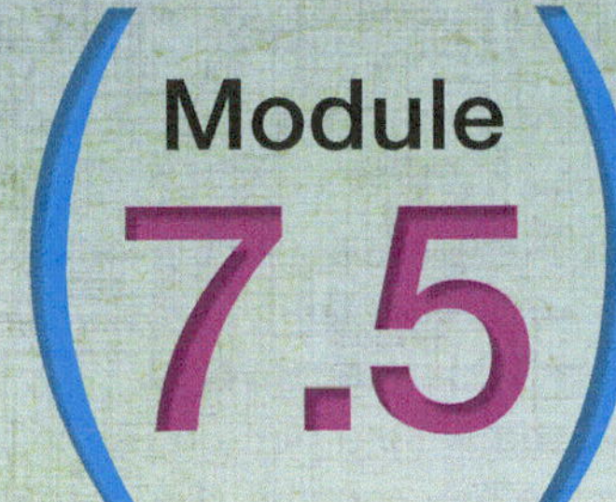

Module 7.5 Global Inequality

Objective

You will learn that income, wealth, and other valued resources are distributed unequally across the countries of the world.

Aleena Ferrante

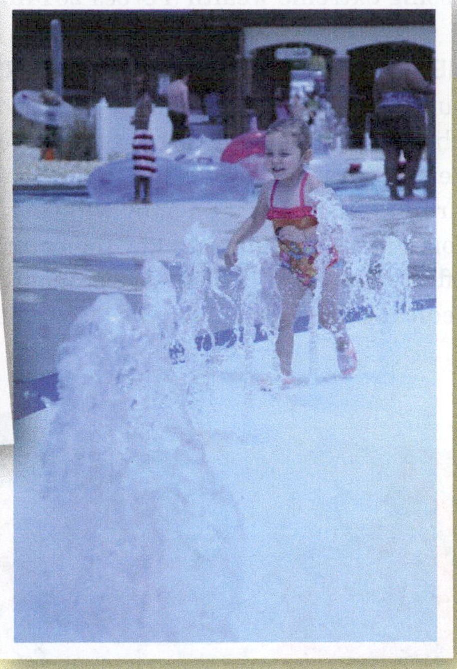

Chris Rasmussen, *Fort Jackson Leader*

What do these photos say about children's ability to access water?

Clearly, we have no control over which of the world's 200-plus countries we are born, yet where people live affects their life chances in dramatic ways. The chances that children will spend a large portion of their day carrying water from a community water well to their home, which may be miles away, are high in many places like Tanzania. The chances that young children in the United States have opportunities to play in water are very high. In assessing global inequality, sociologists look to see how wealth, income, and other valued resources—including access to water—are distributed among the 7.2 billion people living on planet earth.

Poverty and Wealth

It is difficult to define what it means to live in poverty, except to say that it is a situation in which people face great hardship in meeting basic needs for food, shelter, and clothing. Poverty can be thought of in absolute or relative terms.

Absolute poverty is a situation in which people lack the resources to satisfy the basic needs no person should be without. A state of absolute poverty is sometimes quantified in terms of money needed to meet basic needs. The World Bank (2014d) has set that threshold at US$1.25 per day and estimates that 1.22 billion people worldwide live in a state of absolute poverty. **Relative poverty** is a situation in which a person is disadvantaged when compared with a person in an average or more advantaged situation. Thus, relative poverty is a concept that considers not just survival needs but also goods and services that allow people to participate in society (e.g., Internet access, access to reliable transportation).

In contrast to poverty, **extreme wealth** is the most excessive form of wealth, in which a very small proportion of people have money, material possessions, and other assets (minus debts) in such abundance that a small fraction of it, if spent appropriately, could provide adequate food, safe water, sanitation, and basic health care for the 1 billion poorest people on the planet (United Nations 2006). There are about 1,700 billionaires in the world with an estimated combined wealth of $6.4 trillion, an amount that certainly qualifies this tiny group as possessing extreme wealth (*Forbes* 2014b).

INEQUALITIES ACROSS COUNTRIES. When studying inequalities across countries, sociologists identify a valued resource and estimate the chances of achieving or acquiring it by country. For example, parents everywhere want their children to survive pregnancy, childbirth, and beyond. Sociologists compare the chances that a baby will survive the first year of life by country. Infant mortality is considered a key indicator of the overall well-being of a nation, as it reflects maternal health, socioeconomic conditions, and quality of and access to medical care (Centers for Disease Control 2011a).

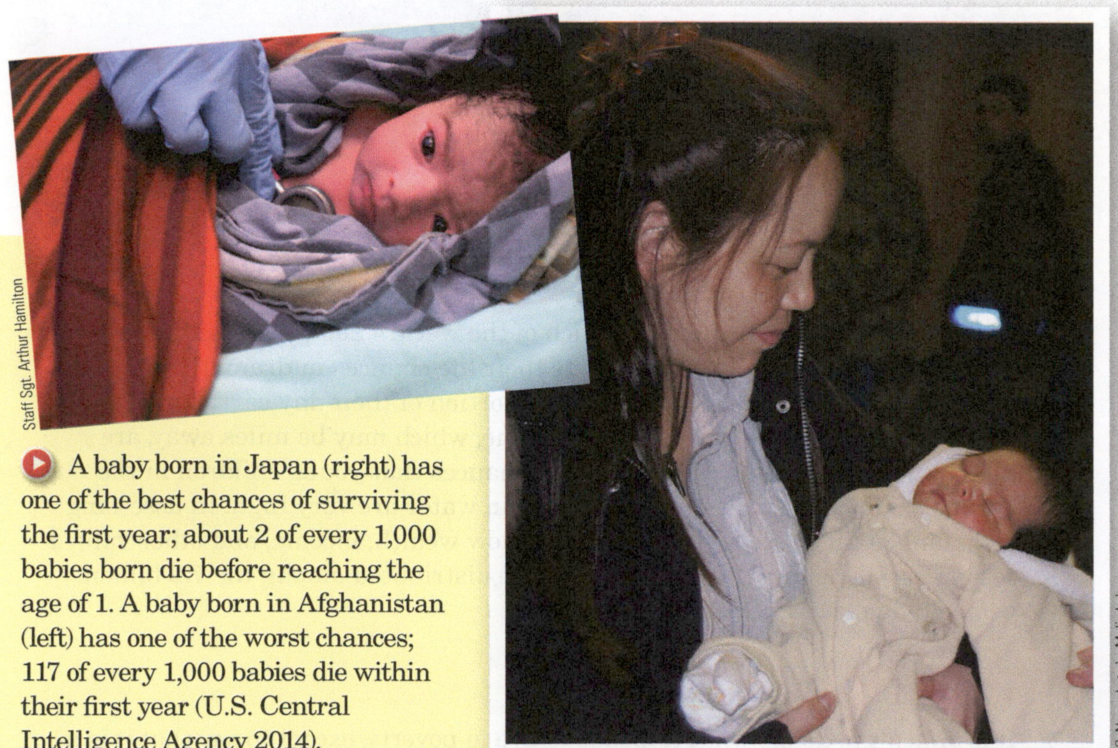
Staff Sgt. Arthur Hamilton

U.S. Navy photo by Melinda Larson

A baby born in Japan (right) has one of the best chances of surviving the first year; about 2 of every 1,000 babies born die before reaching the age of 1. A baby born in Afghanistan (left) has one of the worst chances; 117 of every 1,000 babies die within their first year (U.S. Central Intelligence Agency 2014).

Other startling examples of how life chances vary across countries emerge when we compare the least and most safe countries to become a mother.

Figure 7.5a: 20 Riskiest and 20 Safest Countries to Be a Mother and Baby
The map highlights the 20 riskiest and safest countries to be a mother. Note the safest countries are almost all European and the riskiest are almost all African countries. The chart below the map offers some statistics that show why Finland is the safest country to be a mother and baby and why the Democratic Republic of the Congo is the riskiest. Note that the United States ranks 30 out of 176 countries evaluated by the Save the Children Organization.

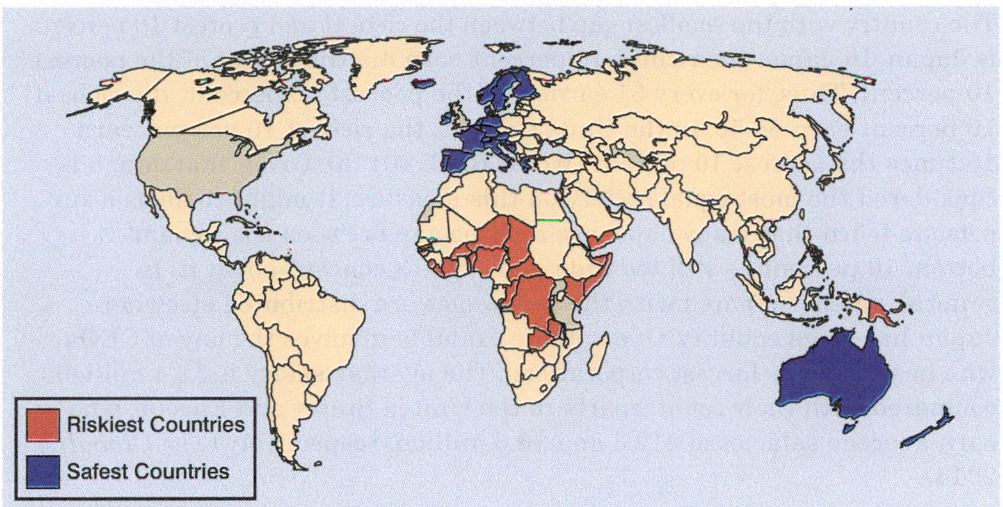

Table 7.5a: Statistics on Finland, Democratic Republic of Congo and United States that Capture Level of Risk to Mothers and Babies

Country	Lifetime Risk of Maternal Death	Under-5 Mortality Rate (per 1,000 live births)	Expected # of Years of Formal Schooling	Gross National Income per Capita (US$)	Participation of Women in National Government (% seats held by women)	Rank (out of 176 countries)
Finland	1 in 12,200	2.9	16.9	47,770	42.5	1
Democratic Republic of Congo	1 in 30	167.7	8.5	190	8.3	176
United States	1 in 2,400	7.5	16.8	48,620	18.2	30

Source of data: Save the Children (2014)

INEQUALITIES WITHIN COUNTRIES. Sociologists look *within* countries to consider how wealth, income, and other valued resources are distributed. They ask questions like: How does the income of the richest 10 percent compare with that of the poorest 10 percent? The country in which the gap between the richest 10 percent and bottom 10 percent is greatest is Sierra Leone. There, the richest 10 percent earn 87.2 times that of the poorest 10 percent. To put it another way, for every $1 earned by the poorest 10 percent, the richest 10 percent earn $87.20.

Sierra Leone is a mineral-rich country where a substantial portion of people live in poverty. Its mineral resources include diamonds, titanium, bauxite, and gold. A former colony of Great Britain, Sierra Leone gained independence in 1961. Control of the country's resources fueled a decade-long civil war from 1991 until 2001, in which virtually the entire population was displaced. The atrocities of this civil war have been documented and include amputations and systematic abuses of women too horrific to mention here (Ben-Ari and Harsch 2005; U.S. Central Intelligence Agency 2014).

The country with the smallest gap between the richest and poorest 10 percent is Japan. In Japan the richest 10 percent earn 4.5 times that of the poorest 10 percent. Thus, for every $1 earned by the poorest 10 percent, the richest 10 percent earn $4.50. In the United States, the richest 10 percent earn 16 times the poorest 10 percent (World Bank 2011a). Given that Japan is considered the most equal society on this measure, it might come as a surprise to learn that many Japanese see this gap between the top and bottom 10 percent as still too wide and express concern about it. In general, when compared with the way wages are distributed elsewhere, Japan has wage equality. One notable example involves the pay of CEOs who head Japan's largest corporations. The average salary is $2.4 million, compared with their counterparts in the United States and Europe, who earn average salaries of $12.3 and $6.6 million, respectively (*The Globalist* 2014).

Responses to Global Inequality

One obvious way to reduce global inequality is to redistribute wealth by transferring some of it from the wealthiest to the poorest. While there is no plan in the works to take from specific rich people per se, since 2000 a plan known as the Millennium Declaration has been in place to redistribute wealth from the richest countries to the poorest ones. That plan revolves around two major commitments from the world's 22 richest countries:

1. to increase current levels of foreign aid to seven-tenths of 1 percent of GDP, and
2. to develop an open, nondiscriminatory trading system.

INCREASE FOREIGN AID. Under this plan the United States, the richest country as measured by its gross domestic product (GDP) of $16.7 trillion, must increase its current level of foreign aid from $34.5 billion to $119 billion. The 22 richest countries, with a combined GDP of $36.7 trillion, should increase foreign aid from $102.6 to $256 billion (U.S. Central Intelligence Agency 2014; United Nations Millennium Development Project 2013). To date the richest countries have not met this commitment.

Chief Mass Comm. Spc. James G. Pinsky

U.S. Air Force photo by Staff Sgt. Joseph L. Swafford Jr./Released

Among other things, the United Nations Millennium Development Project has devised a plan for reducing extreme poverty in developing regions of the world by half by 2015. The latest report proclaims that this goal has been met. One component of the plan asks that the wealthiest countries in the world increase the amount of foreign aid they contribute. While emergency food aid (including supplying bottled water) is important and necessary, it has no lasting effect on development. In contrast, aid dedicated toward building a water well or roads makes a significant contribution to helping a community become self-sufficient. The point is that it is not just aid that is important to realizing the UN plan, but the kind of aid delivered.

END SUBSIDIES, TARIFFS, AND QUOTAS. The wealthiest countries have agreed in principle to eliminate tariffs, subsidies, and quotas on products imported from the poorest countries. However, the United States, Japan, the European Union, and other high-income countries continue to subsidize agriculture and other sectors, such as steel, so that producers in these countries are paid more than world market value for their products. Considerable attention has been given to agricultural subsidies, which give farmers in wealthy economies an estimated $376 billion in support (United Nations 2010b). It is well documented that those subsidized are large agricultural corporations, not small farmers. For example, Riceland Foods (the top recipient) received $554.3 million in subsidies between 1995 and 2010 (Environmental Working Group 2012).

In addition, the wealthiest countries apply tariffs and quotas to many imported items, thereby increasing their cost to consumers. Consider sugar. The U.S. government sets quotas limiting the amount of raw sugar, which sells for about 12.2 cents per pound on the world market, imported from Brazil and about 40 other countries. The U.S. government also adds a tariff to the sugar that is imported. These policies limit competition for U.S. sugar growers, who can sell their raw sugar to processors in the United States for 20.8 cents per pound, 8 cents higher than the global competition. Of course, this means that sugar prices in the United States are artificially high (*Tampa Bay Times* 2012).

We should note that there are often good reasons why the U.S. government applies subsidies and tariffs. For example, the United States government had good reason to impose a 35 percent tariff on tires from China after a surge in tire imports from that country lowered tire production in the United States, from 218.4 million to 160.3 million tires per year. The U.S. government argued that China was subsidizing its domestic tire production by undervaluing its currency. Its low labor costs also gave China an unfair competitive advantage (Chan 2010). Despite these complications, as of 2010, 81 percent of the goods imported from the poorest countries by the richest countries now enter duty-free. The UN goal is 97 percent (UN Millennium Development Project 2013).

Chris Caldeira

The wealthiest countries are not the only ones imposing tariffs on imports. Anyone who visits Vietnam, especially its urban areas, quickly observes that it is a land of motorbikes. The widespread use of motorbikes can be explained in part by the tariff Vietnam places on imported cars and small trucks, which was once 90 percent added to the price. The tariff is to be lowered to between 47 and 70 percent by 2017. The high tariffs make it too expensive to purchase trucks and cars (*Thanh Nien News* 2006; Business-in-Asia.com 2012).

Criticism of the Millennium Declaration

Critics of the UN plan argue that there are other factors that make it difficult for poor economies to compete in the global economy. As a case in point, multinational and global corporations take advantage of those who will work for less. The estimated 188 million Apple iPhones sold in 2013 were manufactured overseas, primarily in China and other Asian countries where employees live in company dorms, work 60 hours or more each week, and earn $17 or less per day. Millions compete for these factory jobs. The low-wage labor is the backbone of the operation known as Apple, which employs 63,000 workers and earns $400,000 in profit per employee (*Forbes* 2014c, Duhigg and Bradsher 2012).

A second reason why critics say the UN plan is not sufficient relates to **brain drain**, the emigration from a country of the most educated and most talented people, including actual or potential hospital managers, nurses, accountants, teachers, engineers, political reformers, and other institution builders (Dugger 2005). The rich economies support brain drain with immigration policies that give preference to educated and skilled foreigners.

Photographer's Mate 2nd Class Timothy Smith

The concept of brain drain is particularly relevant to health care workers. In fact, the British Medical Association (Mayor 2005) has grown so concerned about the shortage of health care workers around the world and the migration of such workers from poor to rich countries that it has called for efforts to reduce this trend. Special immigration policies lure these Indonesian and other nurses away from their home countries to work in the richest countries.

Finally, critics argue that more focus should be placed not on the UN Millennium Project but on the hundreds of creative and successful grassroots efforts to reduce poverty. One is the Grameen Bank microlending project, which was first piloted in 1976 in Bangladesh, a country whose economy is among the poorest. The goal was to examine the possibility of extending tiny loans to the poorest of rural poor women. The goals were to eliminate the high interest rates moneylenders charged the poor and to create opportunities for self-employment among the unemployed living in rural Bangladesh. Today, the bank that piloted microlending has 2,500 branches, with an estimated 7.9 million borrowers living in 83,000 villages. An estimated 90 percent of borrowers repay the loans (Grameen 2014).

What Do Sociologists See?

Sociologists see two meat counters—one in the United States with special cuts of pork (left), the other in Laos with pig's feet. The chances of consuming large quantities of meat on a regular basis are high in the United States—the per capita (per person per year) consumption of meat (beef, veal, chicken, and pork) is 262.21 pounds (577 kilograms). The United States, which has 4.6 percent of the world's population, consumes 21 percent of the meat consumed in the world. In Laos, the annual per capita consumption of meat is 15 pounds (32.2 kilograms), and cuts of meat such as feet, heart, and intestines are consumed (U.S. Census Bureau 2012g).

Chris Caldeira

Chris Caldeira

Critical Thinking

Have you ever experienced absolute or relative poverty? Give examples.

Key Terms

absolute poverty

brain drain

extreme wealth

relative poverty

Module 7.6 Applying Theory: The World's Billionaires

Objective

You will learn how sociologists think about the world's richest people.

Do you ever dream about being a billionaire? What would you do with such wealth? How much would you set aside to invest in social causes that matter to you?

NKU Sociology, Missy Gish

What does it mean to be on the *Forbes* list of the world's billionaires? Those on the *Forbes* list of the world's billionaires are among an elite group of 1,645 people. At the top of the list is Bill Gates, with a net worth of $76 billion and the source of wealth listed as Microsoft. Just making the list is Michael Kors, whose wealth made from retail is valued at $1 billion.

These 1,645 billionaires hold an estimated combined wealth of $6.4 trillion and their combined wealth makes the billionaires of the world the equivalent of the fourth largest economy after the United States ($16.7 billion), European Union ($15.9 billion), and China ($13.4 billion). That means that the combined wealth of these 1,645 people is larger than the GDP of India ($5.0 billion), Japan ($4.7 billion), Germany ($3.2), and each of the other 250 or so countries. Before we look at some of the ways sociologists from each of the four perspectives see billionaires, it is important to understand what a billion is, relative to a million and also to understand what a trillion means since the combined wealth of these 1,645 people is $6.4 trillion.

David Bedard

One way to conceptualize the relative difference between a million, a billion and a trillion is to think about a clock ticking—a second hand moving one increment every second of time. Now imagine listening to each tick of time go by. How long would you have to listen before the second hand ticked a million times? The answer: 11.5 days to hear a million ticks. How long would you have to listen to hear a billion ticks of time go by? The answer: 11.5 years. Now how long would you have to listen to hear a trillion ticks of time? The answer: 32,500 years.

Chris Calderia

From a functionalist perspective, one way to support and reward economic growth is to allow innovators to earn unlimited wealth. A review of the list of billionaires and the major sources of their wealth shows that the founders of Microsoft, Oracle, Google, Amazon, Twitter, Facebook, Dell, and What's Up are among those on the list. These corporations also employ tens of thousands or millions of workers directly and indirectly (Royal Geographical Society 2014). It is in these ways the billionaires contribute to order and stability in society.

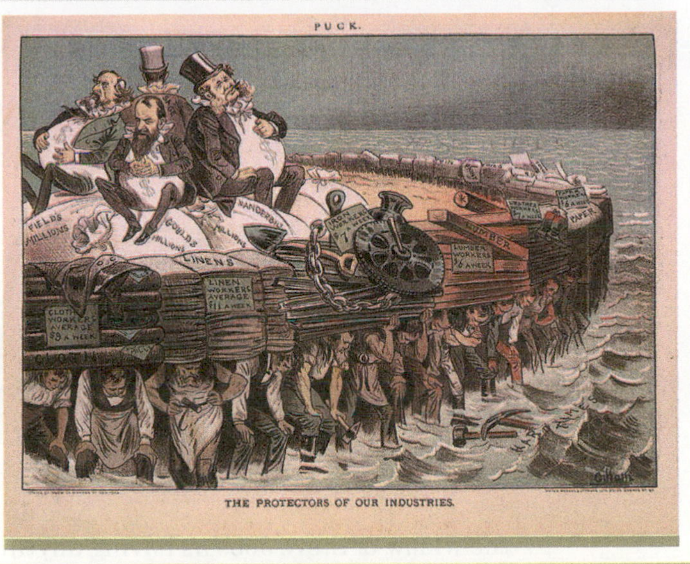

Library of Congress Prints and Photographs Division [LC-USZC4-3108]

This 1883 political cartoon captures the conflict vision of how billionaires secure wealth. It shows four of the wealthiest capitalists who have ever lived—Cyrus Field, Jay Gould, Cornelius Vanderbilt, and Russell Sage. These men were wealthier in real dollars than any considered among the world's richest today. The four are seated on bags of money atop a large raft, which is being held afloat by hundreds of thousands of workers. Today we might imagine (among others) the Walton family, whose wealth comes from retail giant Walmart, sitting atop this raft. There are six Walton family members on the list of billionaires, accounting for $147.8 billion (*Forbes* 2014a). One can argue that the Walton's wealth depends upon the labor of millions of workers worldwide—most of whom work for low wages and no benefits.

Katie, Courtesy of Joan Ferrante

Among other things, symbolic interactionists are interested in how billionaires "market" themselves and what they choose do with their wealth. Some billionaires engage in philanthropic giving with high symbolic value but low financial value relative to their total net worth. The Walton family's net worth is $147.8 billion, and their lifetime giving is estimated at $4.6 billion. By contrast, Warren Buffet plans to donate all of Berkshire Hathaway's value ($58 billion) before or upon his death, to be invested within 10 years for social good. Buffett notes that he has "never given a penny away that had any utility to me," and argues that there are millions of people in the world who sacrifice by giving money "that's important to them" (*Forbes* 2014a).

NKU Sociology, Boni Li

Feminists would surely notice that only 10 percent of the world's billionaires are female. Feminists would also notice that many of the female billionaires on the list are presented as securing their wealth through marriage, divorce, or inheritance. When you think of Apple do you think of Laurene Powell, the widow of Steve Jobs as having anything to do with the success of Apple? It is interesting that billionaires—both men and women—are presented on the Forbes list as individuals with no reference to partners (even lifelong partners) as co-owners of that wealth. This assumes that partners, even lifelong partners, played no role in a billionaire's monetary success. This assumption is problematic, especially if billionaires are parents and their partners assume roles as primary caregivers freeing them up to focus on achieving career and financial goals. For this reason alone many spouses play an important role in billionaires' personal lives, and that contribution is key to their "having it all."

How do we think about such vast amount of wealth and the people who hold it? As we have learned, when *functionalists* think about extreme wealth, they think of the contribution billionaires make to order and stability. *Conflict theorists* focus on how such a vast amount of wealth is derived from the labor of others. *Symbolic interactionists* think about the way billionaires use wealth to present themselves. *Feminists* emphasize gender inequalities and the gendered distribution of this vast amount of wealth.

Chapter 7

Summary: Putting It All Together

Social stratification is the systematic process of categorizing and ranking people on a scale of social worth where the various rankings affect life chances in unequal ways. Every society in the world stratifies its people according to ascribed and achieved statuses. Associated with each status are varying levels of prestige and esteem. Stratification systems fall on a continuum between two extremes—caste and class systems, neither of which exists in a pure form.

Sociologists use the term *social class* to designate a person's overall status in a system of social stratification. The views of Karl Marx and Max Weber are central to class analysis. For Marx, the key variable in determining social class (and life chances) is source of income. Weber defined social classes as being composed of those who hold similar life chances, determined not just by their sources of income but also by their marketable abilities, access to goods and services, and ability to generate investment income. Social class is complicated by status groups (lifestyle) and power.

When sociologists examine social stratification on a global scale, they look to see how wealth, income, and other valued resources are distributed unequally among the 7.2 billion people living in 200+ countries. The contrasts in wealth, income, and life chances are startling. In 2000, United Nations member countries endorsed the Millennium Declaration, an ambitious plan to significantly reduce global inequality by 2015. This plan revolves around increasing the amount of foreign aid the world's wealthiest countries contribute to helping the poorest countries and reducing, even eliminating tariffs, quotas and subsidies. To date the world's wealthiest countries have not increased foreign aid to hoped for levels and progress toward achieving free trade has been uneven. Critics of this plan believe more efforts should be made to reduce brain drain and increase wages.

Sociologists draw upon four perspectives to explain inequality and to understand how it is manifested in daily life. Those perspectives are functionalism, conflict, symbolic interaction, and feminism. Feminists offer the concept of intersectionality, a term that captures the interconnections among race, class, gender, sexual orientation, religion, ethnicity, age (generation), nationality, and disability status. Sociologists recognize these traits as interlocking categories of analysis that, when taken together, profoundly shape how we see ourselves, how others see us, and the structure of our relationships.

Sociologists also look to structural changes in the economy to understand how they create advantaged and disadvantaged groups. They give special attention to how structural changes disrupt networks of occupational contacts, so that large segments of society such as the urban poor cannot connect with those who are steadily employed. Other related structural factors include the creative destruction and turbulent unpredictability of capitalism, the economic "need" for low wage labor, and economic growth fueled by debt.

Photo by Chris Caldeira; Mural by Paul Ygartua/www.ygartua.com

Chapter 8

Race and Ethnicity

Each of the children depicted in this mural painted by Paul Ygartua is meant to represent one of the five so-called universal races. Do you know the names of those five races? In the United States, there are five officially recognized races—"Asian," "Black or African American," "White," "American Indian and Alaskan Native," and "Native Hawaiian and Other Pacific Islander." Can you match the face of each person on the mural with just one of these five races? Your ability to do so (or not) speaks volumes about the idea of race and its usefulness for classifying humanity. You might be surprised to learn that race, and by extension racial categories, makes no logical sense. Still, race has assumed great significance in structuring human affairs. This contradiction is what makes race an especially fascinating topic for sociological analysis.

Module

8.1 Race

Objective

You will learn why sociologists define race as a social construction with real consequences.

What race is President Barack Obama?

Obama is considered the first black president of the United States. Do you think that odd, considering he described his father as a Kenyan immigrant who was "black as pitch" and his Kansas-born mother as "white as milk" (Obama 2004)? The race we have assigned President Obama requires that we forget about his mother and his biological connections to her side of the family, including her ancestors. What race do you consider his wife to be? Does it matter that Michelle Obama carries "within her the blood of slaves and slave owners" (Obama 2009)? The Obamas' complex ancestries help us understand why most biologists and social scientists have come to agree that race is not a biological fact. The reason is that parents who are considered different racial categories can produce offspring. Each offspring, by definition, is a blend of the two races and cannot, in a biological sense, belong to just one racial category.

Defining Race

Sociologists define **race** as human-created or -constructed categories that have come to assume great social importance. Although on some level we can say that race has something to do with skin shade, hair texture, eye shape, and geographical origins of ancestors, it is so much more than that. When sociologists study race, they study its social importance—the meanings assigned to physical traits, racial categories, and the effect race has on opportunities in life. We know that race is human-created if only because racial categories vary across time and place. This variation suggests that it is people who "determine what the categories will be, fill them up with human beings and attach consequences to membership" (Cornell and Hartmann 2007, 26).

What race is the baby in this photo? Because of his appearance, this baby is seen as belonging to the race of just his mother. In the United States people look to skin shade and hair texture to determine a person's race, but those physical markers hide the connections this baby has with the history and ancestors associated with his father's "race."

U.S. Air Force photo by Staff Sgt. Laine McNeal

Race as Illusion

When sociologists proclaim race to be an illusion, they mean that the most basic assumptions we hold about race—what we believe to be true—are in fact distorted. Simply consider that it may seem natural to divide people into racial categories, but upon close analysis it is illogical. First, there are no sharp dividing lines specifying the physical boundaries that distinguish one racial category from another.

Some of the children pictured are brothers and sisters; others are cousins. Notice the continuum of skin shades. Which of the children pictured appear white; which appear black? Where is the line that marks the point at which skin shade is considered black on one side of that line and white on the other side?

NKU Sociology, Missy Gish

A second problem with trying to place people in racial categories is that millions of people in the world are products of sexual unions between people of different races. Obviously, the offspring of such unions cannot be one biological race. Even if we are able to classify each child as a single race, the biological reality does not support a clear-cut classification.

Third, the diversity of people within any one racial category is so great that knowing someone's race tells us little about that person. For example, the U.S. government defines as Asian any person having origins in any of the original peoples of the Far East, Southeast Asia, or the Indian subcontinent. So people expected to identify as Asian include those who have roots in very different places, including Cambodia, India, Japan, Korea, Malaysia, Pakistan, Siberia, the Philippines, Thailand, and dozens of other countries. Similar diversity exists within populations labeled racially as "Black or African American," "White," or "American Indian and Alaskan Native."

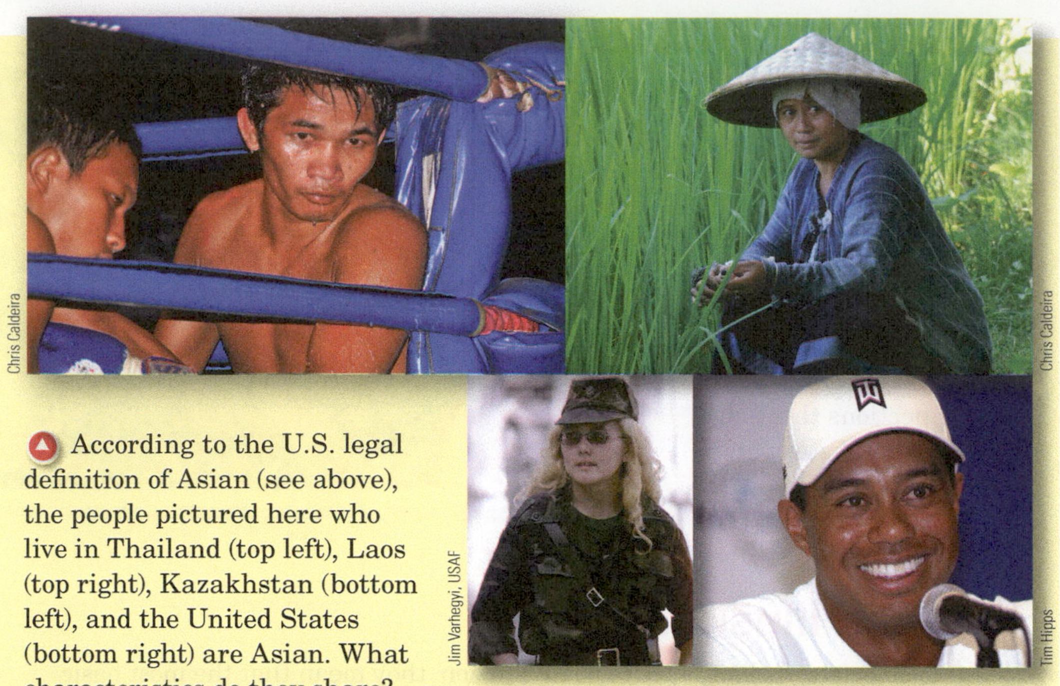

Chris Caldeira

Chris Caldeira

Jim Varhegyi, USAF

Tim Hipps

According to the U.S. legal definition of Asian (see above), the people pictured here who live in Thailand (top left), Laos (top right), Kazakhstan (bottom left), and the United States (bottom right) are Asian. What characteristics do they share? They do not share a language, a country of origin, or a culture. The only trait those pictured have in common is that each has at least one blood relative or ancestor considered as being from a country designated as Asian.

Finally, the assumptions about race and ideas about racial categories and who belongs in each vary by time and place. In Brazil almost everyone, regardless of physical appearance, thinks of themselves as multiracial. Americans think of race in categorical terms and apply the label "biracial" or "multiracial" only to those who appear "almost white."

NKU Sociology, Missy Gish

Moreover, most Americans would not think of the boy on the right as multiracial. They would be more likely to label the boy on the left as multiracial. Since these two brothers are offspring of the same mother and father, they are both multiracial.

Racial Formation Theory

With regard to race, sociologists are interested in how something that cannot be supported by logic has come to assume such great importance. They are interested in the strategies people use to make people, like the two brothers pictured above, fit into existing racial categories. One answer is that people make rules that make the categories work. For example, in the United States, depending on time in history and place, the government has used different rules to classify children born to parents considered different races into ONE racial category, including classifying the child as the "race of mother," "race of father," and "the race they named first" (e.g., if a person said "I am Asian and white," that person was classified as Asian, but if the person said "I am white and Asian," that person was classified as white).

Sociologists Michael Omi and Howard Winant (1986) offer **racial formation** theory as a way to understand the social importance of race. Omi and Winant argue that for race to exist in its present form, people must learn to see race—that is, they must learn to give arbitrary biological features, such as skin color and hair texture, social significance. Moreover, they must develop what the two sociologists call **racial common sense**, ideas people share about race that they believe to be so obvious that they do not even think to question their validity. Omi and Winant (2002) maintain that racial common sense informs our expectations and interactions that involve race and that this commonsense understanding of race persists even when it is challenged. Many Americans assume (believe it to be common sense) that ALL enslaved appeared "black" and were of African descent, and that the masters appeared "white" and were of European descent, never considering that an unknown but significant proportion of slaves were offspring of master and slave.

Isaac & Rosa, Slave Children from New Orleans.
Photographed by Kimball, 477 Broadway, N. Y.
Ent'd accord'g to act of Congress in the year 1863, by Geo. H. Hanks, in the Clerk's Office of the U.S for the So. Dist. of N. Y.
Library of Congress

Isaac and Rosa, pictured here, illustrate that the enslaved—a group most imagine as having dark skin and tightly curled hair—could appear "white." Slave narratives tell us that forced and unequal sexual relationships between master and enslaved were commonplace. As just one example, in her autobiography, *Incidents in the Life of a Slave Girl*, Linda Brent (who appears almost white) tells of her master being the father of at least 11 children by enslaved women. In an effort to avoid a sexual encounter with her master, Linda has an affair and two children with a white lawyer. The couple's two children can pass as white and, by law, are owned by Linda's master (as children of enslaved women, no matter the fathers' status, followed the condition of the mother). How does knowing the complex history of slavery change "commonsense understanding" of black and white as separate racial categories?

Sociologists see race as socially constructed. This means that the characteristics we have come to believe define race are products of social beliefs and values imposed by those who had (or have) the power to create the labels and categories. Once those labels and categories were put in place, it became easy to reify them. "Reify" means to treat them as if they are real and meaningful and to forget that they are made up. When we reify categories, we act as if people are those categories. When people do things or appear in ways that don't fit their assigned category, we act as if something is wrong with them or as if they are exceptions to the rule, rather than questioning the category scheme.

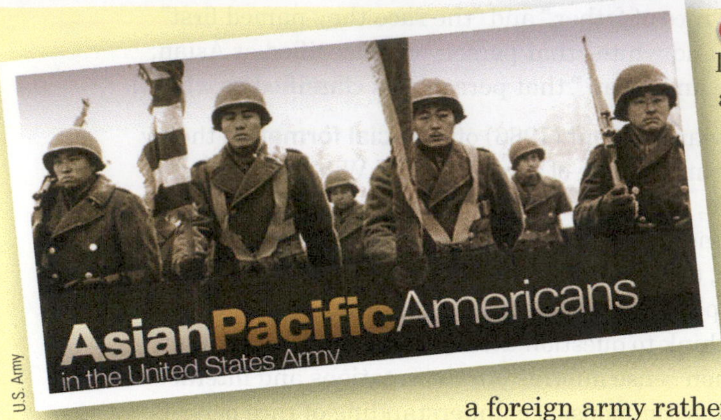

Asian Pacific Americans
in the United States Army

U.S. Army

Many Americans have learned to think of people who appear Asian as "outsiders" or as recent immigrants to the United States who were born elsewhere and do not speak English, and if they do speak the language they expect to hear an accent. Thus, when most Americans glance at this photo, it brings to mind a foreign army rather than the thought that the people of Asian descent pictured are part of the U.S. military (which they are).

What Do Sociologists See?

Sociologists would relate Omi and Winant's racial formation theory to the situation of this father and daughter. Many people would say that this man has a black daughter or that this little girl has a white father. Hair texture and skin color are the key characteristics we use to make this assessment, setting aside the biological connection. Such an assessment suggests that we have learned to see race, and to give clear cut racial categories great social importance. It is likely that the father and daughter also see themselves as belonging to different racial categories.

DoD photo by Petty Officer 2nd Class Stephanie Tigner, U.S. Navy

Critical Thinking

Think of your living and deceased relatives on the maternal and paternal sides of the family. Can you identify a relative who is considered a different race from you? Explain.

Key Terms

race racial common sense racial formation

Module

8.2 Ethnic Groups

Objective

You will learn the processes by which ethnicity assumes social and personal significance.

Can you tell the ethnicity of the woman holding the fish? If yes, what clue did you use to make that determination?

NKU Anthropology, Sharyn Jones

If you guessed that this woman's ethnicity as Fijian, then you have correctly named the ethnicity by which she is known and identifies. One way this woman announces and shares her ethnicity is through the food she values and eats and the way she catches and prepares it. Sociologists are interested in the processes by which ethnicity takes on personal, symbolic, and social significance.

Defining an Ethnic Group

An **ethnic group** consists of people who share, believe they share, or are believed by others to share a national origin, a common ancestry, a place of birth, or distinctive social traits (such as religion, style of dress, or language) that set them apart from other ethnic groups. Distinguishing between race and ethnicity is complicated because racial and ethnic identities are intertwined. In the United States, for example, Fijians are considered to belong to the racial category "Native Hawaiian and Other Pacific Islanders." However, many Fijians self-identify as "black" as they believe that their ancestral roots can be traced to East Africa (specifically to what is now Zanzibar, Tanzania). In fact, most Fijians present themselves as descendants of one of the Lost Tribes of Israel.

It is impossible to list all the ethnic groups that exist in the United States and elsewhere in the world. It is also difficult to specify the unique social characteristics and markers that place people into a particular ethnic group. For example, does a diet centered around fish make someone Fijian? Or is it the ability to speak Fijian? Because language and cultural traditions are imprecise markers of ethnicity, one way to determine someone's ethnicity is to simply ask, "What is your ethnicity?" Self-identification is problematic, too, because people's sense of their ethnicity can range in intensity from no awareness of an ethnic identity to an all-encompassing awareness (Verkuyten 2005).

For some people, their claimed ethnicity is a complete lifestyle that involves being born in a particular place, speaking the language, dressing a particular way, and interacting primarily with others in that ethnic group. For others, their ethnicity is symbolic; that is, they feel a nostalgic allegiance to, love for, or pride in the cultural traditions of a specific immigrant group or a home country but they do not incorporate ethnic traditions into their everyday life. In other words, people embrace the most visible and recognizable symbols that are "easily expressed and felt" (Gans 1979). They may express their ethnicity by celebrating holidays (St. Patrick's Day), buying signature products (Italian beer), eating certain food (tacos), naming a child an ethnic-sounding name (Mario), watching films or television programs with ethnic characters, or rooting for a soccer team from the home country. None of these activities "take much time" or effort.

Selective Forgetting and Ethnic Renewal

The ethnicity with which a person identifies is also affected by **selective forgetting**, a process by which people forget, dismiss, or fail to pass on a connection to one or more ethnicities. This "forgetting" is affected by larger societal forces. For example, in the United States, people assigned to some racial categories have more freedom than others in claiming an ethnic identity. Americans classified as white have a great deal of freedom to claim an ethnic identity as long as it is one associated with the category of white. But it is difficult for people who appear white to claim an ethnicity associated with races considered nonwhite (e.g., Kenyan) because others are likely to dismiss such a connection as meaningless (e.g., you don't look Kenyan). Americans classified in racial terms as black have less choice; they are expected to identify as simply "black" or as being of African descent, even though they know they have ancestors of specific African and other ethnicities or feel a special connection to ancestors from countries other than African (Waters 1994).

Library of Congress

In taking a long view of American history, it should come as no surprise that people of African ethnicities (Igbo, Geb, Yoruba) and Native American ethnicities (Seminole, Choctaw) produced offspring as their paths crossed in complex ways. Some Native Americans owned black slaves; some slaves sought refuge in and outside of Native American communities; and Native Americans were part of the Underground Railroad movement. The Black Seminoles, descendants of free blacks and escaped slaves, formed communities just outside of Seminole communities. It is rare to hear of someone claiming Igbo–Seminole ethnicity.

People's sense of ethnicity, even when they are free to define it, can also shift through a process known as **ethnic renewal**, a process by which people take it upon themselves to find, learn about, and claim an ethnic heritage. This occurs when people discover a once-hidden ethnic past, as when an adopted child learns about and identifies with newly found biological parents or when someone learns they have ancestors of a certain ethnicity and then take steps to revive lost traditions.

Chris Caldeira

It is not usual for Korean-born children raised in families classified as white to identify as "white." When adoptees reach college age and move away from home to go to school, they may find that others consider and treat them as Asian (not Korean per se). This experience may launch a desire to explore and connect with ethnic roots by joining a Korean or Asian support group or enrolling in Korean language or a college course that focuses on Asian history (Shiao and Tuan 2008).

In light of the information presented thus far about ethnicity, is it any wonder sociologist Max Weber (1922a) argued that "the whole conception of ethnic groups is so complex and so vague that it might be good to abandon it altogether"? So why study ethnicity? One answer is that sociologists are interested in studying the processes by which people make ethnicity important (or not important).

Dominant Group Ethnic Identity

Sociologist Ashley W. Doane (1997) defines a **dominant ethnic group** as the most advantaged ethnic group(s) in a society; it is the ethnic group that possesses the greatest access to valued resources, including the power to create and maintain the system that gives it these advantages. Dominant status is achieved over a long history that includes conquest, colonialism, and forced or invited labor migrations that left some groups and the offspring that followed with more advantages than the descendants of those exploited. Those who are part of the dominant ethnic group, however, tend to dismiss that history as irrelevant to any of its advantages. In the United States, the dominant ethnic group began as Anglo-Americans, then expanded to encompass Protestant European Americans and eventually to encompass European Americans. In comparison to other ethnic groups, Americans from Western European ancestries several generations removed are the least likely to recall incidents in which they have personally faced prejudice, discrimination, or disadvantage because of ethnicity. In other words, because ethnicity is viewed as a largely insignificant factor in their lives. It is important to clarify that not everyone who is part of the dominant ethnic group holds a powerful or advantaged status. The term *dominant* refers to the fact that European Americans are overrepresented among those holding advantaged statuses.

Those in the dominant ethnic group take it for granted that they will see their "ethnic group" predominantly displayed in stores and in advertisements around the world. Estée Lauder markets its beauty products in 150 countries using essentially the same appeals, messages, and photographs. This advertisement is in a subway station in Beijing, China. Why do you think a woman who appears as shown is used to sell cosmetic products to people of color around the world?

NKU Sociology, Boni Li

Doane (1997) refers to the dominant ethnic group as possessing a **hidden ethnicity**, a sense of self that is based on no awareness of an ethnic identity. This is because the dominant group's culture is considered normal, normative, or mainstream. In the case of the United States, Western European standards dominate the form and content of its educational and judicial systems and permeate its mass media and political institutions. The normalization of the dominant culture promotes the belief that European Americans are cultureless; they are simply being American. This normalization creates a framework that makes it difficult for those in subordinate ethnic groups to challenge the existing system. Specifically, those who work to hold onto their ethnicity are portrayed as unwilling to give up a "foreign" culture and become "Americans." This failure to let go is labeled as undesirable. Those—even Native Americans—who seek to hold onto their cultural identity, advocate for cultural pluralism, or ask that their cultural experiences be recognized as significant to American history are viewed as divisive, guilty of political correctness, or asking for special treatment.

Involuntary Ethnicity

Then there is the phenomenon of **involuntary ethnicity**. In this situation, a government or other politically advantaged group creates an umbrella ethnic category and assigns people from many different cultures and countries to it. The category becomes the label by which diverse peoples are known and with which they are forced to identify. The category "Hispanic" qualifies as an example. In the United States the only officially recognized ethnic categories are "Hispanic" and "non-Hispanic." The Hispanic category, created in 1970, includes anyone who has roots in a Spanish-speaking country, of which there are at least 19. As one contrast to the U.S. system, China officially recognizes 56 different ethnic groups, including the Han (which encompasses 92 percent of the population), Tibetans, Uighur, and Yoa (Lilly 2009).

THE HISPANIC/LATINO CATEGORY. In the mid-1970s, the U.S. government chose to count two official ethnicities: (1) Hispanic or Latino, and (2) Not Hispanic or Latino. The U.S. government defines Hispanic/Latino as "a person of Mexican, Puerto Rican, Cuban, Central or South American culture or other

Spanish culture or origin, regardless of race" (U.S. Office of Management and Budget 1997). Note that a person from Brazil is not considered Hispanic or Latino because that country is a former Portuguese, not Spanish, colony. When the Census Bureau first classified the Hispanic population in the 1970s, it represented less than 5 percent of the total population. Today, that ethnic category includes 16.3 percent of the population and is expected to account for 24.4 percent of the total U.S. population by 2050. Approximately 50.4 million people classified as Hispanics live in the United States.

Figure 8.2a: Hispanic or Latino Population as a Percent of Total Population by County

About half (52 percent) of the Hispanic population identifies as white; 36.7 percent identify as "other." Just 2.5 percent identify as black (Humes et al. 2011). Notice that the Hispanic population is concentrated in the western United States, in the area that was once part of Mexico.

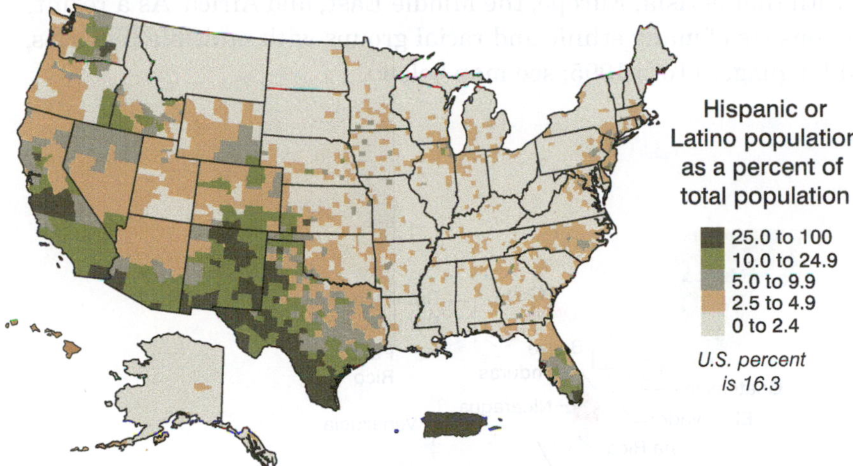

Source: Ennis, Rios-Vargas, and Albert (2011)

In the United States, people who are ethnically classified as Hispanic can be of any race. (Note: This is also true for people classified ethnically as non-Hispanic.) The people pictured in these photos are all considered Hispanic. What race would you assign to each?

THE TERM *HISPANIC*. Most people we think of as Hispanics have to learn to define themselves as such (Novas 1994). In addition, many so-called Hispanics reject that label because it forces them to identify with conquistadors and settlers from Spain, who imposed their culture, language, and religion on their ancestors. "For Latin Americans, who, like North Americans, fought hard to win their independence from European rule, identity is derived from their native lands and from the heterogeneous cultures that thrive within their borders" (Novas 1994, 2).

Panethnicity is a process by which various people with distinct histories, cultures, languages, and identities are lumped together and viewed as belonging to one catchall category, such as Hispanic/Latino. *Hispanic/Latino* is a panethnic category because it applies to people from, or with ancestors from, 19 Central and South American countries (excluding Brazil) that were once under Spanish control. To complicate matters even further, the history of these 19 countries is intertwined with that of Asia, Europe, the Middle East, and Africa. As a result, each country consists of many ethnic and racial groups with multiple histories, cultures, and languages (Toro 1995; see map below).

It is important to point out that the U.S. government also groups together people with many ethnicities into a single racial category. As one example, it classifies 2,000 tribes, each with distinct cultures, as Native American. The United States has always grouped diverse peoples into single racial and ethnic categories for administrative, statistical, surveillance, and other purposes. The people classified as such often do not see the grouping as natural and must learn to see themselves as belonging to a government-imposed category. Sometimes, as a result of widespread discrimination in the society, a significant number of members of this imposed category identify with the label, transcend differences, and organize as a political force to advocate for political inclusion, needed services, and basic rights.

What Do Sociologists See?

Sociologists see a photo of two people who are both considered Native American or First Nation peoples. The girl is from Tewa tribe of New Mexico and the man is considered Sioux. Of course, the Tewa and Sioux are only two of the more than 2,000 ethnic groups that make up the category Native American. This vast diversity raises questions as to the value of the category Native American for knowing about someone.

Library of Congress

Library of Congress

Critical Thinking

Is there an ethnicity with which you identify? Explain. How do you express your ethnicity?

Key Terms

dominant ethnic group
ethnic group
ethnic renewal
hidden ethnicity
involuntary ethnicity
panethnicity
selective forgetting

Module 8.3 Chance, Choice, and Context

Objective

You will learn that a person's race and ethnicity is a product of chance, choice, and context.

What race are Dora the Explorer and the child touching her?

Rich McFadden

You likely classified Dora as Hispanic/Latino, which, in the context of the United States is considered an ethnic category,
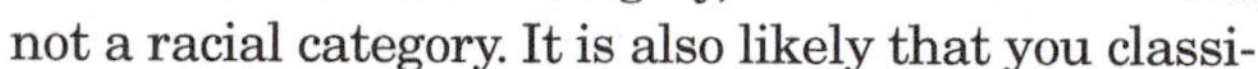
not a racial category. It is also likely that you classified the little girl as black or African American. But what if you lived in Brazil? In Brazil, Dora and this little girl touching her would likely be seen as brown, or even white. Both most certainly would be viewed as multiracial.
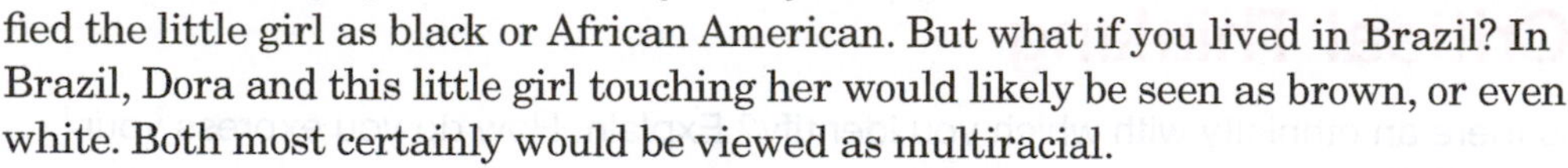

When sociologists study race, they study its social importance—the meanings assigned to physical traits such as skin color and hair texture, the rules for assigning people to racial categories, and the choices people make to present themselves as a particular race or to challenge people's perceptions of their race. The differences in the way people around the world view race force us to think about the processes by which race assumes social importance. These processes involve chance, choice, and context.

Chance, Choice, and Context

Chance is something not subject to human will, choice, or effort. We do *not* choose our biological parents, nor can we control the physical characteristics we inherit from them. **Context** is the social setting in which racial and ethnic categories are recognized, created, and challenged. **Choice** is the act of choosing from a range of possible behaviors or appearances. The choices one makes about whether to identify or present oneself as a particular racial or ethnicity are constrained by chance and context (Haney Lopez 1994).

In evaluating the relative importance of chance, context, and choice, consider the case of a highly visible American: Tiger Woods, who is by *chance* the son of a mother who is considered half Thai, a quarter Chinese, and a quarter "white."

Woods's father is considered half "black," a quarter Chinese, and a quarter American Indian (Page 1996, 285). Simply by *chance*, Woods inherited physical features that people in the United States associate with the category "black."

Theoretically, Tiger Woods could choose to publicly identify as white (or specifically as Dutch ethnicity) but Woods lives in a country (context) where few people would attribute any significance to that part of his ancestry. Likewise, in the *context* of United States, Woods' Asian ancestry is viewed as having much less significance than his African ancestry. When Woods first came on the golf scene, he tried to present himself as "Cablinasian"—a mixture of Caucasian, black, Native American, and Asian—but as he has pointed out, "In this country, I'm looked at as being black. When I go to Thailand, I'm considered Thai. It's very interesting. And when I go to Japan, I'm considered Asian. I don't know why it is, but it just is" (Barkley 2006). If Tiger Woods lived in another context such as Brazil, people living there would assume he was a Brazilian golfer first and foremost, and if asked his race they might classify him as brown or white, but certainly as multiracial.

By chance, this baby was born with a skin shade that is much lighter than that of her mother who is holding her. In the context of the United States, most see a mother who appears "black" holding a "white" child.

Lisa Southwick

Race in the Context of the United States

The U.S. government officially recognizes five racial categories plus a sixth, "some other race" (a category of last resort for those who resist identifying, or cannot identify, with one of the five official categories):

- American Indian or Alaskan Native—a person having origins in any of the original peoples of North, Central, or South America and who maintains tribal affiliation or community attachment;
- Asian—a person having origins in any of the original peoples of the Far East, Southeast Asia, or the Indian subcontinent;
- Black or African American—a person having origins in any of the Black racial groups of Africa;
- Native Hawaiian or Other Pacific Islander—a person having origins in any of the original peoples of Hawaii, Guam, Samoa, or other Pacific Islands; and
- White—a person having origins in any of the original peoples of Europe, the Middle East, or North Africa (U.S. Census Bureau 2011g; see Table 8.3a).

It is significant that "Black or African American" is the only racial category that does not refer to original peoples. Notice that the words *original peoples* appear in every other definition. If the words original peoples were included in the definition of Black or African American (rather than *Black racial groups of Africa*), everyone in the United States would have to claim "black" as their race. We know from existing archaeological evidence that all humans evolved from a common African ancestor.

Table 8.3a: Percent of Total U.S. Population by Official Racial Category

About 79 percent of the 318 million people in the United States self-classify as white. If people identify with a race other than the recognized five, they can check "other" on the census form. About 5 percent of the American population identify as such. Notice that Hispanic is missing from the table it is considered an ethnic group (see Module 8.2).

Official Racial Category	Percentage of Population
White	79.1%
Black/African American	12.3%
Asian	3.6%
American Indian/ Native Alaskan	0.9%
Native Hawaiian/ Pacific Islander	0.1%
Other	5.5%

Source: U.S. Census Bureau (2011g)

Race in the Context of Brazil

The U.S. system of racial classification may seem a natural way to divide humanity until we contrast it with another system such as that of Brazil. The Brazilian idea of race holds that Africans, indigenous peoples, and Europeans had mixed to the point that race was no longer important. For administrative purposes, the Brazilian government uses three broad categories (more like segments on a continuum)—*branco* (white), *pardo* (brown), and *preto* (black), with *branco* and *preto* considered ends on that continuum (see figure below). These three categories apply to 99 percent of the country's population. Two other categories that apply to the remaining 1 percent are *amarelo* (yellow) and *indigena* (indigenous).

Because the Brazilian system recognizes a continuum of hair textures and skin shades, it does not specify a place on the continuum that marks the point at which the white (*branco*) segment gives way to the brown (*pardo*) and brown gives way to the black (*preto*) segment.

NKU Philosophy, Rudy Garns

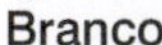

Branco

NKU Philosophy Rudy Garns

Preto

NKU Philosophy Rudy Garns

The continuum-like view of race is reflected in the popular language of Brazil. Sociologist Edward E. Telles (2004) found that when presented with an open-ended question asking people their race, Brazilians answered with 135 distinct terms, 45 of which were named only once or twice. Ninety-eight percent of respondents used one of seven popular terms to describe their color: *branco*/white (42 percent), *moreno*/no clear race (32 percent), *pardo*/brown (7 percent), *moreno claro*/light of no clear race (6 percent), *preto*/black (5 percent), Negro (3 percent), and *claro*/light (3 percent). Telles was intrigued by the fact that more than one-third (38 percent) used the term *moreno* or *moreno claro*. *Moreno*, a term not used by the Brazilian census, means a "colored person" of ambiguous or no clear race. *Moreno claro* means a person of light color of ambiguous race.

Changing Contexts and Race

Until the 2000 Census, the U.S. system of racial classification required that an individual identify with only one of its official racial categories. While Americans have always acknowledged racial mixture unofficially by using (often derogatory) words like *mixed*, *mulatto*, *half-breed*, *mongrel*, and *biracial*, the government still forced anyone who has more than one racial background to choose only one racial category. This practice changed with the 2000 Census, when for the first time in its history, the United States allowed people to identify with two or more of its official racial categories. This change is monumental because until the 2000 Census, the United States had never officially recognized intermixture.

While the U.S. government now allows people to identify with more than one official race, it has yet to decide what to call people who do so. One thing is clear: to date the government has stated that it will not classify them as multiracial (Schmid 1997). It also appears that, even when given the choice of identifying with more than one racial category, almost 98 percent of Americans still identify with only one. Because the U.S. government now allows people to identify with more than one race, there are officially 63 race categories, including the six official racial categories and 57 combinations of those official categories (e.g., "Black-Asian" or "White-Black-Asian").

Like the United States, Brazil's system of racial classification is undergoing change after supporting multiracial identities for hundreds of years. In an effort to acknowledge and remedy discrimination against those with the darkest skin shades, Brazil's public universities have instituted affirmative action policies that now require applicants to identify with one of two racial categories—white or black ("Negro"). This system of racial classification is a two-category scheme advanced by those in Brazil's black consciousness movement—someone is either *preto* (Negro) or *branco*. That movement seeks to dismantle ideas of race as a continuum, to destigmatize "blackness," and to challenge the unspoken assumption that brown (which is treated as "not black") is superior to black. Black consciousness movement activists argue that the emphasis on multiracialism has discouraged browns and blacks from mobilizing to fight well-documented discrimination and prejudices. Thus, the movement encourages all people who see themselves as *moreno* and *pardo* to identify as Negro (Telles 2004; Bailey 2008).

Lance Cpl. Cory D. Polom

Study the faces of the people in this photograph. Using Brazil's new two-category system, decide who is *preto* (Negro) and *branco*. Imagine that you had grown up thinking that any amount of "white blood," no matter how dark one's appearance, would make someone "not black." With what category would you identify—*preto* or *branco*?

What Do Sociologists See?

These athletes happened to be Brazilian. In studying the faces of these athletes, sociologists recognize that American ideas of race do not apply in Brazil—that the athletes are likely to think of themselves as Brazilian and multiracial, not as black or white or Asian athletes playing for Brazil.

U.S. Department of State

Critical Thinking

How do you think the systems of racial classification in the United States and Brazil have shaped family relationships in the two countries?

Key Terms

chance

choice

context

Module 8.4

Minority Groups

Objective

You will learn the criteria sociologists use to determine minority group status.

Can you identify the minority in this photograph? What criteria did you use to make your selection?

Sgt. Mark Fayloga

Minority groups are populations within a society that are regarded and treated as inherently different from those in the mainstream. They are systematically excluded (whether consciously or unconsciously) from full participation in society and are denied equal access to power, prestige, and wealth.

Characteristics of Minority Groups

Depending on the context, anyone can hold minority status. If we think on a national scale about which groups are most likely to be excluded from full participation in society, what groups come to mind? Likely a long list of racial, ethnic, and religious groups comes to mind, along with women, those in the LGBT community, the very old and the very young, and the physically impaired. But in the context of applicants seeking employment as a daycare provider, we would have to list males of every race as a minority because of widespread and unfounded suspicions about their motives for wanting to be around young children.

Although we focus on ethnic and racial minorities, the concepts described here can apply to any minority. Sociologists have very specific criteria in mind when evaluating minority status. Sociologist Louis Wirth (1945) identified a number of essential traits that are characteristic of minority groups: involuntary status, lack of control of valued resources, exclusion from societal advantages, isolation from the dominant group, and minority group status overshadowing personal characteristics.

First, according to Wirth, membership in minority groups is *involuntary*. In fact, people are generally born into them. He argues that as long as people are free to join or leave a group, they do not constitute a minority. The idea that one cannot be free to join or leave a group to qualify for membership in a minority group raises questions about the meaning of being free to join or leave.

For example, one might be tempted to argue that these Muslim teenagers who wear headscarves are not a minority because they are free to remove their headscarves and blend in. The problem with this line of reasoning is that, for those who are devout Muslims, doing so violates some deeply held religious convictions.

U.S. Army photo by Sgt. Catherine Threat/Released

Second, *minority status is not based on size*. A minority may be a numerical majority in a society. In the United States, legally recognized racial-ethnic minorities are the ethnic category Hispanic and the four nonwhite racial groups—American Indian/Alaskan Native, Asian, Black/African American, and Native Hawaiian/Other Pacific Islander. All five are numerical minorities. However, there are geographic areas of the United States—counties, cities, neighborhoods—where nonwhite groups are the numerical majority.

Third, minority groups are *excluded from full participation* in the larger society. That is, as a group, minorities do not enjoy the advantages that members of a dominant group take for granted. **Advantaged** refers to symbolic support (positive images) and access to valued resources that give one group more opportunities relative to another, including the opportunity to live a long life, to make a good income, to survive the first year of life, and much more (see Table 8.4a).

Table 8.4a: Differences in Life Chances by Race and Sex (United States)

Life chances are a critical set of potential opportunities and advantages, including the chance to survive the first year of life, to grow to a certain height, to receive medical and dental care, to avoid a prison sentence, to graduate from high school, to live a long life, and so on (Gerth and Mills 1954). Why do you think Asian females live almost 17 years longer on average than black males? This dramatic difference speaks to the way in which race affects life chances. Which racial category has the highest chance of dying before age one? The lowest chance? Why might this be the case?

Chance of . . .	Highest Chance	Lowest Chance	Difference
Living a long life (average life expectancy)	Asian female 86.7 years	Black male 70.2 years	**16.5 years**
Graduating from high school on time*	Asian 91.4%	Black 61.5%	**29.9%**
Going to prison	Black male 32.5%	White female 1%	**31.5%**
Earning a high median weekly income (average salary working full time)	Asian male $936	Hispanic female $508	**$428 per week**
Dying before reaching one year of age	Black 1,100 per 100,000	Asian 389 per 100,000	**711 babies per 100,000**
Living in poverty*	Native American/Black 25.8%	White (non-Hispanic) 12.3%	**13.5%**
Having no health insurance	Native American 35%	White 11.9%	**23.1%**

Sources of data: U.S. Bureau of Labor Statistics (2012b); Stillwell (2010); U.S. Department of Justice (2007); U.S. Census Bureau (2011j); Murphy, Xu, Kochanek (2012)
*Data not available for sex categories.

In addition, minorities do not enjoy the freedom or the privilege to move within the society the way those in the dominant group do. Peggy McIntosh (1992), professor of women's studies, has identified a number of such **privileges**, special taken-for-granted advantages and immunities or benefits enjoyed by a dominant group relative to minority groups, including the following:

- Most of the time—when I am at school, work, or just out walking—I am in the company of people of my race or ethnic group.
- I feel confident that I can rent or purchase housing in any area in which I can afford to live.
- I can go shopping and not worry that I am being followed or targeted for surveillance by store detectives.
- I can do poorly on a test without worrying that my classmates or professor will attribute it to my race or ethnicity.
- Whether I use checks, credit cards, or cash, I can count on my skin color not to work against the perception that I am financially reliable.
- If I perform well at some activity that is associated with my racial or ethnic group, people will still recognize my hard work and not attribute my success to natural abilities. (McIntosh 1992, 73–75)

Royal Air Force photo by Sgt. Mitch Moore/Released

Brian Prechtel, USDA

If you look at these photos and think that being Muslim is not compatible with U.S. military service or that Asians are naturally good at science, then you have failed to accord them the taken-for-granted benefits accorded to those in dominant groups. The Muslim soldier (unlike a white soldier and assumed Christian counterpart) faces questions about loyalty to the United States. The Asian scientist is often not recognized for the hard work of learning science when others believe Asians are naturally good at the subject.

A fourth essential trait of disadvantaged status is that minorities are *socially and spatially isolated* from those in the dominant group. This isolation manifests itself in

- segregated residential arrangements (ethnic neighborhoods and/or gated communities),

- portrayals of specific minorities as particularly dangerous,
- laws prohibiting certain racial/ethnic groups from marrying those in dominant groups (miscegenation),
- acts of profiling that target minority groups for special surveillance, and
- underrepresentation of minorities in key political and economic positions. (Wooddell and Henry 2005)

NKU Philosophy, Rudy Garns

Many Americans hold irrational fears about black men being dangerous, to the point that they are nervous when in their presence. Many black men know this is the case, as illustrated by the following observation: "I am a large black man, and when I enter an elevator, I can see and sense people's fear. . . . I hate to see this fear and do everything possible to set them at ease. When I enter an elevator, I immediately smile, nod my head, go to the far corner and avoid any eye contact. I do not know whether that is proper elevator etiquette, but it is how I handle the situation" (Meeks 2011).

The fifth essential trait is that those who belong to minority groups are "*treated as members of a category*, irrespective of their individual merits" (Wirth 1945, 349). In other words, people focus on those physical and cultural characteristics that identify someone as belonging to a minority. Those characteristics are considered so important to the individual's identity that they overshadow other characteristics that person might possess.

The five characteristics of minority group status apply especially to the situation of **involuntary minorities**, those who did not choose to be a part of a country (nor did their ancestors); rather, they were forced to become part of it through enslavement, conquest, or colonization. Those of Native American, African, Mexican, and Hawaiian descent are examples. There are also some groups considered white today (most notably the Irish) that came to the United States involuntarily as indentured servants (Painter 2010).

Unlike peoples who voluntarily immigrated to the United States expecting to improve their way of life, those forced to become part of the United States had no such expectations. In fact, their initial forced incorporation involved a loss of freedom and status (Ogbu 1990). As a case in point, Native Americans (often referred to as First Nations people) and Native Hawaiians represent minority groups that were pulled into the United States against their will. In 1993, the Congress of the United States issued the Apology Resolution, in which it acknowledged that "the overthrow of the Kingdom of Hawaii occurred with the active participation of agents and citizens of the United States" and that "the Native Hawaiian people never directly relinquished to the United States their sovereignty as a people over their national lands."

The Queen of Hawaii in 1893 at the time of the overthrow was Liliuokalani. Her overthrow is not forgotten by those in the Hawaiian Sovereignty Movement who see the 1993 apology as not insufficient for addressing wrongs committed against the Hawaiian people. For one thing, the apology included no plan to make reparations of any kind.

Hawaii State Archives Digital Collections, PPWD-16-4-014

What Do Sociologists See?

In a racial/ethnic sense this woman possesses a kind of privilege that comes with being considered white. That characteristic allows her to shop most, if not all the time, without feeling store security is following or watching her.

NKU Sociology, Missy Gish

Critical Thinking

Describe a specific context in which you were a minority (not necessarily a racial/ethnic minority). Use Wirth's characteristic traits to inform your description of the experience.

Key Terms

advantaged
life chances
privileges
involuntary minorities
minority groups

Module 8.5 Racism

Objective

You will learn that racism relies on flawed logic to explain and justify differences and inequalities.

Do you view one of these three players as possessing more natural talent for the game of basketball?

NKU Philosophy, Rudy Garns

Is your prediction the athlete in the middle? Is your prediction based on a belief that blacks are superior athletes by nature and that whites are just not as good at basketball? Such a prediction qualifies as racist thinking.

Racism is a set of beliefs that uses biological factors to explain and justify inequalities between racial and ethnic groups. Generally people who hold racist beliefs treat them as accurate explanations of the existing state of affairs. On closer analysis, however, race-based explanations fall apart. Racism is structured around two notions:

- people can be classified into racial categories according to physical traits such as skin color, hair texture, and eye shape; and
- these physical attributes are deemed so significant that they are used to explain and determine behavior and inequalities. There is no other possible explanation for these inequalities.

Origins of Racism

Racism, or some variation on racism, as a way of explaining differences between groups of people has probably always existed. Modern racism, however, emerged as a way to justify European exploitation of people and resources in Africa, Asia, and the Americas. Between 1492 and 1800, Europeans learned of, conquered, and colonized much of North America, South America, Asia, and Africa, setting the tone for international relations for centuries to come. Among other things, this exploitation took the form of enslavement and **colonialism**, which occurs when a foreign power uses superior military force to impose its political, economic, social, and cultural institutions on an indigenous population in order to control its resources, labor, and markets (Marger 1991).

When Europeans' demands for low-cost and slave labor could not be met by native-born populations of places they colonized in the western hemisphere (the United States, Brazil, and the Caribbean, for example), they imported slaves from Africa or indentured workers from Asia and Europe. In fact, an estimated 11.7 million enslaved Africans survived their journey to the Americas between the mid-fifteenth century and 1870 (Chaliand and Rageau 1995). After slavery ended, the Europeans colonized the African continent. By 1914, nearly all of Africa had been divided into European colonies. The Europeans forced local populations in Africa to cultivate and harvest crops and to extract minerals and other raw materials for export to the colonists' home countries and beyond.

Racism helped justify this exploitation of nonwhite peoples and their resources by pointing to the so-called superiority of the white race. More precisely, the exploitation was justified by **scientific racism**, the use of faulty science to support systems of racial rankings and theories of social and cultural progress that placed whites in the most advanced ranks and stages of human evolution.

Library of Congress Prints and Photographs Division [LC-USZ62-136241]

This 1870 illustration shows caricatures of five different racial categories. Which female figure appears to be the standard against which other women are judged? The white-appearing woman in the center of the illustration is the ideal representation/standard of liberty against which the other women are compared. If the woman in the middle is the "standard," the caricatures of Irish and black women "deviate" the most from the ideal and are thus depicted as the least acceptable representations of liberty (and the nation).

Racial rankings such as those pictured above shaped how people, and not just in the United States and Europe, saw themselves. In Brazil, for example, many of the political elite embraced the doctrine of white superiority and instituted policies to "whiten" the population, arguing that in the end miscegenation would eliminate, or at very least dilute, the black population.

Flaws in Racist Arguments

Anyone who takes the time to look will find that race-based explanations of differences among people fall short. To take one example, people who argue that blacks are naturally superior athletes usually point to physiological differences that give black athletes an advantage over other athletes. As evidence that such differences exist, they point to the disproportionately high number of blacks participating in the money sports—the most visible, best-paid, and most televised sports (basketball, football, and boxing). Such evidence does not convince sociologists that blacks are naturally superior athletes.

In evaluating the argument that blacks are superior athletes, we must ask why black athletes from African countries do not dominate international competitions, such as the Olympic Games. In fact, athletes from the 55 African countries have competed in the Olympics for more than 50 years, and during that time, they have won a total of 347 Olympic medals out of about 17,500 awarded. During this time the United States won 2,681 (Wikipedia 2014).

Sociologists look to the social processes that channel athletes of a certain racial classification into a sport considered the domain of that race. With regard to black athletes, note that athletes and entertainers have always been the most highly publicized black achievers, and they have arguably been just as influential in shaping black identity as Martin Luther King Jr., Malcolm X, and other black leaders. That high visibility has surely played some role in channeling black athletes' talent toward the money sports (Hatfield 2006).

Sociologists would also point to other factors that channel black, white, and other athletes' talents toward specific sports and away from others. These factors include financial resources to pay for equipment, lessons, and playing time; encouragement from parents and peers; perceptions that a sport "belongs" to a particular race; and geographic location related to warm- and cold-weather sports.

Tim Hipps, U.S. Army Installation Management Command

Lance Cpl. Corey A. Blodgett

Finally, can we make the case that black basketball players are really better athletes than Olympic luge athletes or Koreans who excel in martial arts? Are speed, strength, and quick reflexes qualities that water polo and martial arts athletes also possess? What about hockey, gymnastics, swimming, soccer, cycling, sailing, rowing, archery, volleyball, skiing, and other less lucrative sports dominated by athletes of other races? Can we explain white dominance in water polo or Asian dominance in martial arts only as products of athletes' race?

What Do Sociologists See?

When looking at these American athletes from the 2014 Winter Olympics, sociologists notice the seeming absence of diversity. They would explain the relative absence of nonwhite athletes by the limited numbers living in cold-weather environments and the high costs associated with playing and training for these sports relative to, say, track and basketball. It is also not surprising to learn that nonwhite Olympians often cross over from track and field or football teams into bobsled teams.

U.S. Army photo by Gary Sheftick

Critical Thinking

Has your understanding of what constitutes racist thinking changed as a result of reading this module? Explain.

Key Terms

colonialism

racism

scientific racism

Module 8.6 Prejudice and Discrimination

Objective

You will learn about prejudice and discrimination and their consequences.

Imagine that you are standing alone late at night waiting for the elevator in an underground parking garage. Are you thinking about who might be on the elevator when the doors open? Are you hoping a person of a particular sex, race, age, social class, or nationality is on the elevator? Or is not?

Lisa Southwick

Prejudice

If you are hoping a person of a particular sex, race, age, social class, or nationality is on the elevator (or not), you likely hold prejudice toward those groups. A **prejudice** is a rigid and, more often than not, unfavorable judgment about a category of people that is applied to anyone who belongs to that category. Prejudices are supported by **stereotypes**—generalizations about people who belong to a particular category that do not change even in the face of contradictory evidence. Stereotypes give holders an illusion that they know the other group and that they possess the right to control images of the other group (Crapanzano 1985).

Stereotypes are supported and reinforced in a number of ways. First, in **selective perception**, prejudiced persons notice only the behaviors that support their stereotypes. These people use stereotypes as facts supported by their own observations (Merton 1957a). Many people stereotype Asians as being naturally good at math and sciences and think that their natural ability, in turn, explains why 25 percent of engineers in the United States are Asian despite the fact that Asians are less than 5 percent of the total population. Yet these same people do not point to natural ability to explain why almost 88 percent of airline pilots in the United States are classified as white males (U.S. Bureau of Labor Statistics 2013a).

Second, stereotypes persist in another way: when a prejudiced person encounters someone who contradicts a stereotype associated with their race, the former sees the latter as the exception to the rule; such a conclusion serves to support stereotypes. Third, prejudiced individuals keep stereotypes alive when they evaluate the same behavior differently depending on the race or ethnicity of those involved (Merton 1957a). For example, incompetent behavior of racial and ethnic minority members is often attributed to their race; in contrast, incompetence exhibited by someone from the advantaged group is almost always treated as a personal shortcoming.

Finally, stereotypes flourish through a process known as the **self-fulfilling prophecy** (Thomas and Thomas 1928, 572). A self-fulfilling prophecy begins with a false definition of a situation that is assumed to be accurate. That is, people behave as if their definitions are true. In the end, the misguided behavior produces responses that confirm the false definition (Merton 1957a). With regard to race, a self-fulfilling prophecy occurs when teachers, coaches, and parents channel a child's interests into areas they believe are appropriate to that child's race. Over time, real differences in quantity, quality, and content of instruction create seemingly race-based differences in talent. The cycle of self-fulfilling prophecies can be broken when the original assumption is challenged, changing the definition of situation.

NKU Sociology, Missy Gish

As an example of a self-fulfilling prophecy, many colleges and universities offer diversity scholarships, and they are disproportionately awarded to students who are classified as nonwhite or Hispanic/Latino. Many Americans have come to think of diversity scholarships as reserved for minorities, engineering scholarships for Asians, and so on. This thinking creates a self-fulfilling prophecy. As a result, parents of students classified as black and Hispanic may push their children to "go" for diversity scholarships, not thinking them eligible for other kinds of academic scholarships.

Discrimination

In contrast to prejudice, which is an attitude, discrimination is a behavior. **Discrimination**, intentional or unintentional, is the unequal treatment of racial or ethnic groups without considering merit, ability, or past performance. Discrimination blocks access to valued experiences, goods, and services. Sociologist Robert K. Merton argues that knowing whether people are prejudiced does not help predict whether they will discriminate. This is because prejudiced people do not always discriminate and unprejudiced people sometimes do. To illustrate this point, Merton describes four types of people.

1. **Nonprejudiced nondiscriminators** (all-weather liberals) accept the creed of equal opportunity, and their conduct conforms to that creed. They represent a "reservoir of culturally legitimized goodwill" because they not only

believe in equal opportunity but also take action against discrimination (Merton 1976, 193).

2. **Nonprejudiced discriminators** (fair-weather liberals) accept the creed of equal opportunity but discriminate because they simply fail to consider discriminatory consequences or because discriminating gives them some advantage. For example, whites decide to move out of their neighborhood after a black family moves in—not because they are prejudiced against blacks per se, but because they are afraid of declining property values.
3. **Prejudiced nondiscriminators** (timid bigots) reject the creed of equal opportunity but refrain from discrimination, primarily because they fear possible sanctions or being labeled as racists. Timid bigots rarely express their true opinions about racial and ethnic groups and often use code words such as "inner city" or "those people" to camouflage their true attitudes.
4. **Prejudiced discriminators** (active bigots) reject the notion of equal opportunity and profess a moral right, even a duty, to discriminate. They derive significant social and psychological gains from the conviction that anyone from their racial or ethnic group is superior to other such groups (Merton 1976). Prejudiced discriminators are the most likely to initiate hate crimes, actions aimed at humiliating someone in the target group or destroying their property or lives. Nonprejudiced discriminators are vulnerable to going along with this suggestion.

Sociologists distinguish between individual discrimination and institutionalized discrimination. **Individual discrimination** occurs when a person acts to block another's opportunities or does harm to life or property. **Institutionalized discrimination** is the established, customary way of doing things in society—the unchallenged laws, rules, policies, and day-to-day practices established by a dominant group that keep minority groups in disadvantaged positions (Davis 1978). Such discrimination is difficult to identify and rectify because the discrimination results from simply following established practices that seem on the surface to be impersonal and fair or part of the standard operating procedures.

Segregation

Racial and ethnic **segregation** is the physical and/or social separation of people by race or ethnicity. Segregation may be legally enforced (de jure) or socially enforced without the support of laws (de facto). The segregation may be spatial or hierarchical. Spatial segregation occurs when racial or ethnic groups attend different schools, live in different neighborhoods, and use different public facilities, such as restaurants and even drinking fountains. It also occurs when people of different racial or ethnic groups are in the same buildings but sit in different places for lunch or work on different floors or in different rooms (in the kitchen versus the dining area of a restaurant). Segregation is hierarchical when people in advantaged categories occupy the most prestigious positions while those in the disadvantaged categories are concentrated in the least prestigious positions, such as servants, maintenance workers, and laborers.

In the United States, the Jim Crow laws enforced racial segregation between whites and nonwhites from 1880 to 1964. These laws resulted in the establishment of separate but unequal race-specific bathrooms, recreational facilities, hospitals, and drinking fountains.

NKU Sociology, Missy Gish

If you use Facebook or other social media, do your "friends" reflect the racial diversity of the United States? Although racial segregation in public spaces is no longer supported by law in the United States, we must acknowledge that segregation exists in fact. For most Americans, all the important and meaningful events in our lives—weddings, funerals, graduation parties, and holiday gatherings—are experienced and shared with people considered the same racial category, or limited to two racial categories.

Redlining refers to institutionalized practices that deny, limit, or increase the cost of services to neighborhoods because residents are low-income and/or minority. Redlining can affect access to financial services (loans, checking accounts, credit cards, mortgages), insurance, health care, and grocery stores. The term *redlining* refers to a 1960s practice when banks actually marked red lines on maps, highlighting the communities in which they would not invest. The term was later applied to systematic discrimination against a geographically based population that possesses a characteristic labeled as not good for business. A National Community Reinvestment Coalition (2008) study found that lenders were more likely to issue higher-interest loans to African Americans and Hispanics than to whites with comparable credit histories (see Table 8.6a).

- In 71 percent of the 165 metro areas studied, middle- and upper-income blacks were at least twice as likely as their white counterparts to receive high-interest loans (such as 9.25 percent vs. a low-interest rate of 6.25 percent).
- In 22 percent of the 165 metro areas studied, middle- and upper-income Hispanics were at least twice as likely as their white counterparts to receive high-interest loans.
- In 47.3 percent of the 165 metro areas studied, lower-income blacks were at least twice as likely as their white counterparts to receive high-interest loans.

Table 8.6a: Monthly and Total Home Mortgage Payments by Interest Rates
This table shows the economic consequences of discriminatory lending patterns. A $140,000, 30-year mortgage financed at 6.25 percent translates to an $862 monthly payment, or $310,000 over the life of the loan. By contrast, a 30-year loan at 9.25 percent translates to a $1,152 monthly payment and a total of $415,000 over the life of the loan.

Cost of House	Interest Rate	Monthly Payments	Total Payments After 30 Years
$140,000	6.25%	$862	$310,000
$140,000	8.25%	$1,052	$379,000
$140,000	9.25%	$1,152	$415,000

Source: National Community Reinvestment Coalition (2008).

Ethnic Cleansing

Ethnic cleansing is an extreme form of forced segregation. It is a process by which a dominant group uses force and intimidation to remove people of a targeted racial or ethnic group from a geographic area, leaving it ethnically pure, or at least free of the targeted group. Ethnic cleansing also involves the destruction of cultural artifacts associated with the targeted groups, such as monuments, cemeteries, and churches. One example of ethnic cleansing was the forced removal of Native Americans from their ancestral lands.

Genocide is the calculated and systematic large-scale destruction of a targeted racial or ethnic group. The destruction can take the form of killing an ethnic group en masse, inflicting serious bodily or psychological harm, creating intolerable living conditions, preventing births, "diluting" racial or ethnic lines through rape and forced births, or forcibly removing children to live with another group (United Nations 1948).

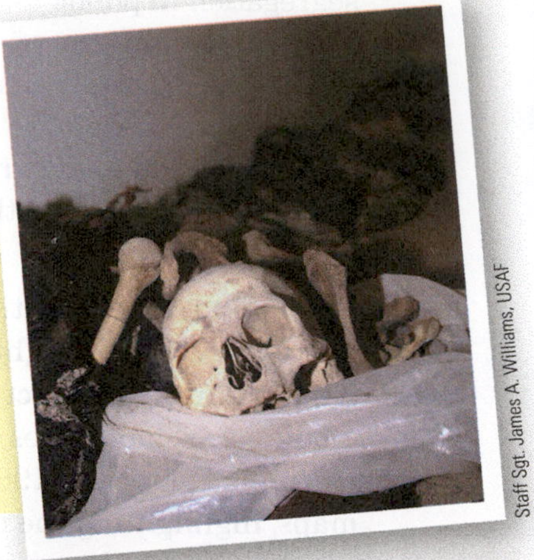
Staff Sgt. James A. Williams, USAF

By one estimate, 38 million people around the world died as a result of genocide in the 20th century (Oberschall 2000). Genocide is more often than not state-sponsored, in that a dominant group uses state apparatus (police, military, surveillance) to eliminate those targeted.

What Do Sociologists See?

Notice the sign to the left directing white women and "colored women" to move to different public spaces to get water and access to restrooms. This photograph was taken during the Jim Crow era; it represents physical and social segregation enforced by law. By law, "coloreds" and whites had separate bathrooms and drinking fountains.

Library of Congress Prints and Photographs Division [LC-DIG-fsa-8a26761]

Critical Thinking

Consider the people at the last important and personally meaningful social gathering you attended, whether it was a funeral, baptism, or wedding. What was the racial composition of the people attending this event? Explain.

Key Terms

discrimination
ethnic cleansing
genocide
individual discrimination
institutionalized discrimination
nonprejudiced discriminators
nonprejudiced nondiscriminators
prejudice
prejudiced discriminators
prejudiced nondiscriminators
redlining
segregation
selective perception
self-fulfilling prophecy
stereotypes

Module 8.7 Assimilation, Integration, and Pluralism

Objective

You will learn about the process by which racial and ethnic distinctions disappear, blend, or coexist.

Study these before-and-after photographs. Do you see the changes as merely physical or do you think this man has changed as a person?

Cumberland County Historical Society

Cumberland County Historical Society

These before-and-after photographs were taken in the mid- to late 1800s to showcase how Carlisle and other boarding schools "effectively" changed Native Americans from an uncivilized to a civilized state of being. They capture what sociologists call **assimilation**, a process by which ethnic, racial, and/or cultural distinctions between groups disappear because one group is absorbed, sometimes by force, into another group's culture or because two cultures blend to form a new culture. Two main types of assimilation exist: absorption assimilation and melting pot assimilation.

Types of Assimilation

In **absorption assimilation**, members of a subordinate ethnic, racial, and/or cultural group adapt to the ways of the dominant group, which sets the standards to which they must adjust (Gordon 1978). According to sociologist Milton Gordon, absorption assimilation has at least seven levels, in which a subordinate group

1. abandons (by force or voluntarily) its culture for that of the dominant group,
2. enters into the dominant group's social networks and institutions,
3. intermarries and procreates with the dominant group,

4. identifies with the dominant group,
5. encounters no widespread prejudice by those in the dominant group,
6. encounters no widespread discrimination by those in the dominant group, and
7. has no value conflicts with those in the dominant group.

A subordinate group is completely absorbed into the dominant group once all seven phases are achieved. Gordon maintains that level 1 assimilation is likely to take place before the other six are achieved. He also states that even when a group "abandons" its culture (level 1), it does not always lead to the other levels of assimilation.

Library of Congress Prints and Photographs Division [LC-USZ62-107203]

In the mid- to late 1800s and well into the next century, there were schools established to Americanize (and Christianize) Native American peoples including the Potawatomi, Winnebago, Chippewa, and Mesquakie, shown here, who attended the Mesquakie Day School, near Toledo, Iowa. These schools sought to achieve level 1 assimilation—that is, to force students to abandon their cultures by cutting their hair, assigning them "white" names, and requiring them to speak only English.

Gordon proposes that if those in the subordinate group are able to join the advantaged groups' social circles on a large enough scale (level 2 assimilation), a substantial number of interracial or interethnic marriages are bound to occur (level 3 assimilation) because of social interactions between the groups (Gordon 1978). Of the seven levels of assimilation, Gordon believes that gaining access to the advantaged racial group's social networks and institutions is the most important; if that occurs, the other levels of assimilation inevitably follow. Yet, in practice, gaining such access is very difficult.

Assimilation need not be a one-sided process in which a minority group is absorbed into the dominant group. Ethnic and racial distinctions can also disappear through amalgamation or **melting pot assimilation** (Gordon 1978). In this process, previously separate groups accept many new behaviors and values from one another, intermarry, procreate, and identify with a blended culture. The term *Blackanese* is used in reference to those who seek to identify as both black and Japanese. Those who see themselves as Blackanese seek to embrace both cultures and maintain that each culture has equal influence in their lives.

Integration

Assimilation, by definition, involves some level of integration. The term *integration* is often used in conjunction with the legal term **desegregation**, the process of ending legally sanctioned racial separation and discrimination. Desegregation often involves removing legal barriers to interaction and offering legal guarantees of protection and equal opportunity. **Integration** occurs when two or more racial groups interact in a previously segregated setting. Such integration may be in violation of the law or be court-ordered, legally mandated, or the natural outcome of people crossing the "color line" once legal barriers have been removed.

U.S. Air Force

When people celebrate the entry of a minority into social circles previously closed to them, they are celebrating a first and necessary step toward integration (e.g., first Muslim elected to the House of Representatives, first black chosen Miss America, first Asian selected as a first-round draft pick in the NFL, first white to attend a historically black college). Shown here is Dr. Mae Jemison, the first African American woman to travel into space. Jemison, born in 1956, was admitted into the astronaut training program in 1987 and eventually traveled into space in 1992 to conduct experiments on weightlessness and motion sickness.

When evaluating the extent of integration in society, it is important to ask this question: With whom do you live, learn, pray, celebrate, and mourn? If the answer involves only people of a single race or ethnicity, then one must conclude that, in fact, he or she lives a segregated life. Many Americans believe that racial and ethnic integration has been realized and that the United States is a color-blind society. This belief may be the result of a phenomenon known as **virtual integration**, in which simply seeing other racial groups on television and in advertisements gives "the sensation of having meaningful, repeated contact with other racial groups without actually having it" (Lynch 2007).

Pluralism is a situation in which different racial and ethnic groups coexist in harmony; have equal social standing; maintain their unique cultural ties, communities, and identities; and participate in the economic and political life of the larger society. These groups also possess an allegiance to the country in which they live and its way of life. In a pluralistic society, there is no one race or ethnic group considered to be the standard to which other races should aspire. Rather, cultural differences are respected and valued.

Although it is difficult to find an example of a country that practices pluralism in every way, it is possible to find descriptions of that ideal. The United States presents its ideal of pluralism as a melting pot in which the country welcomes immigrants from all over the world, who bring a vitality and energy to the country's way of life.

Sgt. Jasmine Chopra

Arguably the best example of pluralism in action is the U.S. military, which recruits people from every racial and ethnic group into its ranks. In assessing the extent to which pluralism is practiced, we might ask: To what extent do the soldiers have equal opportunity to advance and to maintain cultural identities?

The usefulness of the melting pot analogy must be assessed in light of the complex history of the United States. That history involved the European conquest of Native American peoples; the enslavement of African peoples; the annexation of Mexican territory, along with many of its inhabitants (who lived in what is now New Mexico, Utah, Nevada, Arizona, California, and Texas); and an influx of voluntary and involuntary immigrants from practically every part of the world. In addition, Puerto Ricans, American Samoans, Hawaiians, and other peoples all became part of the United States through a form of domination known as conquest or colonialism. The most celebrated group is voluntary immigrants—the millions of people who more or less chose to move to the United States.

One of the most interesting, significant, and long-lasting aspects of this global story is the U.S. government's establishment of a racial and ethnic classification scheme that applied to all who lived in and immigrated to the United States. The categories to which people were assigned reflected and reinforced the prejudices and discrimination of the times and set the tone for race relations then and far into the future. Perhaps as many as 2,000 distinct groups of Native American peoples, speaking seven different language families, were placed in a single category: "Indian." The millions of voluntary and involuntary immigrants from Europe eventually became "white." The peoples from all of Latin America became "Hispanic." Those from the Far East, Southeast Asia, and the Indian subcontinent were lumped into the category "Asian." The peoples from Hawaii and other Pacific islands (such as American Samoa and Guam) were eventually lumped into the category of "Native Hawaiian and Other Pacific Islander." Those of African descent became "black."

The Civil Rights Movement

Shortly after slavery was abolished in the United States, a state-sanctioned system of racial discrimination, known as Jim Crow (1877), was put into place. Under Jim Crow, blacks (and other minorities) were denied the right to vote and sit on juries; subjected to racial segregation (separate and unequal facilities); disadvantaged with regard to employment opportunities; and subjected to widespread, systematic discrimination, including violence against persons and property. After decades of struggle and resistance, that system was overturned with the ratification of the Civil Rights Act of 1964, the Voting Rights Act of 1965, and the Fair Housing Act of 1968. The civil rights movement was a response to that systematic discrimination, not just in the South but across the nation. The civil rights movement reached its most organized phase in the late 1950s and 1960s. It encompassed other related movements as well, including the American Indian Movement, La Raza Unida (the Unified Race), the antiwar (Vietnam) movement, and the women's movement.

In the popular imagination, the civil rights movement involved confronting blatant white supremacists such as the Ku Klux Klan. However, activists also confronted institutional discrimination as embodied in local, state, and federal agencies; judicial systems; and legislative bodies, including police, the National Guard, judges, and all-white citizens' councils. Most notably, police departments, especially in the South, arrested civil rights activists on false or trumped-up charges, and all-white juries found whites who murdered blacks not guilty. Some officials, such as Alabama Governor George Wallace, used the National Guard and state police to prevent school integration.

Library of Congress Prints and Photographs Division [LC-USZ62-107203]

College and high school students played key roles in the civil rights movement. They participated in bus boycotts, sit-ins, freedom rides, and school integration.

Black churches played a key role in the civil rights movement as well. In an atmosphere of profound discrimination and inequality, the black church had become not just a place to worship but also served as a community clearinghouse, a credit union, a support group, and a center of political activism (National Park Service 2009). Churches were the context for the emergence of key civil rights leaders such as the Reverends Martin Luther King Jr., Ralph Abernathy, Bernard Lee, and Fred Shuttlesworth.

Church and student groups founded organizations to coordinate their efforts, many of which were in place before this civil rights movement. These included the Southern Christian Leadership Conference (SCLC), the Student Nonviolent Coordinating Committee (SNCC), the Congress of Racial Equality, the National Association for the Advancement of Colored People (NAACP), and the National Urban League.

Rachel Ellison

Dr. Martin Luther King Jr. is the face and charismatic leader of the civil rights movement. He was deeply influenced by Gandhi's teachings and wrote: "Since being in India, I am more convinced than ever before that the method of nonviolent resistance is the most potent weapon available to oppressed people in their struggle for justice and human dignity." Over the course of the movement and up until his death, King was arrested 30 times.

The Selma-to-Montgomery March was an especially significant moment for the civil rights movement because ABC television broadcast it, making it the first event of the activist movement to be televised. What Americans saw created enough outrage to lend national support to the movement. It began on what is known as Bloody Sunday, March 7, 1965, when state and local law enforcement agents stopped 600 demonstrators six blocks into the march and attacked them with clubs and tear gas. Civil rights leaders sought protection from the courts to march and it was granted. The escalating intensity of this movement pushed the federal government to become involved. Most notably, President John F. Kennedy used the power of his office to enforce desegregation in schools and public facilities. Attorney General Robert Kennedy filed suits against at least four states to secure the right to vote for blacks. When Congress passed the Civil Rights Act of 1964 and the Voting Rights Act of 1965, President Lyndon Johnson signed them into law knowing that it might cost him the next presidential election and severely weaken the Democratic Party's chance of winning in the next election cycle. Of course, federal and Supreme Court judges also played key roles in ruling against segregation and discrimination (National Park Service 2009).

What Do Sociologists See?

Lacey Justinger, USAG Fort Bliss, Texas

The racial diversity reflected in this photograph is a legacy of the civil rights movement and longstanding efforts to integrate U.S. schools. That integration was accompanied by desegregation, the process of ending legally sanctioned racial separation. Since these children pictured are sons and daughters of U.S. military, we can point to the military efforts to support pluralism, a situation in which different racial and ethnic groups coexist in harmony and in which racial and cultural differences are respected and valued.

Critical Thinking

Assess the degree to which your life is integrated with people of other races and ethnicities by asking: With whom do I live, learn, pray, celebrate, and mourn? Does your life experience best identify with assimilation, integration, or pluralism?

Key Terms

absorption assimilation
assimilation
desegregation
integration
melting pot assimilation
pluralism
virtual integration

Module 8.8 Applying Theory: Racial Classification

Objective

You will learn how the four theoretical perspectives help us think about the meaning and purpose of racial classification.

With what race do you think this person identifies—American Indian/Native Alaskan, Asian, Black/African American, Native Hawaiian/Pacific Islander, white, or some other race?

NKU Anthropology, Sharyn Jones

According to the U.S. official definitions of race, the man pictured here is Native Hawaiian/Pacific Islander because he is Fijian. While he acknowledges his Fijian roots, he views himself as a descendant of one of the Lost Tribes of Israel (Ham) and as such considers himself "black."

Since 1790, the U.S. government has required that someone in each household across the United States complete a survey which has always included a question asking about race. Why does the United States do this? Each of the theoretical perspectives—functionalist, conflict, symbolic interaction, and feminism—offers a framework for thinking about this U.S. policy. Keep in mind that while the United States now allows people to identify with more than one race, less than 3 percent of the population does so (U.S. Census Bureau 2011g).

Courtesy of Joan Ferrante

Upon immigrating the United States, this woman of Pakistani descent learned she was "Asian" while filling out the required paperwork and her uncle told her to check "Asian," and, as she said, "it stuck." From a functionalist perspective, racial classification is a tool the U.S. government uses to count and manage its population. Answering the race question on surveys, census forms and applications can be thought of as rituals that reinforce the racial reality the government has constructed. Think about how often you have been asked to reveal your race on an application or survey. In doing so you are supporting the existing system of racial classification.

Chris Caldeira

Conflict theorists would point out that today's system of racial classification cannot be separated from its historical roots, which can be traced to the exploitation of people and resources in Africa, Asia, and the Americas. Racism and the racial categories that form the cornerstone of racist thinking helped to justify this exploitation of nonwhite peoples and their resources. Conflict theorists maintain that when the United States divides the 318 million residents of the United States into five racial and two ethnic categories, it reinforces differences and competing interests. As one example, conflict theorists question why the U.S. government maintains that Hispanics can be of any race and why Hispanic is the only named ethnic category in the United States. If, for example, Hispanics were classified as black, then together the two would represent 28.6 percent of the population. Separately, however, Hispanic represent 16.3 and blacks 12.3 percent of total population.

Interactionist Perspective

NKU Sociology, Missy Gish

Among other things, symbolic interactionists are interested in self-referent terms—the words we use to describe the self and the words we believe other people use to think of us. Symbolic interactionists would recognize that the system of racial classification gives people words to think about self and others. The two children pictured are cousins, but one appears white and the other black. Symbolic interactionists are interested in how each incorporates white- or black-appearing family members into their identity. For example, how might this little boy see himself relative to his father, who appears black, and his mother, who appears white. Or how might this young girl think of herself relative to her white-appearing cousin?

Library of Congress

In studying the connection between a system of racial classifications and gender relations, feminists believe that it cannot be studied apart from the larger political structures, including the government, that make and enforce rules governing racial classification. Feminists are compelled to ask: What does it mean to live in a society that created a categorical system of racial classification such that biological parents can be considered a different race from their children (e.g., a "white" father can have a "black" child)? The categorical system can be traced to enslavement, when the offspring of master and enslaved followed the condition and race of the mother. Frederick Douglass, the great orator and abolitionist pictured here is viewed as black. Yet in his autobiography he writes that he knew his father to be white and the "opinion was . . . whispered that my master was my father; but of the correctness of this opinion I know nothing." He could not know for sure because enslaved women were forbidden by law to name the father of their children.

The four perspectives taken together help us to think about the context in which this father-son relationship is embedded. From a *functionalist* point of view, if this man and son identify as different races they are supporting the social order that treats race as categorical. From a *conflict* point of view when people view themselves as belonging to a distinct racial category it reinforces racial boundaries and cultivates differences. Such a view discourages people—like this little boy—from embracing the racial history and ancestors that do not match up with his physical appearance. A *symbolic interactionist* point of view encourages us to think about ways the language of race shapes the father and son's racial identity (e.g., white son of a black father) and also the way others think about their relationship. In a society that has treated "black" and "white" as distinct categories there is no model for how this son can incorporate his black heritage into his sense of self? Finally the *feminist* perspective alerts us to a system of racial classification that recognizes one gender (in this case the mother) as determining the race of the child while dismissing the other parent (in this case the father) as unimportant.

NKU Sociology, Missy Gish

Summary: Putting It All Together

Chapter 8

Sociologists make a distinction between the concepts of race and ethnicity. Race is a physiologically based classification scheme in which racial categories are typically associated with selected physical traits, such as hair texture and skin shade, that have been assigned extraordinary social and political significance. The fact that Brazil and the United States have very different ideas about race tells us that race is human created. That is, people make the categories, attach meanings to them, and give them social significance with real consequences.

An ethnic group consists of people who share, believe they share, or are believed by others to share a national origin; a common ancestry; a place of birth; and/or distinctive social traits that set them apart from other ethnic groups. Ethnic identification is not just a matter of individual choice. People selectively remember and forget some ethnicities that make up their heritage. Some racial groups have more freedom than others in claiming or disclaiming an ethnic identity. Sometimes ethnicity is imposed, and sometimes it is revived and renewed. The United States recognizes six official racial categories and two ethnicities (e.g., Hispanic and nonHispanic). Like all such classification schemes, the U.S. categorical system is characterized by fatal flaws in logic. For one thing, there is no sharp dividing line to mark the boundaries that distinguish one race from another. Second, millions of people are products of sexual unions between people who appear to be different races so their offspring by definition cannot be a single category. Finally, the diversity within racial categories is so great that knowing someone's racial category tells us little about them as a person.

Disadvantaged racial and ethnic groups are considered minority groups. Among other things, minorities do not choose that status and those who occupy that status are systematically excluded from full participation in society and are denied equal access to opportunities to acquire power, prestige, income, wealth and other valued resources. Disadvantaged status is justified and perpetuated by racism, prejudice, and discrimination. Prejudice, discrimination, and racism are the cornerstones of segregation, ethnic cleansing, and genocide.

Assimilation is a process by which ethnic and racial distinctions between groups disappear because one group is absorbed into another group's culture (absorption assimilation) or because two cultures blend to form a new culture (melting pot assimilation). Assimilation, which involves some level of integration, is facilitated by desegregation. Pluralism is a situation in which different racial and ethnic groups coexist in harmony; have equal social standing; maintain their unique cultural ties, communities, and identities; and participate in the economic and political life of the larger society.

Chris Caldeira

When out in a public setting, did you ever go into the "wrong" bathroom by accident? How did you feel? Have you ever been out with someone—an older child but one still too young to go to bathroom by themselves or an adult who needs help going to the bathroom—but you were the "wrong" gender to accompany them? Or do you know someone, including yourself, who identifies as transgender and prefers going to a bathroom that corresponds to the gender with which they identify? Do you think there might be some males (children, teens, adults) who feel uncomfortable using a urinal? These questions get at the problems associated with sex-segregated bathrooms. Asking these kinds of questions forces us to imagine another way of organizing an essential human activity. Talk of changing the way people use public bathrooms is explosive because change means we must think about, acknowledge, and accommodate populations who we have asked to make do and to keep quiet about an arrangement that discriminates and that makes many feel uncomforable to the point of believing that they are committing a serious crime if they must go into or accidently venture into the "wrong" bathroom (Greenaway 2009).

Module

9.1 Sex and Gender

Objective

You will learn that sex is an anatomical distinction and that gender is a social construction.

Chris Caldeira

Can you tell if this baby is a boy or a girl? If so, how? If not, why?

The decision to cover the baby boy in pink, blue, green, yellow, or some other color relates to the concept of gender, a society's beliefs about how boys, girls, women, and men are expected appear and behave. While this baby seems happy and not to care that it is covered with a green blanket, many parents believe it is very important to dress and surround their babies in colors so people know whether their baby is a girl or a boy.

Sex

A person's **sex** is based on **primary sex characteristics**, the anatomical traits essential to reproduction. Most cultures classify people at birth into two sex categories—male and female—based on what are considered to be clear anatomical distinctions. Biological sex, however, is not clear-cut, if only because some babies are born intersexed. The medical profession uses the broad term **intersexed** to classify people with some mixture of male and female primary sex characteristics. Although we do not know how many intersexed babies are born each year, medical experts who treat intersexed children estimate that one or two in every thousand babies born undergo surgery to "correct" their genitals in some way (Dreifus 2005; Intersex Society of North America 2014).

If some babies are born intersexed, why does society not legally recognize an intersexed category? In the United States and most other countries, no such category exists because such children are typically treated with surgery and/or hormonal therapy. The rationale underlying medical intervention is the belief that the condition "is a tragic event" about which something must be done (Dewhurst and Gordon 1993, A15). Why does no clear line exist to separate every newborn into one of two categories, male or female? One answer lies in the biological sequence of events that occur in the first weeks after conception, at which point the human embryo develops the potential to form either ovaries or testes. Approximately eight weeks into development, the ovaries or the testes disintegrate. About a week later, the outer appearance we come to associate with male or female begins to develop. This complex chain of events does not always occur as theorized (Lehrman 1997, 49).

Do you question whether this female wrestler carrying Mrs. Obama at a Team USA event is genetically female? Female athletes who exceed expectations—whose performance far exceeds that of female opponents—face the risk of competitors challenging their status as female. Such tests take several weeks to complete. The test involves a medical examination and reports from a gynecologist, an endocrinologist, a psychologist, an internal medicine specialist, and an expert on gender (Dreger 2009).

Official White House photo by Sonya N. Hebert

In addition to using primary sex characteristics to distinguish one sex from another, we also use what are called **secondary sex characteristics**, physical traits not essential to reproduction, such as breast development, quality of voice, and distribution of facial and body hair, that supposedly result from the action of so-called male (androgen) and female (estrogen) hormones. We use the term so-called because all people produce androgen and estrogen (Garb 1991). Like primary sex characteristics, there is no clear line to mark any secondary sex characteristic as distinctly male or female. For example, there is no demarcation that separates a male voice from that of a female or a female pattern of body hair distribution from that of a male.

Gender

Gender is the socially created and learned distinctions that specify the physical, behavioral, and mental and emotional traits believed to be characteristic of the recognized sexes, males and females. Thus the terms **masculinity** and **femininity** refer to traits believed to be characteristic of males and females. Masculinity and femininity are concepts that are taught, learned, emulated, and enforced (Lorber 2005).

Ideas about what constitutes masculinity and femininity are often expressed as **gender ideals**. A gender ideal is at best a caricature, in that it exaggerates the characteristics believed to make someone the so-called perfect male or female. In fact, gender ideals may not exist in reality. Consider that few, if any, women have feet that are 4–6 inch long or 18-inch waistlines. Yet at one time, 4–6 inch feet were considered the *ideal* size for women in China—an impossible standard that no female could achieve without enduring foot binding as a young girl. Likewise, an 18-inch waist was once an ideal in the United States.

Library of Congress Prints and Photographs Division [LC-USZ62-101143]

Library of Congress Prints and Photographs Division[LC-USZ62-104036]

The two photos speak to the power of gender ideals to set a standard of beauty with regard to waist and foot size that few, if any, women can achieve without taking extreme measures. Without the help of a corset, and sometimes surgery to remove the two lower ribs, few women have a wasp-like waist. Likewise, few, if any, women have feet that are 4 inches in length without breaking the four smallest toes on each foot and binding them under the heel. The X-ray shows the bone structure of a foot that has been subjected to binding. Both images were created between 1888 and 1920.

In thinking about masculinity and femininity, it is important to realize that there are many forms of masculinity and many forms of femininity and what is defined as such varies by "region, religion, class, national culture, and other key social factors" (Gendered Innovations 2014). People behave and think in ways that depart from what they and others perceive their gender to be. Professional tennis player Serena Williams may present herself as feminine in dress and hairstyle, but she plays tennis in ways that many label as masculine. The point is that "any one person—woman or man—engages in many forms of femininity and masculinity," depending on context, expectations of others, age, and other factors (Gendered Innovations 2014).

To this point, we have drawn a distinction between sex and gender. Although sociologists maintain that there is no sharp line that distinguishes so-called male traits from so-called female ones, they would never argue that there are no biological differences. Sociologists are, however, interested in the extent to which people accentuate biological differences as a way of meeting gender ideals. They are also interested in ways gender ideals shape practically every aspect of life—influencing, among other things, how people dress, how they express emotions, and what occupations they choose. In this regard we can look at the ways gender ideals are used to sell products.

Commodification occurs when economic value is assigned to things not previously thought of in economic terms such as an idea, a natural resource (water, a view of nature), or a state of being (youth, sexuality). The **commodification of gender ideals** occurs when companies create products for people to buy so that they can express themselves in ways that meet ideals their society holds about some gender category. These products are accompanied by advertising campaigns capitalizing on gender insecurities and promising consumers that if they buy and use the products they achieve relief. The list of available products is endless, especially to achieve ideals of femininity. These products are aimed to feminize from the tip of toes (nail polishes) to the top of the head (hair dye). Products on the market for men include erectile dysfunction drugs, touch of gray hair products, body gels, body hair removers, and muscle-building drinks.

The trend toward marketing cosmetics to preteens has led critics to claim that young girls are being hypersexualized and in the process are joining the ranks of older women, many of whom have been socialized to believe that they fall short in their natural state (Wiseman 2003).

NKU Sociology, Missy Gish

Transgender

Not everyone can or wants to meet gender ideals associated with masculinity and femininity. Like the concept of sex, there is no line separating one gender from another. Simply consider those who identify or are labeled as transgender as further evidence that no clear line separates male from female. In the most basic and broadest sense, the term **transgender** includes those who identify with a gender that is at odds with their socially or officially recognized sex, as determined at birth. Some "refuse" to do the work required to comply with ideas of masculinity and femininity; others challenge existing ideals and seek to create alternative gender categories with which they feel more comfortable. Even when people resist gender ideals, those ideals are pointed to as what they are not. It should not surprise us, then, that some societies recognize and allow (although they may not completely accept) people to be something else.

There are a variety of popular and academic terms used to label those who do not meet, or are perceived as not meeting, the gender expectations associated with being male and female. Terms like "third sex," "other gender," and "transsexual" are among the broad concepts most widely used to label those self-defined or socially defined as being neither man nor woman but something in between, or as being completely distinct from male and female. Some societies (past and present) have specific labels for other genders including the following:

- Kathoey of Thailand—those born anatomically male but "having a female heart."
- Whakawahine/Wakatane of New Zealand (Maori culture). Whakawahine applies to those born anatomically male who prefer the company of females and pursue occupations that are considered female. Wakatane applies to those born anatomically female who pursue occupations associated with males, most notably warriors.
- Mamluk of Egypt—a term used in AD 1200–1700 in reference to girls perceived to have masculine characteristics who were raised as boys and accorded legal status and advantages of males (Independent Lens 2013).

- Fa'afafine of American Samoa and other Polynesian cultures. According to the Samoa Fa'afafine Society, "to be a fa'afafine you have to be a Samoan, born a man, feel like you are a woman, be sexually attracted to males and, importantly, proud to be called a fa'afafine" (Montague 2011). *Fa'afafine* are people who are not biologically female but who have taken on the "way of women" in dress, mannerism, appearance, and role.

It is important to keep in mind that those who make up third gender populations are very diverse and that labels such as those used in the United States such as gay, transvestite, drag queen, or even transgender do not capture the essence of being *fa'afafine* in Samoa, *kathoey* in Thailand, or *mamluk* in Egypt (Fraser 2002; VanderLaan et al. 2012).

Courtesy of Asia and Pacific Transgender Network Development

This photograph is striking in that it shows people, once considered biologically male, who transitioned to females and are part of the Asia and Pacific Transgender Network (APTN), an organization devoted to saying "no" to discrimination and marginalization and to championing the health, legal, and social rights of those who are transgender.

What Do Sociologists See?

Sociologists see an advertisement promoting a gender ideal—that femininity can be achieved by wearing shoes like those in the advertisement. Many women who wear such shoes complain that they are uncomfortable, even painful to wear. The average woman can stand in them about one hour before pain "kicks in." One study by the Society of Chiropodists and Podiatrists found 40 percent of women admit enduring pain for the sake of fashionable shoes, compared to 12 percent of men. Moreover, 90 percent of women claim to have experienced some foot problem, ranging from blisters to muscular strains, from wearing uncomfortable shoes (*New York Daily News* 2013).

NKU Sociology, Boni Li

Critical Thinking

Can you think of a time in your life when you worked particularly hard to achieve a gender ideal or to resist a gender ideal? Explain.

Key Terms

commodification

commodification of gender ideals

femininity

gender

gender ideals

intersexed

masculinity

primary sex characteristics

secondary sex characteristics

sex

transgender

Module 9.2 Gender Socialization

Objective

You will learn the process by which people, especially children, learn gender expectations.

Scott Hansen, Northwest Guardian

Nathan Pfau

Think back to the activities you engaged in as a child. Did you choose those activities or did your parents choose them for you? Did they support your choices?

Masculinity and femininity are not innate or natural qualities. People learn to be masculine or feminine and to appear and to do the things that their society considers masculine and feminine. When a baby is presented to the world as male or female, just about everyone who comes in contact with that child begins to treat them as such. When a baby is presented in an ambiguous way, people work to learn the baby's gender.

With prompting and encouragement from others, children learn to talk, walk, and move in gendered ways (Lorber 2005). They also learn **gender roles**, the behavior and activities expected of someone who is male, female, or some other gender. These expectations channel energies in gender-appropriate directions. As children learn their society's expectations about how boys, girls, and transgender should look and behave, most will, by their own behavior and appearance, reproduce and perpetuate those expectations. Children learn that when they fail to behave in what are considered gender-appropriate ways, there are consequences (Lorber 2005).

Children are taught to be a gender when parents and others intentionally convey (or choose not to convey or resist) the larger societal expectations to be one of two categories. Children also learn about gender in indirect ways when they observe others' behaviors toward those are considered male, female, or transgender; hear jokes, comments, and stories; and see how men, women, and transgender are portrayed in the media.

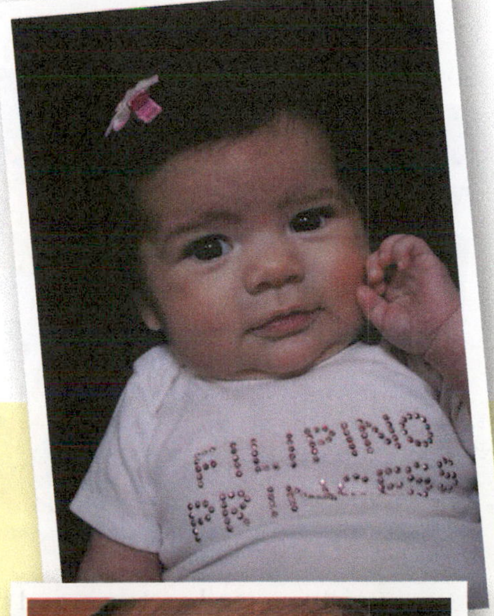

Courtesy of Missy Gish

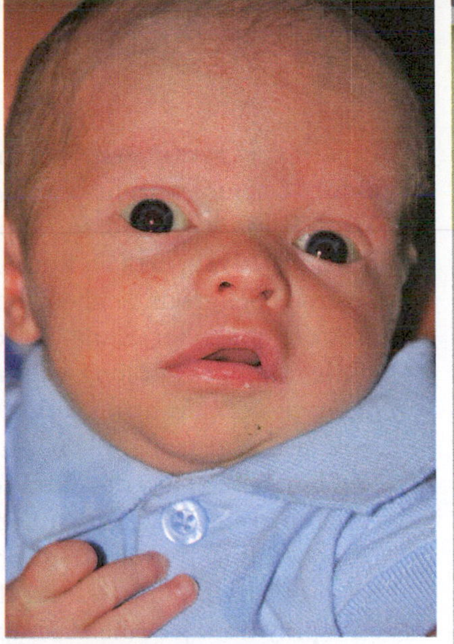
Chris Caldeira

Socialization theorists argue that an undetermined, yet significant, portion of male-female-transgender differences are products of the ways in which people are treated. These two children pictured are too young to care or choose on their own whether they want to wear clothes considered gender-appropriate, but their parents have invested money and time to present their babies in clothes and colors symbolic of a particular gender.

Agents of Socialization

Agents of gender socialization are the significant people, groups, and institutions that act to shape our gender identity—whether we identify as male, female, or something in between. Agents of socialization include family, classmates, peers, teachers, religious leaders, popular culture, and mass media. There have been a number of studies documenting ways caregivers reward some behaviors and ignore others depending on whether a child is known or thought to be a boy or a girl. For example, early childhood teachers are more accepting of girls' cross-gender behaviors and explorations than they are of boys'. According to this research, teachers believe that boys who behave like "sissies" are at greater risk of growing up to be homosexual and psychologically ill-adjusted than are girls who behave like "tomboys." This finding suggests that while American society has expanded the range of behaviors and appearances deemed acceptable for girls, it has not extended the range for boys in the same way (Cahill and Adams 1997).

Likewise, children's toys and celebrated images of males and females figure prominently in the socialization process. Barbie® dolls, for example, have been marketed since 1959 with the purpose of inspiring little girls "to think about what they wanted to be when they grew up." The dolls are available in 67 countries. An estimated 95 percent of girls between ages 3 and 11 in the United States have Barbie® dolls, which come in several skin colors and 45 nationalities (Mattel 2010).

NKU Sociology, Missy Gish

For boys, G.I. Joe was the first action figure toy on the market, launched in 1964, and it was followed by a long line of action figures, including Transformers™, Micronauts™, Star Wars™, Power Rangers™, X-Men™, Street Fighter™, Bronze Bombers™, and Mortal Kombat™. The popularity of such toys is boosted by comic books, motion pictures, and cartoons, and they appear on school supplies, video games, card games, lunch boxes, posters, and party supplies (Hasbro Toys 2010; Son 1998).

It is important to point out that the gender socialization process is not uniform, if only because parents vary in how they raise their children. In *The Gender Trap: Parents and the Pitfalls of Raising Boys and Girls*, sociologist Emily Kane (2012) reports on findings from 42 in-depth interviews with parents from diverse backgrounds about how they go about socializing their children in matters of gender. Kane found that some parents give little to no thought to, while others think long and hard about, gender socialization. Kane identified five types of parenting styles: naturalizers, cultivators, refiners, innovators, and resisters. Naturalizers are parents who believe that biology is destiny, that girl and boy babies are destined to be feminine and masculine, respectively. Naturalizers are inclined to strongly discourage gender-atypical behavior and feel a responsibility to raise their children in gender-appropriate ways. Cultivators, on the other hand, understand that gender is taught and learned and that parents play a major role in shaping gender identity. Consequently, they cultivate gender-appropriate behavior because they believe it is in the best interest of their children to do so.

Refiners are parents who believe biology and socialization are key factors in determining gender and will permit, even encourage their children to deviate in modest ways from gender expectations, but will not push to make significant changes in the gendered norms and expectations that shape their children's lives outside the home. Innovators place greater emphasis on socialization than on biology and are optimistic in their belief that they can raise children in gender-neutral ways, and they are committed to addressing the oppressive elements of gender. Finally, resisters not only encourage nonnormative behavior, they actively reject raising their children to meet traditional gender expectations as well.

Kari Hawkins, USAG Redstone

Cpl. Timothy Childers

While it is difficult to label a parenting style by the one moment in time that each of these photographs captures, what clues might you use to predict the parenting style of this military mother holding her daughter and of the father encouraging his son to give it his best shot? Use sociologist Emily Kane's typology described above to narrow each parent's style down to one or two of the five styles.

Norms Governing Body Language

Learning to be male or female involves learning norms governing the way males and females present themselves. That includes learning the sex-appropriate norms governing body language. Norms governing male body language suggest power, dominance, and high status, whereas norms governing female body language suggest submissiveness, subordination, vulnerability, and low status. These norms are learned, and people give them little thought until someone breaks them, at which point everyone focuses on the rule breaker.

Lisa Southwick

"In this typical office scene, the man in the photo holds power with an authoritative stance—one hand in pocket and the other at mid-chest, straight posture, and head high; the woman is submissive with smile, arms and hands close to her body. Note the man's wide, stable stance and the woman's unstable stance. Many women tend to slip into a posture similar to that shown when talking to a shorter male authority figure." (Mills 1985, p. 8)

"In [this] photo, the man defers to authority by assuming a feminine, subordinate posture—with scrunched-up spine, constricted placement of arms and legs, canted head, and smiling attentiveness" (Mills 1985, 8).

Lisa Southwick

Such norms governing appropriate body language for males and females can prevent women from conveying a sense of security and control when they are in positions that demand these qualities, such as a lawyer, politician, or physician. In this regard, women face a dilemma: to be perceived as feminine and nurturing, a woman needs to appear "passive, accommodating, affiliative, subordinate, submissive, and vulnerable." To be perceived as a competent manager, a woman needs to appear "active, dominant, aggressive, confident, competent, and tough" (Mills 1985, 9).

What Do Sociologists See?

The photograph captures but a moment in time. It does suggest the child has at least one parent whose parenting style might be labeled as a refiner, innovator, or even a resister. It is clear that the child has access to toys girls and boys play with and feels free to play with both kinds of toys.

Terra Schultz, Courtesy of Joan Ferrante

Critical Thinking

Describe the parenting style(s) that best characterizes the one to which you were exposed as a young child.

Key Terms

agents of gender socialization

gender roles

Module 9.3

Sexualities and Sexual Orientations

Objective

You will learn the meaning of sexuality and sexual orientation.

Do you think your parents wondered before you were born what your sexual orientation might be?

Chris Caldeira

We are bombarded daily with messages about sexuality. They may come from sex education classes that warn of the health dangers of unprotected sexual activity; from fairy tales; from song lyrics; or from commercials, movies, and news events. Other messages come from those close to us: friends who come out to us or parents who kid us about having a boyfriend or girlfriend. Finally, messages come from observing the treatment of people around us: we notice uncomfortable reactions toward women who breast-feed their babies in public and toward a man who appears feminine or a woman who appears masculine; we take notice of the boy and girl everyone wants to date or not date; we take note of reactions and facial expressions when someone says he or she is from San Francisco.

Sexuality

Sexuality encompasses all the ways people experience and express themselves as sexual beings. The study of sexuality considers the range of social activities, behaviors, and thoughts that generate sexual sensations and experiences and that allow for sexual expression. Sexuality is not an easy subject to present, for several reasons.

NKU Anthropology, Sharyn Jones

First, for most of us sex/sexuality education focused on abstinence and the dangerous consequences of sexuality, most notably on sexually transmitted diseases. Those lessons were usually uninformed by any discussion of what to make of sexual excitement, sexual attraction, or the relentless messages regarding sexuality all around us.

Second, people who have had difficult sexual experiences—those molested or raped as children, men who cannot achieve erections or orgasms, and women and men who have been sexually assaulted—may be uncomfortable with the topic (Davis 2005). Third, it is very difficult to discuss human sexuality in all its dimensions when heterosexuality and all that it entails is presented as normal and legitimate and any sexuality outside that norm is considered deviant and in need of fixing.

Sexual Orientation

Sexual orientation is an expression of sexuality. While sociologists are interested in the topic of sexual orientation, the American Sociological Association does not issue a statement about what sexual orientation means. According to the American Psychiatric Association (APA) (2009), **sexual orientation** refers to "an enduring pattern of emotional, romantic, and/or sexual attractions to men, women, or both sexes. Sexual orientation also refers to a person's sense of identity based on those attractions, related behaviors, and membership in a community of others who share those attractions." The word *enduring* suggests that one encounter does not make someone gay or lesbian. This caveat speaks to the fact that many people have experienced at least one same-sex sexual encounter at some point in their lives. Results from the most recent survey conducted by the Centers for Disease Control (2011b) found that 1 in 8 women and 1 in 16 men ages 15–44 have had a sexual experience with someone of the same sex.

Sexual orientation falls along a continuum, with its endpoints being exclusive attraction to the other sex and exclusive attraction to the same sex. In the United States, we tend to think of sexual orientation as falling into three distinct categories: heterosexual (attraction to those of the other sex), gay/lesbian (attraction to those of one's own sex), and bisexual (attraction to both men and women). It is important to realize that there are other labels that cultures apply to expressions of human sexualities (APA 2009).

Sexual orientation should not be confused with other related and intertwining terms that shape the experiences of sexuality and sexual orientation, including:

biological sex—the physiological, including genetic, characteristics associated with being male or female;

gender role—the cultural norms that guide people in enacting what is considered to be feminine and masculine behavior;

gender identity—the awareness of being a man or woman, of being neither, or something in between (gender identity also involves the ways one chooses to hide or express that identity); and

transgender—the label applied to those who feel that their inner sense of being a man or woman does not match their anatomical sex, so they have to undergo medical procedures and behave and/or dress in ways that actualize their gender identity.

People enact sexual orientation in relationships with others. Thus, according to the APA (2009), "sexual orientation is closely tied to the intimate personal relationships that meet deeply felt needs for love, attachment, and intimacy." Based on what we know to date, the core attractions that emerge in middle childhood through early adolescence prior to sexual experiences are the foundation of adult sexual orientation. The experiences of coming to terms with sexual orientation vary. People can be aware of their sexual orientation even if they are celibate or have yet to engage in sexual activity. Others come to label their orientation after a sexual experience with a same-sex and/or other-sex partner. Still others ignore, suppress, or resist pulls toward those of the same sex because of widespread social disapproval (APA 2009).

Sociology, Missy Gish

Serina Beauparlant, Courtesy of Joan Ferrante

In addition to sexual behaviors, sexual orientation includes "nonsexual physical affection between partners, shared goals and values, mutual support, and ongoing commitment. . . . One's sexual orientation defines the group of people in which one is likely to find satisfying and fulfilling romantic relationships that are an essential component of personal identity for many people" (APA 2009).

Every group has established **sexual scripts**, responses and behaviors that people learn, in much the same way that actors learn lines for a play, to guide them in sexual activities and encounters. These scripts are gendered in that males and females learn different scripts about the sex-appropriate responses and behavioral choices open to them in specific situations (Stein 1989). Even if they resist following the script, people know the script they are expected to follow and must come to terms with accepting or rejecting that script. The sexual scripts of the dominant culture call for behaviors and responses that support its definitions of what it means to be heterosexual. Other sexual scripts constructed by those in lesbian, gay, bisexual, and transgender (LGBT) communities and other cultures are dismissed as deviant.

Social Movements

If we simply think about the men and women we encounter every day, we quickly realize that many people cannot, do not, or outright refuse to express their sexuality in idealized ways. Social movements occur when enough people organize to make a change, resist a change, or undo a change in some area of society.

It is not easy to piece together the complete history of LGBT revolutionary social movements and the reactionary movements that emerged to oppose any gains made. In the United States, LGBT movements can be dated to the establishment of the first gay rights organization in Chicago in 1924 (the Society for Human Rights). Over time, the movements have involved various players and assumed various names, including gay liberation, lesbian feminism, the queer movement, and the transgender movement (Bernstein 2002).

Sociologist Mary Bernstein describes the movements' goals as both cultural and political: the "cultural goals include (but are not limited to) challenging dominant constructions of masculinity and femininity, homophobia, and the primacy of the heterosexual nuclear family (**heteronormativity**). Political goals include changing laws and policies in order to gain new rights, benefits, and protections from harm" (2002, 536). Strategies to achieve these goals include building communities, lobbying legislators, voting for politicians sympathetic to LGBT issues, holding street marches of celebration and protest, and promoting LGBT culture through international, national, and community events, magazines, films, literature, and academic research.

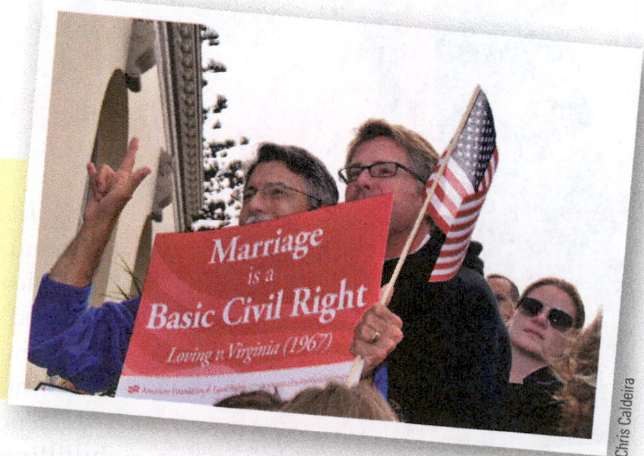

Demonstrations, such as this gathering in San Francisco to oppose a proposed constitutional amendment banning same-sex marriage, represent another strategy to achieve political goals.

The relative successes of the LGBT movements, and by extension opposition movements, can be gauged by examining each of the 50 states' position on same-sex marriage and unions. Currently 28 states enforce same-sex marriage or union bans. The various LGBT movements have met resistance from social movements broadly labeled the Religious Right and from state, local, and federal politicians who seek to preserve the heterosexual nuclear family and reserve marriage as a right granted only to a man and a woman. In 2013, the Supreme Court struck down the federal Defense of Marriage Act, ruling that married same-sex couples were entitled to federal benefits. The decision had the effect of recognizing gay marriages at the federal level regardless of the state the federal employee lives in.

What Do Sociologists See?

Sociologists see that sexuality encompasses all the ways people experience and express themselves as sexual beings. One expression of sexuality is sexual orientation. The image suggests that both parties express their sexualities through heterosexual attraction, which is considered to be normal and legitimate. At one time in the United States, heterosexual attraction between parties classified as different races did not comply with dominant norms, and such marriages were banned.

Cpl. Patrick Fleischman

Critical Thinking

Have you had the chance to vote on a state constitutional amendment to ban same-sex marriage? If so, how did you vote? If not, imagine how you would vote. Explain.

Key Terms

- biological sex
- gender identity
- gender role
- heteronormativity
- sexual orientation
- sexual scripts
- sexuality
- transgender

Module 9.4 Life Chances and Structural Constraints

Objective

You will learn the extent to which a person's gender determines his or her life chances.

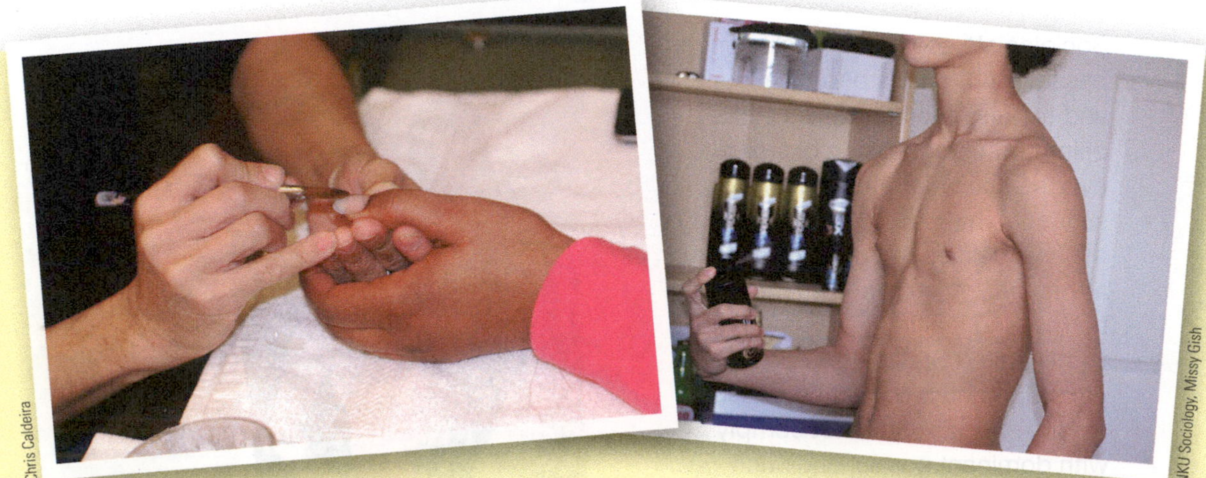

Chris Caldeira

NKU Sociology, Missy Gish

In a typical week, how much time do you take and how much money do you spend to present yourself as a particular gender?

Life Chances

Sociologists define **life chances** as the probability that an individual's life will turn out a certain way. Life chances apply to virtually every aspect of life—the chances that a person will become an airline pilot, play T-ball, major in elementary education, spend an hour or more getting ready for work or school, or live a long life.

Sociologists are interested in the processes by which being seen and presenting oneself as male, female, or transgender increases the probability that a person's life will be a certain way. Ideas about what men, women, and other genders should be shape every aspect of life, including how people dress, eat, and carry themselves; the time they wake up in the morning, what they do after they wake up, the social roles they take on, the things they worry about, and even ways of expressing emotion and experiencing sexual attraction (Bem 1993; Jones 2011).

To understand the power of gender in shaping life chances, consider the now-classic research by Alice Baumgartner-Papageorgiou (1982). She asked middle school students how their lives would be different if they were the other gender. Their responses reflected culturally conceived and learned ideas about sex-appropriate behaviors and appearances and about the imagined and real advantages and disadvantages of being male or female. The boys generally believed that their lives would change in negative ways if they became girls. Among other things, they would become less active and more restricted in what they could do. In addition, they believed that they would become more concerned about their appearance, finding a husband, and being alone and unprotected in the face of a violent attack—"I'd use a lot of makeup and look good and beautiful" and "I would not be able to help my dad fix the car and his two motorcycles" (2–9).

The girls, on the other hand, believed that if they became boys they would become less emotional, their lives would be more active and less restrictive, they would be closer to their fathers, and they would be treated as more than "sex objects"—"My father would be closer, because I'd be the son he always wanted" and "People would take my decisions and beliefs more seriously" (5–13).

Baumgartner-Papageorgiou's findings about how one's life is shaped by gender seem to hold up across time and age groups. When I asked my students how their lives would change if they were the other gender, their responses were remarkably similar to those described above. Decisions about how early to get up in the morning, which subjects to study, whether to show emotion, how to sit, and what sports to play are influenced by society's ideals of masculinity and femininity rather than by criteria such as self-fulfillment, interest, ability, or personal comfort.

Baumgartner-Papageorgiou's findings that men would feel less powerful as women are also supported by sociologists Jill E. Yavorsky and Liana Sayer's (2013) research on male-to-female transsexuals (transwomen). The two sociologists interviewed 25 transwomen about perceptions of personal safety, before and after making their transition. Most reported becoming more fearful and believed that now as females they could not fight off an attacker "even though most were large statured and were socialized as males." The sociologists concluded that cultural norms and messages presenting women as the weaker sex overrode decades of prior experiences and exposed the socially constructed nature of fear and bodily agency (Yavorsky and Sayer 2013).

Chris Caldeira
Chris Caldeira

Among other things, transwomen (far left) must "come to terms with gender rituals that positioned women as frail and men as able." Transmen (right) must come to terms with the view that they are better able to control of their emotions (Schilt and Connell 2007). This transman's wave suggests such control when we consider how women might wave to crowds.

Structural Constraints

How do we explain the fact that, for the most part, nurses are females and carpenters are males? Sociologists believe that one answer to this question lies with **structural constraints**, the established and customary rules, policies, and day-to-day practices that channel behavior in a certain direction and that shape a person's life chances. Structural constraints are at work, for example, when students choose a major, even if unconsciously, that matches up with expectations about what kinds of work are appropriate for someone of their sex. Note that about 85 percent of bachelor's degrees in engineering and in computer-information sciences are awarded to males, whereas 94 percent of bachelor's degrees in library sciences are awarded to females. Other majors dominated by women include education, health professions, and public administration/social services. Approximately 80 percent of all bachelor's degrees awarded in these fields go to women (National Center for Education Statistics 2011a).

Structural constraints are also at work when women choose or are "pushed" into work roles that emphasize personal relationships and nurturing skills or that pertain to products and services labeled as family-oriented and feminine. Men are more likely to be pushed into work roles that emphasize decision making and control and that pertain to machines, products, and services considered masculine. Sociologists argue that we must also consider how the jobs men and women "choose" channel behavior in stereotypically male and female directions. The point is that it is not the day care worker per se that is feminine; it is the skills needed to do the job well that make the day care worker behave in ways we associate with femininity. Presumably, anyone holding the job of day care worker will display those "feminine" characteristics (Anspach 1987).

To be successful at the job of drill instructor, a person must be aggressive, relentlessly critical, and forceful—qualities we associate with masculinity. All people who take on this job—male or female—will find themselves displaying "masculine" characteristics.

Sgt. Jose Nava

The concept of structural constraints offers insights as to how institutions are **gendered**. Sociologists look for ways in which gender is embodied and embedded in institutional arrangements. Institutions are gendered when there is an established pattern of segregating the sexes into different workspaces or jobs, of disproportionately assigning one sex to positions of power, and of otherwise disadvantaging one sex relative to the other (Britton 2000; Martin 2004). To determine whether an institution is gendered, sociologists ask the following kinds of questions: Do work institutions hold expectations about whether a male or female is best suited for a particular job? Are some occupations disproportionately occupied by females (administrative assistant) and others by males (vice president)? Are women the social studies teachers and men the science teachers? Do employers offer female employees maternity leave but do not offer male employees paternity leave? If the answer to any of these questions is "yes," then the institution is gendered.

Have you ever played a sport? Were your coaches male or female? No matter your gender, it is likely that your coach was male. For the most part it is commonplace for men to coach females, but not for females to coach men. Simply consider that 58 percent of women's college teams have a head coach who is male; less than 2 percent of men's teams have a female head coach. This state of affairs qualifies as a gendered pattern (Rhode and Walker 2008; Brennan 2013).

NKU Philosophy, Rudy Garns

What Do Sociologists See?

Sociologists see a sign that announces a public bathroom that is welcoming of all genders. The existence of this facility is a response to structural constraints that prevent people of the "wrong" gender from accompanying young children, disabled persons, and others of another gender in need of assistance going to the bathroom.

Chris Caldeira

Critical Thinking

How do you think your life would change if you presented yourself as another gender?

Key Terms

gendered

life chances

structural constraints

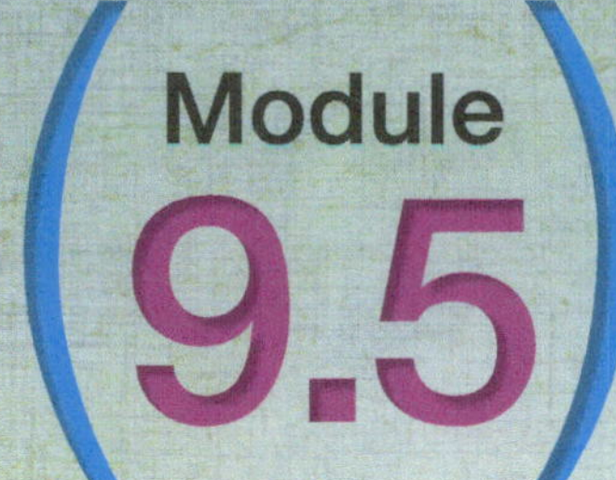

Module 9.5 Gender Stratification

Objective

You will learn that sociologists seek to understand situations that put one sex at a disadvantage relative to the other.

Mass Comm. Spc. 2nd Class Erick S. Holmes

Mass Comm. Spc. 2nd Class Ronald Gutridge

What would a completely equal relationship between a man and a woman look like?

If the two held the same occupation, were physically similar in size, and made the same incomes as the man and woman pictured above suggest they do, then the relationship would be essentially equal. Each year the World Economic Forum publishes a report on the global **gender gap**, defined as the disparity in opportunities available for men and women. The report considers the situation of women relative to men in 135 countries with regard to four areas: economic participation and opportunity, health and survival, educational attainment, and political empowerment. In doing so, the World Economic Forum is considering **gender stratification**, the extent to which opportunities and resources are unequally distributed between men and women.

According to this report, there is no country in which women have more *overall* opportunities than their male counterparts, but there are countries in which men and women share more equally in the available resources and opportunities. According to the measures used, Iceland ranks first as the country with the least inequality, the United States ranks 17th, and Yemen ranks 134th or last with the greatest inequality (see Table 9.5a).

Table 9.5a: Indicators Used to Rank Countries on Gender Equality/Inequality
Notice that female life expectancy exceeds that of males in all three countries. With regard to political power, the data show that women in Yemen have virtually no political power. Compare Iceland and the United States on all indicators. On which indicators is Iceland ahead of the United States? On which indicators, if any, are Iceland and the United States similar? Why do you think Iceland is ranked ahead of the United States with regard to gender equality?

	Iceland		United States		Yemen	
	Females	Males	Females	Males	Females	Males
Indicator	Economic Opportunity					
% Working age in labor force	81	86	62	78	26	74
Median income	$29,535	$40,000	$38,388	$40,000	$1,064	$3,890
% of Professionals/ workers	56	44	55	45	15	85
% of legislators, senior managers/ officials	33	67	43	57	2	98
Length of paid maternity leave	26 weeks @ 80% pay		12 weeks @ no national policy		60 days @100%	
Health and Survival						
Life expectancy (in years)	75	73	72	68	55	43
Political Empowerment						
% of Parliament/ Congress	40	60	18	82	0	100
Number of years of last 50 with female or male head of state	20	30	0	50	0	50

Source of Data: World Economic Forum (2013).

Explaining the Gender Gap

Using a complex formula, the World Economic Forum estimates that in Iceland 87 percent of the overall gender gap has been closed. In the United States, 71 percent of the gap has been closed, and in Yemen 46 percent has been closed. Sociologists seek to identify the social factors that put one sex at a disadvantage relative to the other. Inequality exists when one sex relative to the other:

1. faces greater risks to physical and emotional well-being,
2. possesses a disproportionate share of income and other valued resources, and/or
3. is accorded more opportunities to succeed.

YEMEN. In Yemen there are many customs and traditions that work to keep women in positions inferior to men. In particular, Yemen is a **patriarchy**, an arrangement in which men have systematic power over women in public and private (family) life. Male power is supported by law, and in the case of Yemen, those laws are Islamic (*shari'a*) and govern all aspects of life. In addition, "women are prohibited from interpreting the religious texts that define Islamic laws, and they cannot serve as family court judges. One can speculate that if women had this right, they would interpret Islamic texts differently than men, who for the most part have defined a woman's duty as obeying her husband" (Freedom House 2008). Because 55 percent of Yemeni women are not literate, most are unable to read Islamic texts (World Economic Forum 2013).

These Yemeni women are participating in a U.S. military-sponsored animal husbandry training program with the goals of improving women's position in society and the country's overall livestock health and productivity. Here a U.S. staff sergeant is showing a thermometer and explaining its uses.

Tech. Sgt. Carrie Bernard

ICELAND. Iceland has one of the highest per capita incomes in the world. In 2013, the country ranked number one in the world with regard to gender equality. Gender discrimination is prohibited in Iceland, and laws mandating gender equality in schools and education have been in place since 1976. In 2008, a new legal mandate—the Act on the Equal Status and the Equal Rights of Women and Men—was passed. The act promotes gender equality in all spheres of society. Among other things, the act "stipulates that equal participation of women and men shall be promoted in committees, boards and councils under the auspices of the government and local authorities, the gender proportion being not less than 40 percent where there are more than three members." In addition to an administrative function, the Centre for Gender Equality is the national bureau that provides counseling and other support education to governments, corporations, and nonprofits (Gudmundsson 2008). Among other things, Iceland is known for its family leave policy, which guarantees both parents the opportunity to care for newborns, and for its extensive system of day care and child development centers.

THE UNITED STATES. Of the 134 countries studied for the 2013 rankings, the United States ranked 17th in closing the gender gap. In comparison to women in Iceland, American women are much closer to their male counterparts in terms of income. The United States falls short in the areas of political empowerment and female labor force participation. Unlike Iceland, the United States does not mandate equal representation in government and other public organizations. As a result, 99 of 535 seats (only 18.5 percent) of the U.S. Congress are held by women (see Figure 9.5a).

Figure 9.5a: The Gender Gap: 10 Most Equal and 10 Least Equal Countries in the World
The map shows the 10 countries in which women's overall opportunities relative to the men in that country are most and least equal. The 10 countries with greatest equality for women are Norway, Finland, Sweden, Iceland, New Zealand, the Philippines, Denmark, Ireland, the Netherlands, and Latvia. The 10 least equal are Nepal, Oman, Benin, Morocco, Côte d'Ivoire, Saudi Arabia, Mali, Pakistan, Chad, and Yemen.

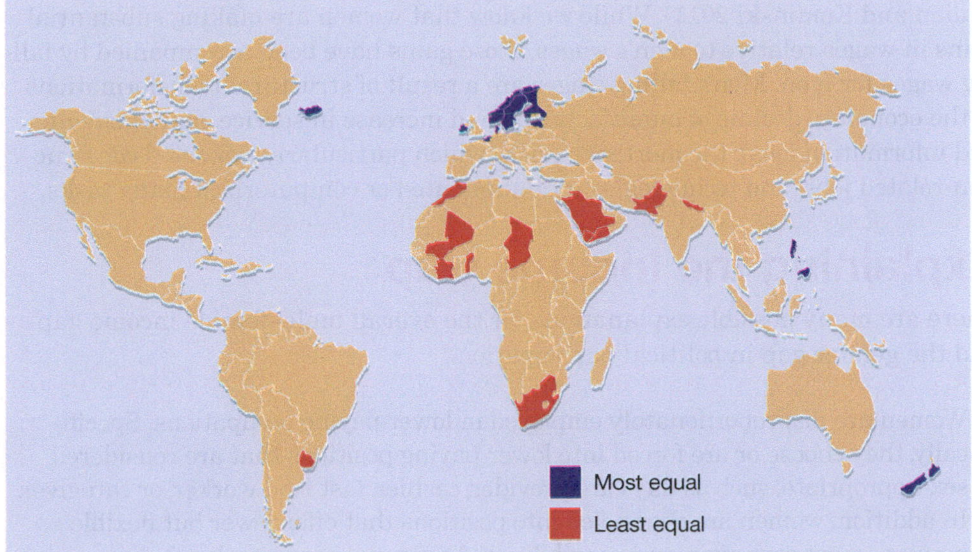

Source of data: World Economic Forum (2011)

While the income gap between men and women is significantly smaller in the United States than in Iceland, there is still income inequality between men and women. Figure 9.5b offers a graphic depiction of gender inequality in pay as it relates to full-time wage and salary workers in the United States.

Figure 9.5b: Women's Earnings as a Percentage of Men's, 1979–2012 (Full-Time Wage and Salary Workers)
In 1979 women working full-time earned about 63 cents for every dollar earned by their male counterparts. In 2010 that gap had decreased by 29 cents on the dollar, so women earned about 81 cents for every dollar earned by men.

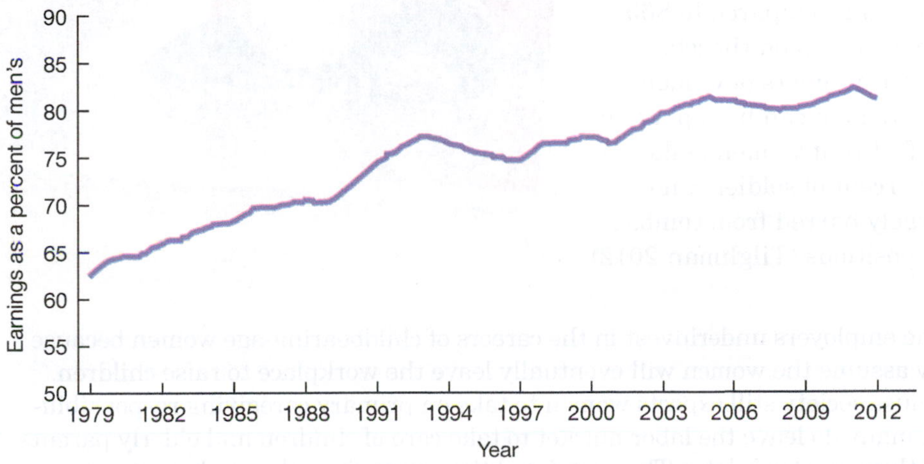

Source: U.S. Bureau of Labor Statistics (2013b)

These income differences between men and women vary by age group. Women ages 25–34 earned 90 cents for every dollar earned by men, and women 45 and older earned 75 cents for every dollar earned by men (U.S. Bureau of Labor Statistics 2013b). If we take a broader view and examine total earnings of men and women over an extended period of time, such as a 40-year-long work career, we find that the average woman earns $434,000 less than her male counterparts. That estimate varies by educational level, with high school–educated women earning $270,000 less and college-educated women and women with professional degrees earning $713,000 less than their male counterparts over a 40 years (Julian and Kominski 2011). While we know that women are making substantial gains in wages relative to men's wages, those gains have been accompanied by falling wages for men. Men's falling wages are a result of structural transformations in the economy (decline of manufacturing and increase in service, management, and information jobs); the mortgage crisis, which particularly impacted construction-related jobs; and technologies that automated or computerized many tasks.

Explaining the Income Gap

There are many possible explanations for the overall male–female income gap and the gender gap in political participation.

- Women are disproportionately employed in lower-paying occupations. Specifically, they choose or are forced into lower-paying positions that are considered sex-appropriate, such as day care provider, cashier, fast food worker, or caregiver. In addition, women are channeled into positions that offer fewer but flexible hours to meet care giving responsibilities. As a group, women also choose not to work in or are forced out of higher-paying occupations that require them to relocate or take on dangerous assignments in the fields of construction, logging, mining, and the military.

Lance Cpl. Chelsea Flowers

The different kinds of jobs men and women hold explain why men suffer 92 percent of workplace fatalities. The man pictured here is one of 100,000+ men wounded on the job as a soldier in Iraq and Afghanistan, compared to 865 women wounded on the job. The lower numbers of women relative to men can be explained by the fact that women make up 15 percent of soldiers and are largely barred from combat career positions (Tilghman 2012).

- Some employers underinvest in the careers of childbearing-age women because they assume the women will eventually leave the workplace to raise children. Because society still expects women to take on primary caregiving responsibilities, many do leave the labor market to take care of children and elderly parents and then reenter it later. The associated time away from the workplace puts them behind with regard to wages and promotions.
- Employers often view women's salary needs as less important than men's and pay women accordingly. Unfortunately, many still consider women's earnings as supplemental to a presumed male partner—earnings that can be used to buy "extras"—when in reality many women are heads of households. When

negotiating for salaries, women underestimate their worth to employers and ask for less than their male counterparts. Some employers steer males and females into different gender-appropriate assignments (such as sales clerks in baby clothes departments rather than hardware) and offer them different training opportunities and chances to move into better-paying jobs (Love 2007).

- Women encounter a **glass ceiling**, a term used to describe a barrier that prevents women from rising past a certain level in an organization, especially for women who work in male-dominated workplaces and occupations. The term applies to women who have the ability and qualifications to advance but who are not well-connected to those who are in a position to advocate for or mentor them. With regard to men who work in female-dominated professions, they encounter the **glass escalator**, a term that applies to the invisible upward movement that puts men in positions of power, even within female-dominated occupations. In this case, management singles out men for special attention and advancement such that men are encouraged to move from school teacher to assistant principal to principal or from social worker to program director.

What Do Sociologists See?

Sociologists see that this woman represents one of the 89.4 percent of all registered nurses who are female. The male represents one of the 10.6 percent of all nurses who are male. The median weekly income for female nurses is $1,086. The median weekly income for male nurses is $1,189, $103 per week more than female counterparts (U.S. Bureau of Labor Statistics 2013b).

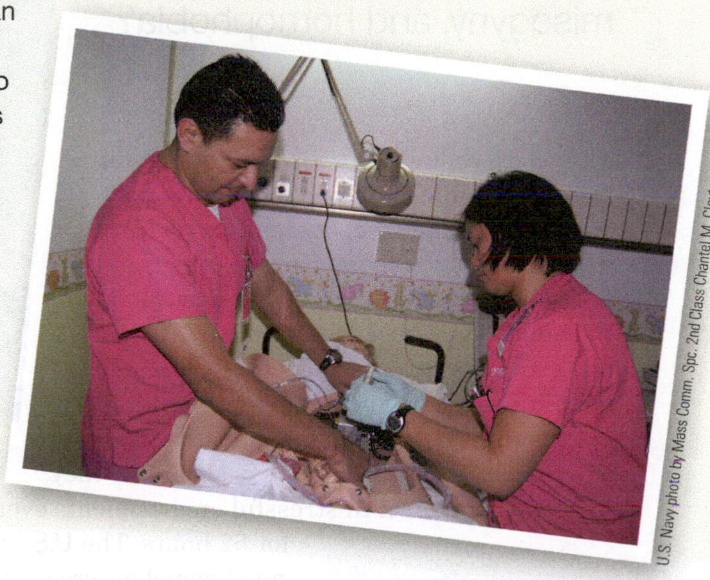

U.S. Navy photo by Mass Comm. Spc. 2nd Class Chantel M. Clayton

Critical Thinking

Is there an area of your life where you feel advantaged relative to the other sex?

Key Terms

gender gap

gender stratification

glass ceiling

glass escalator

patriarchy

Module 9.6

Sexism and Feminism

Objective

You will learn about the meaning of sexism and feminism.

How might environments that cultivate hypermasculinity also cultivate sexism, misogyny, and homophobia?

U.S. Navy photo by Mass Comm. Spc. 2nd Class Shauntae Hinkle (Released)

The photo shows Navy SEALs undergoing rigorous training. Their hands are bound as they negotiate deep water. Navy SEALS are one of four elite military forces that also include Army Rangers, Delta Forces, and Green Berets. These forces operate in small, self-contained teams. The success of elite forces depends on each member's ability to overcome severe physical and mental hardship. Their physical stamina is such that they are able to perform in stressful environments even after they have been awake for 52 hours. The U.S. military faces a dilemma: It must somehow create soldiers, and especially elite forces, who are willing to put their lives on the line, but meeting this need also puts them at high risk for cultivating an environment that supports hypermasculinity, sexism, misogyny, and homophobia.

Sgt. Joel A. Chaverri

Hypermasculinity involves exaggerating the traits and behaviors believed to be characteristic of males by placing excessive emphasis on strength to the point that a man's muscles and reproductive organs are presented as impossibly large.

Sexism

Sexism is the belief that one sex—and by extension, one gender—is innately superior to another, justifying unequal treatment of the sexes. Sexism revolves around three notions:

1. People can be placed into two categories—male and female.
2. A close correspondence exists between a person's reproductive organs (e.g., primary sex characteristics) and other characteristics such as emotional state, body language, personality, intelligence, the expression of sexual desire, and athletic capability.
3. Primary sex characteristics are viewed as so significant that they explain and determine behavior and the inequalities that exist between the sexes.

Sexism rationalizes unequal treatment of men, women, and those who are transgender as natural and even just. Sexism can be so extreme that it can involve a hatred for one sex. That hatred is called **misogyny** when it is directed at women; it is called **misandry** when it is directed at men. Some of the most publicized charges of misogyny are leveled against those areas of life where hypermasculinity is cultivated, including:

- that segment of hip-hop/rap artists who portray themselves as pimps and the women around them as prostitutes and sex objects who must obey them;
- football teams, especially professional and Division I; and
- the U.S. military, especially special forces

In its 2012 report on sexual assault, the Pentagon estimated that 26,000 service members experienced unwanted or forced sexual contact. Forty-seven percent of unwanted contacts were directed at females, who account for 15 percent of military personnel. Fifty-three percent involved unwanted sexual contact toward men by men. While it might seem tempting to point to the repeal of the "Don't Ask, Don't Tell" policy, which allowed the openly gay to serve in the military, the Pentagon report suggests that men who identify as heterosexual engage in sexual violence against other men aimed at humiliating the victims or as part of a hazing ritual. Some experts suggest that the "Don't Ask, Don't Tell" policy had the effect of actually suppressing complaints from male victims because the violated men believed that they would be viewed as gay and discharged (Dao 2013). Based on the Pentagon findings, one might speculate that hypermasculine environments such as the military promote violence not only toward women but also toward men.

The Connection Between Homophobia and Hypermasculinity

The term **homophobia** is used in at least two ways. On one level, the word refers to an irrational fear held by some heterosexuals that any same-sex person will make a sexual advance toward them. On another level, homophobia applies those who are so afraid or offended by the possibility of an unwanted advance by a same-sex person that they engage in or support violence toward gays and others who do not conform to prevailing notions of masculinity and femininity. Other discriminatory acts include banning gays from certain occupations (military service, child care).

Chris Caldeira

The sign commemorates the life of Matthew Shepard. In 1998, at age 21, Matthew was tortured and murdered in an anti-gay hate crime in Wyoming. Shepard's case brought international attention to extreme homophobia and hate crimes against gays. His tragic death became a catalyst driving state and federal legislation.

Hypermasculinity and all that it entails point to another dimension of sexism—the belief that people who behave in ways that depart from ideals of masculinity or femininity are considered deviant and in need of fixing and should be subject to negative sanctions. This ideology was reflected in U.S. military policy toward gay men and lesbians when "Don't Ask, Don't Tell" (DADT) was the law. At that time the U.S. Department of Defense (1990) maintained that homosexuality was "incompatible with military service" and that the "presence of such members adversely affects the ability of the Armed Forces to maintain discipline, good order, and morale" (25). Over the course of a decade, the U.S. military discharged 12,500 servicemen and women for homosexuality (Bumiller 2009).

U.S. Navy photo by Mass Comm. Spc. 2nd Class Patrick Gordon

Whether taking part in training exercises or relaxing afterward, soldiers often make physical contact with other soldiers. Would the presence of openly gay men and lesbians in the military disrupt these kinds of bonding activities? Is physical contact of any kind sexual? Might the presence of two soldiers who share a strong friendship also be disruptive?

People who were opposed to the presence of gay men and lesbians in the military stereotyped them as sexual predators just waiting to pounce on heterosexuals while they shower, undress, or sleep. Opponents seem to believe that any same-sex person is attractive to a gay man or lesbian. But as one gay ex-midshipman noted, heterosexuals "have an annoying habit of overestimating their own attractiveness" (Schmalz 1993, B1). In December 2010, Congress passed legislation repealing DADT, officially ending the policy on paper. U.S. servicemen and women were already serving with coalition partners who allowed gays to serve openly. In fact, openly gay soldiers serve in the British (a major coalition partner) and 23 other militaries without any significant problems. The British military even recruits soldiers at gay pride events (Lyall 2007).

Feminism: A Response to Sexism

When women living in the United States are asked, "Do you consider yourself a feminist or not?" only one in four answers yes. When asked the same question accompanied by this definition of feminist—someone who believes in social, political, and economic equality of the sexes—65 percent of women answer yes. One possible explanation for the difference is that few people consider the label of feminist a compliment (CBS News Polls 2005).

Feminism is a perspective that seeks to understand the position of women in society relative to that of men in the context of the economic, political, and cultural structures in which their lives are embedded. In addition, feminists advocate for equal opportunity. Questions about what that equality looks like and how equality should be achieved distinguish feminist camps from one another.

Library of Congress Prints and Photographs Division [LC-DIG-ppmsc-05887]

This early-20th-century political cartoon speaks to a widespread misperception that feminists as a group disdain men and see them as the weaker sex. Here four women are inspecting what appears to be a bug. It is actually someone of the "male species" on his knees begging for mercy.

Many feminists believe that any inequality between genders, including that which gives females an advantage over males, needs to be addressed. The following quotations from well-known feminists demonstrate the range of concerns and positions feminists hold:

> It's important to remember that feminism is no longer a group of organizations or leaders. It's the expectations that parents have for their daughters, and their sons, too. It's the way we talk about and treat one another. —Anna Quindlen

> If divorce has increased by one thousand percent, don't blame the women's movement. Blame the obsolete sex roles on which our marriages were based. —Betty Friedan

> Women do not have to sacrifice personhood if they are mothers. They do not have to sacrifice motherhood in order to be persons. Liberation was meant to expand women's opportunities, not to limit them. The self-esteem that has been found in new pursuits can also be found in mothering. —Elaine Heffner

> No one sex can govern alone. I believe that one of the reasons why civilization has failed so lamentably is that it had one-sided government. —Nancy Astor

> I myself have never been able to find out precisely what feminism is: I only know that people call me a feminist whenever I express sentiments that differentiate me from a doormat, or a prostitute. —Rebecca West

Remember, Ginger Rogers did everything Fred Astaire did, but she did it backwards and in high heels. —Faith Whittlesey

Feminism's Activist Roots

The history of feminism cannot be separated from efforts to bring about change in the lives of women, and by extension the lives of children and men. The following very selective list highlights some major events in feminist history. They include gaining the right to vote, securing fair labor standards, and opening doors that were previously closed. These are rights that most of us—male and female—have come to take for granted.

1920 The 19th Amendment to the U.S. Constitution becomes law, guaranteeing women the right to vote.

1943 With many men fighting in World War II, over 6 million women hold factory jobs as welders, machinists, and mechanics.

1963 The Equal Pay Act, signed by President John F. Kennedy, prohibits the practice of paying women less than men for the same job.

1964 President Lyndon B. Johnson signs the Civil Rights Act, outlawing discrimination in unions, public schools, and the workplace on the basis of race, creed, national origin, or sex.

1965 Under Title VII of the Civil Rights Act of 1964, the Equal Employment Opportunity Commission is established to prohibit discrimination in the workplace on the basis of sex, religion, race, color, national origin, age, or disability.

1966 The National Organization for Women (NOW) is founded by Betty Friedan with the purpose of challenging sex discrimination in the workplace.

1972 The U.S. Senate approves the Equal Rights Amendment, as well as Title IX, making sex discrimination in schools that receive federal funding illegal and requiring schools that receive such funds to give females an equal opportunity to participate in sports.

1973 The U.S. Supreme Court rules that a Texas law restricting abortion in the first trimester is unconstitutional. As a result, anti-abortion laws in nearly two-thirds of the states are declared unconstitutional, legalizing abortion nationwide.

1975 President Gerald Ford signs a defense appropriations bill allowing women to be admitted into U.S. military academies.

1993 President Bill Clinton signs the Family Medical Leave Act, allowing eligible employees to take up to 12 weeks of leave for reasons of illness, maternity, adoption, or a child's serious health condition.

1996 U.S. women's successes in the Summer Olympics (19 gold medals, 10 silver, 9 bronze) are attributed to the Title IX legislation that supported and encouraged girls' participation in sports.

1997 The Supreme Court rules that college athletics programs must actively involve men and women in numbers that reflect the proportions of male and female students.

2009 The Lilly Ledbetter Fair Pay Act extends the 180-day statute of limitations for filing an equal-pay lawsuit to every paycheck issued rather than tying the statute to the first discriminatory pay check earned. (Adapted from Boxer 2007)

A feminist viewpoint emphasizes the following kinds of themes:

- the right to bodily integrity and autonomy;
- access to safe contraceptives;
- the right to choose the terms of pregnancy;
- access to quality prenatal care, protection from violence inside and outside the home, and freedom from sexual harassment;
- equal pay for equal work;
- workplace rights to maternity and other caregiving leaves; and
- freedom for both men and women to make choices in life that defy gender expectations.

What Do Sociologists See?

Sociologist see that football revolves around hypermasculinity or exaggerating the traits and characteristics of the "ideal" male, giving emphasis to physical attributes such as muscles which are portrayed as impossibly large. The size and muscle mass along with the football uniform, which includes shoulder and thigh pads, serves to exaggerate the male physique. Men communicate power and dominance over one another through their body and its ability to both take and give hard hits.

U.S. Navy photo by Damon J. Moritz/Released

Critical Thinking

Identify some example of popular culture—such as a movie, a song, a sport, a cartoon, a toy—that exhibits misogyny, misandry, homophobia, or hypermasculinization.

Key Terms

feminism
homophobia
misandry
misogyny
sexism

Module 9.7

Applying Theory: Sex Testing

Objective

You will learn how the theoretical perspectives help us think about the uses of sex testing.

Did your mother know from the results of an ultrasound or other test your sex before you were born?

L.A. Shively

If your parents knew your sex before you were born, why do you think it was important to them to know beforehand? If your parents did not, why do you think they resisted knowing? The sociological theories—functionalist, conflict, symbolic interaction, and feminism—help us go beyond individual cases and think about the larger consequences and issues associated with sex testing months before babies are born.

Ms. Jennifer M. Caprioli (Drum)

Sex testing allows parents to "know" the sex of the baby around 20 weeks after conception. From a functionalist point of view this knowledge contributes to order and stability as the parents have time to prepare for a boy or girl, including buying gender-appropriate clothes and other items. In short, it allows parents the time to make a plan for raising children to conform to gender expectations of their society.

Conflict theorists argue that ALL babies have the same needs—to be cared for, kept warm, and fed. In reality, conflict theorists say, sex tests simply allow parents to plan out the baby's future in very gender-specific and unequal ways that narrow or expand their children's opportunities in life. Expectant parents may project their sons to be football players and their daughters to be dancers. Knowing the baby's sex 20 weeks into the pregnancy may encourage some parents, disappointed by the baby's sex, to choose abortion. In fact, the availability of ultrasound technologies to determine a baby's sex 20 weeks after conception is believed to be responsible for sex ratio imbalance in places where there are clear cultural preferences for boys (Dhar 2012).

Lisa Southwick

Phil Grout

Staff Sgt. Malcolm McClendon

Symbolic interactionists are interested in how knowing (or not knowing) the sex of the baby 20 weeks into a pregnancy affects a parent's self-awareness or sense of self. Do parents feel differently about themselves when they learn they are having a baby girl versus a baby boy? Are parents' projections of future interactions with the child different depending on whether the baby tests as a boy or a girl? Symbolic interactionists are also interested in how parents and other involved parties negotiate interactions where the question of the baby's sex comes up. How do some parents, who choose not to sex test their baby, withstand questions like "Do you know the sex of the baby yet? Why not?" How do they explain their decision not to know? After the baby is born, do these parents interact in less gender-specific ways with their baby?

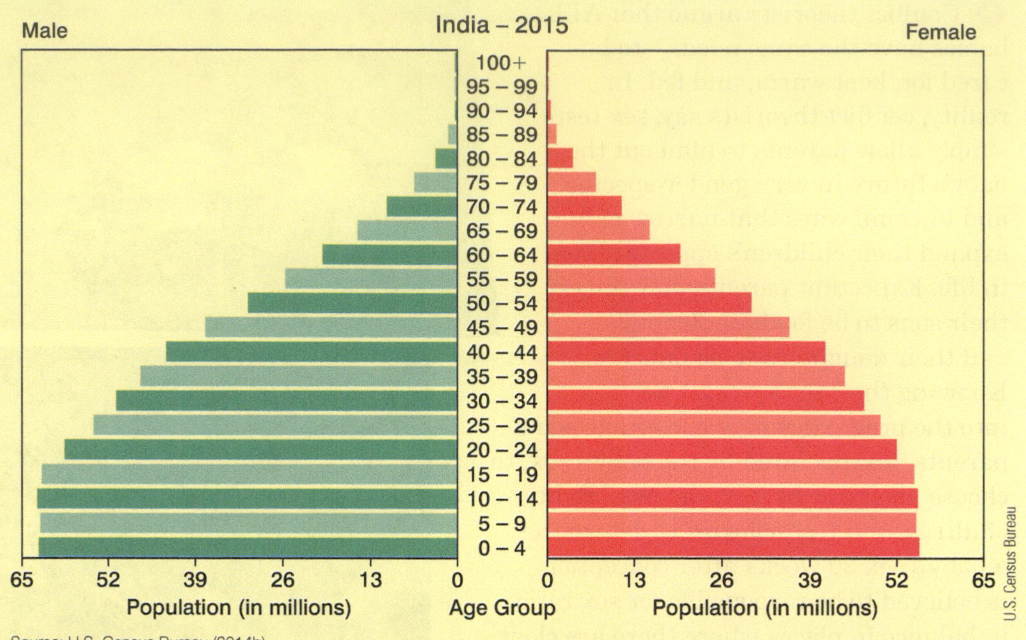

Source: U.S. Census Bureau (2014b)

Like conflict theorists, feminists are concerned about gender inequality that sex testing promotes and they also see a worrisome connection between sex testing and female infanticide, the targeted abortion of female fetuses because of a cultural preference for males and corresponding low status assigned to females. Feminists point out that there seems to be few, if any, societies with a cultural preference for female babies. The widespread practice of female infanticide is believed to be the cause when there is a dramatic imbalance between the number of boys and girls in a society. Notice that the population pyramid for India shows about 62 million boys age 0–4 and about 55 million girls. Other age groups also show significantly more boys than girls.

Summary: Putting It All Together

Chapter 9

Sex is a biological distinction determined by the anatomical traits essential to reproduction. While most cultures classify people into two categories—male and female—sex should not be considered a clear-cut category. Gender is a social distinction about how males and females should be; it is something that is carefully constructed, taught, learned, and enforced. Not every society divides people into so-called opposite genders. For example, American Samoans and other Pacific island peoples accept a third gender known as *fa'afafine*.

Sociologists are especially interested in gender ideals. Often, gender ideals do not exist in reality, yet that does not stop people from trying to attain them. The commodification of gender ideals is the process of introducing products to consumers through advertising campaigns that promise that those who buy and use the products will achieve masculine or feminine ideals.

Sex, and by extension gender, affects people's life chances, the *probability* that an individual's life will take a certain path or turn out a certain way. The effect a person's gender has on his or her life becomes evident when we ask people to imagine life as another gender. Structural constraints are the established and customary rules, policies, and day-to-day practices that channel behavior in a certain direction and that shape a person's life chances. One example relates to the structural constraints that push many men and women into careers that correspond with society's ideals about sex-appropriate work. Institutions are gendered when there is a pattern to the relationships, practices, images, and belief system that supports segregating the sexes and empowering or subordinating one sex relative to the other. An institution is gendered when there is an established pattern of segregating the sexes into different workspaces or jobs, of disproportionately assigning one sex to positions of power, and of otherwise disadvantaging one sex relative to the other.

When sociologists study inequality between males, females, and those who are transgender, they seek to identify the social factors that put one gender at a disadvantage relative to the other. Inequality is justified by sexism, which can be so extreme that it takes the form of misogyny and misandry. Sexism also encompasses homophobia and hypermasculinization. Feminism, a response to sexism, is a perspective that advocates equality between men and women. Most feminists believe that any gender inequalities, including that which gives females an advantage over males, needs to be addressed.

Chapter 10

Economics and Politics

Chris Caldeira

When sociologists study economies, they focus on employment opportunities; the labor force; and how goods and services are produced, distributed, and consumed. When they study politics they focus on who has power over, and access to, scarce and valued resources and who has the power to make laws, policies, and decisions that affect others' lives. Sociologists also seek to understand how the economy and politics are interconnected. As one obvious example, consider that governments enact tens of thousands of laws that affect income. Some of the most controversial are federal laws governing the tax rates on investments, most notably stocks. Right now the maximum rate is 20 percent, up from 15 percent during the George W. Bush and the first term of the Obama administrations. But that rate was as high as 39.9 percent during the Carter administration (Brookings Institution 2012). There are also laws setting minimum wage levels. At the time of this writing the federal minimum wages was $7.25, with some state and city governments mandating higher minimum wages. San Francisco mandates a minimum wage above $10.00; the state of Connecticut's minimum wage is $8.70 and by law it must increase to $10.10 by January 2017 (Stoller 2014).

Module

10.1 Economic Systems

Objective

You will learn about the economic systems of socialism, capitalism, and the welfare state.

Would you be willing to pay more taxes if the government guaranteed child care, health care, public transportation, a free college education with a monthly stipend for 55 months, and other benefits from cradle to grave?

NKU Sociology, Missy Gish

You would receive these kinds of benefits if you lived in Finland. The catch is that you would live in a smaller house, pay more taxes (51.2 percent), and consume less. You would also see fewer people who are extremely poor or extremely wealthy.

Economic systems are social institutions that structure employment opportunities and opportunities to earn income and create wealth. These systems also regulate the production, distribution, and consumption of products and services. We can classify the economies of the world as falling somewhere along a continuum that has capitalism and socialism as its extremes. Keep in mind, however, that no economy fully realizes capitalist or socialist principles and that, in practice, economic systems are some combination of the two.

Capitalism

Capitalism is an economic system in which the raw materials and the means of producing and distributing goods and services are privately owned. That is, individuals (rather than employees or the government) own the raw materials, land, machines, tools, trucks, buildings, and other inputs needed to produce and distribute goods and services. In theory, this economic system is profit-driven and free of government interference. Being profit-driven is the most important characteristic of capitalist systems. In such systems, business owners and employers are motivated to increase and maximize profits by seeking maximum return on investments and using labor and resources in cost-efficient ways. Theoretically, consumer demand and choice drives the amount and kinds of goods and services offered. In addition, the best businesses survive and thrive because consumers "vote" with their purchases.

The laws of supply and demand drive capitalist systems. This means that as the demand for a product or service increases, the price rises. Businesses respond to increased demand by increasing production, which in turn "increases competition and drives the price down" (Hirsch et al. 1993, 455). Although most economic systems in the world are classified as capitalist, in reality no system fully realizes capitalist principles. Simply consider that the U.S. government ignores capitalist principles any time it intervenes to regulate an industry, stimulate the economy, prevent a recession, or when it makes laws that affect how income and wealth is distributed. Consider also that the U.S. Postal Service, national parks, the public school system, Medicaid, and libraries are publicly owned enterprises that are, in theory, not profit-driven.

The United States is considered as having a capitalist economy. For the most part, privately owned businesses decide what products to make and what services to offer. Individuals decide what they want to buy. In a capitalist system the government contracts with businesses for the products and services its needs. As one example, Royal Dutch Shell, one of the largest private suppliers of fuel to the U.S. government, including to the military, has been awarded contracts worth billions of dollars (Reuters 2014; U.S. Department of Defense 2013). In contrast, about 75 percent of the oil companies in the world are government-owned and control 90 percent of proven oil reserves. Thus governments, not corporations, control most of the world's oil production (World Bank 2011b).

Karl Marx believed that capitalist systems ignore too many human needs and exploit human labor for the sake of profit. Marx believed that if systems of production were in the right hands—those of socially conscious people motivated not by a desire for profit or self-interest but by an interest in the greatest good to society—public wealth would be more than abundant and could be distributed according to need.

The search for the lowest-cost—even free—labor underlies exploitation of workers. The bottom image shows the enslaved from Africa in severely crowded quarters sailing the Atlantic to the United States, Brazil, the Caribbean, and other slave-holding societies. The top photo shows a man from Mexico scaling a fence. The fence marks a boundary separating a society where farm workers earn $5 per day from one where workers earn $81.36 per day (U.S. Department of Agriculture 2014).

Petty Officer 1st Class Matthew Tyson

Library of Congress Prints and Photographs Division[LC-DIG-ppmsca-05933]

Competition and the drive for profit push capitalists to cut the costs of production by introducing labor-saving machinery, laying off workers, and/or finding workers who will work for less. Marx argued that at some point the drive to lower production costs would result in so many jobs lost and workers displaced that the system would collapse. For Marx, one contradiction of capitalism is that the businesses must lay off and lower the wages of workers who they also depend on to buy their products (Kilcullen 1996).

Socialism

The term *socialism* was first used in the early 19th century in response to excesses of capitalism, specifically the poverty and inequality that accompanied the capitalist-driven Industrial Revolution. In contrast to capitalism, **socialism** is an economic system in which raw materials and the means of producing and distributing goods and services are collectively owned. In other words, public ownership—rather than private ownership—is an essential characteristic of this system. Socialists reject the idea that what is good for the individual and for privately owned businesses is good for society as a whole. Instead, they believe the government or some worker or community organization should play the central role in regulating economic activity on behalf of the people as a whole. Socialist ideals celebrate the common worker, who labors for the good of society. "From each according to his ability, to each according to his needs" (Marx 1875).

The sculpture of John Lennon is in Cuba. The World Socialist Movement considers Lennon's song *Imagine* as its anthem, the lyrics conveying "the socialist vision." The words urge listeners "to imagine a world without property, without religion, without nations, living in peace. It postulated an economic order in which both greed and hunger would be impossible. Socialists also share this vision . . . a classless economic order in which wage labor, money and buying and selling have been replaced by free people working together to meet their needs without the constraints imposed by the market system, in short a world of peace, equality, abundance and ecological sustainability. You may think that we are dreamers, but we are not the only ones. I hope some day you'll join us. And the world will live as one" (World Socialist Movement 2014).

Socialists maintain that goods and services important to human welfare like oil, financial services, health care, transportation, and the media should be state- or government-owned. In socialism's most extreme form, the pursuit of personal profit is forbidden. In less extreme forms, profit-making activities are permitted as long as they do not interfere with larger collective goals. As with capitalism, no economic system fully realizes socialist principles. The People's Republic of China, Cuba, North Korea, and Vietnam are all officially classified as socialist economies, but they permit varying degrees of profit-making activities that generate personal wealth.

Welfare States

The term **welfare state** applies to an economic system that is a hybrid of capitalism and socialism. In this economic model, the government (through taxes) assumes a key role in providing social and economic benefits to some or all of its citizens, including unemployment benefits, supplemental income, child care, social security, basic medical care, transportation, education (including college), or housing. Under one welfare state model followed by the United States (with the exception of Social Security and Medicare, to which everyone over age 65 who has contributed to the system is entitled), such benefits are provided to those who fall below a set minimum standard, such as a poverty line or a certain income level. Under a second welfare state model, the benefits are awarded in a more comprehensive way (e.g., all families with children, all college-age students, universal health care). Most European countries follow the second model. Finland, for example, funds all schools equally. In Finland day care and preschools are free, uniformly high in quality, and staffed by teachers who have received extensive training. Finland's investment in equality of educational opportunities explains in part why its students are among the best-performing in the world (Kaiser 2005).

What Do Sociologists See?

Sociologists inspired by Marx see a social program—food stamps—where government intervenes to offset the shortcomings of a free market system. There must be a safety net for those who lose their jobs or for whom the free market system does not pay a living wage. Recipients of foods stamps purchase products from private businesses; in that sense food stamps spending benefits businesses.

FOOD STAMPS
GROW GARDENS!
SNAP EBT
SNAP EBT benefits (food stamps)
can be used to purchase
food-producing plants and seeds.
For info: www.SNAPgardens.org or call 888-963-SNAP
To volunteer, visit www.SNAPgardens.org/volunteer

Rachel Ellison

Critical Thinking

What are some examples of government-run (local, state, federal) programs where government intervenes in workings of the "free market"?

Key Terms

capitalism

economic system

socialism

welfare state

Module 10.2 The U.S. Economy and Jobs

Objective

You will learn some characteristics that define the U.S. economy.

Mr. Douglas Demaio (IMCOM)

Mass Comm. Spc. 2nd Class Eric C. Tretter/Released

Which occupation, veterinarian or postal clerk, is expected to be among the fastest growing jobs between now and 2022?

Veterinarians (along with personal care aides, biomedical engineers, and physical therapists) are projected to be among the fastest growing occupations in this decade. Postal clerks (along with switchboard operators, word processors, and typists) are projected to be among jobs in decline (U.S. Bureau of Labor Statistics 2014a). What forces are driving occupational growth and decline?

The aging of the population is one factor driving a demand for personal care aides, physical therapists, biomedical engineers (for medicines and medical technologies), and veterinarians (to care for animals who serve as companions and helpers). Advances in digital and voice-driven technologies are forces that are lowering demand for postal clerks and switchboard operators. In this module we examine some of the characteristics of the U.S. economy that are shaping career opportunities.

Job Growth by Sector

We can think of an economy as comprising three sectors: primary, secondary, and tertiary. The **primary sector** includes economic activities that extract raw materials from the natural environment. Mining, fishing, growing crops, raising livestock, drilling for oil, and planting and harvesting forest products are examples. The **secondary sector** consists of economic activities that transform raw materials into manufactured goods such as houses, computers, and cars.

The **tertiary sector** encompasses economic activities related to delivering services (such as health care, entertainment, sales) and to creating and distributing information. One way to identify the relative importance of each sector of an economy is by determining how much it contributes to total employment. Chart 10.2a shows the 10 occupations that employ the largest numbers of people employed and are also considered the fastest growing between now and 2022. These 10 categories account for 18 percent of all projected job growth in the next decade. Nine of these 10 occupations are part of the tertiary sector.

Chart 10.2a: Number of People Employed (in thousands) in 10 Largest Occupational Categories and Median Income, 2012

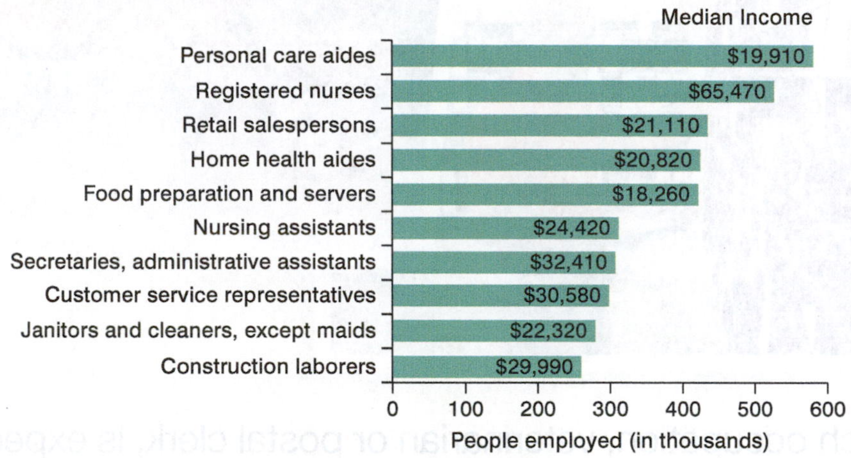

Source of data: U.S. Department of Labor (2014)

Look at the bar graph to see two pieces of information about each of the occupations listed: (1) the number of people employed in each job and (2) the median income. For example, notice that there are almost 600,000 employed as personal care aides with a median income of $19,910.

A Two-Tier Labor Market

Over the past 40 years or so (since 1975), a two-tier labor market has emerged in the United States, such that those who lack formal education beyond high school or technical skills are at a severe economic disadvantage relative to those with strong educational and professional credentials. As a result, those without a college education are likely to hold occupations that pay lower wages and offer few benefits (U.S. Central Intelligence Agency 2014). Such occupations include personal care aides and nursing assistants, with median incomes of $24,420 and $19,910, respectively. Median means 50 percent of workers in each of these occupational categories earn salaries lower than each amount and 50 percent earn higher wages. While we might argue that it is the lack of formal education that explains low wages, there are also ideological factors at work.

In a comparative study of low-wage labor in selected high-income countries, we see that the United States has the highest percentage of low-wage workers, followed by Ireland and the United Kingdom (see Chart 10.2b). Italy, Finland, Portugal, and Belgium have significantly smaller percentages of their workforces earning low wages.

Chart 10.2b: Percentage of Low-Wage Workers in the Workforce, 2011

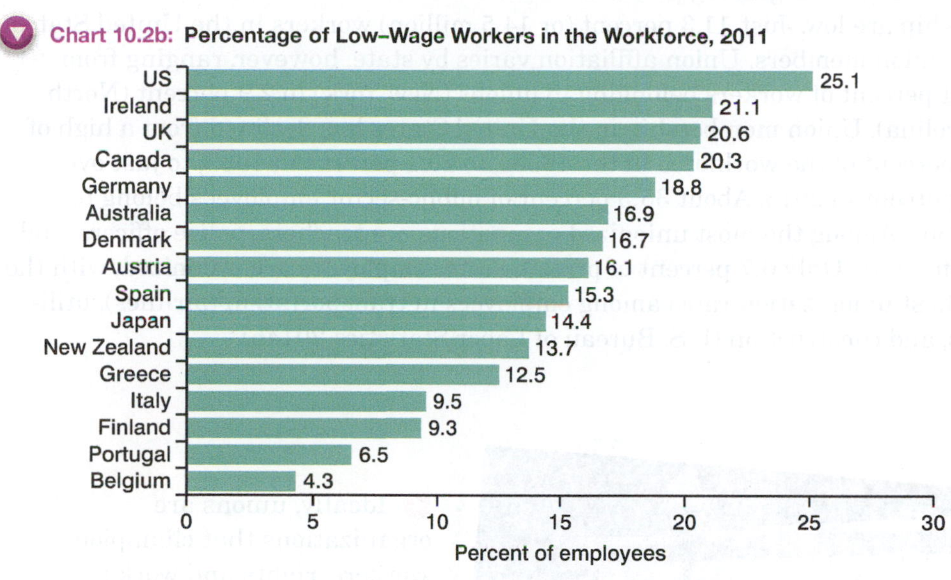

Source of data: OECD (2014a)

Researchers Eileen Appelbaum and John Schmitt (2009) wanted to understand why such a high percentage of workers in the United States earn low wages relative to workers in other high-income countries. To answer this question they focused on nursing assistants. In the United States, 38 percent of nursing assistants earn $12.40 or less per hour. In the United Kingdom, that figure is 21 percent; in Germany, France, and the Netherlands, less than 10 percent of nursing assistants earn the U.S. equivalent of $12.40 or less. Why are nursing assistants more likely to be counted among the ranks of low-wage workers in the United States than in the other five countries? The researchers discovered that countries have different levels of tolerance regarding wage inequality. In addition, European countries place a high value on quality health services and see that quality as dependent on health care workers at all levels being well paid and trained. The United States, by contrast, seems to believe that some health care–related tasks can be done by the unskilled and low-paid workers with minimal training (Appelbaum and Schmitt 2009).

As a cost-saving strategy, U.S. hospitals assign seemingly routine tasks such as monitoring temperature, taking blood pressure, applying sterile dressings, and drawing blood to minimally trained nursing assistants (who receive about six weeks of training), not to registered nurses or LPNs. In the Netherlands, by contrast, a hospital nursing assistant must complete a 34-month program that includes 56 weeks of theory-oriented coursework. France has taken steps to eliminate unqualified nursing assistants by giving them a choice to resign or enroll in a strong vocational training program to upgrade their knowledge and skills (Appelbaum and Schmitt 2009; Schmitt 2012).

Other Factors That Fuel the Low-Wage Tier

Relative to comparable economies (the wealthiest countries in the world), a number of factors work to increase the size of the low-wage labor force in the United States—factors that act to suppress wages and also to reduce benefits. For one, the collective bargaining power of workers is reduced when rates of union membership are low. Just 11.3 percent (or 14.5 million) workers in the United States are union members. Union affiliation varies by state, however, ranging from 24.4 percent of workers belonging to unions (New York) to 2.9 percent (North Carolina). Union membership in the United States has declined from a high of 35 percent of the workforce in the 1950s, to 20.1 percent in 1983, to just over 11 percent in 2013. About 35.3 percent of public-sector employees belong to unions. Among the most unionized occupations are teachers, police officers, and firefighters. Only 6.7 percent of private-sector employees are unionized, with the highest unionization rates among employees in transportation (airlines), utilities, and construction (U.S. Bureau of Labor Statistics 2014b).

Chris Caldeira

Ideally, unions are organizations that champion workers' rights and work to guard and improve conditions of employment including workplace safety; negotiating wages; establishing rules for hiring and firing; and promoting and securing benefits. Unions use the power of numbers to pressure employers to pay and treat workers fairly.

In addition to low union membership, U.S. minimum wage laws set the minimum so low that these laws actually have the effect of increasing the number of working poor. At the time of this writing, the federal minimum wage was $7.25 per hour. Keep in mind that employees working full time at $7.25 per hour earn $14,780 per year. That hourly rate varies by state, with some states having no minimum wage law in place and other states mandating a wage that exceeds the federal minimum. It is worth noting that no state law has set the minimum wage higher than $8.25 per hour. If we compare the purchasing power of those earning minimum wages in high-income countries, we see that the purchasing power of $7.25 in the United States is exceeded by an added $2.75 in Belgium, by an added $3.45 in Luxembourg, and by an added 60 cents in Canada (OECD 2014b). In making such comparisons we must remember that low-wage workers in the United States are also less likely than low-wage workers in comparably wealthy countries to be paid for sick days, family leave, vacation and holidays, and to have other benefits such as retirement and life insurance.

In January 2014 President Obama signed an executive order requiring any businesses awarded federal contracts to pay their employees a minimum of $10.10 per hour. The executive order also requires these employers to pay "health and welfare" benefits valued at $3.81 per hour to its minimum-wage employees. Contractor-operated fast-food establishments on military bases such as this Subway must also comply (*Military Times* 2014).

Kari Hawkins, USAG Redstone

There are some policies that lift people out of poverty. The Earned Income Tax Credit (EITC) is the third largest social welfare program ($63 billion) in the United States after Medicaid ($402 billion) and food stamps ($78 billion). In 2011, the last year for which data are available, 27 million households received $56 billion in EITC, adding an average of $2,074 to their household incomes. The Census Bureau estimated that EITC has lifted 5.4 million above the poverty line (U.S. Department of Health and Human Services 2014).

Chart 10.2c: Estimated Number of Americans Who Did Not Fall into Poverty Because of Government Safety Net Program

Number Kept out of Poverty by ....

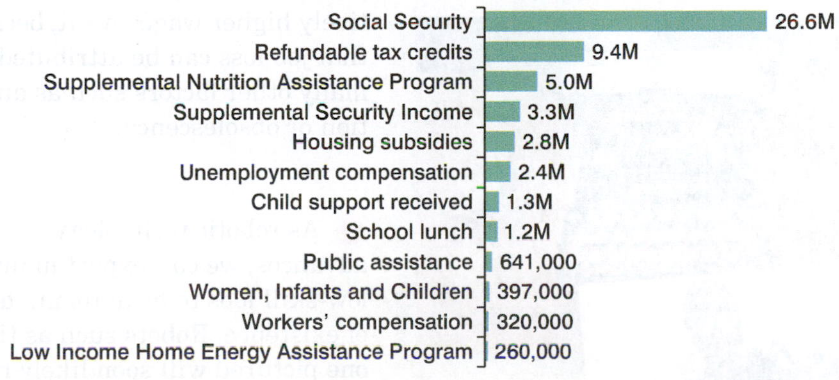

Chart 10.2d: Estimated Number Who Fell into Poverty By Event

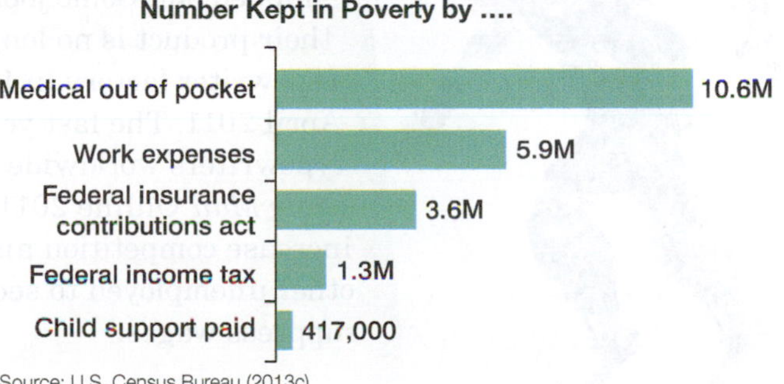

Source: U.S. Census Bureau (2013c)

Outsourcing as a Factor

What role, if any, does outsourcing play in creating the two-tier labor market? It is impossible to know the number of office, information, and other jobs that have been outsourced to foreign countries. It is also impossible to calculate the effects outsourcing has on wages. This is because the U.S. Bureau of Labor Statistics does not collect data on jobs outsourced, and corporations rarely share outsourcing plans with the public or government policy makers. To complicate matters, the United States government does not collect data on the impact of outsourcing, positive or negative (Hira and Hira 2008; Levine 2012b).

When we think of outsourcing, we often think of China as the destination for manufacturing jobs and India for technology and service jobs. Although these two countries are the best-known destinations, manufacturing and service jobs are outsourced to just about every country in the world.

Lisa Southwick

Outsourcing is only one factor behind the net loss of 8 million U.S. factory jobs since mid-1979, the year the number of manufacturing jobs peaked at 19.6 million (Norris 2012). We cannot offer precise accounts of where those 8 million factory jobs with relatively higher wages went, because that job loss can be attributed to many other factors such as automation or obsolescence.

Chris Caldeira

DARPA

As robotic technology advances, we can expect many low-skill jobs to be automated out of existence. Robots such as the one pictured will soon likely run across warehouse floors gathering items to be shipped out. Some jobs will just disappear because their product is no longer used. The world's last typewriter factory in India closed its doors in April 2011. The last year it was open it sold 800 typewriters worldwide (*Huffington Post* 2011; *Daily Mail* Online 2011). These kinds of job losses increase competition among the low-skilled and other unemployed to secure work and act to further suppress wages.

The outsourcing of factory jobs from the United States to elsewhere has been going on for more than 50 years. For example, in 1965 RCA was the first U.S.-based corporation to set up what is known as a *maquiladora* or foreign-owned manufacturing operations on the Mexican side of the border with the United States. Today there are about 4,500 such factories along the border, 85 percent of which are U.S.-owned (Connolly 2011; Export.gov 2012). Advances in digital technologies, however, have led to a second stage of outsourcing that has put a wider range of jobs up for international competition—those jobs are in the areas of office tasks and IT. At first, the outsourced jobs involved routine office work that required little training and little direct or face-to-face contact with customers or coworkers—such as bill processing, bookkeeping, data entry, and payroll. But then, the not-so-routine, high-skilled jobs became targets for outsourcing, including work done by architects, radiologists, and lawyers.

What Do Sociologists See?

Chris Caldeira

From a profit-making point of view, the 1.6 million drivers who haul large loads and the 1.3 million who drive delivery trucks in the United States (U.S. Bureau of Labor Statistics 2013a) represent "costs" of delivering a product from one destination to another. As driverless technology is perfected, we can anticipate that the jobs of truck drivers will eventually disappear.

Critical Thinking

Have you ever worked for minimum wage or lower? Describe the job you held in terms of hours worked and benefits. Did you ever feel your labor was worth more than that amount? Explain.

Key Terms

primary sector

secondary sector

tertiary sector

Module 10.3 Transnational and Global Corporations

Objective

You will learn a way to think about the size and power of the world's largest corporations.

NKU Sociology, Missy Gish

What does it mean to be the largest corporation in the world and have $481.7 billion in annual revenue? Royal Dutch Shell's annual revenue of $481.7 billion in 2013 made it the 30th largest economy in the world, just larger than that of Nigeria, which has a GDP of $478 billion.

Transnational Corporations

A **transnational corporation,** sometimes referred to as a multinational or global corporation, has operations in more than one country. Transnationals compete, plan, produce, sell, recruit, acquire resources, and do other activities on a multi-country scale. A transnational can range in size from fewer than 10 to millions of employees.

While world's largest transnationals are often referred to as global corporations, theoretically a truly global corporation should have established some kind of economic relationship in every country in the world. By that criterion, probably no corporation is yet global. As a case in point, McDonald's and UPS are considered global corporations but McDonald's has a presence in 119 countries and UPS in 200-plus countries (McDonald's 2014; UPS 2014). To be truly global both would have to have operations in 260 territories and countries.

Transnationals establish operations in foreign countries for many reasons, including to obtain raw materials (such as oil and diamonds), to avoid paying taxes, to employ a low-wage labor force, and to manufacture goods for consumers in a host country (as does Toyota Motor North America, Inc.). Transnationals are headquartered disproportionately in the United States, Japan, and Western

Europe. These global enterprises make "the key decisions—about what people eat and drink, what they read and hear, what sort of air they breathe and the water they drink, and ultimately what societies will flourish and which city blocks will decay" (Barnet 1990, 59).

When you read the names of the world's largest transnational corporations, the 10 largest of which are listed in Table 10.3a, it is difficult to imagine their size and power of influence without some basis for comparison. We can get some idea of their size by comparing the annual revenues of a corporation to a country's GDP. A corporation's annual revenue is the total amount of money it receives for goods sold or services provided over the course of a given year. Taken together, the annual revenue of the top 10 global corporations is $3.671 trillion. GDP is the total value of all goods and services produced *within* the country over a year's time. Only four countries in the world—the United States, China, India, and Japan—have a GDP that exceeds $3.671 trillion. The annual revenue of the world's largest corporation, Royal Dutch Shell, is $481.7 billion. Only 30 countries have a GDP larger than that amount; those countries include Australia, Brazil, Canada, China, Colombia, India, Iran, Italy, Japan, Mexico, Poland, South Africa, and the United States.

Table 10.3a: The World's Largest Transnational or Global Corporations, 2013
Notice that 6 of the 10 largest corporations in the world extract or refine oil. In what ways might petroleum industries slow down national efforts to use less oil or to make the transition to sustainable sources?

Rank	Global Corporation	Industry	Revenues (in billions)	Profits (in billions)	Headquarters
1	Royal Dutch Shell	Petroleum	$481.7	$26.6	Netherlands
2	Walmart	General merchandisers	$469.2	$17.0	United States
3	ExxonMobil	Petroleum	$444.9	$44.9	United States
4	SINOPEC	Petroleum	$273.4	$7.6	China
5	China National Petroleum	Petroleum	$408.6	$18.2	China
6	BP	Petroleum	$388.3	$11.6	UK
7	State Grid	Electricity (power)	$298.4	$12.3	China
8	Toyota Motors	Automobile	$265.7	$11.6	Japan
9	Volkswagen	Automobile	$247.6	$27.9	Germany
10	Total	Petroleum	$234.3	$13.7	France

Source of data: CNN Fortune Global 500 (2014)

Criticism and Support for Transnationals

Critics of transnational corporations maintain that they are engines of destruction. That is, they exploit people and natural resources to generate profits. They take advantage of desperately poor labor forces, lenient environmental regulations, and sometimes almost nonexistent worker safety standards. Supporters of transnational corporations, by contrast, maintain that these companies are agents of progress. Most obviously, transnationals employ millions and distribute goods, services, technology, and capital across the globe. In addition, they praise the transnationals' ability to raise standards of living, increase employment opportunities, transcend political hostilities, transfer technology, and promote cultural understanding. As one measure of their contributions, consider that Royal Dutch Shell and other oil producers provide the energy source that underlies the world's transportation structure.

On another level, however, transnationals' operations can aggravate problems related to climate change, obesity, poverty, mass unemployment, and overall inequality. Still, one can also argue that transnationals are not responsible for creating or solving social problems such as obesity. After all, nobody forces people to eat fast foods, choose Walmart over a locally owned store, or drive cars.

Chris Caldeira

McDonald's and other fast-food establishments sell healthy items such as real fruit smoothies. As one former McDonald's CEO pointed out, "You can get a balanced diet at McDonald's. It's a question of how you use McDonald's" (Greenberg 2001).

Moreover, corporations claim that they merely respond to consumer demand. For example, virtually all the major fast-food companies have introduced healthy items on their menus, and most have proven unpopular with consumers. Salads, for example, make up to 2–3 percent of all McDonald's revenue, and McDonald's CEO reports throwing out more than they sell (Jargon, 2013). Nevertheless, critics question whether corporations should have the right to ignore the long-term effects of their products and practices on people and the environment, even if they are responding to consumer demand. Profitable products may benefit a corporation's bottom line and shareholders, but they can also be costly for a society due to **externality costs**—the hidden costs of using, making, or disposing of a product that are not figured into the price of the product or paid for by the producer. Such costs include those associated with cleaning up the environment, treating obesity-related health problems, injured and chronically ill workers, consumers, and others. These costs are not paid by the corporations but ultimately by taxpayers and consumers.

While transnationals and other corporations are very powerful, consumer advocacy organizations have demonstrated that they can hold corporations in check. One thing is clear: when corporate executives feel real pressure from consumers, they act. However, if only a small number of consumers speak out, their claims are often dismissed. Consider comments from a McDonald's CEO after as many as 2,000 protesters trashed McDonald's restaurants and other businesses during four days of protest in Seattle against the World Trade Organization. The CEO noted that while 2,000 people protested, 17.5 million other people visited a McDonald's restaurant to eat (Greenberg 2001). The point is that there is no need to be concerned about the voices of 2,000 activists when 17.5 million consumers are voting with their feet—or mouths. McDonald's has changed some of its practices in response to pressure from other organizations, however.

Greenpeace

Greenpeace is one example of an organization dedicated to mobilizing people to hold governments and corporations responsible for crimes against the environment or for failing to protect the environment. It has recorded a number of successes, including pressuring McDonald's Corporation to agree not to use chickens that have been fed soya, a feed made from soy beans grown in the Amazon rainforest, and not to feed chickens with genetically engineered feed. Here Greenpeace staged a protest in which Ronald McDonald resigns from his job with McDonald's. For a list of Greenpeace victories, see http://www.greenpeace.org/international/about/victories.

What Do Sociologists See?

The Coca-Cola Company can be classified as a transnational corporation that is global in scale. The company introduced its famous soft drink in 1886 (Atlanta, Georgia), and by 1895 it was sold throughout the United States and its territories. Today the product is sold in 196 countries (Coca-Cola 2014).

Robert K. Wallace

Critical Thinking

Review the most current list of the world's 500 largest corporations, which can be found online by using the search term "Global 500." Identify one that has had a positive or negative impact on your life. Explain.

Key Terms

externality costs

transnational corporation

Module

10.4 The Global Economy

Objective

You will learn about the forces creating a global-scale economy and why some countries are rich and others poor.

What does the term *global economy* mean to you? How is this woman part of the global economy? How are you part of a global economy?

Chris Caldeira

The woman in the photo is part of the global economy if only because she rows products out to tourists on cruise ships docked in Vietnamese ports. Among other things, her floating "convenience store" stocks Oreo cookies and Ritz crackers. World system theory explains how the global-scale economy came to be. This theory falls under the category of global society theories. Global society theorists focus on human activity that is embedded in a larger global context. They emphasize that seemingly local events are interconnected with events taking place in other countries and regions of the world. These interconnections are part of a phenomenon known as **global interdependence**, a situation in which human interactions and relationships transcend national borders and in which social problems within any one country—such as unemployment, water shortages, natural disasters, or drug addiction—are shaped by events taking place outside the country.

Global interdependence is part of a dynamic process known as **globalization**—the *ever-increasing* flow of goods, services, money, people, technology, information, and other cultural items across political borders. This flow has become more dense and quick-moving as constraints of space and time that once separated people who lived far apart geographically seemingly dissolve.

World System Theory

Immanuel Wallerstein (1984) is the sociologist most frequently associated with world system theory. He has written extensively about the ceaseless 500-year-plus expansion of a single market force—capitalism—that created the world economy. The world economy, which encompasses more than 260 countries and territories and uncountable cultures, is interconnected by the division of labor. In this world economy, economic transactions cross national boundaries. Although governments seek to shape the global market in ways that benefit their interests, no single political structure such as a world government holds authority over the global economy.

How has capitalism come to dominate the global network of economic relationships? An answer to this question can be found in the ways businesses respond to changes in the economy, especially to economic stagnation. One response is to find ways to lower labor-associated production costs by moving production facilities out of high-wage zones and into lower-wage zones outside a country, introducing technologies that save labor, and forcing people to work for little to no wages (enslavement, indentured servitude).

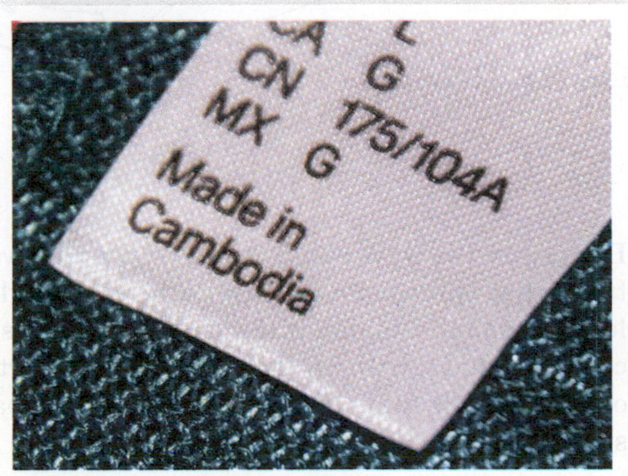

The "Made In" label symbolizes the practice of moving production from high-wage to low-wage labor zones.

A second response to economic stagnation is to secure at the lowest possible price the raw materials needed to make products; these products may be rubber, sugar cane, or coltan (for mobile phones). Karl Marx wrote that the drive for profit is a boundless thirst that chases the capitalist "over the whole surface of the globe" in search of not only low-cost labor but also inexpensive raw materials (Marx and Engels 1848). A third response to economic stagnation is to create new markets that expand the boundaries of the world economy. As one example, Figure 10.4a shows a map of the new markets McDonald's has yet to reach. McDonald's grew from one Illinois-based restaurant in 1955 to a global giant. The corporation began its expansion outside the United States in 1967 when it opened a unit in Canada.

Figure 10.4a: Countries Where There Are No McDonald's
This figure highlights those countries where there are no McDonald's. Why do you think a country might not have a McDonald's? The answers vary by country. Notice that Bolivia is the only country in South America without a McDonald's. A little research suggests that McDonald's once tried to establish roots in Bolivia but was not successful. One reason is that Bolivians like hamburgers but prefer to buy them from indigenous street women that they know, called cholitas, rather than from an anonymous source (Holt 2013).

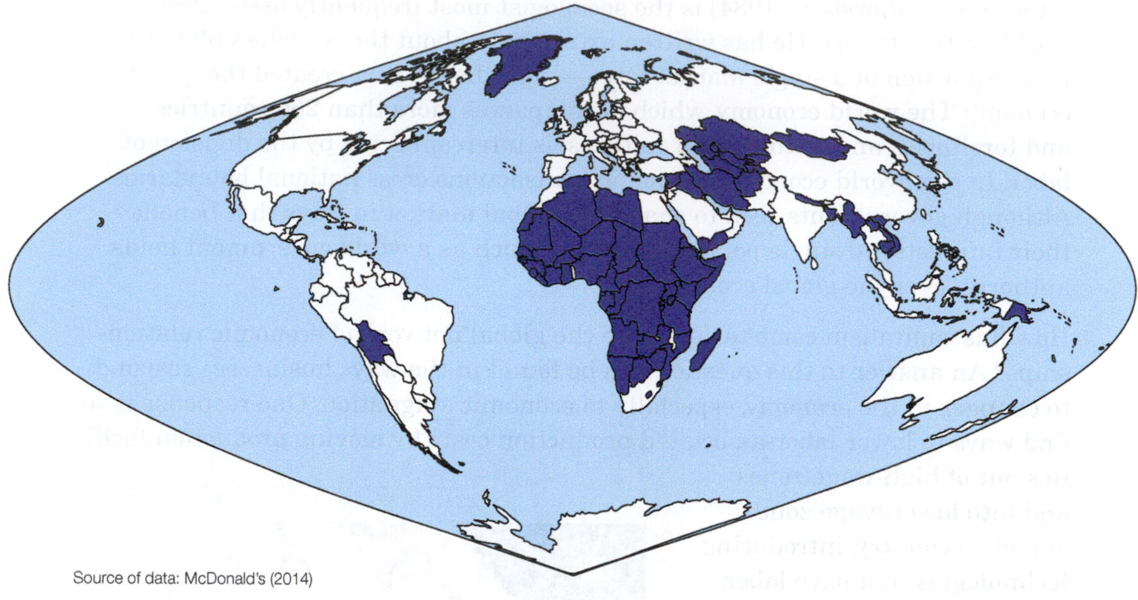

Source of data: McDonald's (2014)

Because of the ways corporations respond to avoid economic stagnation, capitalism has spread steadily across the globe and facilitated connections among local, regional, and national economies. Wallerstein (1984) argues that the 260 countries and territories that have become part of the world economy play one of three different and unequal roles in the global economy: core, peripheral, or semiperipheral.

Core Economies

Core economies include the wealthiest, most highly diversified economies with strong, stable governments. Examples of core economies include the United States, Japan, Germany, France, Canada, and the United Kingdom. The 20 percent of the world's population that constitutes the core economies accounts for 86 percent of total private consumption expenditures (World Bank 2008). Consequently, when economic activity weakens in core economies, the economies of other countries suffer, because exports to core economies decline and prices fall. In spite of their large size and relative stability, core economies have weaknesses, one of which is their dependence on foreign sources for raw materials, such as oil needed to fuel cars and palm oil needed to make soaps and other products.

Much of the processed food people who live in core economies consume depends on palm oil—which is produced in Indonesia, Malaysia, and West African forests. Palm oil is an ingredient in cake mixes. It is also an ingredient in soaps, ice creams, cookies, crackers, and icings. Palm oil is used in many non-food products such as cosmetics, personal care products, and detergents, though the ingredient is rarely listed as "palm oil" on labels. Some of the most used synonyms are glycerin, stearic acid, and vegetable oil (Schuster Institute for Investigative Journalism 2014).

Peripheral Economies

Peripheral economies are built around primary commodities, or even one primary commodity such as coffee, peanuts, or tobacco, or on a natural resource, such as petroleum, tin, copper, or zinc (United Nations Development Programme 2011). As a result, price fluctuations due to bad weather, a bumper crop, or lowered consumer demand have a dramatic impact on people's livelihoods and affect the incidence of poverty, as the vast majority of the poor depend on primary commodities for their livelihoods. Of the 141 countries known as developing, at least 95 depend on primary commodities for 50 percent or more of revenue generated from exports (U.S. Central Intelligence Agency 2014; United Nations Development Programme 2011). For example, coffee accounts for almost 70 percent of the export revenue that Uganda, Ethiopia, and Burundi earn (United Nations Development Programme 2011).

Extreme dependence on one or even a few primary commodities causes many problems. First, fluctuations in the price of that commodity can leave a government flush with money when prices rise and poor when prices fall, leading to cycles of wild government spending followed by drastic spending cuts. Second, corruption and political rivalries inevitably arise when a government or a few powerful people control a country's natural resources, the distribution of those resources, and the revenues from them (Birdsall and Subramanian 2004).

Peripheral economies have a dependent relationship with core economies that traces its roots to colonialism. Peripheral economies operate on the so-called fringes of the world economy. Most of the jobs that connect their workers to the world economy pay little and require few skills. Amid widespread and chronic poverty, pockets of economic activity may exist, including manufacturing zones and tourist attractions.

NKU Anthropology, Sharyn Jones
NKU Sociology, Missy Gish

A country whose major exports include things like sugar, garments, gold, timber, fish, molasses, and coconut oil qualifies as a peripheral economy. This list of items suggests that the economy revolves around human or animal muscle to grow, harvest, and extract products. It also suggests that, in contrast to core economies, people consume unprocessed food such as fresh coconut and not the processed coconut found in, say, a Mounds candy bar.

Semiperipheral Economies

Semiperipheral economies are characterized by moderate wealth (but extreme inequality) and moderately diverse economies. Semiperipheral economies exploit peripheral economies and are in turn exploited by core economies. According to Wallerstein (1984), semiperipheral economies play an important role in the world economy because they are politically stable enough to provide useful places for capitalist investment if employee wage and benefit demands in core economies become too great. The countries known as emerging markets—Mexico, Brazil, Argentina, South Africa, Poland, Turkey, India, Indonesia, China, and South Korea—qualify as semiperipheral.

There are two major theories that offer explanations for this dramatic inequality among the world economies: modernization theory and dependency theory.

Modernization Theory

Modernization is a process of economic, social, and cultural transformation in which a country "evolves" from preindustrial or underdeveloped status to a modern society as did the United States and the countries of Western Europe. According to Hirai Naofusa (1999), a country is considered modern when it possesses the following eight characteristics.

1. A high proportion of the population lives in and around cities such that the society is urban-centered.
2. Energy to produce food, make goods, and provide services does not derive from physical exertion (human and animal muscle) but from oil, coal, and other inanimate sources of energy.
3. There is widespread access to a variety of goods and services, a feature of economies with high standards of living.

4. People have a voice in economic and political affairs.
5. Literacy is widespread and there is a scientific, rather than secular, orientation to solving problems.
6. A system of mass media and communication is in place that offsets the influence of the family and local cultures.
7. There are large-scale, impersonally administered organizations such as government, businesses, schools, and hospitals that reduce dependence on family for child care, education, and social security.
8. People feel a sense of loyalty to a country (a national identity), not to an extended family and/or tribe.

This village school in Afghanistan enrolls more than 600 students. It has no modern conveniences, and these girls are allowed to attend on the condition that their father allows it. Modernization theorists would argue that a country will not modernize until the decision is out of the fathers' hands and becomes a government mandate.

Scott McLaren, Courtesy of Joan Ferrante

Modernization theorists identify the conditions that launch underdeveloped countries on the path to modernization and specify five stages through which countries must pass to reach modernization. Stage 1—a prelaunch stage—the point at which a society is still tradition-oriented; that is, it is dominated by kinship-related obligations and loyalties that modernization theorists claim discourage change and personal mobility. Productivity is limited by a lack of access to tools of modern science and a frame of mind characterized by a long-running fatalism, or the widespread belief that "the range of possibilities open to one's grandchildren" will be no different from those open to their grandparents (Rostow 1960).

According to modernization theory, the next stages are the pre-takeoff (stage 2), the takeoff (stage 3), the drive to maturity (stage 4), and the age of high mass consumption (stage 5). Western countries can jump-start modernization (stage 2) through foreign aid, investment, and technology transfers that include fertilizers, pesticides, birth control programs, loans, cultural exchange, and medical interventions (e.g., inoculation programs). Ideally, these interventions "shock" the traditional ways of doing things so that the country "takes off." Such interventions set into motion the ideas and sentiments so that a "modern alternative to the traditional society" evolves "out of the old culture" (Rostow 1960). Western countries can further hasten modernization by demanding appropriate government reforms and policies. Eventually—perhaps as many as 60 years later—the developing country will reach a final state of modernization characterized by technological maturity and high mass consumption.

Tech. Sgt. Scott T. Sturkol

The U.S. government sent school buses to Afghanistan. How might this change traditional Afghan culture? Consider that, before school buses, schoolchildren walked to school with family members. How might the influence of the family decline when children now spend time on a bus with just their peers?

According to Rostow, modernization involves a transformation of cultural beliefs and values that supposedly support fatalism and collective orientation into those that support a work ethic, deferred gratification, future orientation, ambition, and individualism (important attitudes and traits believed to be essential to the development of a free-market economy or capitalism). As the country modernizes, "the idea spreads, not merely that economic progress is possible, but that economic progress is a necessary condition for some other purpose . . . be it national dignity, private profit, the general welfare, or a better life for the children" (Rostow 1960).

Dependency Theory

Dependency theorists challenge the basic tenet of modernization theory—that poor countries fail to modernize because they reject free-market principles and because they lack the cultural values that drive modernization. Rather, dependency theorists argue that poor countries are poor because they have been, and continue to be, exploited by the world's wealthiest governments and transnational corporations. This exploitation began with colonialism.

Colonialism is a form of domination in which a foreign power uses superior military force to impose its political, economic, social, and cultural institutions on an indigenous population so it can control their resources, labor, and markets (Marger 2012). The age of European colonization began in 1492, with the voyage of Christopher Columbus. By 1800, Europeans had learned of, conquered, and colonized much of North America, South America, Asia, and coastal parts of Africa, setting the tone of international relations for centuries to come. During this time, European colonists forced local populations to cultivate and harvest crops and to extract minerals and other raw materials for export to the "mother" countries. When local populations could not meet the colonists' labor needs, the colonists imported slaves from Africa or indentured workers from Asia and Europe. In fact, an estimated 11.7 million enslaved Africans survived their journey to the "New World" between the mid-16th century and 1870 (Chaliand and Rageau 1995, Holloway 1996).

Between 1880 and 1914, the pursuit of and demand for raw materials and labor increased dramatically. This period, known as the Age of Imperialism, saw the most rapid colonial expansion in history. During this time, rival European powers (such as Britain, France, Germany, Belgium, Portugal, the

Netherlands, and Italy) competed to secure colonies and influence in Asia, the Pacific, and especially Africa. Consider as one measure of the extent of colonization that, during the 20th century, 130 countries and territories gained political independence from their "mother" country.

That process of gaining political independence is known as decolonization.

Decolonization is the process of undoing colonialism such that the colonized country achieves independence from the so-called mother country. Decolonization can be a peaceful process in which the two parties negotiate the terms of independence, or it can be a violent disengagement that involves civil disobedience, insurrection, or armed struggle (war of independence). Once independence is achieved, civil war between rival factions often takes place as each seeks to secure the power relinquished by the colonizer. Some scholars argue that the Americas (which include the United States, Canada, and Central and South America) are technically still colonized lands because the indigenous peoples were not the ones to revolt and declare independence; rather, the colonists and/or their descendants revolted and declared independence. In other words, those who assumed power after independence simply continued exploiting the land and resources that once belonged to indigenous peoples (e.g., Native Americans or First Nation peoples) and exploiting the labor of enslaved and indentured servants (Cook-Lynn 2008; Mihesuah 2008).

Figure 10.4b: African Countries and Year of Independence from Colonial Power
This map of Africa shows the decade in which each country gained independence from its "mother" country. The colonizing country exploited the labor and resources of the colony. After independence, the ties between the two countries did not end, however.

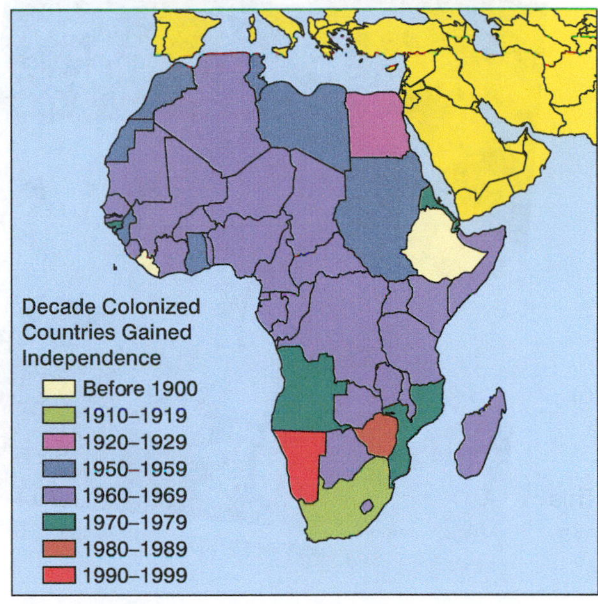

Source: Data from U.S. Central Intelligence Agency (2014)

Gaining political independence does not mean, however, that a former colony no longer depends on or is exploited by its colonizing country. In *How Europe Underdeveloped Africa*, Walter Rodney (1973) argues that in the end, the economies of the African continent—90.4 percent of which was once controlled by colonial powers—assumed roles as the producers of primary products for the benefit of transnational corporations and former colonizers. Examples of primary products include products that are mined (gold) or extracted (oil) from the earth, as well as fish and agricultural products. This continuing economic dependence on the former colonial powers is known as **neocolonialism**. In other words, neocolonialism is a new form of colonialism in which more powerful foreign governments and foreign-owned businesses continue to exploit the resources and labor of postcolonial peoples.

Specifically, resources still flow from the former colonized countries to the wealthiest countries. An example of these dynamics can be found in the United States' and the world's other wealthiest countries' reliance on foreign-trained medical staff, especially physicians.

What Do Sociologists See?

The people pictured in the photo are nursing students attending Notre Dame of Jolo College in the Philippines. The Philippines just happens to be the number one "exporter" of nurses to the United States, Britain, and Canada, and the third largest "exporter" to Australia. U.S. demand for foreign-trained health care professionals will only increase as its population ages (World Health Organization 2013; Tulenko 2010). Currently, 4 percent of nurses (one in every 25) and 22 percent of physicians in the United States were foreign-trained (World Health Organization 2013). This is just one example of the many ways the United States (a core economy) depends on peripheral and semiperipheral economies.

Photographer's Mate 2nd Class Erika N. Jones

Critical Thinking

What country does the coffee or tea you drink come from? Look up information to learn about the extent to which export revenue depends on coffee.

Key Terms

colonialism
core economies
decolonization
global interdependence
globalization
modernization
neocolonialism
peripheral economies
semiperipheral economies

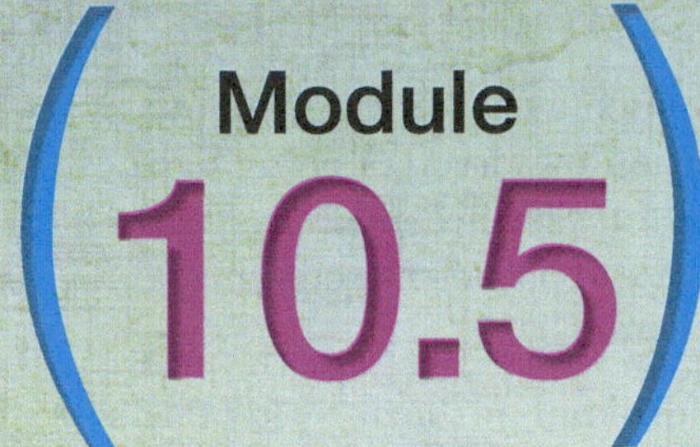

Power and Authority

Objective

You will learn under what circumstances people can exert their will, even in the face of opposition.

Do you recognize the man to the right in this photograph? How might you describe the power he has to attract attention and to have his voice heard?

U.S. Marine Corps photo by Sgt. Tyler L. Main

There is no doubt that Prince Harry of Wales possesses a kind of power over others, a power that derives in part from tradition (he is the son of Charles, Prince of Wales, and Diana, Princess of Wales; his paternal grandparents are Elizabeth II and Prince Philip, Duke of Edinburgh). But there is something else about him that has the power to influence, and that lies with the power of his personality.

A society's **political system** is the institution that regulates access to and use of power that can be used to take control of scarce and valued resources and to make laws, policies, and decisions that affect others' life chances. **Power** is the probability that an individual can achieve his or her will, even against opposition (Weber 1947). That probability increases when an individual has the means to compel people to obey his or her commands, but it increases even more when the individual possesses authority over others. **Authority** is legitimate power—power that people believe is deserved, just, and proper. A leader has authority to the extent that people view him or her as being legitimately entitled to it.

Types of Authority

Max Weber identified three types of authority: traditional, charismatic, and legal-rational. People can possess more than one kind of authority. **Traditional authority** is grounded in the sanctity of time-honored norms that set the terms for how someone comes to hold a powerful position, such as chief, king, queen, or emperor. Usually, the person inherits such a position by virtue of being born into a family that has held power for some time. The masses accept that system because it has always been like that, and to abandon the past is to renounce a heritage and collective identity (Boudon and Bourricaud 1989).

Ultimately, Prince Harry's power is grounded in traditional authority—his father is Prince Charles of Wales, and his grandmother is Queen Elizabeth II. By tradition, after his father Charles, his brother William, and his nephew George, Harry is fourth in line to be king of the United Kingdom and its independent states.

Charismatic authority is grounded in exceptional and exemplary personal qualities. Charismatic leaders are obeyed because their followers believe in and are attracted irresistibly to their vision. These leaders, by virtue of their special qualities, can persuade followers to behave in ways that depart from rules and traditions. Charismatic leaders often emerge during times of profound crisis, such as economic depressions and wars. During such times people are susceptible to a vision of a new order. A charismatic leader is more than popular, attractive, likable, or pleasant. A merely popular person, "even one who is continually in our thoughts," is not someone for whom we would break all previous ties and give up our possessions (Boudon and Bourricaud 1989, 70). Charismatic leaders successfully persuade their followers to make extraordinary personal sacrifices, cut themselves off from ordinary worldly connections, or devote their lives to achieving a vision that the leaders have outlined.

Remembering
Nelson Mandela
July 18, 1918 – December 5, 2013
United States African Command

Civil rights activist and first black president of South Africa Nelson Mandela qualifies as a charismatic leader because of "his fierce dignity and unbending will to sacrifice his own freedom for the freedom of others. . . . In transforming South Africa *he moved all of us*. . . . And the fact that he did it all with grace and good humor, and an ability to acknowledge his own imperfections, only makes the man that much more remarkable" (Obama 2014). At his trial in 1964, Nelson Mandela declared, "I have fought against white domination, and I have fought against black domination. I have cherished the ideal of a democratic and free society in which all persons live together in harmony and with equal opportunities. It is an ideal which I hope to live for and to achieve. But if needs be, it is an ideal for which I am prepared to die."

The source of a charismatic leader's authority, however, does not rest with the ethical quality of his or her vision. Adolf Hitler, Franklin D. Roosevelt, Mao Zedong, Winston Churchill, and the Reverend Martin Luther King Jr. were all considered charismatic leaders. Each assumed leadership during turbulent times. Likewise, each conveyed a powerful vision (right or wrong) of his country's destiny. Charismatic authority results from the intense relationships between leaders and followers. From a relational point of view, charisma is a highly unequal relationship between a guide who inspires and followers who believe wholeheartedly in that guide's promises and visions (Boudon and Bourricaud 1989).

Over time, charismatic leaders and their followers come to constitute an emotional community devoted to achieving a goal and sustained by a belief in the leader's special qualities. Weber argues, however, that eventually the followers must be able to return to a normal life and to develop relationships with one another based on something other than their connections to the leader. Attraction and devotion cannot sustain a community indefinitely, if only because the object of these emotions—the charismatic leader—is mortal.

Legal-rational authority is grounded in a system of impersonal and formal rules that assign power to a position (such as CEO, president). In addition, these rules specify the qualifications for occupying positions, the reach of that power, as well as the terms under which power can be exercised. It is the position that gives the occupant the power to command others to act in specific ways, a power backed by the force of law and organizational structure.

The White House is the residence and workplace of the person who holds the office of president of the United States. Anyone who holds the position of U.S. president possesses legal-rational authority. The Constitution of the United States specifies how the position of president is filled and the powers associated with that position.

Rachel Ellison

The Power Elite

Sociologist C. Wright Mills wrote about the connection between government, industry, and the military in *The Power Elite* (1956). The **power elite** are those few people who occupy such lofty positions in the social structure of leading institutions that their decisions affect millions, even billions, of people worldwide. For the most part, the source of this power is legal-rational—residing not in the personal qualities of those in power, but rather in the positions that the power elite have come to occupy.

The power elite use their positions—and the tools of their positions in a bureaucracy—to rule over, control, and influence others. These tools include surveillance systems, communication structures, and weapons. In writing about the power elite, Mills focuses on those who occupy the highest positions in the leading U.S. institutions. Those institutions are the military, corporations (especially the 200 or so largest), and the government (Mills 1963, 27).

The origins of the power elite that Mills describes can be traced to World War II, when the political elite mobilized corporations to produce the supplies, weapons, and equipment needed to fight that war. U.S. corporations, which were left unscathed by the war, were virtually the only corporations capable of providing the services and products that war-torn countries needed for rebuilding. The interests of the U.S. government, the military, and corporations became further intertwined when the political elite decided that a permanent war industry was needed to contain the spread of communism. Thus, over the past 60 years, these three institutions (and by extension the U.S. workforce) have become deeply and intricately interrelated in hundreds of ways.

Air Combat Command, photo courtesy/Lockheed Martin Communications

As one measure of the ties between the U.S. military, the government, and industry, consider that in 2013 the U.S. Department of Defense alone did business with more than 20,184 contractors and coordinated 349,000 active contracts. The value of these contracts ranges from less than $25,000 to billions (Defense Contract Management Agency 2014). The photo shows the Super Hercules aircraft produced by the Lockheed Martin Aeronautics Company. Other defense contractors include BP, Dell, FedEx, and Krispy Kreme. In 2013 the top five military contractors based on revenue earned from contracts were Lockheed Martin ($43.9 billion), Northrop Grumman ($8.2 billion), Boeing ($81.7 billion), General Dynamics ($31.5 billion), and Raytheon Co. ($24.4 billion) (Defense News 2014).

Simply think about Lockheed Martin's dependence on government contracts. About 94.1 percent of its sales are to the U.S. Department of Defense (DoD), the Department of Homeland Security, and other U.S. government agencies. These government contracts support 115,000 employees working in 572 facilities in 500 cities and 50 states throughout the United States and in 75 other nations and territories (Lockheed Martin 2014).

Because the military, the government, and corporations are so interdependent and because decisions made by the elite of one sector affect the elite of the other two, Mills believes that everyone has a vested interest in cooperation. Shared interests cause those who occupy the highest positions in each sector to interact with one another. Out of necessity, then, a triangle of power has emerged. Mills cautions, however, that we should not assume that the alliance among the three sectors is untroubled, that the powerful share exactly the same interests, that they know the consequences of their decisions, or that they are joined in a conspiracy to shape the fate of a country or the globe. At the same time, it is clear that they know what is on each other's minds. Whether they come together casually at their clubs or hunting lodges or more formally as board members, they are definitely not isolated from each other. There is a community of interest and sentiment among the elite (Hacker 1971).

Mills gives no detailed examples of the actual decision-making process at the power elite level. Rather, he focuses on understanding the consequences of this alliance. Mills acknowledges that the power elite are not free agents but are subject to controls, such as whistleblowers, congressional investigations, and budget constraints.

Pluralist Model

A pluralist model of power views politics as an arena of compromise, alliances, and negotiation among many competing special-interest groups, and it views power as something dispersed among those groups. **Special-interest groups** consist of people who share an interest in a particular economic, political, or other social issue and who form an organization or join an existing organization to influence public opinion and government policy. Special-interest groups are very diverse. They include grassroots groups and well-established groups such as Boeing Corporation, which had $81.6 billion in active government contracts in 2014 (*Defense News* 2014).

Interest groups are often represented by **lobbyists**, people whose job it is to solicit and persuade state and federal legislators to create legislation and vote for bills that favor the interests of the group they represent. There are an estimated 106,000 people employed as lobbyists, about one-third of whom work in Washington, D.C. Lobbyists can represent states, foreign governments, industries, universities, or nonprofit agencies (Lee 2014). Some examples of organizations that lobbied to shape the Affordable Care Act reform legislation and the amount each spent are the Pharmaceutical Research and Manufacturers Association ($20.1 million), Blue Cross and Blue Shield ($16.1 million), the American Medical Association ($12.3 million), and the American Hospital Association ($12.6 million) (Center for Responsive Politics 2009).

Some special-interest groups form **political action committees** (PACs), which raise money to be donated to the political candidates who seem most likely to support their special interests. There are more than 4,500 registered PACs, some of which are called Super PACs. Super PACs, the products of federal court cases (*SpeechNow.org v. Federal Election Commission*; *Citizens United v. Federal Election Commission*), are allowed to raise and spend unlimited amounts in support of or against political candidates. In addition, corporations and unions can contribute unlimited amounts of money directly or though other organizations without complete or immediate disclosure, making it difficult for the public to know the source of funding behind political messages (Center for Responsive Politics 2014). As of March 2014, 967 groups were registered as Super PACs and reported holding $141.2 million in contributions (Center for Responsive Politics 2014).

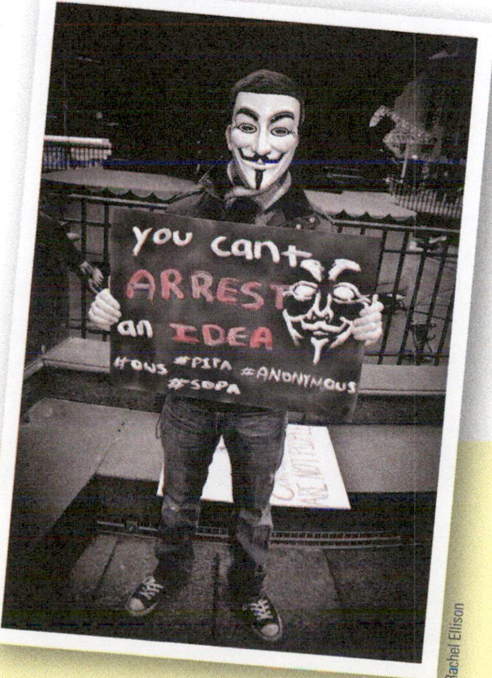

According to the pluralist model, no single special-interest group dominates the U.S. political system. Rather, competing groups thrive and can express their views through opinion polls, unions, protests, e-mails, and PACs. One problem with the pluralist model is that we cannot conclude that every special-interest group has enough resources to represent and defend its interests.

What Do Sociologists See?

Why does this marine "agree" to stand at attention while his instructor yells at him? He agrees to comply because the drill instructor occupies a position that gives her power (legal-rational authority) over him while she is on duty. That authority does not apply should she run into this man in the grocery store when both are off duty and out of uniform.

Sgt. Pamela Shelley

Critical Thinking

Think of someone who has authority over you. Which of the three types of authority—traditional, charismatic, legal-rational—does the person possess?

Key Terms

authority

charismatic authority

legal-rational authority

lobbyists

political action committees (PACs)

political system

power

power elite

special-interest groups

traditional authority

Module 10.6 Forms of Government

Objective

You will learn about various forms of government and how they direct and coordinate political and economic activity.

Which forms of government do the two men represent?

Official White House photo by Pete Souza

Government is the organizational structure that directs and coordinates people's involvement in the political activities of a country or some other territory, such as a city, county, or state. It is also one mechanism through which people gain power and exercise authority over others. Governments make laws and create policies that shape every aspect of life. Governments are often divided into branches of power, such as executive, judicial, and legislative. In this module we consider five forms of government: democracy, totalitarianism, authoritarianism, monarchy, and theocracy.

Democracy

Sir Winston Churchill, prime minister of the United Kingdom during World War II, once said that "democracy is the worst form of government except for all others that have been tried" (1947, 756). **Representative democracy** is a system of government in which power is vested in citizens who vote into office those candidates they believe can best represent their interests. Because it is not feasible for citizens to vote on every matter that affects them, they make their voices heard through those they elect. In a democracy, political candidates can be from more than one party, and those in minority parties can challenge the party holding power. In addition, when a majority of voters elect to change the party in power, the transition is orderly and peaceful. In democracies, elected representatives make laws, vote on taxes, establish budgets, and support or challenge the party in power.

Democratic forms of government extend basic rights to citizens and legal residents. These rights include freedom of speech, movement, religion, press, and assembly (that is, the right to form and belong to parties and other associations), as well as freedom from unwarranted arrest and imprisonment. Other characteristics of democracies include free and fair elections, access to a free press, an educated or informed citizenry, and a constitution that sets limits on executive and other powers (Bullock 1977).

Successful democracies depend on informed voters. Although the United States is classified as a representative democracy, not everyone who is eligible to vote does so. In U.S. presidential elections, typically 65 percent of all those registered to vote actually vote. Thus, the winner is usually awarded victory with the support of about 33 percent of all voters (U.S. Census Bureau 2013d). In light of these statistics, we might question whether elected officials represent everyone or just those who vote. Furthermore, we might question whether elected officials pay most attention to those powerful constituents who donate generously to their campaigns. In assessing whether a form of government is a true democracy, it is important to consider who has the right to vote. All democracies have at one time or another excluded some from the political process based on race, sex, income, property ownership, criminal status, mental health, religion, age, or other characteristics.

Totalitarianism

Totalitarianism is a system of government characterized by (1) a single ruling party led by a dictator, (2) an unchallengeable official ideology that defines a vision of the "perfect" society and the means to achieve that vision, (3) a system of social control that suppresses dissent, and (4) centralized control over the media and the economy. Ideological goals vary but may include overthrowing capitalist and foreign influences (as in Cuba's 60-year revolution to resist U.S. capitalist influence) or creating the perfect race (as in Germany under Hitler). Whatever the government's goals, the political leaders, the military, and the secret police intimidate and mobilize the masses to help the state meet its goals.

Totalitarian governments are products of the 20th century, because by that time technologies existed that allowed a few people in power to control the behavior of the masses and the information the masses could hear. Many of the governments labeled as totalitarian have followed communist principles. Traditionally, communist governments have outlawed private ownership of property, supported the equal distribution of wealth, and offered status and power to the working class, or proletariat. Communist leaders also mobilize the masses to bring about change.

Chris Caldeira

Using the definition of totalitarianism above, Fidel Castro, the once prime minister (1959–1976) and president of Cuba (1976–2008), led a totalitarian government. Like all leaders of governments considered totalitarian, Castro is a controversial figure praised by supporters as a champion of equality, anti-imperialism, and sustainability, and criticized by opponents and rivals as a perpetrator of human rights abuses and Cuba's impoverishment.

Authoritarianism

Under an **authoritarian** government, no separation of powers exists; a single person (a dictator), a group (a family, the military, a single party), or a social class holds all power. No official ideology projects a vision of the "perfect" society or guides the government's political or economic policies. Authoritarian leaders do not seek to mobilize the masses to help realize a vision or meet ideological goals. Instead, the government functions to serve those in power, who may or may not be interested in the general welfare of the people. Common to all authoritarian systems is a lack of checks and balances on the leader's power (Chehabi and Linz 1998).

How does a single person, a group, or a social class gain control of an entire country? Authoritarian leaders typically receive support from a foreign government that expects to benefit from their leadership (Buckley 1998). Saddam Hussein, the former president of Iraq, is one example of an authoritarian leader whom the U.S. government both supported and opposed. The U.S. government supported Hussein during the Iran–Iraq War (1980–1988), even though he invaded Iran, abused the human rights of Iraqi citizens, used chemical weapons on both Iranians and his own people, and sought to produce nuclear weapons. Then the U.S. government opposed Hussein during and after the 1991 Gulf War.

In 2003 the United States military invaded Iraq and deposed Saddam Hussein on the grounds that the dictator had been secretly developing and building weapons of mass destruction. When no evidence of such weapons turned up, the U.S. government justified the invasion on the grounds that Hussein had abused the human rights of Iraqi citizens and used chemical weapons against his own people. This photo shows an American soldier draping the U.S. flag around a statue of Saddam Hussein after taking the capital city.

Derrick Jensen

Monarchy

A **monarchy** is a form of government in which the power is in the hands of a leader (known as a monarch) who reigns over a state or territory, usually for life and by hereditary right. Typically, the monarch expects to pass the throne on to someone who is designated as the heir, usually a firstborn son. The monarch's power may range from absolute (total power) to nominal (in name only) (U.S. Central Intelligence Agency 2012).

Master Sgt. Jerry Morrison

Saudi Arabia qualifies as a hereditary monarchy. Since 1932 it has been ruled by the sons and grandsons of the first king of Saudi Arabia, Abd al-Aziz ibn Saud, shown in these portraits. Saudi Arabia has no official political parties and no national elections. The king's power is tempered by consultation with the royal family and with an appointed consultative body known as the Majlis al-Shura, which debates, rejects, and amends proposed legislation; holds oversight hearings over government ministries; and initiates legislation. While the head of state and government is a king and not a religiously trained leader, Saudi Arabia's constitution and legal system are grounded in Islamic laws and principles (U.S. Central Intelligence Agency 2012).

Theocracy

Theocracy, which means rule by divine guidance, is a form of government in which political authority rests in the hands of religious leaders or a theologically trained elite group. Thus, there is no legal separation of church and state. Government policies and laws correspond to religious principles and laws. Contemporary examples of theocracies include the Vatican under the pope, Afghanistan when it was under the Taliban, and Iran under the ayatollahs. In other forms of theocracy, power is shared by a secular ruler (such as a king) and a religious leader (such as a pope or an ayatollah). A theocracy can also be composed of a secular government whose leaders are devoted to and guided by the principles of the dominant religion. At one time, England was dominated by the Anglican Church, France by the Roman Catholic Church, and Sweden by the Lutheran Church.

What Do Sociologists See?

Sociologists see an inked-stained finger, which has come to be associated with democratic elections, most notably elections taking place in places that have had no long-standing tradition of democracy such as Iraq and Afghanistan. The stained finger marks people so they do not try to vote again.

U.S. Army photo by Staff Sgt. Jacob Caldwell

Critical Thinking

Have you voted in local, state, and/or national elections? What does your voting record say about democracies?

Key Terms

authoritarian

government

monarchy

representative democracy

theocracy

totalitarianism

Module

10.7 Applying Theory: The Power and Reach of the U.S. Military

Objective

You will learn about how sociologists inspired by each of the four perspectives describe the military power of the United States.

Navy photo by Petty Officer 2nd Class Andrea Perez

These planes are part of the U.S. military structure, which spends $612 billion on defense each year, or 40 percent of all military spending in the world. How might you explain this spending if someone from another country asked?

In 2013 worldwide military expenditures approached $1.7 trillion. The United States budgeted $612.5 billion, followed by China ($126 billion). U.S. military spending accounts for 40 percent of all military spending worldwide (SIPRI 2012). The amount spent increases if we consider money spent on other military-related costs not counted as defense spending: costs associated with ongoing involvement in Afghanistan and Iraq, arms transfers to foreign governments, and veterans' benefits ($127 billion). One assessment describes the power of the United States this way: "Throughout the history of mankind certainly no country has existed that so thoroughly dominates the world with its policies, its tanks, and its products as the United States does" (*Der Spiegel* 2003).

Americans don't want to be part of a country that is an empire. But if it is one, they want to think of the United States as a kind and just empire. "Talk of 'empire' makes Americans distinctly uneasy" (Judt 2004, 38).

DoD photo by Fred W. Baker III

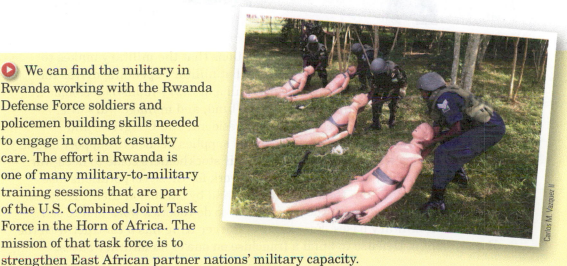

Lee Craker

For one measure of the U.S. military's global reach, simply consider that the three soldiers pictured are among the 40,000 foreign-born servicewomen and men who currently serve in the military. The three are participating in a naturalization ceremony. Naturalization ceremonies can range in size from 25 members from 15 different countries to 256 service members from 76 different countries. At one such ceremony the presiding officer proclaimed "Our nation's unique quality is that it weaves the world's cultures into a great American tapestry, and our military benefits from their strengths" (U.S. Military.com 2014).

We can find the military in Rwanda working with the Rwanda Defense Force soldiers and policemen building skills needed to engage in combat casualty care. The effort in Rwanda is one of many military-to-military training sessions that are part of the U.S. Combined Joint Task Force in the Horn of Africa. The mission of that task force is to strengthen East African partner nations' military capacity.

Carlos M. Vazquez II

Mass Communication Specialist 2nd Class Colby K. Neal

We can also find U.S. military aircraft carriers patrolling in any of the world's major waters including Atlantic Ocean, Pacific Ocean, the Suez Canal, South China Sea, Arabian Gulf, Subic Bay, Indian Ocean, Strait of Hormuz, Baltic Sea, Black Sea, Gulf of Oman, Celtic Sea, Red Sea and the Arctic Ocean (U.S. Navy 2014).

How might sociologists from each of the four perspectives assess the United States as a military power?

DARPA

Rachel Ponder, APG News

Functionalists recognize the contributions that the military makes to the economic order. There is no question that military spending drives the U.S. economy through what sociologists call a military-industrial complex, a relationship between those who declare, fund, and manage wars (the Department of Defense, the office of the president, and Congress) and corporations that make the equipment and supplies needed to support military forces and wage war. Corporations and their stockholders come to need war to maintain profits, and the millions who work for these corporations also rely on war to maintain their employment (and by extension their lifestyles). Corporations contract with the government to do robotics work and to sell fast food on military bases. For example, Burger King has a contract with the Department of Defense to build a franchise on one of its military bases.

African Countries Where U.S. Military Has an Operation or Mission

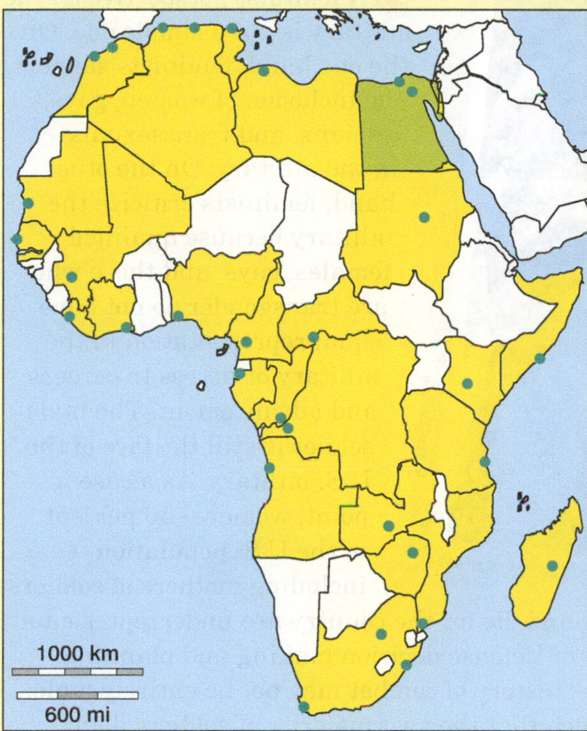

Source: U.S. Department of Defense (2014)

While conflict theorists certainly focus on the corporations that benefit from military contracts, they would also ask: How does the U.S. benefit from military deployments and at whose expense? Conflict theorists point out that it is no coincidence that the U.S. military is stationed in parts of the world that hold needed resources such as oil. In that sense, the military deployments are about gaining control over scarce and valued resources. Given that 21 percent of U.S. oil imports come from African countries, it is no surprise that the United States has a significant and wide-ranging military presence including "base construction, security cooperation engagements, training exercises, advisory deployments, special operations missions, and a growing logistics network" (Turce 2013).

Symbolic interactionists focus on any interactions that take place in military settings and settings where military training, deployments, conflicts, and actions impact lives. Clearly the potential interests are enormous and could range in focus to include a study of babies whom U.S. servicemen/women have conceived with locals overseas. This photo shows children of U.S. military personnel, known as Amerasians, who live in orphanages in Guam, Hawaii, Okinawa, Thailand, South Korea, Vietnam, and the Philippines. Another topic might include how the use of drones and other robotic technologies allowing soldiers to fight at a distance shapes their understanding of the "enemy" versus face-to-face encounters.

CMSgt. Don Sutherland

Mr. Steve L. Morgan (IMCOM)

A feminist perspective on the military is not a simple one. On the one hand, feminists support the inclusion of women, gays, lesbians, and transsexuals in the military. On the other hand, feminists criticize the military because qualified females, gays, and those who are transgender do not have equal representation in the military or access to careers and advancement. The male soldier is still the face of the U.S. military. As a case in point, women—50 percent of the U.S. population—including mothers of soldiers who fight and die for the country are underrepresented and marginalized in Department of Defense decision making and planning. Feminists also recognize that the "history of combat may not be entirely male, but it is overwhelmingly male," such that the vast majority of soldiers doing the fighting and killing have been and are still male (Berlatsky 2013). In this regard, feminists criticize the military as a patriarchal system that values violent conflict over dialogue and compromise (Solaro 2010; Berlatsky 2013).

Summary: Putting It All Together

Chapter 10

Economic systems coordinate human activity to produce, distribute, and consume goods and services. There are two major types of economies—capitalist and socialist. Although the two are treated as extremes along a continuum, in practice, economies are blends of these two systems. The term *welfare state* applies to an economic system that is a hybrid of capitalism and socialism.

Over the past 40 years or so (since 1975), a two-tier labor market has emerged in the United States, such that those who lack formal education beyond high school or technical skills are at a severe economic disadvantage relative to those with strong educational and professional credentials. This emergence of this two-tier labor market is tied to ideological factors that tolerate inequality, minimum wage laws that set the minimum too low, weak forces of collective bargaining and the loss of high-paying jobs because of outsourcing and automation.

The United States, along with Western European countries, China, and Japan, is headquarters to most of the world's largest transnational or global corporations. Transnational corporations, which wield considerable influence over how we live, have facilities and operations in countries outside the one in which they are headquartered. Together, the 10 largest corporations in the world constitute the fifth largest economy in the world.

Political systems distribute power over resources and decision making. Power is the likelihood that an individual can achieve his or her will, even against opposition. That probability increases when people possess authority or legitimate power, of which there are three types: traditional, charismatic, and legal-rational. There are two broad models for the ways power is distributed: power elite and pluralist. The power elite model describes power as being concentrated in the hands of those few people who occupy lofty positions in the social structure of the military, corporations (especially the 200 or so largest), and the government. The pluralist model views politics as an arena of compromise, alliances, and negotiation among many competing special-interest groups, and power is therefore dispersed among those groups.

There are five major forms of government: democracy, totalitarianism, authoritarianism, monarchy, and theocracy. The United States is part of a world economy, which currently encompasses 260 independent countries and innumerable cultures interconnected by the division of labor. The forces driving the creation of this global economy are actions businesses take to prevent economic stagnation and to create growth and increase profit. Over the course of 500 years, these profit-generating responses have created the global economy, assigning each country to core peripheral or semiperipheral status.

NKU Sociology, Missy Gish

Chapter 11

Families

Does it matter who is holding this five-year-old boy's hands? If the adults holding this boy's hands were his two dads, would it change how you see the boy and his life? What if the adults were two uncles, or just two people in the community? What if one or both of the adults happened to be a woman? No matter who is holding this boy's hands, it is obvious that there is a strong bond and emotional tie among the three that is unaffected by any departure from what we define as the norm. Sociologists consider the family a social institution that binds people together through blood, marriage, law, and/or social norms. When sociologists study family, they do not have a particular family structure in mind as an ideal. Instead, they consider the social forces that affect the ever-changing structure of families. Those social forces include, but are not limited to, life expectancy, women's ability to control pregnancy, and employment opportunities for males and females.

Module 11.1 Defining Family

Objective

You will learn about the challenges of defining family and what being part of a family means.

Does this scene evoke any memories of a time spent with your own family?

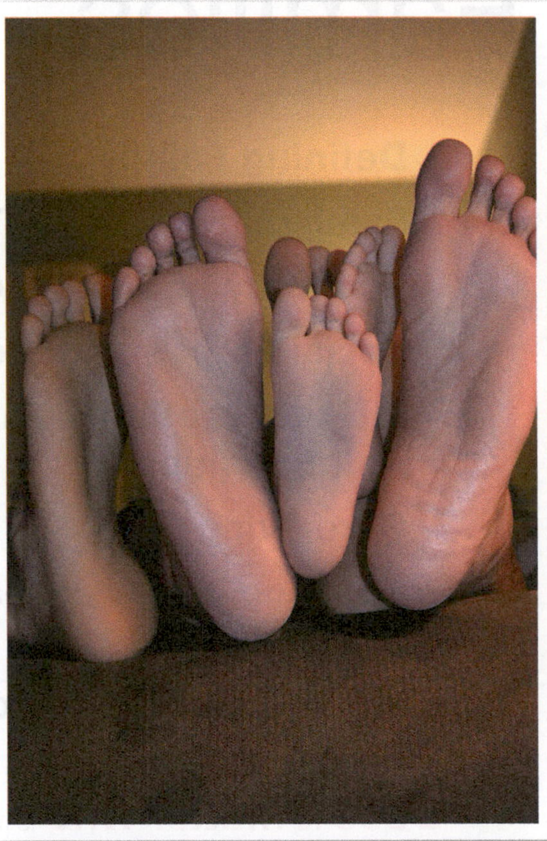

NKU Sociology, Missy Gish

Family is a social institution that binds people together through blood, marriage, law, and/or social norms. It is important to point out that "there is no concrete group which can be universally identified as 'the family'" (Zelditch 1964, 681). An amazing variety of family arrangements exists in the United States and worldwide—a variety reflected in the many norms that specify how two or more people enact family life. These norms govern who can marry, the number and kind of partners one can have, the connection to paternal and maternal relatives, and the living arrangements (see Table 11.1a). In light of this variability, we should not be surprised that when people think of family, they often emphasize different dimensions, such as kinship and other criteria for membership, including legal ties.

Table 11.1a: Norms Governing Family Household Structure and Composition

Number of Partners/Spouses	
Monogamy	One partnership/union/spouse
Serial monogamy	Two or more successive partnerships/unions/spouses
Polygamy	Multiple partners/unions/spouses at a time
Polygyny	One husband, multiple wives at one time
Polyandry	One wife, multiple husbands at one time
Single	Person not in a committed relationship, partnership, or union
Choice of Spouse	
Arranged	Parents select/approve marriage partners for children
Romantic	Person selects partner based on love
Endogamy	Partnership/union/marriage within one's social group
Exogamy	Partnership/union/marriage outside one's social group
Authority	
Patriarchal	Male-dominated
Matriarchal	Female-dominated
Egalitarian	Equal authority between partners
Descent	
Patrilineal	Traced through male lineage
Matrilineal	Traced through female lineage
Bilateral	Traced through both male and female lineage
Multilateral	Lineage traced through one or both biological parents and/or through other social parents (artificial insemination, foster, adopted, surrogacy) (Parke 2013)
Living Arrangements	
Nuclear	Partner/spouse plus children
Extended	Three or more generations living in one household
Single-parent	One parent/guardian living with children
Domestic partnership	People (with or without children) committed to each other and sharing a domestic life but not joined in marriage or civil union
Civil union	A legally recognized partnership (with or without children) providing same-sex or transgender couples with some of the rights, benefits, and responsibilities associated with marriage
Married partners	People joined in marriage (with or without children) with rights, benefits, and responsibilities associated with marriage
Two or more residences	Parents separated, divorced, never married share custody/relatives or other caretakers recognized as family care for children and others who need care

KINSHIP. Conceptions of family emphasizing kinship view families as composed of members linked together by blood, marriage, or adoption. Using these three criteria, the size of any given family network is incalculable, because one person has an astronomical number of living and deceased kin. For this reason every society finds ways to exclude some kin from their idea of family. Some societies, for example, trace family lineage through the maternal or the paternal side only. In addition, people make conscious or unconscious decisions about which kin they will remember or "forget" to tell offspring about (Waters 1990).

NKU Sociology, Missy Gish

These men and boys are all linked together as family through blood or marriage. You might think you *know* which of the adult males is father to each boy, but the man who appears white is not a father to either child. All pictured share family ties to ancestors and blood relatives of African ancestry, yet when you think of the "Black Family Reunion" celebration, do you think of it as attracting people of such diverse appearances? The fact that most Americans likely do not see everyone pictured as "belonging" at the Black Family Reunion celebration speaks to a pattern of "selective forgetting" of family members on a national and cultural level.

MEMBERSHIP. Conceptions of family emphasizing membership often focus on the ideal members. One well-known ideal that defines family membership centers on the belief, even the legal requirement, that marriage be a voluntary union of a man and a woman in a lifelong covenant welcoming of children. However, in reality, few families match this so-called ideal arrangement.

Does this photograph represent the kind of household in which you grew up? Did your household stand out for being larger, blended, or otherwise different from what was considered the mainstream?

Lisa Southwick

LEGAL RECOGNITION. When legal recognition is the defining criterion, a family is defined as two or more people whose living and/or procreation arrangements are recognized *under the law* as constituting a family. Legal recognition means that laws define and enforce the kinds of people who can marry and be recognized as a family. Until recently, federal law in the United States defined *spouse* as "a person of the opposite sex who is a husband or wife" and *marriage* as "a legal union between one man and one woman as husband and wife" (Defense of Marriage Act 1996, section 7). In 2013 the Supreme Court, in a 5-4 decision, struck down part of the Defense of Marriage Act of 1996 when it overturned a section denying federal benefits to same-sex couples. While that decision did not overturn state bans against gay marriage, it did guarantee the right to same-sex marriage and guaranteed employees of the federal government in same-sex relationships the same federal benefits as employees in heterosexual marriages. This decision acknowledges that many people (including gay, lesbian, and heterosexual couples) form lasting, committed, caring, and faithful relationships—and that they are entitled to the legal protections, benefits, and responsibilities associated with marriage. Without these benefits and protections, such people might suffer numerous obstacles and hardships. As a case in point, the U.S. General Accounting Office (2004) identified 1,138 federal statutory provisions in the U.S. Legal Code "in which marital status is a factor in determining or receiving benefits, rights, and privileges."

Functionalist View of Family Life

From a functionalist point of view, family should be defined in terms of the functions it fulfills—that is, in terms of the functions that support order and stability in a society. Four central functions include: (1) regulating sexual behavior, (2) replacing those who die, (3) socializing the young, (4) providing care and emotional support and (5) conferring social status.

REGULATING SEXUAL BEHAVIOR. Marriage and family systems include norms that regulate sexual behavior. These norms may prohibit sex outside of a marriage or specify *who* should engage in sexual partnerships. Such norms can take the form of laws prohibiting marriage and sexual relationships between certain blood relatives (such as first cousins), genders (same), age groups (such as an adult and a minor), and racial or ethnic groups. Other laws regulate the number of people who may form a marriage.

REPLACING MEMBERS OF SOCIETY WHO DIE. For humans to survive as a species, society must replace those who die. Marriage and family systems provide a socially and legally sanctioned environment into which new members can be born or adopted and nurtured.

SOCIALIZING THE YOUNG. The family is the most significant agent of socialization, because it gives society's youngest members their earliest exposure to relationships and the rules of life (see Chapter 3).

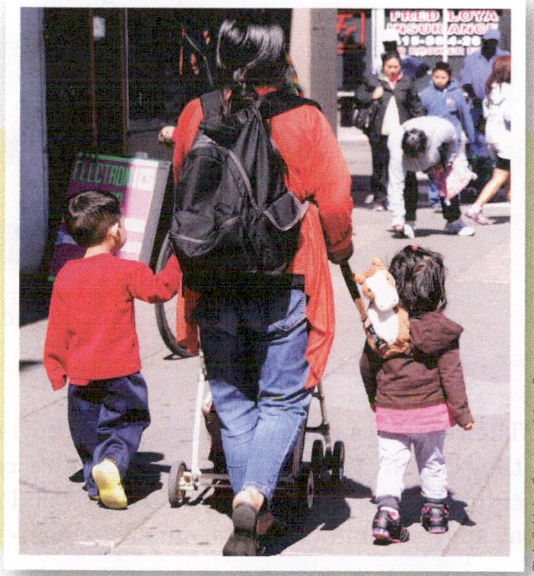

Chris Caldeira, Courtesy of Joan Ferrante

PROVIDING CARE AND EMOTIONAL SUPPORT. A family is expected to care for the emotional and physical needs of its members. Without meaningful social ties, people of all ages deteriorate physically and mentally. The life cycle is such that humans experience a time of extreme dependency in infancy and early childhood and at the end of life.

CONFERRING SOCIAL STATUS. We cannot choose our family, the quality of the relationship between our biological parents or other people we come to know as parents, or the economic conditions into which we are born. Among the things we inherit from our parents are genetic endowment and social class. The genetic or physical characteristics we inherit affect the racial category to which we are assigned. Parents' social class as determined by their occupation, income, and wealth are also important predictors of life chances—the probability of engaging in critical experiences that increase the likelihood of obtaining advantaged status, whether that be surviving the first year of life, living independently in old age, living free of disease, eating a balanced diet, having a tutor or private coach, traveling abroad, and attending high school and college without having to work.

Conflict View of Family Life

While everyone might agree that families should fulfill these five functions, conflict theorists point out that family members do not always care for one another and that the family also perpetuates inequalities by passing on social advantages and disadvantages to members. Moreover, marriage and family systems are structured to value productive work and devalue reproductive work, and to foster and maintain divisions and boundaries.

SOCIAL INEQUALITY. Families transfer power, wealth, property, and privilege from one generation to the next. Obviously, income affects the level of investment parents can make in their children. In the United States, 73 percent of children live in a household in which one parental figure has secure employment; that is, at least one parent worked 35 or more hours per week for at least 50 weeks in the past year. Of course, that percentage varied by living arrangement (see Figure 11.1a).

Figure 11.1a: Percentage of U.S. Children with Secure Parental Employment by Living Arrangement

Under which living arrangement are children most likely to have a parent securely employed?

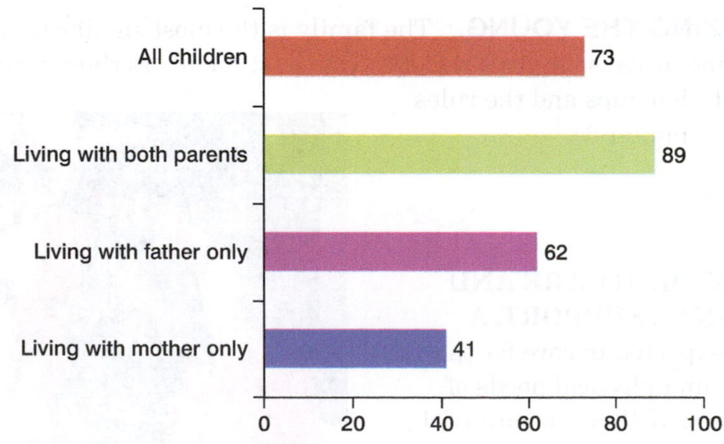

Source of data: ChildStats.gov (2013)

The resources parents possess shape how much personal time and money they can invest in their children. Simply consider the money it costs to enroll children in programs that cultivate social and developmental skills such as sports, summer camps, music lessons, or other activities. Parents act as a bridge connecting their child to a range of experiences that confer social standing—those experiences relate to parents' "choices" about which schools to enroll children in, where to worship, who their children will play with, and the kind of caretakers employed (Parke 2013).

PRODUCTIVE AND REPRODUCTIVE WORK. Friedrich Engels (1884) distinguishes between productive and reproductive work. Productive work is paid work and involves the actual manufacture of food, clothing, and shelter and delivery of services. Reproductive work involves bearing children, giving care, managing a household, and socializing children. While reproductive work is unpaid, it is still "work," and it involves essential tasks that must be done if societies are to thrive. Everywhere women as a group spend more time doing reproductive work than men; and men spend more time working for pay than women.

The difference in time men and women spend doing household/caregiving tasks versus paid work is affected by whether a couple has children and the amount of income each partner contributes to the household. Generally, with dual-career couples, the more equal the two incomes, the less the gender imbalance with regard housework.

The most recent data on heterosexual couples living in the United States shows that on average men spend about 10 more hours a week on the job than women, and women do about nine hours more of household/caregiving work than men. Among couples with children, fathers put in on average a 37-hour work week (down from 42 in 1965). The average number of hours fathers spend doing housework is 10 (up from 4 in 1965). The average mother works 21 hours per week (up from 8 in 1955) and spends 18 hours doing housework (down from 32 hours in 1965). Mothers spend 13.5 hours a week caring for children while fathers spend 7.3 hours (up from 2.5 in 1965) (Parker and Wang 2013).

Since 1965, men in heterosexual relationships, and especially fathers, have increased the amount of time they spend doing housework. But, while fathers still devote more hours to paid labor than mothers, men spend about half as much time as women on housework.

Mark Gish, Courtesy of Joan Ferrante

FOSTERING SOCIAL DIVISIONS AND BOUNDARIES. In the United States, we assume that people choose a partner or mate based on love. Upon investigating who marries, however, we find that people's choices are guided by other considerations as well: a potential partner's age, height, weight, income, education, race, sex, social class, and religion, among other things. When the conditions are right, we allow ourselves to fall in love.

Chris Caldeira

The person with whom we fall in love may not be a person we can legally marry. For example, only couples who meet certain legal criteria can get married at the Chapel of the Bells in Nevada or any other site that performs marriages in the United States. Nevada is one of 31 states banning gay marriage; the state does allow same-sex couples to register for domestic partnerships.

All societies have norms defining who may date or marry whom. These norms may be formal (enforced by law) and/or informal (enforced by social pressure). **Exogamy** refers to the practice of choosing a partner from a social category other than one's own—for example, a partner of the other sex or of another race. **Endogamy** refers to the practice of choosing a partner from the same social category as one's own—for example, a partner of the same race, sex, ethnicity, religion, or social class. In the United States there are about 58 million heterosexual couples living together; 90 percent of all marriages involve partners of the same race. However, in 2010, the last year for which we have data, about 1 in 4 black males married someone outside their race, compared with about 1 in 11 black females. Among Asian newlyweds, about 1 in 3 females married outside their race compared with 1 in 6 Asian males (Wang 2012).

What Do Sociologists See?

Sociologists see a mother and children reunited upon the mother's return from military deployment. This scene reminds us that people do not have to live together to be a family. In fact, there are many cultures where women take jobs outside the country to make a better life for their children, who are cared for by fathers and other relatives while they are gone (Lauby and Stark 1988).

Staff Sgt. Chris A. Durney

Critical Thinking

Look at Table 11.1a, which of the norms governing household structure and composition apply to the family or families you grew up with? Did you experience another arrangement not listed?

Key Terms

endogamy

family

exogamy

Module

11.2 Family Structure in Three Countries

Objective

You will learn about the ways in which the larger social context influences the structures of family.

NKU Sociology, Missy Gish

U.S. Army photo

If you asked the Afghan girls and the American girls whether they would like to have children one day and if so, how many, how do you think each would answer?

If American girls answered "none," they would likely draw the response, "Why not?" If they answered "two," no one would likely raise an eyebrow. If they answered "eight," they would surely draw expressions of disbelief. But imagine this question–answer sequence taking place in Afghanistan. An answer of "none" would quite likely draw a reaction of shock. From such exchanges, children learn what their society considers an ideal number of children. They also learn that they must justify decisions that depart from the ideal.

Family Composition

In this module we compare family life in the United States with that of Japan and Afghanistan. We chose Japan because it has one of the lowest fertility rates in the world (1.25); we chose Afghanistan because it has one of the highest (5.43). Both situations have prompted concern and efforts to either increase low fertility or decrease high fertility. In making comparisons, we will consider three factors that shape the experience of living in a family: (1) women's control over reproductive life, (2) the type and timing of death, and (3) caregiving responsibilities.

FEMALE CONTROL OVER REPRODUCTIVE LIFE. If we look at the statistics presented in Table 11.2a, we can gain some insights about the typical size of families in the three countries. We see that **total fertility**, the average number of children that a woman bears in her lifetime, is about five for women in Afghanistan. In the United States, total fertility is two children, and in Japan it is about one (see Table 11.2a).

Table 11.2a: Fertility-Related Statistics

By age 40-44 almost 20 percent of Japanese and American women remain childless. A woman who is childless at age 40-44 is a rare occurrence in Afghanistan. In which country are chances greatest that a baby will be born to a teenage mother? What percentage of women in each country report using some form of contraception?

	Japan	United States	Afghanistan
Chance of a teen giving birth per year (ages 15–19)	1 in 200	1 in 20	1 in 9
Total fertility (average number of children per woman)	1.24	2.05	5.64
% of women age 15–49 using any method of contraception	80.0%	78.6%	21.2%
% of all women who give birth to a child outside of marriage	2%	35.0%	Virtually zero
% of childless women, age 40–44	18.0%	18.8%	2.6%

Sources: World Bank (2014a, 2014b); Shattuck and Kreider (2014); Eberstadt 2012

DEATH OF FAMILY MEMBERS. Death results in the loss of a family member. We can think of death in terms of the time in life at which it occurs. Specifically, sociologists ask: What are the chances a child will survive the first year of life and beyond? What are the chances a woman will survive childbirth? And what are the chances of living beyond age 65? As we will see, the answers to these questions offer important clues about the size of the family unit and the broad nature of family relationships.

Official U.S. Marine Corps photo by Cpl. Sarah M. Maynard

Journalist 1st Class Lynn Jenkins

U.S. Air Force photo by Master Sgt. Jim Varhegyi/Released

For the most part, parents in Japan and the United States can expect their children to survive them. Look at Table 11.2b on page 411. How many babies in each country die before reaching age 5? In Afghanistan the answer is 198.6 die per 1,000 live births.

When parents can expect a baby to survive, the parents feel more secure about the future and, thus, less need to have additional children. On the other hand, when the mortality rate for babies and children is high and there is no government safety net in place to support people in old age, parents have an incentive to have a large number of children to ensure that at least one survives into adulthood.

Each year in the United States, 28 women die for every 100,000 live births as a result of pregnancy complications. In Japan that number is 6, and in Afghanistan it is 400. High fertility rate coupled with the uncertainty that women will survive childbirth, and with the uncertainty about whether children will survive the first year of life and beyond, speaks to women's lack of control over their lives, but especially over their reproductive lives.

Lack of access to clean water is responsible for many waterborne illnesses such as diarrhea, the number one cause of death among children under 5 years old. The United Nations and other international agencies work to find sources of clean water near where people live so that children, who are often charged with collecting water, do not have to walk far to get it. In fact, many children, especially girls, do not go to school because they spend their days carrying water.

MC1 Monique Hilley

Life expectancy, or the average number of years after birth a person can expect to live, offers insights about the timing of death. Notice in Table 11.2b that 84 percent of people in Japan can expect to reach age 65 and to live, depending on their sex, on average another 17 or 22 years beyond that age. In the United States, 77 percent can expect to live to age 65, and then on average an additional 17 to 20 years. In Afghanistan, 56 percent of people live to age 65 and upon reaching that age can expect to live an average of 10 or 11 years longer.

How might high infant mortality and maternal mortality affect the size of families? The higher infant and maternal mortality, the greater the number of

Table 11.2b: Selected Life Expectancy and Mortality Statistics

	Japan	United States	Afghanistan
Infant mortality (per 1,000 babies born)	2.21	6.0	71
Under age 5 mortality (per 1,000 children)	3.3	7.8	99
Maternal mortality (per 100,000 live births)	6	28	400
Lifetime risk of maternal mortality	1 in 1,800	1 in 12,100	1in 49
Average life expectancy	83.9	78.5	49.7
Percentage of population expected to reach age 65	84.0%	77.4%	56.0%
Average number of years after age 65 one can expect to live	22.5 (females) 17.4 (males)	20.0 (females) 17.2 (males)	11.0 (females) 10.0 (males)

Sources: United Nations (2013); U.S. Central Intelligence Agency (2014); World Health Organization (2014)

children women are likely to have. Conversely, the more confident women are that they will survive childbirth and that their children will also survive, the fewer children they are likely to have.

CAREGIVING AND DEPENDENCY. The family plays an important role in meeting the emotional and physical needs of its members. People's emotional and/or physical needs tend to be more evident when they are very young and very old. The age composition of the population offers broad insights about where family caregiving energies are directed.

Figure 11.2a: Age and Sex Composition of Japan and Afghanistan, 2014
Can you tell by looking at the two population pyramids which country has the older population? The younger population? In Japan, which age group has the most men? The most women? Which age group is the largest in Afghanistan? In which country is the proportion of children under 19 years of age considerably larger than that of adults age 65 and over?

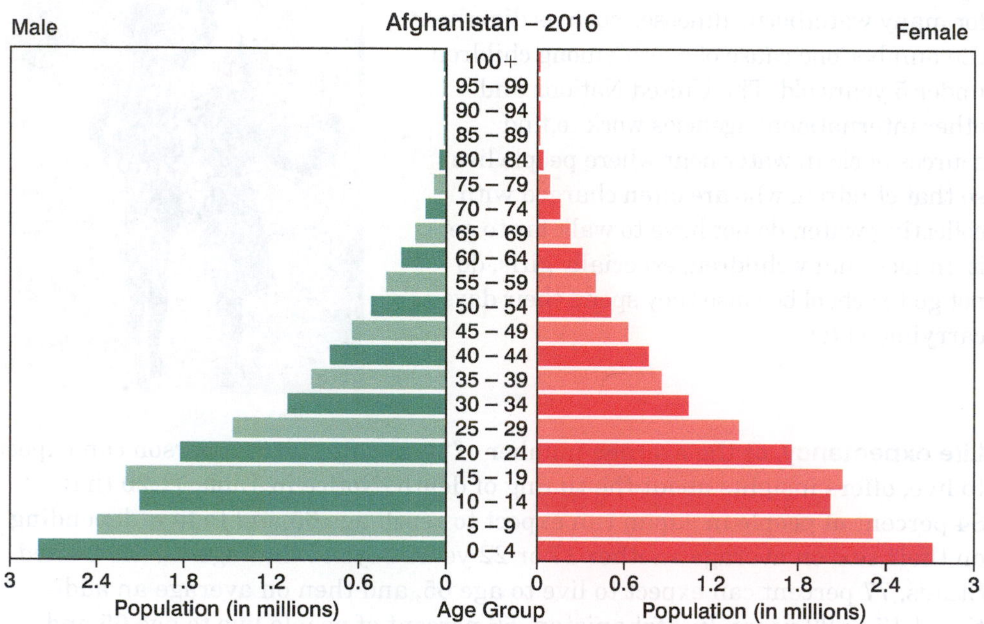

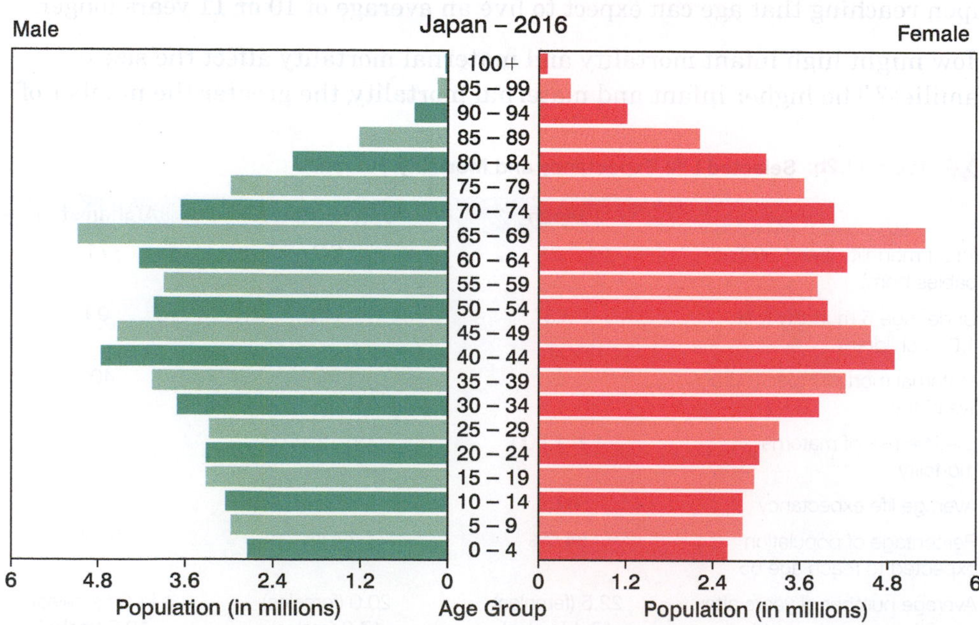

Source: U.S. Census Bureau (2014b)

Population pyramids are a snapshot of a society's age and sex composition. In comparing Japan's population pyramid with that of Afghanistan, it is clear that Japan has an **aging population**, one in which people tend to live longer and in which those in older age cohorts outnumber those in younger age cohorts. As a result, the Japanese can expect to participate in family life for a greater number of years.

Japan's Family System

Japan's very low total fertility rate, combined with its long life expectancy and low immigration rate, means that the country has one of the world's oldest populations. The low total fertility rate is a major national concern, and it has prompted a variety of responses encouraging people to have children. Japan's fertility rate of 1.2 children per woman can be explained by a number of social forces including, but not limited to, the dismantling of the *ie* family structure, a dramatic decline in arranged marriages, and employment barriers.

DISMANTLING OF THE *IE* FAMILY STRUCTURE. Until Japan's defeat in World War II, everyone was officially part of a male-headed multigenerational household known as an *ie*. Relationships between family members were shaped by the Confucian values of filial piety, faith in the family, and respect for elders. Firstborn sons held privileged status, inheriting and controlling family wealth (primogeniture). Legally, women could not choose a spouse or own property. Daughters were viewed as temporary family members until marriage, at which point they moved in with their husband's family (Takahashi 1999). Brides served and obeyed the husband and his parents, including caring for them in old age. After World War II ended in 1945, the United States imposed sweeping changes on Japan's family systems, granting equal rights to women, ending arranged marriage, and abolishing primogeniture. In spite of these changes to the family structure, the belief that problems should be managed within the family has persisted. This persistence is evidenced by the fact that the Japanese government provides only limited public services for the elderly; women are expected to serve as primary caregivers to the young and old. This caregiving responsibility is likely to have a depressing effect on fertility as women anticipate how to balance the needs of children and aging parents.

THE DECLINE OF ARRANGED MARRIAGES. Since 1945 the proportion of all marriages considered arranged has fallen from almost 100 percent to 10 percent. When parents no longer arranged marriages, young people were left to their own devices. Japan has not fully adjusted to this shift, and it has yet to develop a couples culture (Ogawa et al. 2009).

In Japan, 70 percent of working females ages 20 to 39 live with their parents while contributing little to household expenses (Ogawa et al. 2009). Sociologist Masahiro Yamada (2000) coined the phrase *parasite singles* to describe those who work but are free of the financial and emotional pressures associated with parenting and marriage. According to government surveys, among the most stressful pressures these singles wish to avoid is that of guiding children through Japan's highly competitive educational system.

U.S. Air Force photo/1st Lt. Bryan Bouchard

To give their children an edge, many parents in Japan enroll their children in juku (after-school cram schools) and hire tutors at a combined cost of $303 to $433 per month. Keep in mind that in Japan scores on qualifying exams determine acceptance into the best high schools and colleges. GPA, letters of recommendation, athletic ability, and extracurricular activities are not taken into consideration (8Asians.com 2009; *The Economist* 2011). An estimated 40 percent of all elementary students and 75 percent of all high school students attend one of the 50,000 juku across Japan (Sato 2005; Kittaka 2013).

The *juku* pressure is particularly stressful for mothers, who almost single-handedly guide children through the process—many working fathers are largely absent from the home. The workplace emphasizes work over family, as employees work long hours, take little time off for personal reasons, and accept sudden job transfers (Newport 2000).

ENTRENCHED BARRIERS TO EMPLOYMENT. Although Japan's labor laws forbid discrimination against women, it exists at every level, including recruitment, hiring, training, compensation, and promotion. For the most part, Japanese women are expected to quit working when they marry or have children. When mothers do work, they tend to accept low-paying and insecure employment in exchange for the flexibility needed to meet household and caregiving responsibilities. In addition, the tax system offers married women incentives to "choose" part-time employment (Bauwens 2013; Lam 2009). In sum, women must choose between work and family life. As a result, many are not willing to give up the security of employment to have children.

Afghanistan's Family System

In Afghanistan the family is *the* source of economic and social support, because there are very few public services. The typical Afghan family is extended, with members sharing a household or a valley. Extended family households span several generations, including "[t]he eldest female, usually his wife, [who] runs the household, and is in charge of the other women including her daughters, the wives of her sons, any other wives her husband may have (Islam allows each man to have four wives but most are too poor to afford this), and any unmarried or widowed cousins, aunts, etc. who live with the family" (Asian Studies Center 2013). The Afghan family is male-dominated; authority rests with male elders, and inheritance is passed through the male line. As a side note, the practice of excluding women from inheritance is not in keeping with the Qur'an, which states that daughters receive an inheritance that is one-half that of a son (Asian Studies Center 2013). Women are expected to want children, especially sons, as infertility is "a frightening social stigma" for a woman and her family (Blood 2001). About 15 percent of Afghan women are in the labor force compared to 80 percent of the men (United Nations 2013).

In Afghanistan males assume control over women's lives, usually deciding whether they can pursue an education or career. The men also approve the spouse and negotiate the price of the bride for their sons. Independent females are viewed as a threat, and families tolerating such independence are ostracized. The ideal mate for a son is a paternal first cousin. The typical Afghan bride is in her teens and is matched with a man in his mid-twenties. When price is paramount, very young girls are matched with much older men who can afford to pay the price (Blood 2001).

Five percent of Afghan women can expect to complete secondary school, compared to 34 percent of men (United Nations 2013). Although there are some exceptions, education is a gender-segregated experience (Brown 2010). Under the Taliban (pre-2001), who enforced a conservative interpretation of Islamic law, girls could not go to school; women could not hold public jobs and could leave their homes only under rare circumstances, and only if wearing full-body burqas. The Taliban mandated that men grow beards and banned Western-style clothing. Crimes such as theft, adultery, and drinking were subject to severe punishments, including lashings, stoning, and amputation (Shoup 2006). After 2001, the year of the U.S. military intervention, a new constitution was put into place that guaranteed equal rights for women but allowed ethnic minority groups to establish their own family law in keeping with their religious traditions. These family laws tend to place women in subordinate positions (Boone 2009).

What Do Sociologists See?

It is likely that these Afghan girls cannot imagine a life without children as their status in society depends on their being a mother, especially to a son. Their chances of dying in childbirth over the course of their lives are 1 in 49, and there is a 1 in 10 chance that any child they have will die before age 5.

U.S. Air Force photo by Staff Sgt. Brian Ferguson

Critical Thinking

How many children, if any, do you think you will have (including those you might have already)? What factors figure into your decision?

Key Terms

aging population

total fertility

life expectancy

Module 11.3 Economy and Family Structure

Objective

You will learn how the economy shapes relationships among family members, but especially between men and women.

Ingrid Barrentine

Why might someone who suffers physical and/or mental abuse at the hands of a partner not take action to change the situation?

Three answers are when (1) the abuser is physically stronger, (2) the person being abused does not have the financial resources to leave, and (3) the abused lack access to police or other sanctioned agents of violence control. With this kind of question in mind, sociologist Randall Collins (1971) proposed a theory of sexual or **gender stratification**, the system societies use to rank males and females on a scale of social worth such that the ranking affects life chances in unequal ways. This theory is based on three assumptions. First, people tend to use their economic, political, physical, and other advantages to dominate others. Second, any change in the way resources are allocated to men and women alters the structure of domination. Third, ideology is used to support and justify one group's domination of another.

Mass Comm. Spc. 2nd Class Jason Johnston

Under what conditions might Afghan women, or any woman for that matter, be denied access to education, be forbidden to leave the house unless accompanied by an adult male relative, and be required by law to cover from head to toe? According to Collins (1971), when males hold political and economic power and when the society defines males as innately superior to females. How would things change if women were given access to education and to opportunities for earning an income independent of males?

Collins points out that, in general, males tend to be physically stronger than females; thus, the *potential* for coercion by males exists in most encounters with females. He maintains that women have historically been viewed and treated as men's sexual property and that this ideology lies at the heart of gender stratification. The extent to which women are viewed as sexual property and subordinate to men depends on whether women have an institutionalized source of protection outside the home to turn to and whether women's income is equal to that of men's. In this regard, Collins identified four historical economic structures that shape women's relationship with men: (1) low-technology tribal societies, (2) fortified households, (3) private households, and (4) advanced market economies.

LOW-TECHNOLOGY TRIBAL SOCIETIES. Low-technology tribal societies include hunting and gathering societies. These societies do not have technologies that permit the creation of surplus wealth—that is, wealth beyond what is needed to meet basic needs for food and shelter. In such societies, sex-based division of labor is minimal, because the emphasis is on collective welfare and the belief that all members must contribute to ensure the group's survival. Because almost no surplus wealth exists, marriage between men and women from different families does little to increase a family's wealth or political power. Consequently, daughters are not treated as property, in the sense that they are not used as bargaining chips to achieve such aims. Relatively speaking, women in low-technology tribal societies have greater social standing and value than women in other arrangements described below (Cote 1997).

FORTIFIED HOUSEHOLDS. Fortified households are preindustrial arrangements where the household functions as an armed unit and the head of the household acts as a military commander. The household provides protection; there is no place outside the household to turn for protection. Fortified households include within them propertyless laborers and servants. In the fortified household, "the honored male is he who is dominant over others, who protects and controls his own property, and who can conquer others' property" (Collins 1971, 12). Men treat women as sexual property in every sense: daughters are often bargaining chips for establishing economic and political alliances with other households, and male heads of household take sexual liberties with female servants. In this system, women's power depends on their relationship to the dominant men.

Slaveholding households in the United States qualified as fortified households in that they included propertyless servants and enslaved females treated as sexual property. This engraving speaks to the sexual relationship between master and slaves, in which the products of their union were slaves who appeared often as white but assumed the legal status of the mother—the master's property.

HarpWeek

PRIVATE HOUSEHOLDS. Private households emerge with market economies; a centralized, bureaucratic state; and agencies of social control that alleviate the need for citizens to take the law into their own hands. Under the private household arrangement, men monopolize the most desirable and important economic and political positions. Men are still heads of household in that they control the property and assume the role of breadwinner; women are responsible for housekeeping and childrearing. In the context of market economies, people turn to businesses and government agencies for goods and services; workplaces exist apart from households; family sizes are smaller and there is a police force to which women and other victims can appeal in cases of domestic violence. Because the family is "less important," market economies give rise to the notion of romantic love as a key factor in choosing a partner.

Library of Congress Prints and Photographs Division[LC-USZ62-61085]

This 1916 drawing suggests one image of the private household: men offer women economic security because they dominate the important, high-paying positions. Women offer men companionship and emotional support, and they strive to be attractive—that is, to achieve ideal femininity, which might include possessing an 18-inch waist or wearing high-heeled shoes. At the same time, before marriage women try to act as sexually inaccessible as possible, because theoretically they offer sexual access to men in exchange for economic security. Given the dominance of the private household, can you see how this economic arrangement might discourage same-sex partnerships? The question has relevance for understanding the forces that make same-sex partnerships acceptable in the larger society.

ADVANCED MARKET ECONOMIES. Advanced market economies offer widespread employment opportunities for women. Although women remain far from being men's economic equals, some have the financial resources to enter into a relationship with men offering more than an attractive appearance; they can provide an income and other personal achievements. Because they have more to offer, women can make demands on men to be sensitive and physically attractive, to meet the standards of masculinity, and to help with housework and child care. In the United States today, about 30 percent of all full-time employed married women living in dual-income households in which their male partner is also employed full time earn more money than the man (U.S. Bureau of Labor Statistics 2014c). This situation may explain in part why increasing commercial attention has been given to male appearance and sexual performance (e.g., the erectile dysfunction drugs Viagra and Cialis).

Because advanced market economies offer women greater opportunities in the labor market, women are less dependent on men for money and protection. This freedom lessens the incentives to look only to men as potential partners.

Likewise, since men can no longer count on women "willingly" assuming homemaker and caretaker roles, incentives to marry for these reasons decline. It should come as no surprise that same-sex relationships become more common, and even accepted.

Lesbian and Gay Marriages and Partnerships

Gay partnerships, unions, and marriages (hereafter referred to as gay partnerships), like heterosexual relationships, are complex relationships that cannot be analyzed in simple ways. To what extent do same-sex or other unconventional partnerships mirror the inequalities that characterize heterosexual partnerships in general? As you might imagine, there is limited research on so-called unconventional partnerships, if only because it is difficult to identify those who are in such relationships. Most, if not all, of the existing research on unconventional partnerships focuses on gay and lesbian relationships. Those unfamiliar with same-sex commitments often frame them in terms of heterosexual ideals—ideals that assume one person enacts a traditional masculine role and the other enacts the traditional feminine role. The research, however, indicates that most childless lesbian and gay couples tend to reject the traditional heterosexual ideals associated with masculinity and femininity as a template for their relationships (Kurdek 2005). This generalization changes when same-sex couples become parents, at which point the relationship faces pressures to becomes unequal with regard to child care, household tasks, and employment (Biblarz and Savci 2010).

What Do Sociologists See?

Of the four historical economic structures named by Collins, it is the advanced market economy where we would expect to find products like Cialis and other erectile dysfunction medications. In advanced market economies a significant percentage of women have achieved an economic status equal to or greater than that of men. As a result, women can make demands of men, including demands on how they should appear and perform.

NKU Sociology, Missy Gish

Critical Thinking

Identify a couple you know in a marriage, civil union, or other committed partnership. Use Collins's theory to describe the signature qualities of the relationship.

Key Terms

gender stratification

Module 11.4 Social Forces and the Changing Family

Objective

You will learn that family forms change and explore the social forces behind change.

Library of Congress Prints and Photographs Division[LC-DIG-nclc-02070]

What would it be like to grow up in a household of this size and sleep in a bedroom with as many as five siblings?

Sociologists think of family in the broadest sense of the word—that is, they think beyond the "isolated, self-sufficient nuclear family"—to include any arrangement in which people make a commitment to care and support others. In this regard, the data that the U.S. Census Bureau collects on households offers a more complete picture of those who live together in a physical space.

Changing Household Structures in the United States

Table 11.4a shows that the distribution of different household structures in the United States is not static; over the span of 100-plus years, it has changed quite dramatically, especially with regard to single-person and single-parent households.

Table 11.4a: U.S. Households: 100-Plus Years of Change
The table is a profile of household structures for 1900, 1950, and 2013. Which household characteristic and living arrangement has seen the greatest decline? Which household characteristic and living arrangement has seen the greatest increase?

	1900	1950	2013
Household Characteristics			
Average household size (persons)	4.8	3.4	2.6
% of households with seven or more people	20.4%	4.9%	1.5%
% of households with one person living alone	5.0%	9.3%	28.0%
Same-sex couple households	—	—	1.0%
Living Arrangements of *Children* by Family Status			
Two-parent—farm family	41%	17%	2%
Two-parent—father breadwinner, mother homemaker	43%	56%	23%
Two-parent—dual-earner	2%	13%	40.7%
Single-parent	9%	8%	26.5%
Children not living with a parent	5%	6%	4.1%
Child living with stepparent, grandparent, or adoptive parent	—	—	3.7%
Same-sex couple households	—	—	0.005%*

Sources of Data: ChildStats.gov (2012); U.S. Census Bureau (2002, 2013a)
* For the years 1900 and 1950, the phrase "two-parent" encompassed only heterosexual parents because the Census Bureau likely only counted male–female couples as parents. While there were no doubt same-sex couple households in those times, they were not counted as "parents."

The Rise and Fall of the Breadwinner System

As you can see, household composition, and by extension dominant family structures, have changed over time. The forces behind change are complex. But to gain insights about how family structures change, we turn to an analysis of the rise and fall of the breadwinner system by sociologist Kingsley Davis (1984).

Before industrialization—that is, for most of human history—the workplace was made up of the home and the surrounding land, and both men and women worked together to produce for the household. Industrialization destroyed the household economy by moving production from the home to the factory. In the process, women's contribution to the household production declined and a new economic arrangement Davis called the breadwinner system emerged.

This 1887 drawing depicts an image we have come to associate with the breadwinner system. As the breadwinner, men worked not in the home or on surrounding land but with nonkin in factories, shops, and offices. The man's economic role now made him the link between the family and the wider market economy. At the same time, his personal participation in the household diminished. His wife stayed home and performed the parental and domestic duties that women had always performed. Under the breadwinner system, the woman bore and reared children, cooked meals, washed clothes, and cared for her husband's personal needs, but to an unprecedented degree her economic role changed because she did not produce what the family consumed—production had been removed from the home (Davis 1984).

Historically, the breadwinner system is not typical; it is a household arrangement characteristic of the middle and upper, predominantly white, classes, and its heyday was from 1860 to 1920. The breadwinner system emerged around the time that family size reached its peak and infant and childhood mortality began to decline. The breadwinner system came about because many women had too many children to engage in work outside the home.

The breadwinner system did not last long because it placed too much strain on husbands and wives. The strain stemmed from what Davis identified as structural weaknesses:

- never before had the roles of husband and wife been so distinct,
- never before had women not produced what their families consumed,
- never before had men spent most of their waking hours separated from their families, and
- never before had men assumed sole responsibility for supporting the entire family.

Given these weaknesses, the system needed strong normative controls to survive: "The husband's obligation to support his family, even after his death, had to be enforced by law and public opinion." Sexual relations considered "illegitimate," specifically when babies were born outside of marriage, "had to be condemned; divorce had to be punished, and marriage had to be encouraged by making the lot of the 'spinster' a pitiful one" (406). Davis maintains that the strains were too great, and these strict normative controls eventually collapsed.

The breadwinner system's fall was marked by a dramatic increase in the percentage of women age 16 and over in the *paid* labor force. That percentage rose from less than 20 percent in 1900 to 60 percent in 1980. Today the percentage is 57.7 percent (U.S. Bureau of Labor Statistics 2013c). The fall of the breadwinner system was also marked by an increase in the percentage of married women in the labor force, which rose from 15.4 percent in 1900 to 53.1 percent in 1980. Today that percentage is about 58 percent (U.S. Bureau of Labor Statistics 2013c). Davis links women's entry (and especially married women's entry) into the paid labor market to a series of demographic changes that gave women the time and the incentive to work outside the home. Those changes include a decrease in total fertility, an increase in life expectancy, an increase in the divorce rate, and an increase in employment opportunities.

DECLINES IN TOTAL FERTILITY. The decline in total fertility (number of children born over a woman's lifetime) actually began before married women entered the labor force in large numbers. Davis attributes the decline to the forces of industrialization, which changed children from economic assets to economic liabilities. Not only did women have fewer children, but the number of years separating first- and lastborn also decreased, so that women had their last child at a younger age. These changes in number and spacing of children gave women time to work outside the home, especially after their children entered school.

INCREASED LIFE EXPECTANCY. As life expectancy increases and women devote fewer years to bearing and raising children, the time devoted to child care came to occupy a smaller proportion of their lives. Because women's life expectancy exceeds that of men's and because brides tended to be younger on average than their partners, the average married woman could expect to outlive her husband by 8 to 10 years. Although few women would name increased life expectancy as an incentive to search for paid work, the possibility of living many years longer than their husbands changed the way women looked at and planned out their lives.

INCREASED DIVORCE RATE. Davis traces an increase in the divorce rate to the shift of economic production outside the home:

> With this shift, husband and wife, parents and children, were no longer bound together in a close face-to-face division of labor in a common enterprise. They were bound, rather, by . . . the husband's ability to draw income from the wider economy and the wife's willingness to make a home and raise children. The husband's work not only took him out of the home but also frequently put him into contact with other people, including young unmarried women who were strangers to his family. Extramarital relationships inevitably flourished. Freed from rural and small-town social controls, many husbands either sought divorce or, by their behavior, caused their wives to do so. (410–411)

Davis argues that an increase in the divorce rate preceded married women's entry into the labor market by several decades. However, once the divorce rate reached a certain threshold (a 20 percent or greater chance of divorce), more married women seriously considered seeking employment to protect themselves in case their marriages failed. When both husband and wife participate in the labor force, the chances of divorce increase even more. Working outside the home puts both in contact with others, increasing the possibility of extramarital romantic attachments.

INCREASED EMPLOYMENT OPPORTUNITIES FOR WOMEN. Paid employment increased women's economic independence and transformed gender roles. There is "nothing like a checking account to decrease someone's willingness to be pushed into marriage or stay in a bad one" (Kipnis 2004). While women were motivated to seek paid work because of changes in the childbearing experience, increases in life expectancy, and the rising divorce rate, women could act on these motivations only when opportunities for paid employment increased. As industrialization matured, it brought a corresponding increase in the kinds of jobs perceived as suitable for women. In addition, legal protections such as Title IX officially protected women from discrimination in the workplace and opened career opportunities.

While some women have always managed to work in heavily male-dominated occupations, their chances of doing so increase when there is legal support. Dr. Mae Jemison was the first African American woman accepted to the NASA astronaut program. Her unique biography, work ethic, and talents aside, Jemison's successes were helped along by laws such as Title IX, which made discrimination illegal in 1973, the year she enrolled as a chemical engineering at Stanford University at the age of 16 (U.S. Air Force 2012).

U.S. Air Force graphic/Sylvia Saab

Davis does not argue that the dual-career model is a problem-free arrangement. First, it lacked "normative guidelines" such that each "couple has to work out its own arrangement" (1984, 413). Second, even in the two-income system, women tend to remain primarily responsible for domestic matters. One reason women bear this responsibility is that men and women are unequal in the labor force. (At the time Davis wrote his study, females working full time earned 66 cents for every dollar men earned. Today, they earn about 81 cents for every dollar men earn [U.S. Bureau of Labor Statistics 2013b].) Davis believes that as long as women make an unequal contribution to the overall household income, they will do more work around the house.

What Do Sociologists See?

Sociologists see a member of the Afghanistan National Taekwondo Federation Junior Team, which consists of 6 Afghan girls and 11 Afghan boys. Here a female team member is demonstrating her ability to kick hard enough to break boards. Only recently did some Afghan women gain the opportunity to practice and work out with men. Such experiences of physical and social empowerment change the way Afghan women see themselves and extend possibilities in life beyond being a wife and mother. The woman pictured states that "My favorite part is breaking the boards. It makes me feel strong" (Campbell 2011).

U.S. Army Spc. Anthony Murray Jr., International Security Assistance Force Public Affairs

Critical Thinking

Does the breadwinner system apply to the family in which you were raised? If yes, did your parents face the strains Davis describes? If not, what demographic factors shaped the family system in which you were raised?

Module 11.5 Intergenerational Family Relationships

Objective

You will learn about the social forces that affect intergenerational family relationships.

Do you have a great-grandparent in your life? How about a great-great-grandparent?

Tony Rotundo

What does it mean to have a great-grandmother, grandmother, and mother in your life? What does it mean for a grandmother to still have her mother in her life? While great-grandparents and even great-great-grandparents have always been present in some people's lives, the likelihood that most of us will know and remember these kinds of older relatives has increased substantially over the past few decades. Moreover, the likelihood that parents will be in children's lives for 70 or even 80 years has increased substantially as well.

In the context of family, we define a **generation** as people sharing an age category passing through time who are distinguished from those sharing other age categories by cultural disposition (dress, language, preferences for songs, activities, entertainment); posture (walk, dance, the way the body is held); access to resources; and socially expected privileges, responsibilities, and duties (Eyerman and Turner 1998). Each generation is distinguished by titles it occupies such as baby, teenager, parent, and great-grandparent. As people age they pass into and out of life stage categories. In this module we consider how three social forces have broadly affected intergenerational relationships among family members. Those forces are increased life expectancy, declines in parental authority, and changes in the economic status of children.

Dramatic Increases in Life expectancy

Since 1900 the average life expectancy at birth has increased by 33 years in the world's richest countries and by 20 years (or more) in middle- to low-income countries (Duncan 2012). Sociologist Holger Stub (1982) describes at least four ways in which gains in life expectancy have altered the composition of the family in the last century. First, the chance that children will lose one or both parents before they reach 16 years of age has decreased sharply. In 1900 the chance of such an occurrence was 24 percent; today it is less than 1 percent. At the same time, parents can expect their children to survive infancy and early childhood. In 1900, 250 of every 1,000 children born in the United States died before reaching age 1; 33 percent did not live to age 18. Today, 6 of every 1,000 children born die before they reach age 1; less than 5 percent die before reaching age 18. Not only can parents and children be much more secure that the other will survive, but the length of time parents, siblings, and other relatives share each other's lives has increased.

Second, the number of people surviving to old age has increased. In countries such as Japan, Italy, Germany, and the United States, where the total fertility rate is low and declining, the proportion of older people in the population is increasing.

Library of Congress Prints and Photographs Division[LC-USZ62-104928]

This 1896 photographic print shows six generations of women. This portrait indicates that multigenerational and even six-generation families existed in the past. From a societal point of view, however, the number of such families was insignificant. But what happens when four-, five-, and six-generation families become commonplace? It will take some time for societies' institutions to adjust to that reality, because there is virtually no model in place for how members of families with four or more generations are expected to interact with each other.

Third, the potential length of the average marriage and other intimate partnerships has increased. Given the mortality patterns in 1900, newly married couples could expect their marriage to last 23 years before one partner died (assuming they did not divorce). Today, if they do not divorce, newly married couples can expect to be married for 53 years before one partner dies. This structural change may be one factor underlying the currently high divorce rates. When people could expect to live only a few more years in an unsatisfying relationship after retirement or their children left home, they would usually resign themselves to their fate. But today, the thought of living 20 or 30 more years in an unsatisfying relationship can provoke decisive action. Divorce now dissolves marriages, whereas the death of a partner dissolved them 100 years ago.

Fourth, people now have more time to choose and get to know a partner, settle on an occupation, attend school, and decide whether they want children. Now, these areas of experience are occurring later in life than in past generations. Moreover, an initial decision made at any one of these stages is not final. The amount of additional living time enables individuals to change life course. In fact, one might argue that the so-called midlife crisis derives from the belief "that there yet may be time to make changes" (Stub 1982, 12).

Decline in Parental Authority

Around the beginning of the 20th century, children learned from their parents and other relatives the skills needed to make a living. As the pace of industrialization increased, jobs moved away from the home and into factories and office buildings. Parents no longer trained their children, because the skills they knew and possessed were becoming obsolete. Children came to expect that they would not make their living in the same way their parents did. In short, as the economic focus shifted from agriculture to manufacturing, the family became less involved in their children's lives, including their adult children's lives. Ultimately, the transfer of work away from the home and neighborhood removed opportunities for parents and children to work together.

Gradually, values and norms developed that supported privacy and independence—for example, norms that elders should not interfere in the lives of adult children and that parents and adult children should reside in separate households. Popular support pushed governments to establish social security and health insurance programs for the retired, low-income, and disabled. Those policies further reinforced changes in family structure and values such that schooling, work, social, and leisure activities became largely age-segregated experiences, to the point that if intergenerational activities were to take place, they needed to be planned in advance.

The responsibilities once assigned to the family are now performed by organizations serving the public in general, such as day care, nursery schools, and preschools. In 1970 about 20 percent of 3- and 4-year-olds in the United States were enrolled in preschool programs. Today about 50 percent attend a full-time preschool program. While this increase is certainly connected to the increased percentages of mothers in the paid labor force, many parents have also come to believe that early education programs are essential to a child's future educational and social success (Social Issues Reference 2009).

Tonya K. Townsell, Presidio of Monterey Public Affairs

Sr. Airman Bethann Hunt, USAF

Contrast the experience of these American children (top) with that of these Afghan children from the Pashtun tribe sitting together at home in Kabul. While kindergartens exist in Afghanistan, only about 2,000 children attend them, in a country where 7 million children are age 5 and under (Education Encyclopedia 2009; UNICEF 2013). Can you see how parental authority is diminished when outside agencies assume responsibility for a portion of children's care?

The Economic Status of Children

Technological advances associated with the Industrial Revolution and the shift from an agriculture-based economy to a manufacturing-based one also changed children from economic assets to liabilities. Mechanization decreased the amount of physical effort and time needed to produce food and other commodities. Consequently, children lost the opportunity to contribute to household income and as a result lost their economic value. In agricultural and other extractive economies, children represent an important source of free unskilled labor for the family. This fact may partly explain why some of the highest total fertility rates in the world occur in places like Afghanistan, Somalia, the Democratic Republic of the Congo, and Niger—places where labor-intensive agriculture and extractive industries are crucial to the economy.

Demographer S. Ryan Johansson (1987) argues that in most industrialized economies, couples who choose to have children bring them into the world to provide intangible, emotional services—services such as love, companionship, an outlet for nurturing feelings, enhancement of dimensions of adult identity—rather than economic services. These desires seem to hold for all income and age groups (teens vs. adults who delay childbearing until later in life).

Catherine Mulconrey

NKU Philosophy, Rudy Garns

Study the two photographs showing children who are about the same age. When you think of the role of children in the United States today, do you think of children performing labor or do you think of children as having talents that need to be developed? The photo of two children selling goods in the street depicts labor that contributes to family's income. The child in gymnastics class is participating in an activity for which the family must pay.

The shift away from labor-intensive production stripped children of opportunities to make an economic contribution to the family and made children expensive to raise. Today in the United States, depending on income, the average yearly expenses for childrearing can range from $10,200 to $17,000 (U.S. Department of

Agriculture 2013). These estimates cover only the basics; when we include extras, such as summer camps, private schools, sports, and music lessons, the costs go even higher. Of course, for many parents, the cost of raising children does not stop when they turn 18. The high cost of raising children may be one reason that, in one recent survey of 2,200 adults, children were among the least cited contributors to a successful marriage. Faithfulness, a good sexual relationship, household chore-sharing, income, good housing, shared religious beliefs, and similar tastes and interests were cited more often than children (Pew Research Center 2011a).

The Economy and Family Life

As we have seen, the economy shapes relationships among family members. Wages and opportunities for paid employment certainly shape intergenerational relationships (e.g., parent–child, child–grandparent). Simply consider that as the number of married and unpartnered women with children entering the workforce increased, grandparents were called on to care for children while the mother worked. Currently, grandparents are primary caregivers for 24 percent of preschool-aged children (Laughlin 2013). Also consider the increasing proportion of young adults aged 25 to 34 living with a parent or a grandparent. Specifically, 17 percent of men and 12 percent of women live in their parents' homes (up from 13 and 7 percent, respectively, in 2005) (U.S. Census Bureau 2014b). Explanations include the large debt incurred from taking out college loans, a shortage of jobs that pay sufficient wages, and life expectancies that delay entry into committed partnerships, parenthood, and careers.

What Do Sociologists See?

Sociologists see children whose parents have signed them up to be part of a swim team. Sociologists also place these swimmers in the context of an economy where children are a financial liability. To participate on swim teams, parents can expect to pay each year at least $1,000 in membership and coaching fees, not including swim suits and transportations costs to swim meets and practices (Boisey Swim Team 2014).

Mrs. Michelle Kennedy (Drum)

Critical Thinking

Estimate what your parents spent to raise you to age 18. Include in that estimate any contributions relatives such as grandparents made to ease the financial burden.

Key Term

generation

Module 11.6

Caregiving

Objective

You will learn how sociologists think about those who give and receive care.

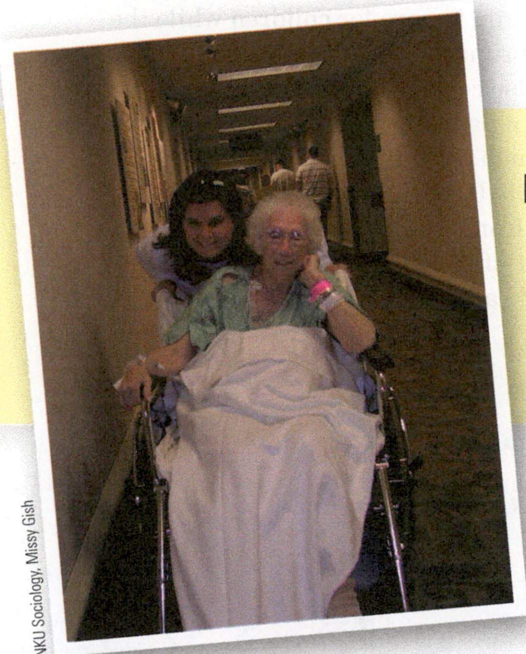

NKU Sociology, Missy Gish

Is there someone in your life for whom you give, have given, or expect to give care? Have you ever needed care?

Obviously we needed care when we were very young and, unless we die suddenly, we will very likely need care at the end of life. In this module we explore caregiving apart from that needed to raise children.

Caregivers

We use the term **caregivers** to mean those who provide assistance to people who, because of illness or physical or mental impairment, cannot manage day-to-day activities on their own. The assistance that family members, neighbors, and friends provide in a home setting is **informal care**. Fee-for-service assistance provided by credentialed professionals in-home or in a dedicated facility is **formal care**. In this module we focus on informal caregiving because most caregiving is informal (Day 2009). By one estimate there are 65.9 million adults (29 percent) in the United States giving care on an informal level to someone. It is believed that 66 percent of caregivers are female and that females are more likely than their male counterparts to provide personal care (16 percent of male caregivers help with bathing compared to 30 percent of females) (Family Caregiver Alliance 2014).

The 65.9 million caregivers in the United States average 25.1 hours of care per week and devote 25.8 billion hours to assisting a partner, relative, or friend (Family Caregiver Alliance 2014). The exact number cared for is not known, but we do know there are 37.6 million people who have a physical or mental impairment. It is difficult to know how many of these 37.6 million people need help with doing day-to-day activities.

Three million (4.1 percent) of noninstitutionalized persons under 18 years of age have a physical, mental, or sensory impairment. Generally, when one thinks of populations labeled as disabled, those age 65 and over come to mind; 15.0 million in that age group are considered disabled. The percentage of the U.S. population (all age groups) considered disabled is 12.2 percent, or 37.6 million (U.S. Census Bureau 2012i).

Ms. Jessica Obermeyer (3rd ID)

With regard to the population age 65 and over in the United States, one in four requires assistance with daily activities such as bathing, walking, dressing, and eating. An estimated 39 percent of all adults 18 and older in the United States provide some kind of informal care to a person age 65 or older. Almost 15 million caregivers care for someone who has Alzheimer's disease or other dementia (Family Caregiver Alliance 2014).

Keep in mind that caregiving is a complex activity that goes beyond time commitments and perceived burden. Caregiving, especially when it is long-term or ongoing, creates strain and tension for both the provider and receiver of care.

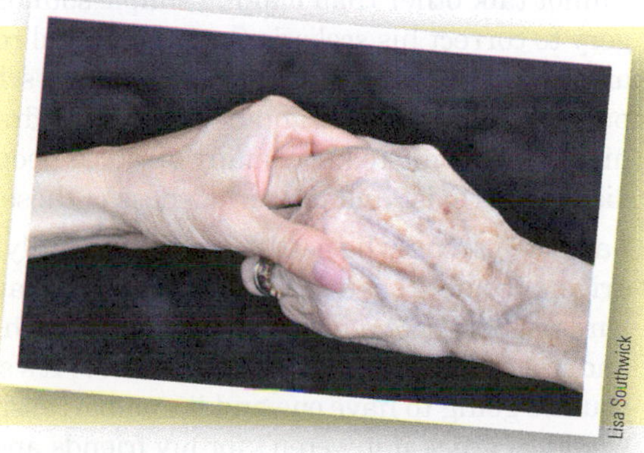
Lisa Southwick

The relationship between provider and receiver is subject to any number of strains and stresses related to

- the amount of time the caregiver is able to devote and the time those receiving care need or require;
- balancing the demands of caregiving with the caregiver's other commitments, including job and family;
- the physical and emotional demands of caring and being cared for; and
- dignity and privacy (changing diapers, using the toilet, bathing) (Day 2009).

People take on caregiving responsibilities for any number of reasons, including

- the desire to pay back recipients for the sacrifices they made;
- the emotional bond between the caregiver and the recipient;
- out of obligation, including a desire to live a life free of regret or guilt;
- a feeling of accomplishment, especially when the recipient expresses appreciation and satisfaction; and
- personal growth (Yamamoto and Wallhagen 1997).

To gain insights about the relationships between those giving and receiving care, I routinely ask my students to describe anyone in their family who needs care because of a physical or mental impairment. The last time I asked my students to write on this topic, 85 students wrote about a total of 109 family members. Four examples follow:

- My uncle has a learning disability. He can barely read and write. His speech is slurred and it is very difficult to understand him. His disability does not affect him very much though; he gets around and goes to work every day just like anyone else. Even with his learning disability he is great at working on cars and fixing miscellaneous things around the house. His disability isn't physical but whenever he has to read or sign something, my mom will have to read it and tell him what it means.
- My mother has severe arthritis. Her bones are deteriorating in her wrist and her ankles. My mother used to be a hairdresser but four years ago she was no longer able to move her hands and stand on her feet. My mom has had two wrist replacements and one ankle operation. She has been on disability for four years now and it's time to reapply. My mom takes many pills every morning and every night, about 20 each time. She often uses her wheelchair. She's able to take care of herself but if she pushes herself too much, she needs help and becomes stiff.
- My younger brother is 19 and was born with several problems. He is mentally retarded and has scoliosis and a club foot. He could not walk until he was 4; he cannot talk other than making simple sounds and he has had one major surgery to correct his scoliosis (he had a metal rod put into his back) and will have another next year after he graduates high school. . . . He uses a handheld computer to communicate, and he just got his first job at Kroger. Growing up with him I have come to know many disabled people. . . . It's unbelievable how many disabilities exist and the seriousness of those disabilities.
- I am the person who needs care in my family. Because I have epilepsy I cannot drive. I cannot consume certain foods that cause migraines, which trigger seizures. When I get a migraine, I have so much pressure on my head my sight blurs and it is hard to see or want to move. . . . It's scary because you don't know when you are going to have one and it's embarrassing when you do. I get sick easily. I would be stuck if it weren't for my friends and family who drive me places.

Social Pressures to Be Caregivers

Sociologists Ingrid Arnet Connidis and Julie Ann McMullin (2002) maintain that men face expectations about their role as caregivers that are different from and less taxing than those women face. Among other things, men are handed a "legitimate excuse" to not engage in caregiving, and when they do, to assume a limited role. That excuse is supported by longstanding beliefs that women are naturally better caregivers. Furthermore, paid employment excuses men from caring for aging parents and others, but for women it does not (Connidis and McMullin 2002, 562).

When you see a group of young children, do you expect a woman or a man to be monitoring their activities? Your expectations say a lot about whether you view men or women as responsible for caregiving.

Chris Caldeira

When women feel socially pressured to assume a disproportionate share of caregiving responsibilities, they can experience ambivalence toward those for whom they must care—an ambivalence intensified by the fact that they are "expected not only to provide the care but also to derive satisfaction and fulfillment" from it (563).

These responsibilities and conflicting feelings can cause tensions in family, work, and other relationships. In response, a woman may "choose" to quit work, to pay a third-party caretaker, to solicit others (e.g., grandparents) to help, and so on. Of course, the response depends on the available resources from which a person can draw.

NKU Sociology, Missy Gish

We know from the U.S. Census Bureau (2012h) that grandparents are primary caregivers for 23 percent of preschool-aged children. In this regard, grandparents, especially grandmothers, are acceptable substitutes for the mother.

In family life men face considerably less pressure to assume caregiving roles. This lack of pressure is reinforced by their public invisibility in such roles. For example, less than 1 percent of day care providers are male; less than 10 percent of nurses are male.

U.S. Navy photo by Mass Comm. Spc. Jason R. Wilson

Women vary in their ability to resist expectations that they assume a disproportionate share of responsibility. Obviously, financially secure women have the option to hire others to provide some or all of the care. Women employed in low-paying jobs may be forced to use vacation and personal days, work fewer hours, change shifts, or drop out of the labor force to meet caregiving obligations. Simply hiring a third party to provide care may not reduce strain and accompanying ambivalence if women do not trust other providers to give their child, parent, or other relative high-quality care.

Connidis and McMullin point out that, even in light of the increasing number of family-friendly workplace policies (e.g., parental leave, job sharing, on-site day care, telecommuting), it is female employees who take the most advantage of them and who sacrifice careers to meet familial responsibilities. Even with family-friendly policies, we must remember that women are still part of a society that defines them as best able to provide care.

One way that men hear the message that they are not naturally suited to provide caregiving relates to the high praise they receive when they do engage in that activity. When you see a man taking care of a child, are you inclined to give him more praise than you would if it were a woman engaging in the same level of care? The low expectations society places on men play a role in eliciting high praise.

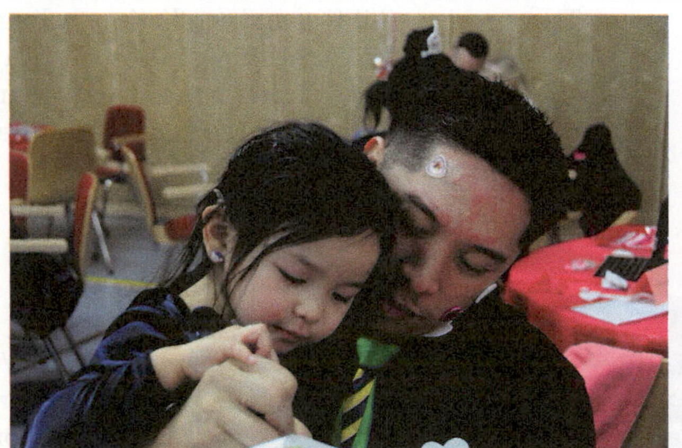

Chelsea Bissell, U.S. Army Garrison Grafenwoehr Public Affairs

Even though men face considerably less pressure to be caregivers, they are likely to experience strain and ambivalence when they do. When men engage in caregiving, their motives are often suspect (e.g., are they sexual predators? gerontophiliacs?). Onlookers may question men's ability to provide such care, especially with regard to the intimate aspects of caregiving—bathing, changing adult diapers, and dressing. Anecdotal evidence suggests that men feel unprepared for the job and isolated from support networks (Leland 2008). One man interviewed in the *New York Times* who is a caregiver to his elderly mother shared his feelings:

> She doesn't know if I'm her husband or her boyfriend or her neighbor. She knows she trusts me. But there are times when it's very difficult. I need to keep her from embarrassing herself. She'll say things like, "I adore you." I don't know who she's loving, because she doesn't know who I am. Maybe I'm embarrassed about it—it's my mom, for Christ sakes. But it's weird how the oldest son becomes the spouse. (Nicholson 2008)

Connidis and McMullin point out that people employ many strategies, including humor, confrontation, substance abuse, acceptance, and abandonment, to deal with strain and stresses. The two sociologists hypothesize that the fewer perceived or real options available to caretakers, the more likely they are to resign themselves to the situation. However, when substantial numbers of people experience tensions and ambivalence, there is a potential for fundamental change to occur.

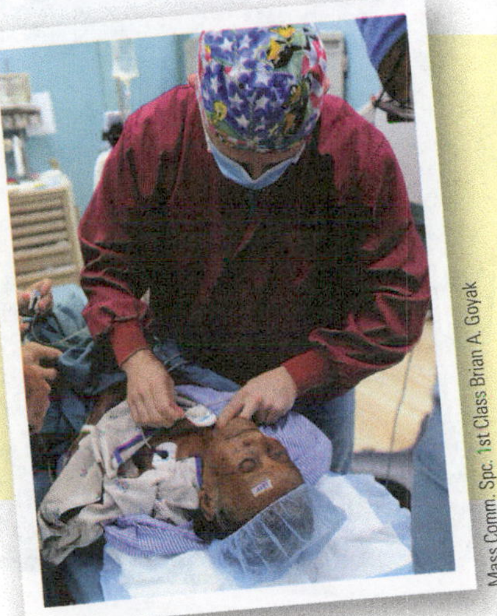
Mass Comm. Spc. 1st Class Brian A. Goyak

This woman, age 101, is being prepared for a routine surgery. Her needs for postoperative care are likely complicated by her advanced age. As the numbers of people aged 80 to 100+ years old increase, the pressures on younger generations to provide care are bound to increase. This demographic pressure has the potential to change current norms that assign primary responsibility to women, most notably daughters.

Lance Cpl. Cory Yenter, USMC

One should not conclude that caregiving is only a burden. It can be a very rewarding experience. But it is also a labor-intensive experience. When social structures define only half the population, women, as naturally suited for caregiving, it eliminates an important source of support. When social structures portray men as unsuited to caregiving, those structures alienate men from establishing meaningful bonds with others.

Impairment and Disability

The experience of caring and being cared for is shaped by whether the parties think of the condition as being an impairment or a disability. An **impairment** is a physical or mental condition that interferes with someone's ability to perform an activity that the average person can perform without technological assistance or without making changes to the physical environment. To illustrate, most people confined to a wheelchair can read, so they do not have an impairment with regard to reading; they do have an impairment when it comes to walking, for which they need assistance.

From a sociological point of view, a **disability** is something society has imposed on those with a specific impairment because of the ways inventions have been designed and social activities organized to exclude some and accommodate others.

Charles M. Belluomo, for *Soldiers* magazine

In this vein, one might argue that those who invented bicycles designed them with the intent of remove barriers of time and distance only for those people with functioning legs. But what if the inventors had designed bikes thinking about how people without legs could "pedal." This shift in emphasis would have created a bicycle like that pictured here. The point is that the way the bicycle was originally designed excluded those capable of "pedaling" with their hands—designing it without thinking of this possibility had the unanticipated consequence of creating a disability.

Likewise, we may treat people confined to wheelchairs as impaired with regard to cooking, but if stoves were designed to accommodate those who must sit while cooking, the so-called impairment would disappear. Disability is imposed when emphasis is placed on the loss of some mental or physical capacity with no consideration given to ways of reducing barriers to full participation (Barton 1991).

Those with impairments and disabilities often experience what many have called the **tyranny of the normal**—a point of view that measures differences against what is thought to be normal and that assumes those with dishabilles and impairments fall short in just about every way. Pam Evans (1991), a wheelchair-bound woman, further clarifies this tyranny by outlining common assumptions "normals" held about the life of impaired individuals:

- That our lives are a burden to us, barely worth living.
- That we crave to be normal and whole.
- That any able-bodied partner we have is doing us a favor and that we bring nothing to the relationship.
- That if we are particularly gifted, successful, or attractive before the onset of disability our fate is infinitely more tragic than if we were none of these things.
- That our need and right to privacy isn't as important. . . .

While it would be naive to believe that these kinds of thoughts never occur to those with impairments and disabilities, these common assumptions capture the dynamics underlying the tyranny of the normal.

What Do Sociologists See?

Sociologists think of the distinction between impairment and ability. While the men shown playing volleyball may have a disability that limits mobility, they are not impaired with regard to using their hands and arms to serve, set, and spike the ball, provided the net is lowered.

J. D. Leipold

Critical Thinking

Think about the people who have cared for you in some capacity. Do ideas covered in this module apply to your experiences? Explain.

Key Terms

caregivers	impairment
disability	informal care
formal care	tyranny of the normal

Module

(11.7) Applying Theory: Changing Family Forms

Objective

You will learn how the major perspectives shape how sociologists look at changing family forms.

Is there something "different" about your family that departs from what is considered the norm?

Terra Schulz, Courtesy of Joan Ferrante

For many people the word family evokes a mother, a father, and children living in the same household. But we know that the percentage of all households in the United States that meet this criterion is 26 percent. This household arrangement has declined if only because as life expectancy increases and fertility decreases, the time (as a portion of their total life span) people spend taking care of children, if they have them, becomes smaller. Moreover, children are not the only ones who need care. All people need care, support, and affection and derive satisfaction from meeting others' needs and knowing they can count on others to meet their needs. The demographic challenges of our time, most notably an aging population, require that we welcome new forms of family—forms once thought of as odd, dysfunctional, or deviant—as sources of caregiving and emotional support. From a sociological perspective, family is not "static"; it is a "dynamic system" that accommodates fluxes and flows in members and relationships. In other words, the kinds of "new" families formed are adaptations to social and demographic change (Widmer 2010). The four sociological perspectives—functionalist, conflict, symbolic interaction and feminist—offer conceptual tools for evaluating these new forms.

This marriage legally recognizes the relationships between the woman and the man on the right, but it also legally recognizes the groom's son as a member of a family that is sometimes referred to as blended. From a functionalist point of view, the best criterion by which to assess the effectiveness of any "new" family arrangement is not by focusing on whether members meet certain criteria (same race, heterosexual, legally married) but how well the arrangement contributes to order and stability by meeting the basic needs of its members. Those basic needs include socializing the young, caring for members, and regulating sexual activity.

Beth Reinersman, Courtesy of Joan Ferrante

Molly Hayden, U.S. Army Garrison–Hawaii Public Affairs

When evaluating new family forms, conflict theorists' assessment is not clouded by those who have the power to define what constitutes a family and committed partnerships—to make laws that award benefits and privilege to some kinds of relationships and families but not others. Conflict theorists recognize that dominant and legally sanctioned family forms can avoid scrutiny, causing people to overlook strains and abuses within those households. After all, the so-called ideal family arrangements are not problem-free.

Since symbolic interactionists focus on interaction and meaning, they would not limit analysis of family to people who occupy positions like biological parent or legal parent, but would also focus on persons in new "nonfamily" groups considered to be like a mother or father or like a sister or brother. Questions of interest include: How are interactions and relationships like those in a family? When does an interaction make someone "feel" like they are part of a family? Family can be thought of as being in decline, not when the dominant forms of family decline or disappear, but when people have no deep connections or feel no commitment to others.

Khalifa al-Sharif, Courtesy of Joan Ferrante

Like the other three perspectives, the feminist perspective challenges narrow views about what constitutes a family or a marriage. Its adherents would challenge the idea that men should be taller or older than female partners, for example. Feminists recognize that laws and dominant norms are used to justify discrimination against family forms that do not meet legal definitions and expectations. In assessing any new family form, feminists challenge any family form that privileges one gender and on partnership, most notably heterosexual marriages with children, over other genders and partnerships. Feminists also recognize that any partnership can be problematic if a partner who assumes a homemaker and caregiver role feels trapped by a breadwinner partner who ruthlessly controls the household budget and spending.

Chris Caldeira, Courtesy of Joan Ferrante

Summary: Putting It All Together

Family is a social institution that binds people through blood, marriage, law, and/or social norms. When sociologists study family, they do not have a particular family structure in mind as a standard. Instead, they consider the complex set of social forces that affect the ever-changing structure of families. Those family structures are evaluated according to the social functions they perform and the extent to which they perpetuate or address existing inequalities and social divisions.

Sexual or gender stratification is the system by which societies rank males and females on a scale of social worth such that the ranking affects the way family life and partnerships are organized. Historically, there have been four economic structures that have shaped gender relations at the family level: (1) low-technology tribal societies, (2) fortified households, (3) private households, and (4) advanced market economies. Each economic structure is characterized according to women's access to agents of violence control and their position relative to men in the labor market. It is clear that when women have access to agents of violence control and equal opportunity in the labor market, they are more likely to have an equal relationship with men and it is more likely that a variety family forms considered unconventional (at first) will flourish.

In the United States, household structures have changed dramatically since 1900. Those changes involve the rise and fall of the breadwinner system (a private household), followed by a variety of household types, with no one type dominant. Increases in life expectancy, the decline of parental authority, and a shift in the economic status of children from an asset to a liability have also altered the structure of households and family life.

Caregiving, especially when it is informal, long-term, or ongoing, creates strain and tension for both the provider and receiver of care. The relationship between provider and receiver is subject to any number of strains and stresses. In family life men face considerably less pressure to assume caregiving roles than do women. This lack of pressure is reinforced by their lack of public visibility in such roles. When men do engage in caregiving, they are likely to experience a different kind of strain and ambivalence than women do. Women's strain and ambivalence derive from social pressures that define them as naturally suited for caregiving and that award them ultimate responsibility for such tasks. Men's strain derives from beliefs that they are not naturally suited to caretaking and from suspicions about why they might want to do such work. The nature and quality of the caregiving relationship are shaped by whether the condition is viewed/treated as a disability or an impairment.

Chapter 12

Education and Religion

- **12.1 Education and Schooling**
- **12.2 Social Reproduction**
- **12.3 Education in a Knowledge Economy**
- **12.4 Rewards and Costs of Higher Education**
- **12.5 Religion**
- **12.6 Civil Religion and Fundamentalism**
- **12.7 Applying Theory: Private Schools**
- **Summary Putting It All Together**

Education and religion are core social institutions that touch the lives of just about everyone. Schools and religions seek to instill what sociologist Émile Durkheim terms a collective consciousness, "a complex of ideas and beliefs that influence ways of seeing and of feeling" (Durkheim 1961, 860). In the first half of the chapter, we focus on schooling, a deliberate, planned effort that takes place in a brick-and-mortar or virtual classroom to impart specific kinds of skills or information. Schooling can be a liberating or positive experience, but it can also be impoverishing and narrowing as well if it involves indoctrination, brainwashing, or neglect. In the second half, we focus on religion, those human responses that give collectively supported meanings to the ultimate and inescapable problems of existence—birth, death, illness, aging, injustice, tragedy, and suffering (Abercrombie and Turner 1978). Like schooling, religiously inspired thought and activity can be liberating or impoverishing depending on the "uses" to which it is channeled.

Module 12.1

Education and Schooling

Objective

You will learn why education, and schooling in particular, is an important area for sociological study.

What is your earliest memory of being a student in school? More than likely, that memory holds important clues about the meaning of school in your life.

David Ruderman, USAG Vicenza Public Affairs

Education includes any experience that trains, disciplines, and shapes the mental and physical potentials of the maturing person. While every experience has the potential to educate, sociologists who study education tend to emphasize **schooling**, a deliberate, planned effort that takes place in a brick-and-mortar or virtual classroom to impart specific skills or information. We tend to think of schooling as a liberating or positive experience, but it can be impoverishing and narrowing if it involves indoctrination, brainwashing, or neglect. In any case, schooling is considered a success when students acquire the skills or thoughts that those who designed the program of learning seek to impart. Sociologists seek to identify the key processes—"inside and outside schools—that affect student outcomes" (Karen 2005). For insights on these factors and processes, we turn to three sociological perspectives: symbolic interaction, functionalism, and conflict theory.

Symbolic Interaction

Schooling is a vehicle through which students are exposed to a collective consciousness or "a complex of ideas and beliefs that influence ways of seeing and of feeling" (Durkheim 1961, 860). The symbolic interactionist perspective offers a framework for capturing the process by which a collective consciousness is transmitted. Since the school days revolve around learning a curriculum, symbolic interactionists focus on how that curriculum conveys shared meanings about the subject matter and much more.

Teachers everywhere teach two curricula simultaneously: one formal and one hidden. The various academic subjects—mathematics, science, English, reading, physical education, and so on—make up the **formal curriculum**. Students do not learn in a vacuum, however. As teachers present the subject matter and students complete assignments, other activities are occurring around them. Those other activities make up the **hidden curriculum**, the lessons conveyed through the way students are taught, the assignments students are asked to do, the way students are assessed (multiple choice vs. essay), the tone of the teacher's voice, attitudes of classmates toward learning, and the frequency of teacher absences are just some examples. These activities teach lessons, not only about the value of the subject, but also about culturally valued ways of thinking and behaving.

Screen capture Courtesy of Larry Huff

When learning is delivered online, it conveys hidden lessons—a hidden curriculum—that go beyond the subject matter. For one, it sends the message that schooling can take place outside of the classroom and that students need not live nearby. It sends the message that students do not have to set aside specific times of the day to learn; they can make their own schedule. Online learning sends the message that education can be worked in to even the busiest of schedules. Being self-disciplined and motivated to complete assignments also matters.

For further insights on the power of the hidden curriculum to convey lessons beyond the subject matter, we consider the comparative research of Yi Che, Akiko Hayashi, and Joseph Tobin (2007), who observed preschools in the United States and China. The three researchers filmed daily life in a Chinese and a U.S. preschool to learn how teachers in each system teach more than the subject matter; they also socialize children to participate effectively in their respective societies. Though it is impossible to make definitive generalizations about preschools in countries as large and diverse as the United States and China, we can identify some very broad cultural differences. The researchers found that, compared with U.S. preschoolers, Chinese preschoolers (four-year-olds) are taught to constructively critique each other's work and to learn from critique. In contrast, American preschoolers are taught to expect praise for their work, not critique. The following scene illustrates a formal curriculum that teaches children a subject—public speaking—but also a hidden curriculum that simultaneously teaches children to give and accept critique.

Each day in a Shanghai preschool class, 22 children engage in a storytelling activity. One child is designated the "story king" and stands in front of the class to tell a story. Upon finishing, the teacher (Mrs. Wang) makes comments, asks students questions about the story, and then calls for a vote on whether today's storyteller earned the title of story king. Eighteen children vote yes. Mrs. Wang asks the four children who voted "no" for reasons: "A child remarks, 'Some words I could hear, but some I couldn't.' 'Don't think his voice was loud enough' says another." The teacher turns to the storyteller and asks if he agrees with the critique, to which he nods yes. "At that point, the teacher comments, 'Next time, he will be loud and clear'" (Che et al. 2007, 7).

In the United States, preschoolers do not typically critique each other's work, and especially not when that work is considered self-expressive or creative. Early-childhood teachers in the United States tend to believe that their job is to protect and support children's self-esteem and that it is not "developmentally appropriate" to subject a child to peer criticism (Che et al. 2007, 7). In fact, American teachers are reluctant to correct a child's mistakes in front of other students. Rather, teachers in the United States often give empty praise—"that's wonderful"—regardless of quality.

The researchers showed an audience of U.S. teachers video clips of preschoolers in China critiquing each other. Some American teachers were bothered by this Chinese practice because they thought children were too young to do and handle this kind of thing. Others were amazed at how well preschool-age children gave and accepted critiques. One teacher remarked, "I'm amazed how well that boy handled the criticism. I'm an adult and I think I would cry if people criticized me like that in front of a group!" (7). The researchers pointed out that perhaps four-year-olds' self-esteem may not be as fragile as Americans assume.

Staff Sgt. Toshiko L Fraley

Would you allow or encourage this child's classmates to critique her work—to identify things she might do to improve it? If you live in the United States, it is unlikely that you would. But in Chinese preschools, such critiquing is encouraged and welcomed. Can you see how schooling is a vehicle through which students are exposed to a collective consciousness, "a complex of ideas and beliefs that influence ways of seeing and of feeling" (Durkheim 1961, 860)?

Functionalist Perspective

The functionalist perspective focuses on the functions of schooling. These functions include transmitting skills, facilitating personal growth, integrating diverse populations, screening and selecting the most qualified students for what are considered the most socially important careers, and solving social problems.

TRANSMITTING SKILLS. Schools exist to cultivate the skills students need to adapt to the society and world in which they live. To ensure that this end is achieved, employers, parents, government officials, and other stakeholders remind teachers what skills and knowledge must be impressed on students. The things to be impressed might be historical facts, the ability to accept critique, or how to play an instrument.

FACILITATING PERSONAL GROWTH. Education can be a liberating experience that releases students from the blinders imposed by the accident of birth into a particular family, culture, religion, society, and time in history. Education can broaden students' horizons, making them aware of the conditioning influences around them and encouraging them to think independently of authority. In that sense, schools function as agents of personal growth and change.

INTEGRATING DIVERSE POPULATIONS. Schools function to socialize (for example, to Americanize, Europeanize, "Chinesize") people of different ethnic, racial, religious, and family backgrounds. In the United States, schools play a significant role in what is known as the melting-pot process. Recall that the creation of the United States involved the conquest of the native peoples, the annexation of Mexican territory along with many of its inhabitants (territory that is now New Mexico, Utah, Nevada, Arizona, California, and parts of Colorado and Texas), and the voluntary and involuntary influx of millions of people from across the globe (Sowell 1981). Early American school reformers—primarily those of Protestant and British backgrounds—saw public education as the vehicle for Americanizing a culturally and linguistically diverse population, instilling a sense of national unity and purpose, and training a competent workforce.

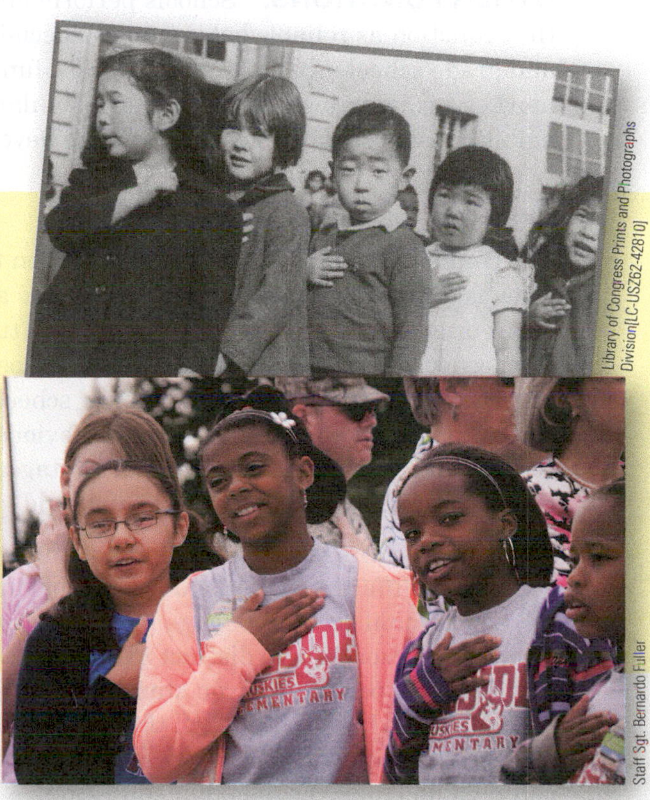

Library of Congress Prints and Photographs Division[LC-USZ62-42810]

Staff Sgt. Bernardo Fuller

These schoolchildren are probably too young to appreciate the meaning of the words in the Pledge of Allegiance, but the act itself reminds them that they are Americans. The fact that the pledge was written in 1892 with schoolchildren in mind speaks to the Americanizing function of schools. The photos show schoolchildren in 1942 (top) and today (bottom) reciting the pledge.

SORTING. Schools use grades earned on tests and assignments to evaluate students and reward or punish them accordingly by conferring or withholding degrees, by rejecting or admitting students into programs of study, and by giving negative or positive recommendations. In theory, these scores and grades channel the most capable and skilled students into the most desirable and important careers, and the least capable and skilled into careers believed to require few, if any, special talents.

SOLVING SOCIAL PROBLEMS. Societies use education-based programs to address a variety of social problems, including parents' absence and lack of involvement, racial inequality, obesity, addictions, malnutrition, teenage pregnancy, sexually transmitted diseases, and illiteracy. Although all countries likely support education-based programs that address social problems, the United States seems to place particular emphasis on education as a primary solution.

These students are part of an education program designed to teach them how to feed and care for a baby. The program also teaches students about the time, resources, and emotional energy that go into taking care of babies and raising children. Such programs are often offered in schools where teen pregnancy is considered a problem.

Arthur McQueen (USAG Miami)

OTHER FUNCTIONS. Schools perform other, less obvious functions. For one, they function as reliable babysitters, especially for nursery school-, preschool-, and grade school–age children. They also function as a dating pool and marriage market, bringing together students of similar and different backgrounds and ambitions whose paths might otherwise never cross.

Conflict Perspective

The conflict perspective draws our attention toward inequalities. Conflict theorists ask questions like: Which schools have access to the most up-to-date computer or athletic facilities? Which types of students are most likely to drop out of high school? Which types of students are most likely to attend college? Conflict theorists argue that, for the most part, schools simply perpetuate the inequalities of the larger society. This point is obvious when we consider that the poorest schools and most economically disadvantaged children usually have the highest dropout rates, lowest graduation rates, and lowest college enrollments.

Among other things, conflict theorists study the extent to which students are exposed to different and unequal kinds of curricula. In this regard, most, if not all, educational systems engage in some form of **tracking**, a process by which students are sorted into distinct instructional groups based on past academic performance, performance on standardized tests, or even anticipated performance. At the high school level the instructional groupings may be college prep, advanced placement, or vocational. At the college level the instructional groups can include honors or remedial education.

Programs for students considered gifted and talented may take place in a classroom setting with seating arranged to make group work and collaboration easy (top photo). In gifted and talented programs, instructors typically do not lecture but guide students in learning activities. Programs for students considered remedial typically take place in a setting with seating arranged in rows (bottom photo) so that the instructor is the focus of attention because students are there to listen, not create knowledge.

Missy Gish, Courtesy of Joan Ferrante

NKU Sociology, Missy Gish

Advocates of tracking offer the following rationales in support of the practice:

- Students learn better when they are grouped with those who learn at the same rate. The brighter students are not held back by the slower learners, and the slower learners receive the extra time and special attention needed to correct academic deficiencies.
- Slow learners develop more positive attitudes when they do not have to compete with the more academically capable.
- Groups of students with similar abilities are easier to teach than groups of students with differing abilities.

There is little evidence to indicate that placing students in remedial or basic courses contributes to intellectual growth, corrects academic deficiencies, increases interest in learning, or narrows the achievement gap relative to students placed in higher tracks (Oakes 1986a, 1986b). Studies on tracking show that

1. Poor and minority students are placed disproportionately in the lower tracks.
2. Different tracks are not treated as equally valued instructional groups. Clear differences exist in classroom climate and in the quality, content, and quantity of instruction. Low-track students are consistently exposed to inferior instruction—watered-down curricula and endless repetition—and to a more rigid, more emotionally strained classroom climate.
3. Low-track students do not develop positive self-images because they are publicly identified and treated as educational discards, damaged merchandise, or even as unteachable. Among average- and low-track groups, tracking seems to foster lower self-esteem and promote misbehavior, higher dropout rates, and lower academic aspirations.

Although many educators recognize the problems associated with tracking, efforts to undo it have collided with demands from parents of high-achieving or gifted students who insist that their children must get something more than the other students (Wells and Oakes 1996). As a result, tracking persists.

What Do Sociologists See?

NKU Sociology, Missy Gish

This sign for an honors program announces a special kind of curriculum often referred to as an "honors experience." The word "honors" evokes images of interactive environments, deep intellectual development, specially designed extracurricular activities, seminar-style classes, classmates with high GPAs, and an environment that permits and encourages intellectual curiosity. The existence of an honors program suggests that honors' students are tracked into an educational experience segregating them from those students not in the program.

Critical Thinking

Do you currently or did you ever experience tracking? Describe how you were tracked and how you believed it shaped your educational experience and sense of self.

Key Terms

education	schooling
formal curriculum	tracking
hidden curriculum	

Module 12.2 Social Reproduction

Objective

You will learn how educational systems perpetuate the social inequalities students bring with them to the classroom.

If a professor asked you to write about this image, would you have difficulty thinking of something to say?

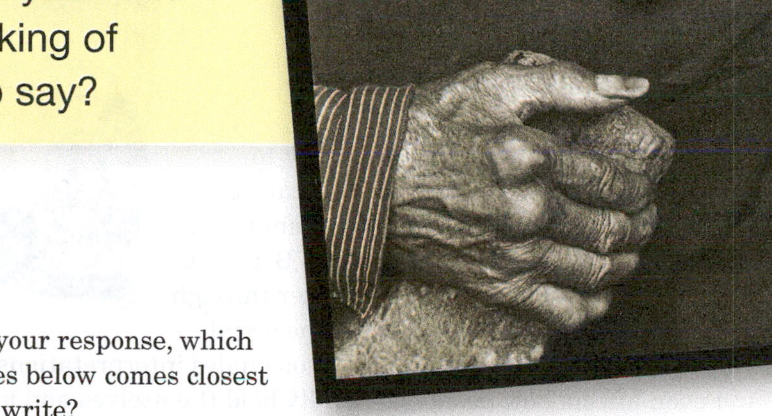

Library of Congress Prints and Photographs Division, Washington, D.C.

In thinking about your response, which of the two responses below comes closest to what you might write?

A. "This man looks like he's got arthritis. His hands are all knotted. I feel sorry seeing that poor old man's hands."

B. "The hands represent a life of toil, of a man who has engaged in labor that left him exhausted at day's end. It makes me think of a time when men worked with their hands (on machines and in fields) and that such hands were quite common as these men approached retirement. It makes me wonder what the hands of those who work in service and information jobs will look like when they retire."

Bourdieu (1984) found that working-class and low-income students tended to use plain, concrete language like that in response A to describe the hands. Students from more advantaged class backgrounds tended to use abstract, aesthetic language that transcends the situation of the particular person pictured, as in response B. These findings led Bourdieu to conclude that students' class positions have an incalculable effect on how they interpret and respond to assignments, including test questions. The class position students occupy is defined by the economic and cultural capital they bring with them to the classroom and upon which they continue to draw (Appelrouth and Edles 2007).

Economic and Cultural Capital

Economic capital refers to material resources—the kind of house in which a student lives, the amount of money parents invest in their child's social and intellectual development, and whether a student has a quiet place to study at home. **Cultural capital** refers to nonmaterial resources, including the kinds of knowledge, social skills, and aesthetic tastes acquired through day-to-day and specially arranged experiences and activities. So children whose parents have the economic capital to send them to science or journalism camps over summer break acquire a kind of cultural capital much different from children whose parents cannot afford to "buy" (or do not even know about) such learning opportunities.

From a middle- to upper-class point of view, parents who allow their children to spend most if not all their free time engaging in unstructured play (top photo) without adult supervision risk put their child at a disadvantage. Parents in more advantaged classes can better afford to invest in or purchase structured learning experiences such as guitar lessons.

NKU Anthropology, Sharyn Jones

Nondice Powell

Bourdieu maintains these kinds of day-to-day activities and experiences, the details of which are often forgotten, have the cumulative effect of shaping the mind or habitus. Bourdieu defined the **habitus** as the mental filter through which people view and understand the social world and their place in it. The habitus guides interpretations of and responses and even affects how people physically hold themselves and move about the world (that is, posture, facial expressions, gestures).

Bourdieu believed that the habitus affects how people physically hold themselves and move about in the world (e.g., posture, facial expressions, gestures). Imagine you are a teacher or school administrator. How might the way each student pictured holds and presents his body affect someone's perception of his academic ability?

Lance Cpl. John Robbart III

Molly Hayden, U.S. Army Garrison Grafenwoehr, Public Affairs Office

According to Bourdieu, no other institution does more than education to reproduce the inequalities that children bring with them to the classroom (Appelrouth and Edles 2007). Through a process sociologists call **social reproduction**, teachers and staff unwittingly perpetuate these inequalities by the unequal ways they treat students and organize their academic experience.

Why Students Drop Out

Bourdieu maintains that most students who drop out make the decision to do so; the school system does not force them out. He believes that tests deserve special scrutiny because the education experience revolves around studying for tests and students' mental energies are organized around taking tests. Bourdieu hypothesizes that it is the prospect of failing tests and studying for tests, and a dislike for the kinds of knowledge one must possess to pass tests, that spurs students to "voluntarily" drop out. The likelihood of failing tests varies by social class and racial category. There are also sharp differences across racial groups and classes with regard to verbal ability, reading comprehension, mathematical achievement, and general knowledge as measured by the standardized tests. The latest data show that, as a group, Asian students tend to score highest, followed by whites, Hispanics, blacks, and Native Americans. (see Table 12.2a).

Table 12.2a: Percentage of Fourth, Eighth, and Twelfth Grade Students Considered Proficient or Above in Reading by Race, 2011

What percentage of Asian students in fourth, eighth, and twelfth grades is considered proficient or above in reading achievement? What percentage of black students in each of three grades is considered proficient or above in reading? (Note that there is still a significant percentage of students in all racial groups who are not proficient in reading.) What does it mean that so few students classified as Hispanic, black, and Native American perform well on reading proficiency tests? Bourdieu would look to the educational system and the way it perpetuates inequalities of the larger society for answers.

Racial Classification	% of Fourth Graders Proficient	% of Eighth Graders Proficient	% of Twelfth Graders Proficient
Asian	49%	47%	48%
Native American	18%	22%	29%
Black	17%	15%	17%
Hispanic	18%	15%	22%
White	44%	44%	46%

Source of data: U.S. Department of Education (2012)

Bourdieu points out that tests give the impression that success in education is meritocratic—that is, that test scores are an objective measure of academic achievement and abilities, when, in fact, test scores may better measure parental coaching and class privilege. We cannot overlook that, as a group, students from advantaged backgrounds are better at taking tests than students from disadvantaged backgrounds. Bourdieu believes that this is because

- test makers are likely to be from academically advantaged backgrounds, and they write questions that reflect their background and point of view; and
- part of doing well on tests is figuring out what test makers want, and it is harder for students from disadvantaged backgrounds to do this because they process the test questions using a different habitus.

Corporation for National and Community Services, Office of Public Affairs

Bourdieu sees the exam as the "clearest expression of academic values" (1984, 142). Tests are the socially accepted, largely unchallenged way to assess whether a specified body of knowledge has been acquired. Bourdieu maintains that the purpose of exams cannot just be to measure academic progress. He views examinations as one of the most powerfully effective ways of impressing upon students the dominant culture and its values.

Bourdieu argues that when students drop out proclaiming that they are lacking in academic ability or that school is not for them, that "decision" places them in a devalorized category. In addition, the act of dropping out both sustains and perpetuates the existing inequalities. Dropouts' beliefs that they cannot make the cut or that "school is not for me" develop as students internalize their objective chances of success. Those objective chances are gleaned from the actual dropout rates of those of the same race and social class standing (see Chart 12.2a).

Chart 12.2a: Educational Survival Rates: Probability of Graduating from High School on Time by Race and Ethnicity

This chart shows the percentages of ninth-grade students by racial classification who graduated from high school in the 2012 school year, four years after entering the ninth grade. The percentage is also a rough indicator of the percentages of high school students who survive the system. The survival rate is highest for those classified as Asian/Pacific Islander (93.5 percent) and lowest for those classified as black (66.6 percent). Keep in mind that race by itself does not explain survival rates; rather, students in black, Hispanic, and Native American categories are disproportionately from low income households and live in neighborhoods where housing values generate lower revenues upon which the schools they attend depend.

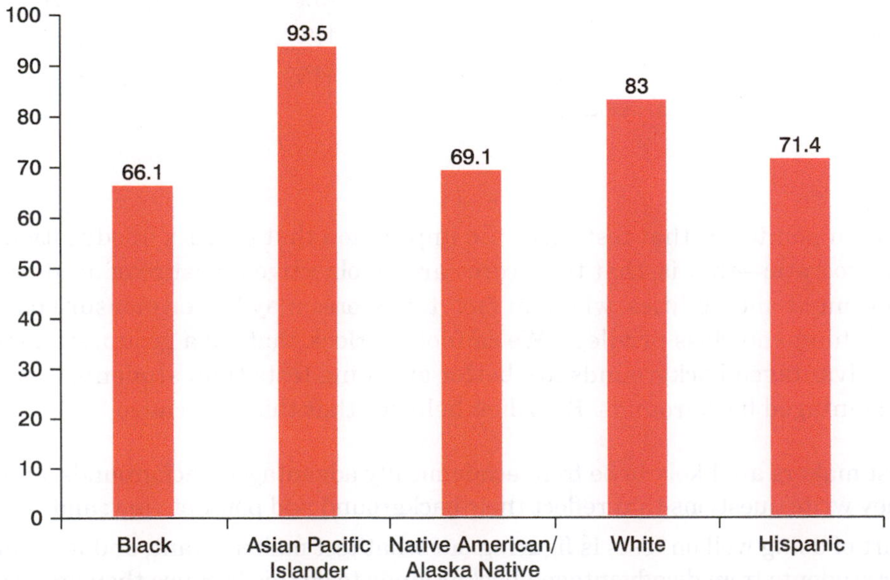

Source of data: National Center for Education Statistics (2013b)

In evaluating Bourdieu's ideas, it is important to keep in mind that he did not say that no one from a disadvantaged background "survives" the educational system—he did say that the probability of surviving is much lower relative to that of advantaged students. The OECD (2011a, 2013) labels students from disadvantaged backgrounds who survive the system as resilient; that is, "they beat the odds stacked against them to outperform peers from the same socioeconomic background to be ranked among the top quarter of students internationally." We can gain insights about the degree to which the U.S. system of education supports resiliency by comparing the percentage of students from economically disadvantaged backgrounds who are counted in the ranks of highest-achieving students (top 25 percent) with percentages of other countries. The educational systems that have the best success are Canada, Finland, Japan, New Zealand, Poland, Portugal, Spain, Liechtenstein, Singapore, and Taiwan. In these countries between 35 and 49 percent of economically disadvantaged students are counted among the high achievers. In the United States, about 28 percent of disadvantaged students are counted as resilient (OECD 2011a, 2013).

What Do Sociologists See?

Sociologists see a smiling mother who is rightfully proud of her son as he gives a piano recital. But this mother is also engaged in a form of concerted cultivation because she has deliberately involved her son in activities designed to cultivate an appreciation for the arts, the self-discipline needed to succeed at a difficult task, and an ability to perform on stage. She has invested financially in her son's development by paying for piano lessons and other related expenses.

Robert K. Wallace

Critical Thinking

Use Bourdieu's theory to frame your educational experiences.

Key Terms

cultural capital

economic capital

habitus

social reproduction

Module 12.3 Education in a Knowledge Economy

Objective

You will learn the meaning of the knowledge economy and the skills that increase chances of success.

Mr. Roger Teel (RDECOM)

Pfc. Gregory Gieske, 2nd HBCT Public Affairs

Which of the students pictured are more engaged with learning?

Think back on your time in school. Which of the two photos that open this module—the one of students conducting an experiment or the one of students listening to someone talk—better captures your academic experience? What kind of student do you think is created through each type of learning experience? Which learning experience better prepares students for success in a knowledge economy?

The U.S. economy, and the global economy of which it is a part, are in the midst of a revolutionary transformation as each moves further away from an industry-dominated economy symbolized by assembly lines rolling out standardized products and toward a knowledge-dominated one symbolized by smart technologies. This transformation has raised questions about what kinds of learning environments best prepare students to compete in a knowledge economy.

The Knowledge Economy

A **knowledge economy** is driven by information-gathering and data-collection activities that can be put to commercial use. People who succeed in knowledge economies—who occupy the highest-paying jobs—are people with the technical, problem-solving, and analytic skills to manage, manipulate, and make meaning from data collected. In knowledge economies, property is intellectual, knowledge is power, and there are corporations that exist for the sole purpose of creating, managing, and disseminating knowledge. In other words, knowledge is something that can be produced, owned, and sold.

A knowledge economy is characterized by an ever-accelerating pace of obsolescence and innovation (Powell and Snellman 2004). *Accelerated pace* refers to the speed at which knowledge is created, transmitted, applied, and revised. *Rapid obsolescence* is a result of ongoing innovation that makes products "old," even as they are first introduced into the market. *Knowledge-intensive activities* refer to jobs that revolve around gathering, collecting, manipulating, and disseminating data and information. Knowledge-intensive also refers to the fact that the majority of people on the planet now have access to digital technologies, most notably computers and smartphones, that allow them to access, create, manipulate, and transmit information that, until just recently, only those occupying positions in the largest organizations (e.g., universities, government agencies, and corporations) could do.

In a knowledge economy, the successful person possesses three critical skills: social networking, adaptability, and entrepreneurship (Hoffman and Casnocha 2012). Networking is considered key, as workers can no longer assume they can learn everything they need to know to analyze an issue or accomplish a task; they have to collaborate and learn from others. Because few will stay in the same job or with the same company over their entire work life, networking is considered a key skill for learning about career possibilities and job openings.

Adaptability is also a key skill, because in a knowledge economy successful people must negotiate the ongoing tsunami of ever-changing information and data. They must "have confidence to dwell in disorder," and flourish "in the midst of dislocation." In a knowledge economy, entrepreneurship is key in that people need to position themselves "in a network of possibilities" rather than paralyzing themselves by counting on one hoped-for job or career opportunity (Bauman 2005) .

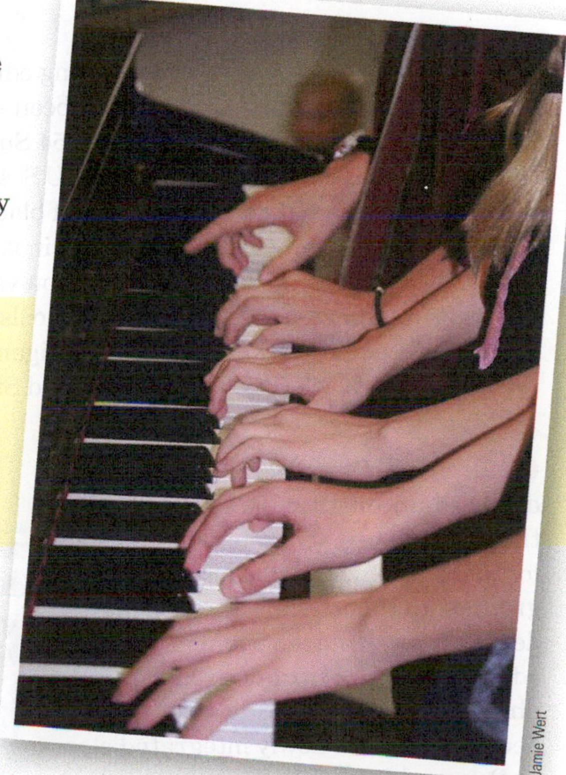

Jamie Wert

Activities that capitalize on social networks, adaptability, and entrepreneurship, like this piano recital involving three pianists, are what prepare one for success in a knowledge economy.

In a knowledge economy, students who learn in environments that cultivate encyclopedic-like understanding and emphasize standardized testing are at a disadvantage relative to students who learn in environments that cultivate flexibility, application, and creativity. While educators are still struggling with the formula for how to best educate students to succeed in a knowledge economy, we can identify some broad characteristics of learning environments that cultivate relevant skills.

Appreciation for Diverse Perspectives

We can argue that the schools that attract students, faculty, and staff who share the same class and racial background are at an educational disadvantage. We know that "embracing diversity is the first item for building teams. Every team building theory states that to build a great team, there must be a diverse group of people on the team, that is, you must avoid choosing people who are only like you. Diversity is what builds teams—a collection of individual experiences, backgrounds, and cultures that can view problems and challenges from a wide-variety of lenses" (Clark 2010). There is much to be gained when all parties believe that diverse perspectives are essential to learning, solving problems, and strategizing.

As one example, people who take natural resources for granted and live in a throw-away culture can learn much from those who are from resource-poor environments. The resource-poor often possess the skills, knowledge, and imagination to build something out of "nothing." Consider the ingenuity required to build this lawnmower out of parts that look to be "junk."

Chris Caldeira

In the United States, progress on creating educational environments that are both diverse and welcoming of diversity has been slow. With regard to race, it has more than a half century since the famous 1954 Supreme Court desegregation decision (*Brown v. Board of Education*), but still U.S. schools are largely segregated. According to the most recent data, 40 percent of blacks attend schools that are 90 percent black, and 40 percent of Hispanics or Latinos attend schools that are 90 percent minority. Of course, the level of segregation varies by place, with 92 percent of blacks in Washington, D.C., attending schools that are 90 percent minority. More than 50 percent of blacks in Illinois, Michigan, and New York attend schools that are 90 percent black (Fry 2007; Orfield and Siegel-Hawley 2012).

Brown v. Board of Education declared unconstitutional state laws that established separate public schools for black and white students on the grounds that such an arrangement denied black children equal access to educational opportunities. This 1955 photograph taken shortly after the decision shows a newly integrated class at Barnard School in Washington, D.C.

Library of Congress, Prints and Photographs Division [LC-U9-183D-20]

Today the average minority student is likely to attend a school that serves students from economically and educationally disadvantaged households located in a low-income neighborhood. That school is likely to serve a student population where relatively few will go on to complete high school, achieve high grade point averages, enroll in advanced placement classes, or feel optimistic about their future. Black, Hispanic, and Native American students are significantly more likely than white and Asian students to find themselves in school environments characterized by high levels of poverty and low levels of academic achievement (see Table 12.3a). The latest data show that 23.2 percent of all elementary students attend high-poverty schools but that percentage varies by race.

Figure 12.3a: Percentage of Public Elementary School Students in High-Poverty Schools, by Race/Ethnicity, 2011–2012

Public schools with more than 75 percent of the students eligible for the free or reduced-price lunch program are considered high poverty. What percentage of students classified as white attend high-poverty schools? What percentage of Hispanics attend high-poverty schools?

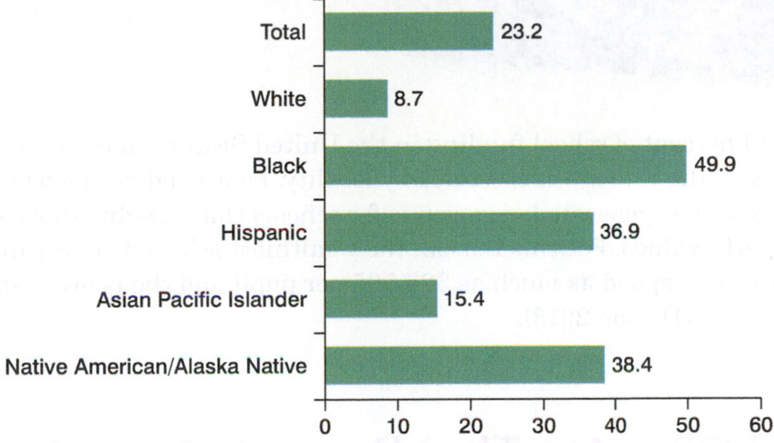

Source of data: U.S. Department of Education (2013b)

In terms of offsetting the effects of inequality, the United States stands out among the 34 highest-income countries as one of three countries (with Israel and Turkey) that spend more money serving advantaged students than serving low-income students. "The bottom line is that the vast majority of the higher-income countries either invest equally into every student or disproportionately more into disadvantaged students." The inequity is due in part to heavy reliance on local property taxes to support schools (Porter 2013).

Rachel Ellison
Lisa Southwick

About 43 percent of school funding in the United States comes from property taxes, the rate of which varies by locality. Poor neighborhoods with low property values generate less revenue for schools than neighborhoods with higher property values. For this reason, the wealthiest school districts in the United States can spend as much as $25,505 per pupil and the poorest spend less than $5,300 (Dixson 2013).

School Climates That Promote Learning

A second characteristic of schools that help predict success in a knowledge economy is school climate. School climate is a term used to capture the kind of environment a school presents. The climate can feel and be inviting, safe, and supportive or it can feel and be exclusionary, unsafe, and unwelcoming.

U.S. Army Courtesy graphic

About one-third of U.S. children report fighting, bullying, and intimidation as a big problem at their schools. About 20 percent indicate that they do not feel safe in school. Forty-six percent of students admit that they "bullied, teased, or taunted someone" and about 48 percent claim that they have been "bullied, teased, or taunted in a way that seriously upset them" (Josephson Institute 2013).

Social climate has at least three interrelated dimensions: physical, social, and academic. The physical dimension includes the building and the classrooms, the size of the school, whether surveillance is obvious (police, metal detectors), and learning resources. The social dimension includes the quality of the interpersonal relationships among students, teachers, staff, and parents; the degree of division based on class, race, and gender; and the reputation of the school in the community and beyond. The academic dimension includes the degree to which students are expected to be active (vs. passive) learners and the quality and type of instruction.

Schools in which the climate feels (and is) uninviting and where keeping discipline takes precedence over learning are likely schools that structure learning in closed ways. That is, assignments and test questions have one correct answer, must be solved a certain way, and are not realistic. Schools with positive and welcoming climates are likely schools that structure learning in open ways. That is, assignments and test questions have no one right answer, elicit multiple interpretations, draw on different perspectives, are not easily solved, take sustained effort (rather than minutes or hours), and involve making choices. Such assignments prepare students for a knowledge economy.

What Do Sociologists See?

When sociologists walk into a school, they assess it on a number of key dimensions such as funding, school climate in all its dimensions, how welcoming it is of diversity, and the ways in which the school reproduces or transcends inequalities.

Benjamin Abel

Critical Thinking

Review the skills needed for success in a knowledge economy. Describe an educational experience you have had that emphasizes one or more of these skills.

Key Term

knowledge economy

Module 12.4 Rewards and Costs of Higher Education

Objective

You will learn about the rewards and costs associated with higher education.

What percentage of students in your high school class graduated in four years? Of those who graduated, what percentage went directly to college? How did they pay for it?

Chris Caldeira

Most Americans are taught to equate more education with increased job opportunities and higher salaries. Chart 12.4a shows that this belief has a basis in reality.

Chart 12.4a: Earnings and Unemployment Rates By Educational Attainment
Chart 12.4a shows that income rises as the level of education increases. Conversely, the unemployment rate decreases as level of education increases. On the other hand, earnings are also affected by factors such as gender. We can see from Chart 12.4b that men earn more than women as a group. We also see that women with four-year college degrees earn more than their female counterparts who have less education. The same is true for males.

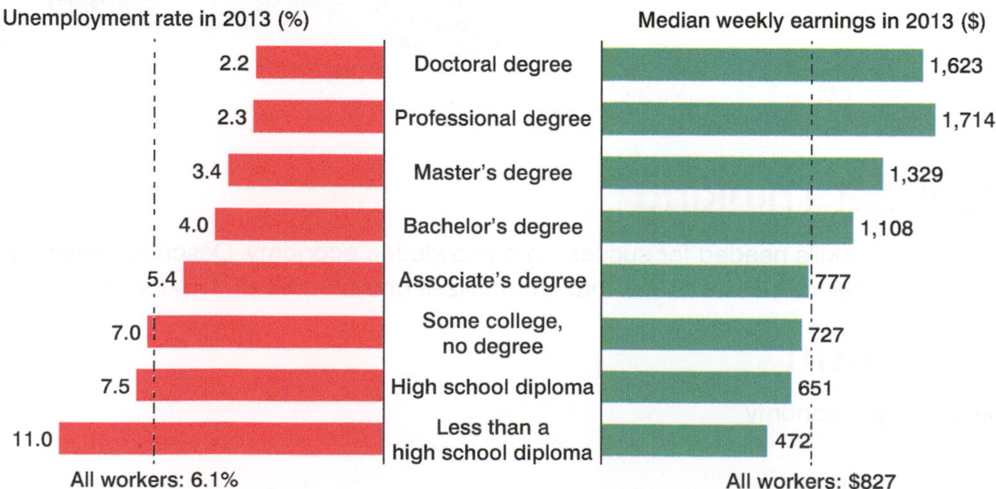

Note: Data are for persons age 25 and over. Earnings are for full-time wage and salary workers.

Source: U.S. Department of Labor, Bureau of Labor Statistics (2014a)

Chart 12.4b: Median Weekly Earning of Full-Time and Salaried Workers Age 25 Years and Over by Sex and Educational Attainment, 2014

At which education level do men most outearn women? Do income differences between men and women increase or decrease as level of education increases?

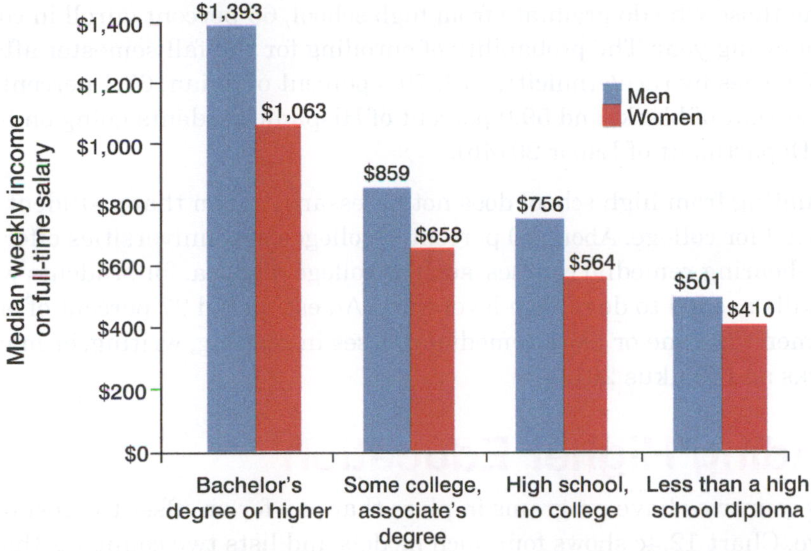

Source of data: U.S. Department of Labor (2014c)

While college graduates possess an earnings advantage over those with less education, it is important to realize that there are some without a college degree who have higher incomes than the college educated. For example, the median weekly income of males with no high school diploma is $501, but wages for this group range from $396 to $1,491 per week. Compare that salary range with the range for males with at least a college degree: that range is $610 to $2,599 (U.S. Department of Labor 2014a).

Who Goes to College?

Since income does rise as level of education increases, it is important to examine who completes high school and goes on to college. In the United States, the on-time, four-year high school graduation rate is 74.5 percent. This means that 74.5 percent of ninth-graders graduate on time from high school four years later. Of course, many who do not complete high school in four years do go on to earn a high school diploma a year or more later, and others eventually earn a general equivalency diploma (GED). The on-time graduation rate varies by school; it can be as low as 5 percent and as high as 100 percent (*The Oregonian* 2011). It also varies for students classified as Asian (93.5 percent), white (83 percent), Native American (69.1 percent), black (66.1 percent), and Hispanic (71.4 percent) (Layton 2012; National Center for Education Statistics 2013a).

Among those who do graduate from high school, 65 percent enroll in college the following year. The probability of enrolling for the fall semester after high school varies by race/ethnicity, with 79.0 percent of Asian, 67.1 percent of white, 59.0 percent of black, and 59.0 percent of Hispanic students going on to college (U.S. Department of Labor 2014b).

Graduating from high school does not necessarily mean that a student is prepared for college. About 70 percent of colleges and universities offer non-credit-bearing remedial courses, such as college algebra, for students who lack the skills needed to do college-level work. An estimated 27 percent of entering freshmen take one or more remedial courses in reading, writing, or mathematics (Sparks and Malkus 2013).

Funding Higher Education

Many countries have programs in place that pay for or offset the cost of going to college. Chart 12.4c shows four such models and lists two countries that exemplify each.

Chart 12.4c: Funding Models for Public Colleges and Universities

The dollar figure in parentheses after each country represents the average annual cost of educating a college student who attends a public university. The chart shows that Model 3 applies to the United States in that there is generous financial support, but students also pay the highest out-of-pocket tuition/fees in the world—on average, $6,312 each academic year (see footnote). This means that on average $23,598 of the total cost per student is supported by government and private sources (e.g., employer tuition reimbursement, federal grants, state grants, and private scholarships).

	Generous Financial Assistance to Students	Less Generous Financial Assistance to Students
Students Pay No or Very Low Tuition/fees (avg. cost paid by student: $0*–$1,850 per year)	**MODEL 1** Iceland ($10,429) Sweden ($20,864)	**MODEL 2** France ($14,945) Ireland ($16,248)
Students Pay High Tuition/Fees (avg. cost paid by student: $3,000–$6,312 per year)**	**MODEL 3** United Kingdom ($15,314) United States ($29,910)*	**MODEL 4** Japan ($16,533) Korea ($10,109)

* Countries in green: students pay no fees or tuition; note that U.S. students on average pay the highest out-of-pocket tuition and fees ($6,312).
** Note that there are no countries where the average cost per student is between $1,850 and $2,999.
Source of Data: OECD (2011a)

We might speculate that students who live in countries where college tuition is free or offered at very low cost are part of a society where education is viewed as a basic right and where people believe a government's role is to ensure access to that right. Note that in three of these countries shaded in green, students pay no tuition or fees, and two of those three also have a generous system of support in place to help students offset costs of living while in college. Students in the third country (Ireland) pay low tuition/fees relative to total cost, but the assistance programs are less generous.

Keep in mind that college students take out loans, even when they live in countries where tuition is free or very low, to offset living expenses. How might the financial burden be reduced if students have access to reliable and efficient systems of public transportation? For the most part, U.S. students who live off campus rely on automobiles to get to school. U.S. students must also find ways to pay for health care, which is considerably more expensive in the United States than elsewhere.

Chris Caldeira

Student Debt after College

College loan programs exist in at least 70 countries. Researchers Hua Shen and Adrian Ziderman (2009) analyzed 44 such programs in 39 countries with two questions in mind: What portion of the loan are students required to repay? And what percentage of money loaned is actually paid back or recovered? The researchers found that many loan programs have "hidden grants" that forgive some of the amount owed depending on actions the borrower takes (working in public service) or on the income borrowers earn after graduation (a graduate's loan payment cannot exceed 10 percent of income). In addition, some loan programs require repayment within as few as five years and others in as many as 20 years. The researchers also found that overall, a high percentage of students around the world who borrow money for college default on those loans, but that the default rate varies by country. In response, some governments have instituted programs to decrease the rate of default, including publishing the names of defaulters, barring defaulters from other credit sources, and deducting money owed from social security payments, tax refunds, and disability payments. Other countries have no such measures in place. In theory, if borrowers pay back loans plus interest, loan programs can sustain themselves. Of the 44 loan programs Shen and Ziderman (2009) studied, the Czech Republic has the best loan repayment ratio, at 108 percent. This high return is possible because the Czech Republic charges at least 12 percent interest on loans, so even if some students default, the high rate offsets money lost from default. Nigeria has the worst loan repayment ratio, at 10.9 percent. Note that about two-thirds of college graduates report having taken out loans while in college. The average graduate who takes out a loan leaves college with $25,250 in debt (Ellis 2011). The U.S. loan repayment ratio is 73 to 78 percent, depending on the source of the loan (Shen and Ziderman 2009).

The Credential Society

The proportion of the population age 25 and over with at least a bachelor's degree has risen from 4.6 percent (about 1 in 20) in 1940 to 30 percent (about 1 in 3) in 2014. What accounts for this increase? Sociologist Randall Collins is associated with the classic statement outlining the reason for the ever-increasing percentages of people with college degrees. Collins begins his argument by pointing out that there have been few, if any, systematic studies examining the role education plays in occupational success. In addition, it is the rare study that examines "what is actually learned in school and how long it is retained" (1961, 39). Regardless, over the decades there has been a steady increase in employer demands for job applicants with a college degree. This demand has created what Collins calls a **credential society**, a situation in which employers use educational credentials as screening devices for sorting through a pool of largely anonymous applicants. Specifically, many employers use the college degree as an indicator that an applicant is responsible, can set goals and follow through, and possesses basic skills. In addition, employers often require a college degree for promotion and advancement, even for employees who have an excellent work record and have demonstrated a high level of competence. In identifying the historical factors behind the emergence of a credential society in the United States, Collins eliminated the ever-increasing need for an educated workforce as a factor, because most jobs created as a result of industrialization and beyond do not require advanced technical knowledge beyond an eighth-grade education.

Table 12.4a: Total Number and Percentage of Labor Force Employed by Level of Education Required for Job in 2012 and Projected Increases by 2022

What percentage of the 145.4 million jobs in 2012 required a bachelor's degree or higher? What percentage required a high school diploma or less? If you answered 23.4 percent and 66.3 percent, respectively, then you are correct. Of the 15.6 million new jobs projected in next 10 years, what percentage require a high school diploma or less? Require a bachelor's degree or higher?

Level of Education	Total Number 2012	%	Number of Jobs Added in Next 10 Years (Projected)	% of All New Jobs	Median Wages
All levels	145,355.8	100.0	15,628.0	100.0	$34,750
Doctoral or professional degree	4,002.4	2.8	638.4	4.1	$96,420
Master's degree	2,432.2	1.7	448.5	2.7	$63,400
Bachelor's degree	26,033.0	17.9	3,143.6	20.1	$67,140
Associate's degree	5,954.9	4.1	1,046.0	7.0	$57,590
Postsecondary non-degree award	8,554.2	5.9	1,337.1	8.6	$34,760
Some college, no degree	1,987.2	1.4	225.0	1.4	$28,730
High school diploma or equivalent	58,264.4	40.1	4,630.8	29.6	$35,170
Less than high school	38,127.6	26.2	4,158.4	26.6	$20,110

Source: U.S. Department of Labor, Bureau of Labor Statistics (2013)

Collins traced the emergence of a credential society to two factors:

1. A long-standing belief going back to the colonial era that associated high economic status with educational attainment. Collins argues that simply "the existence of a relatively small group of experts in high status positions fostered the demand that opportunities to acquire such positions be available on a large scale" (37).
2. The U.S. federal government has always left decisions about what to teach to state and local communities. In addition, it maintains a separation between

church and state (no national church). These two characteristics set the stage for various religious and other special-interest groups to establish their own schools. Collins argues that it was this rivalry among interest groups to educate and socialize the young, and the large number of schools and colleges this spawned, that helps explain, in part, the emergence of the credential society. More specifically, the religious rivalries helped produce the Catholic and Lutheran school systems, and even the public school system. Collins maintains that white Anglo-Saxon Protestant (WASP) elites founded the public school system in response to large-scale Catholic immigration from Europe.

Library of Congress Prints & Photographs Division [LC-USZ62-126152]

This anti-Catholic political cartoon from the 1880s, titled "Wolf at the door, gaunt and hungry—Don't let him in," shows children trying to keep the wolf, symbolizing the threat of Catholic education, from bursting through the door.

Of course, once the system of public mass education was established, the elite founded private schools for their children as a "means of maintaining cohesion of the elite culture itself" (Collins 1961, 38). These rivalries set the stage for religious and elite groups to establish universities as well. As a result, opportunities for education at all levels expanded faster in America than anywhere else in the world, so that today there are an estimated 4,599 two- and four-year degree granting institutions of higher learning (National Center for Education Statistics 2013b).

Collins argues that this large number of degree-granting colleges in the United States helps explain the emergence of the credential society. Employers' demand for credentials, in turn, has fueled and perpetuated a widely held belief that a person must go to college to be successful. The ever-increasing supply of college-educated persons has made a college degree a requirement for many jobs. This may explain why, when surveyed, almost all parents in the United States (94 percent) say they expect their child to attend college (Pew Research Center 2011b).

What Do Sociologists See?

A person with a college education can expect to be employed in occupations where the median weekly income is $1,393 for men and $1,063 for women. New graduates are now part of the 30 percent of those age 22–34 who have a college degree. About 66 percent of college graduates leave with an average debt of $25,250. About 22–27 percent are expected to default.

U.S. Air Force Master Sgt. Jerry Morrison

Critical Thinking

How are you paying for college? What do you project your debt from college loans to total?

Key Term

credential society

Module

12.5 Religion

Objective

You will learn how sociologists approach the study of religion.

Can you identify the religious figure pictured? Do you think it is important to have a framework by which to think about your own and another's religious experiences?

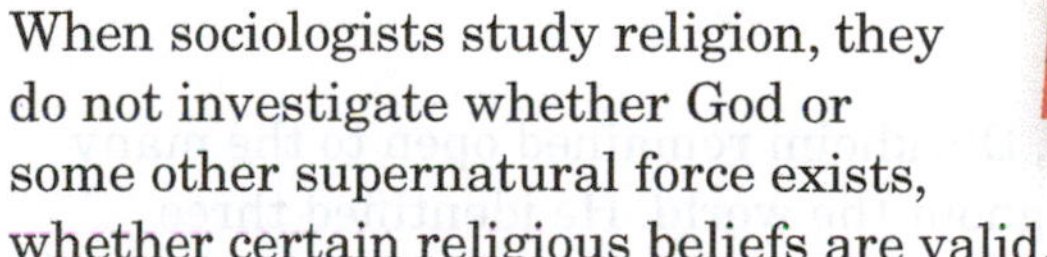

Tony Rotundo

When sociologists study religion, they do not investigate whether God or some other supernatural force exists, whether certain religious beliefs are valid, or whether one religion is better than another. Sociologists investigate the social aspects of religion, which include the ways in which people have used religion to justify the most constructive and destructive behaviors, the ways religion shapes people's behavior and their understanding of the world, and the ways religion is intertwined with social, economic, and political issues.

Defining Religion

Max Weber (1922b) believed that no one definition could capture the varieties and essence of religion. For him, **religion** encompasses human responses to the ultimate and inescapable problems of existence—birth, death, illness, aging, injustice, tragedy, and suffering (Abercrombie et al. 1988). To Weber, the hundreds of thousands of religions, past and present, represent a rich and seemingly endless variety of responses to these problems.

Like Weber, Émile Durkheim (1915) believed that religion is difficult to define. He cautioned that when studying religions, sociologists must assume that "there are no religions which are false" (3). Durkheim maintained that all religions are true in their own fashion; all address the problems of human existence, albeit in different ways. Consequently, he said, those who study religion must first rid themselves of all preconceived notions about what religion should be. We cannot study religion using standards that reflect our own personal experiences and preferences. Preconceived notions and uninformed opinions about the meaning of religious symbols and practices can close people off to a wide range of religious beliefs and experiences.

U.S. Army photo by Spc. Karah Cohen/Released

NKU Sociology, Missy Gish

It is a challenge to see through one's own preconceptions about what is "right" in everyday life. A Western woman may look on the traditional Muslim female head covering, the *hijab*, as a sign of oppression. On the other hand, a Muslim woman may look on Western female makeup and presentation of self as a sign of oppression because many women, even young girls, seem to feel a need to present themselves as sex objects.

In formulating his ideas about religion, Durkheim remained open to the many varieties of religious experiences throughout the world. He identified three essential features that he believed were common to all religions, past and present: (1) beliefs about the sacred and the profane, (2) rituals, and (3) a community of worshipers. Thus, Durkheim defined religion as a system of shared rituals and beliefs about the sacred and the profane that bind together a community of worshipers.

BELIEFS ABOUT WHAT IS SACRED AND PROFANE. The **sacred** includes everything that is regarded as extraordinary and that inspires in believers deep and absorbing sentiments of awe, respect, mystery, and reverence. Sacredness stems from the symbolic power people confer on objects (such as chalices), living creatures (such as cows), elements of nature (such as mountains), places (such as mosques), states of consciousness (oneness with nature), holy days, ceremonies (such as baptism), and other activities (pilgrimages).

The **profane** encompasses everything that is not considered sacred, including things opposed to the sacred (such as the unholy, the irreverent, and the blasphemous) and things that stand apart from the sacred (such as the ordinary, the commonplace, the unconsecrated, and the bodily) (Ebersole 1967). Believers often view contact between the sacred and the profane as being dangerous and sacrilegious, as threatening the very existence of the sacred, and as endangering the fate of the person who makes or allows such contact. Consequently, people take action to safeguard sacred things. For example, some people refrain from speaking the name of God when they feel frustrated; others believe a man must remove his hat during worship.

NKU Anthropology, Sharyn Jones

NKU Anthropology, Sharyn Jones

Some religions define certain animals as sacred. Perhaps the most well known is the Hindu belief that cows are sacred. Mahatma Gandhi once wrote that "[i]f someone were to ask me what the most important outward manifestation of Hinduism was, I would suggest that it was the idea of cow protection" (*Nature* 2007). In some Asian countries such as Thailand, "white" elephants are considered sacred and are revered as symbols of power and good fortune.

RITUALS. In the religious sense, **rituals** are rules that govern how people behave in the presence of the sacred. These rules may take the form of instructions detailing the appropriate day(s) and occasions for worship, acceptable dress, and wording of chants, songs, and prayers. Participants follow rituals to achieve a specific goal, whether it be to purify the body or soul (as through immersion in water, fasting, or seclusion), to commemorate an important person or event (as by celebrating Passover), or to transform profane items into sacred items (e.g., changing water to holy water and bones to sacred relics) (Smart 1976). Although rituals are often performed in sacred places, some rituals are codes of conduct aimed at governing the performance of everyday activities such as sleeping, walking, eating, defecating, washing, and relating to the other sex.

Lance Cpl. Lukas J. Blom

Buddhist monks of the Shingon sect located in Daishoin, Japan, participate in a ritual that involves walking over burning coals as a way of honoring the "Three Awesome Forest Deities" believed to inhabit Mount Misen. These coals were lit by a fire that is believed to have been burning continuously for the past 1,200 years.

According to Durkheim, the manner in which a ritual is performed is relatively insignificant. What is important is that the ritual is shared by a community of worshipers and that in performing it, people feel they are part of something larger.

COMMUNITY OF WORSHIPERS. Durkheim (1915) uses the word **church** to designate those who hold the same beliefs about what is sacred and profane, who share rituals, and who gather in body or spirit at agreed-on times to reaffirm their commitment to those beliefs and rituals. The act of gathering and sharing creates a moral community and cultivates a common identity.

Durkheim (1915) argued that some form of religion appears to have existed for as long as humans have lived. In view of this fact, one might argue that religion must serve some vital function for society. In this regard, one can argue that all religions strive to raise individuals above themselves—to help them achieve a better life than they would lead if left to their own impulses. Religion offers people a code of conduct that can evoke guilt and remorse when violated. Such feelings, in turn, motivate people to make amends.

Religion functions as a stabilizing force in times of severe social disturbance and abrupt change. During such times, many norms that guide behavior may break down. In the absence of such forces, people are more likely to turn to religion in search of a force that will bind them to a group.

In addition, Durkheim observed that whenever any group of people has a strong conviction, that conviction almost always takes on a religious character. Religious gatherings become vehicles for affirming convictions and mobilizing the group to uphold them, especially when those convictions are threatened. People on both sides of controversial issues such as same-sex marriage evoke "God" or a higher power as supportive of their convictions.

The fact that the variety of religions is endless and that religion functions to meet societal needs led Durkheim to reach a controversial but thought-provoking conclusion: the "something out there" that people worship is actually society. In reaching this conclusion, Durkheim asked: How else might we explain the endless variety of religious response? The answer is that people create everything encompassed by religion—images of gods, rites, sacred objects. That is, people play a fundamental role in determining what is sacred and how to act in the presence of the sacred. Consequently, at some level, people worship what they (or their ancestors) have created. This point led Durkheim to conclude that the real object of worship is society itself—a conclusion that many critics cannot accept (Nottingham 1971).

U.S. Army Courtesy photo

Given that Jesus was born in Bethlehem (according to the Bible), a town in the Middle East, is this image an accurate representation of Jesus's physical appearance? Archaeological evidence suggests that the average man at the time of Jesus was 5 feet, 3 inches tall and weighed approximately 110 pounds (Gibson 2004). In addition, this Jesus does not possess the skin shade and hair texture we associate with people living in the Middle East. This presentation of Jesus Christ lends support to Durkheim's idea that the real object of worship is society itself because those who have the power to say what Jesus looks like make him into their image.

The Opiate of the People

Conflict theorists focus on ways in which people use religion to repress, constrain, and exploit others and how religion turns people's attention away from injustice and inequality. This perspective draws inspiration from the work of Karl Marx (1843), who believed that religion was the most humane feature of an inhumane world and that it arose in response to the tragedies and injustices of human experience.

Molly Hayden, U.S. Army Garrison—Hawaii Public Affairs

Marx described religion as the "sigh of the oppressed creature, the heart of a heartless world, and the soul of soulless conditions. It is the opiate of the people." According to Marx, people need the comfort of religion to make the world bearable and to justify their existence. In this sense, he said, religion is analogous to a sedative.

Even though Marx acknowledged the comforting role of religion, he focused on its repressive, constraining, and exploitative qualities. In particular, he conceptualized religion as an ideology that justifies the status quo. That is, religion is used to rationalize existing inequities or downplay their importance. This aspect of religion is especially relevant with regard to the politically and economically disadvantaged. For them, Marx said, religion serves as a source of **false consciousness**. That is, religious teachings encourage the oppressed to accept existing economic, political, and social arrangements that limit their opportunities in this life because they are promised compensation for their suffering in the next world. This thinking inhibits protest and revolutionary change. Marx went so far as to claim that religion would be unnecessary in a truly classless society—one without material inequality—because by definition exploitation and injustice would not exist.

Critics of the conflict perspective argue that it underestimates the power of religion to inspire people to confront inequalities and injustices. Sometimes religion is a catalyst that inspires change and justice.

Historically, African American churches have reached out to millions who have felt excluded from the U.S. political and economic system. For example, African American churches contributed greatly to the overall successes of the civil rights movement. Indeed, some observers argue that the movement would have been impossible if the churches had not been involved (Lincoln and Mamiya 1990).

Library of Congress Prints and Photographs Division Washington, D.C. 20540 USA [LC-U9-30656B-10]

The Protestant Work Ethic

Max Weber wanted to understand the role of religious beliefs in the origins and development of modern capitalism—an economic system that involves the careful calculation of costs of production relative to profits, the borrowing and lending of money, the accumulation of capital, and the drawing of labor from around the world to make products (Robertson 1987). In his book *The Protestant Ethic and the Spirit of Capitalism*, Weber (1958) asked why modern capitalism emerged and flourished in Europe rather than in China or India. He also asked why business leaders and capitalists in Europe and the United States were overwhelmingly Protestant.

To answer these questions, Weber focused on how different religious traditions supported different economic orientations and motivations. Based on his comparisons, Weber concluded that a branch of Protestant tradition—Calvinism—supplied a "spirit" or an ethic that supported capitalism. In particular, Calvinism emphasized **this-worldly asceticism**—a belief that people are instruments of divine will and that God determines and directs their activities. Calvinists glorified God when they accepted a task assigned to them, carried it out in an exemplary and disciplined fashion, and did not indulge in the fruits of their labor (that is, they did not use money to eat, drink, or otherwise relax in excess).

In addition, Calvinists conceptualized God as all-powerful and all-knowing; they also emphasized **predestination**—the belief that God has foreordained all things, including the salvation or damnation of individual souls. According to this doctrine, relatively few people were destined to attain salvation, and people could do nothing to change their fate.

Weber maintained that beliefs in this-worldly asceticism and predestination created a crisis of meaning among adherents. This crisis led them to look for concrete signs that they were among God's chosen people, destined for salvation. Consequently, accumulated wealth became an important indicator of whether one was among the chosen. At the same time, this-worldly asceticism "acted powerfully against the spontaneous enjoyment of possessions; it restricted consumption, especially of luxuries" (Weber 1958, 171). Frugal behavior supported investment and the accumulation of wealth—important actions for the success of capitalism.

This print features men who are considered empire builders in U.S. history, including Andrew Carnegie, Cornelius Vanderbilt, and John D. Rockefeller. The caption (not shown here) reads: "Those Christian men to whom God in his infinite wisdom has given control of the property interests of the country." These words speak to a belief that people are instruments of divine will.

Library of Congress Prints & Photographs Division [LC-DIG-ppmsca-05929]

Do not misread the role that Weber attributed to the Protestant ethic in the rise of a capitalist economy. According to Weber, that ethic was a significant ideological force; it was not the sole cause of capitalism but "one of the causes of certain aspects of capitalism" (Aron 1969, 204). Unfortunately, many people who encounter Weber's ideas draw a conclusion that Weber himself never reached: the reason that some groups and societies are disadvantaged is simply that they lack this Protestant work ethic. Finally, note that Weber was writing about the origins of industrial capitalism, not about the form of capitalism that exists today, which places a heavy emphasis on consumption and self-indulgence. He maintained that once established, capitalism would generate its own norms and become a self-sustaining force. In such circumstances, religion becomes an increasingly insignificant factor in maintaining the capitalist system (Aron 1969).

What Do Sociologists See?

Sociologists see Buddha's feet and notice that they are worn on the top from a ritual in which people who come into the temple and pass the statue rub the feet or kiss them.

NKU Anthropology, Sharyn Jones

Critical Thinking

Whose ideas about religion—Max Weber, Émile Durkheim, or Karl Marx—do you find most compelling? Why?

Key Terms

church
false consciousness
predestination
profane
religion
rituals
sacred
this-worldly asceticism

Module 12.6 Civil Religion and Fundamentalism

Objective

You will learn that civil religion and fundamentalism share some essential characteristics.

Do you recognize the figure looming over actor Laurence Fishburne, who is speaking to an audience on Memorial Day, a national holiday in the United States? Is there a religious quality about this scene?

DoD photo by Mass Comm. Spc. 1st Class Chad J. McNeeley, U.S. Army

The figure behind Fishburne is Abraham Lincoln, the 16th president of the United States, who has been memorialized in countless ways including through a national monument, postage stamps, coins, paper money, and the cities, towns, and streets named after him. In addition, Lincoln is memorialized through often-repeated phrases attributed to him ("government of the people, by the people, for the people, shall not perish from the earth" and "the proposition that all men are created equal").

Sociologist Émile Durkheim defined religion as a system of shared rituals and beliefs about what is sacred and what is profane that bind together a community of worshipers. This combination of characteristics applies not just to religion but to religious-like events such as national holiday celebrations. Thus this combination of characteristics also applies to **civil religion**, an institutionalized set of beliefs about a nation's past, present, and future and a corresponding set of rituals. These beliefs and rituals can take on a sacred quality and elicit feelings of patriotism.

Civil Religion

A nation's values, such as individual freedom and equal opportunity, and its rituals (parades, fireworks, singing the national anthem, 21-gun salutes) often assume a sacred quality. Even in the face of internal divisions centered on race, religion, or gender, national rituals and shared values can inspire awe, respect, and reverence for the country and what it stands for. These sentiments are most evident during times of crisis and war, on national holidays (Veterans Day), and in the presence of national monuments or symbols (the Vietnam Memorial, the flag).

In most societies, leaders defined as great are memorialized and the things they stood for are treated as sacred to the nation. Koreans celebrate King Sejong (top), who is credited with inventing an alphabet known as Hangul, in 1440s, with clear rules governing pronunciation of characters. Sejong created this alphabet with the goal of encouraging literacy among all people, not just the elite. Hangul Day is celebrated on October 9. Che Guevara (bottom) is a revolutionary leader celebrated in Cuba for his role as physician to rebel forces and key strategist in the 26th of July 1959 Movement, after which Cuba was declared a socialist country. Guevara was also active in other "third world" revolutionary movements. He was killed at age 39 while fighting the Bolivian army.

David McNally (USAG-Yongsan)

Chris Caldeira

Sociologist Roberta Cole (2002) argues that America's civil religion found its voice in a 19th-century political doctrine known as Manifest Destiny, a long-standing ideology that the United States, by virtue of its moral superiority, was destined to expand across the North American continent to the Pacific Ocean and beyond (Chance 2002). Manifest Destiny included the beliefs that the United States had a divine mission to serve as a democratic model to the rest of the world, that the country was a redeemer exerting its good influence upon other nations, and that it represented hope to the rest of the world (Cole 2002).

Especially in times of war, presidents offer a historical and mythological framework that gives moral justification for involvement and offers the public a vision and a national identity. We consider statements U.S. presidents have made to illustrate how in times of war the nation's core values can assume a sacred quality and to illustrate the language presidents use to project moral certainty.

> We are Americans, part of something larger than ourselves. For two centuries, we've done the hard work of freedom. And tonight, we lead the world in facing down a threat to decency and humanity. . . . Among the nations of the world, only the United States of America has both the moral standing and the means to back it up. . . . This is the burden of leadership and the strength that has made America the beacon of freedom in a searching world. (George H. W. Bush [1991] on sending 540,000 troops to the Persian Gulf in 1990)

> All of you . . . have taken up the highest calling of history . . . and wherever you go, you carry a message of hope—a message that is ancient and ever new. In the words of the prophet Isaiah, "To the captives 'come out,'—and to those in darkness, 'be free.'" (George W. Bush [2003], on sending troops to Iraq in 2003)
>
> . . . [T]onight, we are once again reminded that America can do whatever we set our mind to. That is the story of our history, whether it's the pursuit of prosperity for our people, or the struggle for equality for all our citizens; our commitment to stand up for our values abroad, and our sacrifices to make the world a safer place.
>
> Let us remember that we can do these things not just because of wealth or power, but because of who we are: one nation, under God, indivisible, with liberty and justice for all. (Barack Obama [2011], upon announcing the death of Osama bin Laden in 2011)

Critics of war liken the moral certainty characteristic of civil religion to "a kind of fundamentalism" and a "dangerous messianic brand of religion, one where self-doubt is minimal" (Hedges 2002).

Fundamentalism

Anthropologist Lionel Caplan (1987) defines **fundamentalism** as a belief in the timelessness of sacred writings and a belief that the words apply to every setting. In its popular usage, the term *fundamentalism* is applied to a wide array of religious groups around the world, including the religious right in the United States, Orthodox Jews in Israel, and various Islamic groups in the Middle East. Religious groups labeled as fundamentalist are usually portrayed as "fossilized relics . . . living perpetually in a bygone age" (Caplan 1987, 5). Americans frequently use the term *fundamentalism* to explain events involving people in the Middle East, especially political turmoil that threatens the interests of the United States. Such oversimplification misrepresents fundamentalism.

Perhaps the most important characteristic of fundamentalists is their belief that a relationship with God, Allah, or some other supernatural force provides answers to personal and societal problems. In addition, fundamentalists often wish to "bring the wider culture back to its religious roots" (Lechner 1989, 51). In this regard, fundamentalists emphasize the authority, infallibility, and timeless truth of sacred writings as a definitive blueprint for life. This characteristic does not mean that a definitive interpretation of sacred writings actually exists. Indeed, any sacred text has as many interpretations as there are groups that claim it as their blueprint.

Second, fundamentalists usually conceive of history as a "process of decline from an original ideal state," which includes the "betrayal of fundamental principles" (18). They see human history as a cosmic struggle between good and evil: the good embodies the principles outlined in sacred scriptures, and the evil results from countless digressions from those principles. To fundamentalists, truth is not relative, varying across time and place. Instead, truth is unchanging and knowable through the sacred texts.

Third, fundamentalists do not distinguish between what is sacred and what is profane in their day-to-day lives. Religious principles govern all areas of life, including family, business, and leisure. Religiously inspired behavior does not take place only in a church, a mosque, or a temple. For example, religious principles guide the business of producing kosher meats, which are prepared according to strict Jewish dietary laws that are considerably stricter than USDA standards. Kosher laws govern how animals are fed, killed, and processed. Kosher laws prohibit, for example, slaughtering cows with broken bones or that are known to be sick.

ACU photo, Ingrid Barrentine, Childhood photo Courtesy of Spc. Simranpreet Singh Lamba

For fundamentalists, religious principles cannot be dismissed out of convenience or because they clash with another set of principles. When U.S. Army Specialist Simranpreet Singh Lamba was a child growing up in India (left), he dreamed of joining the military. When he came to the United States in 2006, he thought he could never join the U.S. military because, as a devout Sikh, he could not cut or shave his hair and he wore a turban. His religious beliefs conflicted with military dress codes. But the U.S. military made an exception for Lamba and two other Sikh soldiers. Sikhs value "the principles of justice, equality and truth" and the religion "emphasizes service to others, particularly in the armed forces" (Petrich 2011).

Fourth, fundamentalist religious groups emerge for a reason, usually in reaction to a threat or crisis, whether real or imagined. Consequently, any discussion of a particular fundamentalist group must include some reference to an adversary—one that may be capitalism, another religious group, feminists, or secularists.

Finally, fundamentalists believe that the trends toward gender equality are symptomatic of a declining moral order and need to be reversed. In fundamentalist religions, women's rights often become subordinated to ideals that the group considers more important to the well-being of the society, such as the traditional family or the "right to life." Such a priority of ideals is regarded as the correct order of things.

ISLAMIC FUNDAMENTALISM. We cannot apply the term *fundamentalism* to the entire Muslim world, especially when we consider that Muslims make up the majority of the population in at least 45 countries (U.S. Central Intelligence Agency 2012). Religious studies professor John L. Esposito (1986) believes that a more appropriate term is *Islamic revitalism* or *Islamic activism*. While the form of Islamic revitalism may vary from one country to another, it seems to be characterized by a belief that existing political, economic, and social systems have failed; a disenchantment with, and even rejection of, the West; soul-searching; a quest for greater authenticity; and a conviction that Islam offers a viable alternative to secular nationalism, socialism, and capitalism.

Esposito asks: Why has religion become such a visible force in Middle East politics? He believes that Islamic revitalism represents a "response to the failures and crises of authority and legitimacy that have plagued most modern Muslim states" (1986, 53). Those crises can be traced to France and Great Britain's division of the Middle East into nation-states after World War I.

The citizens of these Western-created countries viewed many of the leaders who took control as autocratic, corrupt, and "propped up by Western governments and multinational corporations" (1986, 54). Questions of social justice also arose as oil wealth and modernization policies opened up a vast chasm between the oil-rich countries, such as Kuwait and Saudi Arabia, and the poor, densely populated countries, such as Egypt, Pakistan, and Bangladesh. Western capitalism, which was seen as one of the primary forces behind these trends, seemed blind to social justice, instead promoting unbridled consumption and widespread poverty.

For many people, Islam offers an alternative vision for society. According to Esposito (1986), five beliefs guide Islamic activists who follow many political persuasions, ranging from conservative to militant:

1. Islam is a comprehensive way of life relevant to politics, law, and society.
2. Muslim societies fail when they depart from Islamic ways and follow the secular and materialistic ways of the West.
3. An Islamic social and political revolution is necessary for renewal.
4. Islamic law must replace laws inspired or imposed by the West.
5. Science and technology must be used in ways that reflect Islamic values, to guard against the infiltration of Western values.

Muslim groups differ dramatically in their beliefs about how quickly and by what methods these principles should be implemented. Most Muslims, however, are willing to work within existing political arrangements; they condemn violence as a method of bringing about political and social change. However, some do use violence to try to effect change, and in those cases the concept of *jihad* is often evoked. In thinking about the meaning of jihad, it is important to distinguish between religious and political *jihad*. Many Islamic scholars have pointed out that in the religious sense of the word, true *jihad* is the "constant struggle of Muslims to conquer their inner base instincts, to follow the path to God, and to do good in society" (Mitten 2002). But other scholars point out that *jihad* as used by the most radical organizations targets anyone—Muslim or non-Muslim—who stands in the way of their goals.

Secularization

In the most general sense, **secularization** is a process by which religious influences on thought and behavior are gradually removed or reduced. More specifically, it is a process by which some element of society, once part of a religious sphere, separates from its religious or spiritual connection or influences. It is difficult to generalize about the causes and consequences of secularization because they vary across contexts. Americans and Europeans tend to associate secularization with an increase in scientific understanding and in technological solutions to everyday problems of living. In effect, "science and technology give people control over life and death matters once left in the hands of God" (Esposito 1986, 54).

Muslims, in contrast, tend not to attribute secularization to science or to modernization; indeed, many devout Muslims are physical scientists. From a Muslim perspective, secularization is a Western-imposed phenomenon—specifically, a result of exposure to what many people in the Middle East consider the most negative of Western values as reflected in its consumer culture, its violent movies and television programs, and the way its popular culture portrays women and sexualizes young children.

What Do Sociologists See?

The three essential characteristics of religion apply to the funeral of Michael Jackson, which drew a global audience (community of worshippers) estimated by some media sources to be as high as 2.5 billion people. That service assumed a sacred quality. Jackson's signature dance moves and iconic crystal rhinestone-studded glove, as shown and worn by this impersonator, also assumed sacred status and ritualistic focus.

Emily Brainard

Critical Thinking

Can you think of a nonreligious ceremony or event in which participation takes on a religious quality?

Key Terms

civil religion

secularization

fundamentalism

Module

12.7 Applying Theory: Private Schools

Objective

You will learn how sociologists inspired by each of the four perspectives analyze private schools.

Are you one of the 10 percent of students in the United States who went to private elementary, middle, or high school? How do you think it distinguishes you from those who went to public schools?

Lisa Southwick

There are about 60 million students enrolled in pre-K through twelfth grades in the United States. About 10 percent are enrolled in private schools, 80 percent of which have a religious orientation. Private schools with religious orientation are predominantly Catholic (U.S. Department of Education 2013). There are 30,866 private schools in the United States. Almost 70 percent (21,086) of those schools are associated with a religion. Thirty percent are associated with no religious affiliation. The top six religions by number of schools and students across the United States are shown in Table 12.7a below.

Table 12.7a: Six Religions in the United States with Most Schools and Students Enrolled

Religion	Number of Schools	Number of Students	% of All Private School Students
Roman Catholic	6,873	1,928,388	42.9
Christian (unspecified)	4,518	607,130	13.5
Baptist	1,970	203,984	4.5
Amish	1,260	33,419	0.7
Lutheran (Missouri Synod)	994	118,444	2.6
Jewish	954	245,425	5.5

Source: U.S. Department of Education (2013)

NKU Sociology, Missy Gish

NKU Sociology, Missy Gish

At one time many religious schools were all-boy or all-girl schools. If they were co-ed, they segregated the sexes as illustrated by these separate entrances. Today just 1.3 percent (550) are considered all-girl and 2.3 percent (709) are considered all-boy. Ninety-six percent of private schools are co-ed and segregated spaces are a thing of the past (U.S. Department of Education 2013).

The size of private schools vary from fewer than 50 students to more than 750. The average cost of tuition for elementary private school education is $8,148 per year and for private high schools the average is $11,697 per year education (*Private School Review* 2014). The most expensive private elementary school in the United States is about $40,000 per school year; the most expensive high school is $45,000 per school year (this does not include room and board). It is important to point out that private schools serve the most disadvantaged to the most advantaged students across the United States.

Functionalists ask how private schools contribute to order and stability in society. On the one hand, some private school systems have made it their mission to serve distressed communities and disadvantaged students. As one example, the Catholic Inner City School Fund, which consists of eight Catholic elementary schools located in Cincinnati, serves an estimated 1,600 children from the most economically challenged neighborhoods (Archdiocese of Cincinnati 2014). At the other extreme are the most elite private schools with the mission to educate children whose parents can afford high tuition. From a functionalist point of view these schools prepare students to take on the most elite positions in society.

Rachel Ellison

Sociologists inspired by the conflict perspective focus especially on those private schools serving the most elite classes—those students attending the 30 or so elementary and secondary schools with tuition around $40,000 or more per year (Business Insider 2013). From a conflict perspective, the wealthiest parents have the resources to "buy" their children the skills and experiences that will ensure future success. More than 40 percent of students who completed high school at one of the top 30 prep schools in the United States are accepted into the most elite colleges. This is an astounding percentage when we consider that the most elite colleges accept 0.01 percent of all high school graduates. Some elite high schools boast state-of-the-art science and green energy labs and hands-on learning projects in partnership with Ivy League professors (*Forbes* 2010).

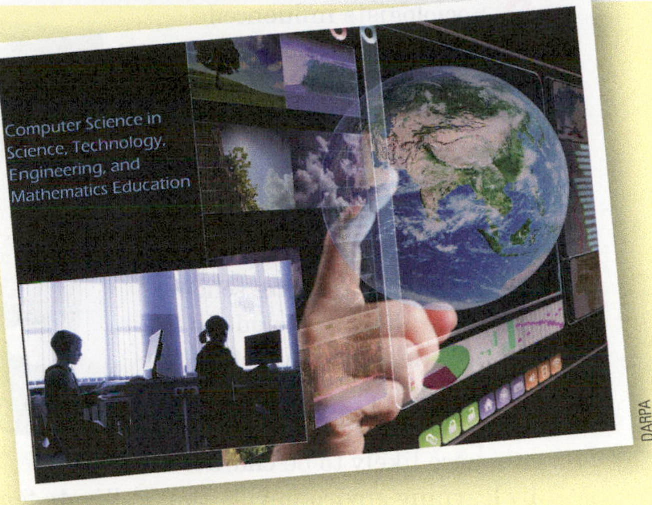

Computer Science in Science, Technology, Engineering, and Mathematics Education

DARPA

Chris Caldeira

You might be surprised to learn that Beyoncé's dance and singing career was launched as a student at a private school—St. Mary's Elementary School in Fredericksburg, Texas, where she enrolled in dance classes and won a school talent show. Symbolic interactionists focus on the experience of going to private schools and how that experience shapes the sense of self and personal biography, not just of Beyoncé. Symbolic interactionists might choose to compare the day-to-day interactions and experiences of one set of students attending a public school versus another set of students attending a private school. Likewise, symbolic interactionists are equally interested in the interaction dynamics by which elite prep schools prepare students for the best careers. These processes may include study abroad experiences in China, France, and South Africa, or incorporate hands-on learning experiences with faculty and graduate students at MIT (Stanger and Jacobs 2013).

Sociologists influenced by a feminist perspective recognize that public schools are very diverse and place focus on identifying the kinds of private schools that are structured to erase or perpetuate gender inequalities. Which kinds of private schools are more likely to include in their curriculum discussion of women as discoverers, inventors, and authors? Which private schools are more likely to be characterized by learning environments that cultivate females' interest in math, science, and technology? How do female students who go to private schools compare to men on measures of academic performance? Today only 2 percent of private schools are gender specific (U.S. Department of Education 2013). Do women who attend one of the 550 all-girl private schools have an advantage over those who attend coed schools? Which kinds of private schools have programs in place to promote male successes when women achieve at higher levels (such as graduation rates, GPA, college enrollment rates, test scores)?

Summary: Putting It All Together

Chapter 12

When sociologists study education they emphasize schooling, a deliberate, planned effort that takes place in a brick-and-mortar or virtual classroom to impart specific skills or information. Sociologists inspired by symbolic interactionists focus on how formal and hidden curriculum convey shared meanings about subject matter and about much more. The functionalist perspective places emphasis on the functions schools perform that contribute to order and stability in society. The conflict perspective draws our attention to inequalities and the way the schools perpetuate existing inequalities of the larger society through a process known as social reproduction.

Because the knowledge economy is assuming dominance, sociologists consider the kinds of learning environments that best prepare students to compete in this new economy. Learning environments that cultivate social networking skills, flexibility, and entrepreneurship are central to success. And so are learning environments that are inviting, supportive, and safe (vs. exclusionary, unsafe, and unwelcoming).

Most Americans are taught to equate more education with increased career opportunities and higher salaries. While it is true that income rises with level of education, this relationship is complicated by gender, race, and other factors such a debt incurred while in college. There are also questions about whether the need for and value of a college degree is inflated by the existence of a credential society.

Religions have three essential features: (1) beliefs about the sacred and the profane, (2) rituals, and (3) a community of worshipers. Because some form of religion appears to have existed for as long as humans have, functionalists maintain that religion must serve some vital social functions. In this regard, Durkheim observed that religious gatherings function as vehicles for affirming convictions and mobilizing the group to uphold them, especially when those convictions are threatened. Marx emphasized religion's repressive, constraining, and exploitative qualities. Weber focused on understanding how norms generated by different religious traditions influenced adherents' economic orientations and motivations. Specifically, Weber concluded that a branch of the Protestant tradition—Calvinism—supplied a "spirit" or an ethic that supported the motivations and orientations of capitalism.

Civil religion is an institutionalized set of beliefs about a nation's past, present, and future and a corresponding set of rituals that take on a sacred quality and elicit feelings of patriotism. The dynamics of civil religion are most notable during times of crisis and war and on national holidays. During these times civil religion can take on qualities we associate with fundamentalism.

David Kamm, NSRDEC photographer

Chapter 13

Social Change and the Pressing Issues of Our Time

When sociologists study social change, they examine the way human relationships and ways of organizing activities are transformed over time. In the process, they work to identify the social forces that trigger change and the consequences of change to individuals, groups, organizations, and society at large. In this chapter we focus on a number of key agents of change including technology, globalization, and social movements. We also focus on specific areas of social change that count among the most pressing issues of our day—the aging of the world's population, the environment, and health care.

Module 13.1 Triggers of Social Change

Objective

You will learn the factors that trigger social change and the consequences of change.

Are you making some plan for the future but find yourself wondering if that plan will ever materialize because you feel some big change is on the horizon?

U.S. Army

On an individual level, social change is experienced when we notice that things are not the same. We may notice that family, friends, or others around us are behaving or thinking in unexpected ways. We may even notice change in our own behavior and thinking. As examples, we may notice that we are spending less time preparing meals, or perhaps that we are texting with friends rather than meeting them in a physical space. We may find that we have less free time or that we are now spending our time in different ways or with different people.

Social change can also be noticed at the community or organizational level. A population moves out and another population moves in. New employees are hired and long-time employees retire. And as demographics in the larger society shift, organizations must adjust to the needs of older clients or a new generation of consumers. A new organization (business, mosque, hospital) appears in a community and residents adjust or resist.

Change can be driven by revolutionary forces. The signature technologies of an era—the wheel, the plow, the factory, gasoline engines, computers, cameras—revolutionize every area of life. People who once worked the land now find themselves working in factories. Some occupations disappear as new occupations emerge. New economies emerge with new winners and losers. Geographic spaces become urbanized; cities expand their influence to encompass suburbs and lands that were once rural. The new arrangements include or exclude people based on their age, gender, race, or religion. For some, the "social movements they join express their desire to obtain a larger share of their society's resources. New institutions emerge to educate, employ, inform, transport, shelter, and care for the health of ever larger populations. . . . Changes like these are a matter of endless fascination to sociologists, and there have been a variety of attempts to explain them" (Karmadonov 2014).

Sociologists define **social change** as any significant modification in the way social activities and human relationships are organized. Sociologists are also interested in **tipping points**, a situation in which what was once a rare practice or event snowballs into something dramatically more common. The process by which the rare becomes commonplace is at first gradual, as the first signs of change go largely unnoticed or are dismissed as insignificant. But at some point a critical phase is reached so that the next increment of change "tips" the scale in a dramatic way.

Chris Caldeira

Chris Caldeira

An example of changing norms: you may notice that there seem to be more adults out walking with pets than with children. This observation corresponds to changes in relationships between pets and their "owners." In fact, many American adults believe they are more likely to define and treat their pets as "children" than when they were growing up. Those with pets often refer to themselves as "pet parents" rather than "pet owners" (Coren 2011).

When sociologists study change, they ask at least three key questions: What has changed? What factors trigger change? What are the consequences of the change?

What Has Changed?

Social change is an important sociological topic. Sociology first emerged as an attempt to understand the dramatic social changes accompanying the Industrial Revolution. As a result, sociology offers a variety of concepts that focus our attention on the kinds of change that have reached into the lives of individuals, groups, communities, and organizations. Those concepts include industrialization, globalization, rationalization, McDonaldization, urbanization, and the information explosion.

INDUSTRIALIZATION. The Industrial Revolution changed society into a **hydrocarbon society**—one in which fossil fuels power machines from automobiles to toasters. In this sense fossil fuels shape virtually every aspect of our personal and social lives. The energy source that lights the night, cools our houses and offices, powers appliances, and charges our mobile phones is likely coal. The energy source that heats our houses, workplaces, and water is likely natural gas; the energy source that enables trains, planes, cars, and buses to move people and goods is most certainly oil. Industrialization wove energy consumption into the fabric of society. Energy consumption is also woven into the fabric of the knowledge economy. Most people fail to realize that using the Internet to call up medical records, send correspondence, shop, take classes, read books, and do countless other activities uses energy. Digitalized information—whether it be a song, photo, video, or spreadsheet—takes up space on servers sitting in one of the 3 million data processing centers worldwide. The servers, running 24/7, are powered by electricity generated from burning fossil fuels (Glanz 2012).

Tony Rotundo

Even the seemingly innocent action of taking a selfie and sharing it with friends and family requires service providers to have their servers fired up and ready to transmit the photo and accompanying text instantaneously. "The complexity of a basic transaction is a mystery to most users: Sending a message with photographs to a neighbor could involve a trip through hundreds of miles of Internet conduits and data centers before the e-mail arrives across the street" (Glanz 2012). All this paperless activity is powered by electricity.

GLOBALIZATION. Sociologists distinguish between global interdependence and globalization. **Global interdependence** describes a situation in which human activity transcends national borders. Examples of such activity are endless but include travel, trade, texting, phone calls, and outsourcing. Global interdependence also applies to social issues such as water shortages, pollution, unemployment, and drug addiction as they are part of the larger global-scale social forces. Because the level of global interdependence is constantly increasing, it is part of a dynamic process known as **globalization**—the ever-increasing capitalist-driven flow of goods, services, money, people, information, and culture across national borders supported by a technological infrastructure consisting of computerized networks, satellite communication systems, and software and hardware (Kellner 2002).

RATIONALIZATION. The term *rationalization* refers to changes in the way daily life has come to be organized since industrialization to accommodate large numbers of people. Rationalization is a process whereby thought and action guided by emotion (such as love, hatred, revenge, or joy), superstition, respect for mysterious forces, and tradition are replaced by thought and action rooted in the most efficient (fastest and/or cheapest) ways to achieve a valued goal.

McDONALDIZATION. This organizational trend refers to a process whereby the principles governing the fast-food industry come to dominate other sectors of the economy (Ritzer 1993). Those principles are efficiency, quantification and calculability, predictability, and control. For example, pharmacies, car washes, banks, and liquor stores have adopted drive-thru services to facilitate their goal of efficiently accommodating customers. Of course, drive-thru service is not all there is to McDonaldization. Whatever the service offered—a college degree in 18 months, a medical clinic in a grocery store, matchmaking with success guaranteed in 6 weeks, a prepaid funeral, or the cheapest plane ticket—we can find one or more of the four McDonaldization principles operating.

U.S. Army photo files

Here a student is learning to drive a car using a driving simulator (virtual robot). Learning through simulation has all the features of McDonaldization. It is *efficient* because the simulator allows the student to experience a range of driving situations in the shortest amount of time. The simulated experience will be the same for everyone (*predictability*), and the time spent practicing can be *calculated* easily. Using this simulator, students can *control* the pace of their learning.

URBANIZATION. **Urbanization** is a transformative process in which people migrate from rural to urban areas and change the ways they use land, interact, and earn a living. Today, urban populations encompass not only city dwellers but also suburbanites and even residents of small towns that have been pulled in by urban sprawl. Urban sprawl spreads development beyond cities by as much as 40 or 50 miles; puts considerable distance between homes, stores, churches, schools, and work-places; and makes people dependent on automobiles (Sierra Club 2007). In addition, the automobile and the highway system have permitted people to live in places that give them access to more space than they need for a comfortable life. In the past 30 years, the average house size has increased from 1,400 to 2,330 square feet (National Association of Home Builders 2011). Urban sprawl has blurred boundaries between cities, suburbs, and rural towns.

THE INFORMATION EXPLOSION. Sociologist Orrin Klapp (1986) defined the **information explosion** as an unprecedented increase in the amount of stored and transmitted data/information from all media sources (including electronic, print, radio, and television). Arguably, the information explosion began with the invention of the printing press. Today, the information explosion is driven by the Internet, a vast fossil fuel–powered computer network linking billions of computers around the world. The Internet has the potential to give users access to every word, image, and sound that has ever been recorded (Berners-Lee 1996). In addition to the Internet, advances in electronics, miniaturization, digitalization, and software programming give people devices that can collect, access, store, transmit, analyze, and manipulate data and other information in nanoseconds. Specifically, these four technologies have sped up old ways of doing things and have changed how people and machines learn and process information.

What Factors Trigger Change?

We usually cannot pinpoint a single factor that triggers a specific social change because change is the outcome of an endless sequence of interrelated events. We can, however, identify key factors that trigger change in general. These triggers include, but are not limited to, technological innovation, revolutionary ideas, conflict, the pursuit of profit, and social movements.

TECHNOLOGICAL INNOVATION. In the broadest sense of the word, **technology** involves using knowledge, tools, applications, and other inventions in ways that allow people to adapt to and negotiate their surroundings. Technology can be material (smartphones, hammers, and ATMs) or conceptual (ways of organizing human activity such as the assembly line or computer software that powers apps). Technologies change the structure of human activity and social relationships: They can save time and dissolve social and physical boundaries, as when X-rays allow physicians to see into the body or when Internet technologies allow education to be delivered 24/7 and outside of traditional brick-and-mortar classrooms.

The University of Phoenix is the largest degree-granting college in the country, with 269,000 students (Hansen 2013). A student body of this size, spread across the United States, could not have been achieved without Internet technology. At its peak, University of Phoenix enrollment was 307,871 in 2011 (National Center for Education Statistics 2013a).

REVOLUTIONARY IDEAS. In *The Structure of Scientific Revolutions*, philosopher Thomas Kuhn (1975) defines paradigms as the dominant and widely accepted theories and concepts in a particular area or field of study that seem to offer the best way of looking at the world at a particular time. A scientific revolution occurs when enough people in the community reject an existing paradigm in favor of a new paradigm. The new paradigm causes converts to see the world in an entirely new light and to wonder how they could possibly have taken the old paradigm seriously. "When paradigms change, the world itself changes with them" (111).

Until recently, the dominant paradigm held that humans were different from, even superior to, animals. The belief was that humans possessed a sense of self, language, and consciousness and animals did not. But now that paradigm is being challenged. Simply consider the prairie dog pictured: Do you think of it as possessing a language? Recent research suggests that these animals have a very specific language system in that they are able to warn other prairie dogs about intruders to the point of describing what intruders are wearing—"Here come a tall human in a blue shirt with a hat on" (Slobodchikoff 2011). How might this paradigm shift change the way humans treat and think about animals?

National Park Service

CONFLICT. Sociologist Lewis Coser (1973) argues that conflict occurs when those who control valued resources strive to protect their interests from those who hope to gain a share. Conflict can be an invigorating force that prevents a social system from becoming stagnant, unresponsive, or inefficient. Conflict—whether it involves military buildups, violent clashes, or public debate—is both a consequence and a cause of change. In general, any kind of change has the potential to trigger conflict between those who stand to benefit from the change and those who stand to lose because of it.

DARPA

As one example of how conflict triggers change, the U.S. military has invested billions in the Legged Squad Support System (LS3), a highly mobile, semi-autonomous legged robot capable of carrying 400 pounds of provisions and equipment over rugged terrain. The robot frees up soldiers from carrying 100+ pounds of gear, thus reducing physical strain, fatigue, and degraded performance. The LS3 pictured is learning to "follow squad members through rugged terrain and interact with troops in a natural way, similar to a trained animal and its handler" (DARPA 2014).

THE PURSUIT OF PROFIT. The capitalist system is a change agent because of the ways it makes profit. To make a profit, employers must find ways to produce goods and services in the most cost-effective manner, and that includes creating new technologies that speed up the production and delivery of goods and services and eliminate labor, such as a drone that delivers packages to doorsteps, thus eliminating labor associated with delivery.

SOCIAL MOVEMENTS. Social movements occur when enough people organize to resist a change or to make a change. Usually those involved in social movements work outside the system to advance their cause because the system has failed to respond. To draw attention to their cause and accomplish their objectives, supporters may strike, demonstrate, walk out, or otherwise disrupt the social order.

What Are the Consequences of Change?

Sociology offers a rich set of concepts, theories, and questions to guide any analysis of social change and its consequences. In particular, the four theoretical perspectives offer basic questions to guide analysis of any social change:

> Functionalist: How does X contribute to order and stability? What are the intended and unintended consequences of X?
> Conflict: Who benefits from X and at whose expense?
> Symbolic Interaction: Does the emergence of X change the way we interact with others? Does it challenge, sustain, or alter existing meanings?
> Feminist: How are gender relations affected by X?

What Do Sociologists See?

This surgeon sits at the Da Vinci Surgical System, a robotic system where the surgeon "looks through two eye holes at a 3-D image of the procedure, meanwhile maneuvering the arms with two foot pedals and two hand controllers" (U.S. Army 2012). The surgeon need not be in the same room as the patient or even in the same building or city. Sociologists would analyze this change by asking questions like: How does this technology change the meaning of the doctor–patient relationship? And might women be less likely to be surgeons now that robotic surgery capitalizes on the skills of those who played video games as children and teens?

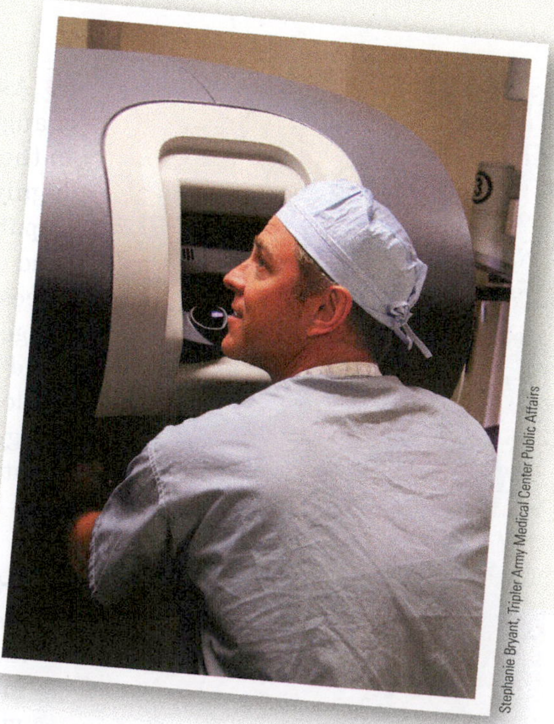

Stephanie Bryant, Tripler Army Medical Center Public Affairs

Critical Thinking

Look over the list of social changes named in the module that have revolutionized human activity. To which revolutionary change do you most relate?

Key Terms

global interdependence	information explosion	tipping point
globalization	social change	urbanization
hydrocarbon society	technology	

Module 13.2 Technology and Human Societies

Objective

You will learn about the relationship between a society's signature technology and the amount of surplus wealth it is able to create.

Can do you think of ways the tractor and plow affect your life, especially the job you do?

USDA

In this module we take a long view and consider six broad types of societies in which humans have lived over their history as a species. Each is defined by a signature technology. The six societies are hunting and gathering, pastoral, horticultural, agrarian, industrial, and postindustrial. Each of the six is distinguished by the amount of surplus wealth that society is able to produce. **Surplus wealth** is a situation in which the amount of available food items and other products exceeds that which is required to **subsist**, or to meet basic needs for human survival.

HUNTING AND GATHERING SOCIETIES. **Hunting and gathering societies** do not possess a signature technology that allows members to create more than they need to survive; people subsist on wild animals and vegetation. As the name suggests, hunters and gatherers do not live in a fixed location; they are always on the move, securing food and other subsistence items. A typical hunting and gathering society is composed of 45 to 100 members related by blood or marriage. The institution of the family is central to people's lives, and emphasis is placed on group welfare (Massey 2002). The division of labor is simple, and most people engage in activities related to survival. The statuses that matter revolve around gender, age, and kinship. Because almost no surplus wealth exists, there is little inequality.

Sociologist Douglas Massey (2002) predicts that by 2020 the last hunter-gatherers on the planet will cease to exist, ending 6 million years of dedication to "the most successful and long-persistent lifestyle in the career of our species" (Diamond 1992, 191). In fact, it may already be impossible to find a society that meets all its needs from hunting and gathering.

PASTORAL AND HORTICULTURAL SOCIETIES. About 10,000 to 12,000 years ago, humans began domesticating plants and animals. Now, instead of searching for and gathering wild grains and vegetation, people planted seeds and harvested crops. Instead of hunting for wild animals, people captured, tamed, and bred them. Domestication (the signature technology) offered a predictable food source and allowed people to secure more grain, meat, and milk than they needed to survive. Domestication of animals offered people another source of power—animal muscle—to transport heavy loads, to guard sheep, and so on. Surplus food gave some people time to pursue other activities beyond securing subsistence, such as making vases to store food.

Domestication is the hallmark of two types of societies: pastoral and horticultural. **Pastoral societies** rely on domesticated herd animals to subsist. Pastoralism was adopted by people living in deserts and other places with limited amounts of vegetation. Those able to acquire and manage the largest herds assumed powerful statuses and passed their advantaged positions on to their children.

Today in Afghanistan there are an estimated 1.2 million people considered pastoral. Here the nomadic Kuchi people are shown as they move goats, donkeys, camels, and cattle through the Panjshir Valley to the high country for the summer.

U.S. Air Force photo/Tech. Sgt. John Cumper

Most pastoral peoples are nomadic, moving their herds when grazing land and/or water sources are depleted. Even though they may be on the move, pastoralists are able to accumulate possessions such as tents, carpets, bowls, and other cultural artifacts, because now they have animals to carry those possessions. In the course of their travels, pastoralists encounter other nomads and settled peoples with whom they trade and/or fight to secure grazing land. The statuses people occupy revolve around gender, age, and kinship, but material possessions and success in conflicts are now also important in determining status.

In contrast to pastoralists, people who live in **horticultural societies** rely on hand tools such as hoes (the signature technology) to work the soil and digging sticks to punch holes in the ground into which seeds are dropped. Horticultural peoples grow crops rather than gather food and employ slash-and-burn technology in which they clear land of forest and vegetation to make fields for growing crops and grazing animals. When the land becomes exhausted, people move on, repeating the process. In contrast to pastoralists, horticultural societies are relatively settled. The horticultural system offers a level of predictability and residential stability that gives people the incentive and means to create surplus wealth, including houses, sculptures, and jewelry. The creation of surplus wealth is accompanied by conflict over available resources and by inequality with regard to its distribution.

AGRARIAN SOCIETIES. The invention of the plow 6,000 years ago triggered a revolution in agriculture and marked the emergence of **agrarian societies** built on the cultivation of crops using plows pulled by animals to achieve subsistence. The plow—the signature technology—made it possible to cultivate large fields and increase food production to a level that could support thousands to millions of people, many of whom lived in cities and/or were part of empires. Although the plow was a significant invention, planting and harvesting food still depended largely on human and animal muscle.

U.S. Navy photo by Lt. Cmdr. Chuck Bell

Library of Congress, Prints and Photographs Division LC-DIG-ppmsca-15046

The major agricultural revolution that launched the agrarian societies occurred around 5000 BC with the invention of the scratch plow (still used in some parts of the world); its forward-curving blade cut deep into the soil, bringing nutrients to the surface and turning weeds under. The plow (right) was a great advance over hoes (left) because it allowed farmers and their descendants to replenish the soil year after year (Burke 1978, 9).

Agrarian societies are noted for dramatic inequality; monarchs (kings, queens, or emperors) hold absolute power over their subjects, who for the most part do not question such power. In agrarian societies there are a small number of landowning elites and large numbers of people known as serfs, peasants, or the enslaved. There are also small numbers of merchants, traders, and craftspeople. In addition to dramatic inequality, agrarian societies are often engaged in wars to protect and/or expand their territory and to exercise control over resources.

One well-known symbol of ancient and medieval empires is the 4,500-mile-long Great Wall of China, built during the Quin, Han, and Ming dynasties. The technological complexity of the wall offers insights into the kinds of human activity—building the Great Wall—that surplus wealth generated by the plow made possible. The plow also allowed people to do something other than spend their day securing food, because there was surplus food that allowed people time to establish governments, build roads, create systems of writing, build palaces, and much more.

NKU Philosophy, Rudy Garns

The plow is believed to have changed the status of women relative to men in dramatic ways. As surplus wealth and populations increased, warfare between peoples fighting over land and resources became commonplace. The invention and proliferation of metal weapons supported military forces created to advance and protect the interests of the political elite. The military excluded women, who spent much of their reproductive life pregnant or nursing, to offset the high death rates and low life expectancy (Boulding 1976).

Women's status was reduced in yet another way. The plow increased the amount of land that could be cultivated, and by extension it increased the amount of food produced. Men operated the plows and managed heavy draft animals in fields away from the home. Women's reproductive lives and caretaking responsibilities, in conjunction with the efficient technology men operated, made it difficult for women to outproduce men. As a result, women's share of the production diminished. Elise Boulding (1976) described women's situation as follows:

> The shift of the status of the woman farmer may have happened quite rapidly, once there were two male specializations . . . plowing and the care of cattle. The situation left women with all the subsidiary tasks, including weeding and carrying water to the fields.

INDUSTRIAL SOCIETIES. **Industrial societies** rely on mechanization or machines to subsist. These signature innovations allowed humans to produce food, extract resources, and manufacture goods at revolutionary speeds and on an unprecedented scale. Mechanization changed everything: Most notably, a small percentage of the population grew the food needed to sustain a society that could include hundreds of millions of people. The products of industrialization improved nutrition and living standards, which increased human life expectancy, decreased fertility, and lowered death rates so that the number of older people eventually came to outnumber the young. Under industrialization, population size increased from about 954 million people in 1800 to 2.5 billion by 1950 (Massey 2002).

Chris Caldeira

These hundreds of bottles of nail polish of varying colors speak to the concept of surplus wealth made possible by the Industrial Revolution. Today people have time to paint their nails and toes because, thanks to the agricultural revolution, they don't have to spend their day securing food. And thanks to the Industrial Revolution they can take time to stand and deliberate over what color nail polish to try instead of using that time to gather food. In *The Paradox of Choice*, Barry Schwartz (2004) raises the point that too much "choice no longer liberates, but debilitates. It might even be said to tyrannize."

The ability to support this level of choice suggests that there is enough surplus wealth to support a diverse economy, including institutions of higher education that offer as many as 350 different majors to choose from, supermarkets that carry as many as 50,000 items, and a pharmaceutical industry through which the 70 percent of Americans on prescription drugs can fill their prescriptions (CBS News 2013). The mass production of goods allowed people not only to buy products that distinguished them from others but also to buy more products than they needed, creating great social differences. On the one hand, the products of industrialization have allowed many more people to experience a high standard of living and social mobility. On the other hand, this level of production has created dramatic differences in material wealth, especially between those in the top 10 percent (who own the means of production or invest in stocks that yield income) and the bottom 20 percent (Massey 2002).

POSTINDUSTRIAL SOCIETIES. Massey (2002) estimates that humans spent 300,000 generations as hunter-gatherers, 500 generations as agrarians, 9 generations in the industrial era, and only 1 generation in the postindustrial era. Sociologist Daniel Bell, who has been writing about the coming of a **postindustrial society** since the 1950s, described it as a society that relies on the intellectual technologies of telecommunications and computers (Bell 1999, xxxvii). According to Bell, these signature technologies encompass four interdependent revolutionary innovations: (1) electronics that allow for incredible speed of data transmission and calculations, which can be done in nanoseconds; (2) miniaturization, or the drastic size reduction of electronic devices; (3) digitalization, which allows voice, text, images, and data to be integrated and transmitted with equal efficiency; and (4) software, applications that allow people to perform a variety of tasks and generate a variety of simulated experiences.

The objects you see embedded inside this snow pea pod are microprocessors and chips powerful enough to do the research once done in a building-sized, fully equipped laboratory. The microprocessor and chips shown power a handheld lab that tests for chemical or biological hazards, detects pollutants, performs medical diagnostics, and performs drug testing (Sandia National Laboratories 2014).

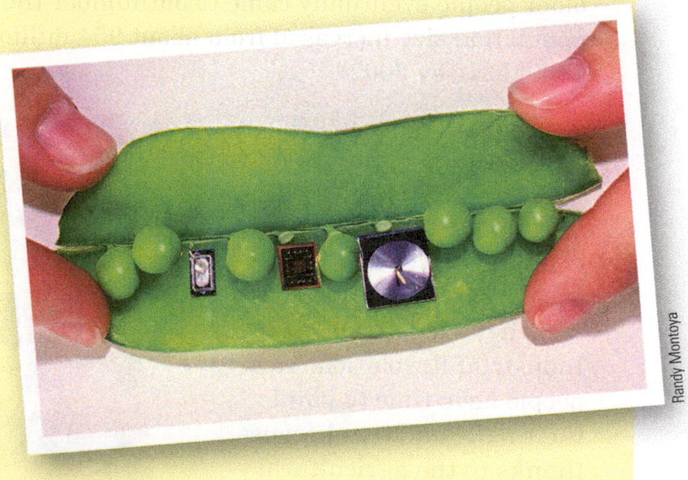
Randy Montoya

Postindustrial societies, built upon these intellectual technologies, are distinguished by the following characteristics:

- a substantially greater share of the working population employed in service-providing occupations (in the United States, from 29 percent in 1950 to 79.9 percent today) (U.S. Bureau of Labor Statistics 2013d);

- an increased emphasis on education as the avenue of social mobility (in the United States, in 1950, 6 percent of the population had at least a four-year today 30 percent do);
- a recognition that capital is not only financial but also social (that is, access to social networks serves as an important source of information and opportunity);
- the dominance of intellectual technology grounded in mathematics and linguistics that takes the form of what are known as apps;
- the creation of an electronically-mediated global communication infrastructure; and
- an economy defined not simply by the production of goods but by applied knowledge and the manipulation of numbers, words, images, and other symbols.

What do these changes mean? In answering this question, keep in mind that Bell does not believe that technology determines the nature of any social change. Rather, it is the ways people choose to use and respond to the technology that shape social change.

U.S. Food and Drug Administration

Although it is virtually impossible to catalog all the changes associated with the intellectual technologies, they have (1) sped up old ways of doing things; (2) given individuals access to the equivalent of personal libraries, publishing houses, and production studios; (3) changed how people learn; and (4) permitted real-time exchange of information on a global scale.

The unprecedented and immediate access to information and people can be overwhelming. Although computer software and telecommunications technologies have increased the speed of information generation and exchange, keep in mind people must still read, discuss, and contemplate the information to give it meaning. These activities are very slow compared with the speed at which the information is generated.

Relative to other types of societies, the postindustrial society presents its members with a distinct set of challenges that are interpersonal in nature. The communication infrastructure and service economy multiply interactions among people, making interpersonal relationships a primary focus. That focus is complicated because we leave permanent records of our transactions with others. The great difference in the postindustrial society is the tremendous change in the number of people one knows or can potentially know (Bell 1976, 48).

What Do Sociologists See?

The oil gusher symbolizes a source of power that allowed for mechanization. In turn, mechanization allowed humans to mass produce food, extract resources, and manufacture goods on an unprecedented scale. Mechanization fueled by fossil fuels such as oil permitted the creation of enough surplus wealth that only a small percentage of the workforce was needed to grow the food needed to sustain hundreds of millions of people, freeing those millions to do other things.

Library of Congress Prints & Photographs Division [LC-USZ62-54453]

Critical Thinking

Do you hold an occupation that creates products or delivers services that are not needed for subsistence? Explain.

Key Terms

- agrarian societies
- horticultural societies
- hunting and gathering societies
- industrial societies
- pastoral societies
- postindustrial society
- subsist
- surplus wealth

Module 13.3 Globalization

Objective

You will learn about how the forces of globalization shape much of what goes on in our day-to-day lives.

Do you recognize any of the balloon characters and in what country this street scene is set?

NKU Philosophy, Rudy Garns

Dora the Explorer, Spiderman, and Minnie Mouse are among the balloon characters a vendor working at Ermou Street in Athens, Greece, sells to tourists and residents. The photo is interesting sociologically because it illustrates two important forces shaping our daily lives: globalization and glocalization. Globalization is the ever-increasing flow of goods, services, money, people, information, and other cultural items across national borders. The photo illustrates globalization because Dora, Minnie Mouse, and Spiderman were "born" in the United States in 2000, 1928, and 1962, respectively. Now, through a process known as globalization, the three are widely known around the globe. While the "migration" of Dora, Spiderman, and Minnie Mouse is an example of globalization, it is also an example of **glocalization**, the process by which the product is transplanted from one local setting to countless local settings around the world. In this case, cartoon characters, which at one time existed only inside the heads of their creators working at Disney Studios and Marvel Comics, found their way to Ermou Street (Athens, Greece), one of thousands of local settings across the globe where these characters can be found. Emphasis should be placed on the word *transplanted* because just as the body adapts to an organ transplant, the people in local settings accept, modify, adapt to, and resist foreign items. This vendor has clearly embraced these characters.

The concept of glocalization draws our attention to the intersection of the local and the global. Globalization and glocalization are intertwined because globalization always involves a series of countless glocalizations. Simply consider that the man on Ermou Street in Athens is just one of the countless people and localities in the world where Dora, Spiderman, and Minnie Mouse have made their way. As we will see, sociology offers a conceptual framework to think about how the twin social forces of globalization and glocalization shape meanings, identities, social relationships, and human activities. Whether or not people recognize it, the forces of globalization shape every aspect of their daily lives. How they respond makes globalization a local (or glocal) event. We showcase five examples of where global forces and local responses intersect.

Chris Caldeira

This driver is challenging the globalization of meat consumption with the glocal plea to "Be Veg . . . Save the Planet." Worldwide meat production has tripled over the last 50 years, such that the average person on the planet consumes 80 pounds (38 kg) of meat per year. Of course, consumption varies by location, with the average American eating 274 pounds (125.4 kg) of meat each year and the average person in India consuming 7 pounds (3.2 kg) (*Economist* 2012). Raising the billions of animals to satisfy this annual level of global-scale consumption involves a tremendous amount of animal waste that releases methane and nitrous oxide (greenhouse gases) into the atmosphere. "The world's supersized appetite for meat is among the biggest reasons greenhouse gas emissions are still growing rapidly" (Engelman 2013).

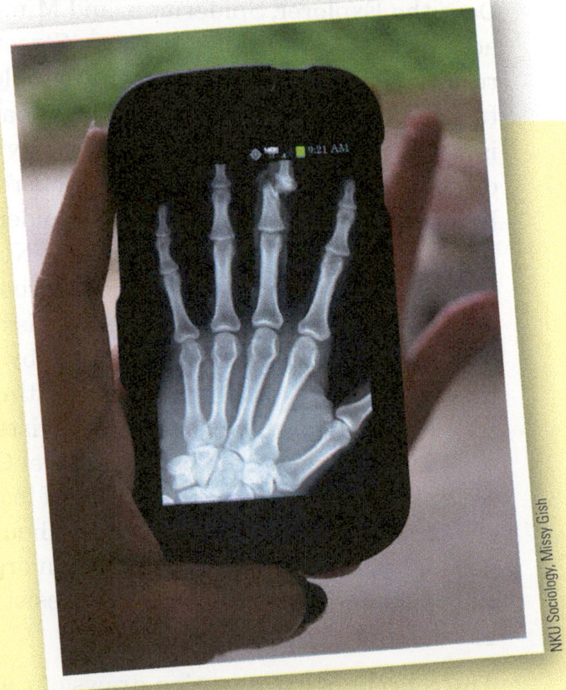
NKU Sociology, Missy Gish

There is no question that the smartphone qualifies as a global-scale social force. Smartphones, first released in 1996, have been embraced by billions of people across the globe in countless locations. Using smartphone technology, doctors and other health care professionals in one location can monitor and treat patients in any locality. One day smartphones will likely be equipped with an app that can take an X-ray on the spot and instantly transmit the image to a doctor or health care professional in another location. This kind of technology will be especially embraced by those living in remote locations who must travel long distances to receive care.

It is very likely that this apple juice is made from apples grown in at least ten countries including Germany, Austria, Italy, Hungry, Argentina, Chile, China, Turkey, Brazil, and the United States. Imagine the global effort involved in coordinating the delivery of apples from ten or more countries to a central location for processing and then delivering that juice by ship, train, and truck to "local" stores, restaurants, schools, and vending machines all over the world. One also has to imagine the hundreds (if not thousands) of U.S. apple growers, who once dominated the global and domestic markets, closing orchards because they cannot compete with China, for example, which now meets 50 percent of the world demand for apple juice.

NKU Sociology, Missy Gish

Chris Caldeira

The globalization of gay pride includes the opening of the first gay bar in Lamai Beach, Thailand, more than 115 years after the first known group to campaign publicly for gay rights was founded in Berlin, Germany, in 1897. By 1922, 25 local chapters had been formed in Europe, only to be suppressed by the Nazis. In the late 1940s and 1950s, support groups formed in the United States (the Mattachine Society in Los Angeles and the Daughters of Bilitis in San Francisco) and in Europe (the Dutch Association for the Integration of Homosexuality in Amsterdam). The Stonewall Rebellion—the event in which patrons of the gay bar Stonewall Inn rioted in protest after a police raid—is widely considered a watershed event propelling the modern gay rights movement. In fact, the Stonewall Rebellion is commemorated annually during Gay and Lesbian Pride Week in many cities around the world.

Sgt. Christopher McCullough

U.S. Drug Enforcement Agency

Anyone who takes prescribed opiate pain medicines takes a drug with opium ingredients likely grown in Tanzania, India, or Turkey. For those who take the opiate known as heroin it is most likely grown in Afghanistan, the largest supplier of this illegal drug in the world. Both Tanzania's and Afghanistan's economies depend on the opium production, but as it stands Afghan farmers are prohibited from producing for the global pharmaceutical market. Someone who becomes addicted to pain killers can turn to heroin.

What Do Sociologists See?

Chris Caldeira

Sociologists think about the forces of globalization underlying this special recognition as the loneliest road. Eureka, Nevada, one of several towns on Highway 50, dubbed "the loneliest road in America," was once part of the globalization trend but has now been bypassed by globalization. Eureka was founded in 1864 as a mining town (gold, silver, lead), and reached its peak population in 1878 at 10,000. Today, 1,100 people live in this town. From the beginning, the town's identity centered on mining, but the value of the ore was determined by global demand. As the value of what Eureka mined declined in the marketplace and its residents, as well as those in surrounding towns, moved away, Eureka's global identity came to center around its location along the "loneliest road in America."

Critical Thinking

Can you think of ways you experience (or have experienced) globalization?

Key Term

glocalization

Module 13.4 Social Movements

Objective

You will learn about types of social movements and some conditions under which people join them.

Is there something about the world you would like to see changed? If so, would you risk imprisonment for that cause?

Tech. Sgt. Anthony Iusi

A **social movement** is formed when a substantial number of people organize to make a change, resist a change, or undo a change to some area of society. For a social movement to form, there must be (1) an actual or imagined condition that enough people find objectionable; (2) a shared belief that something needs to be and can be done about that condition; and (3) an organized effort to attract supporters, articulate the mission, and define a change-making strategy. Usually those involved in social movements work outside the system to advance their cause because the system has failed to address the problem. To draw attention to their cause and accomplish their objectives, supporters may strike, demonstrate, walk out, boycott, go on hunger strikes, riot, or terrorize.

Examples of social movements include the environmental, civil rights, Tea Party, abortion rights, and pro-life movements. Generally, any social movement encompasses dozens to hundreds of specific groups that have organized to address some condition they find intolerable. The environmental movement, for example, consists of thousands of different organizations devoted to reducing air and water pollution, preserving wilderness, protecting endangered species, changing lifestyles, and limiting corporate activities that harm the environment.

Types of Social Movements

Social movements can be divided into four broad categories, depending on the scope and type of change being sought: regressive, reformist, revolutionary, and counterrevolutionary. Definitions and examples of each category follow. Keep in mind that the distinctions between the four categories are not always clear-cut. As a result, you might find that some examples fit into more than one category.

REGRESSIVE MOVEMENTS. Regressive, also called reactionary, movements seek to return to an earlier state of being, sometimes considered a golden era. The International Forum on Globalization (2014) represents such an effort. It is a global alliance of 60 organizations in 25 countries that takes issue with the widely held belief that a globalized economy trickles down to even the poorest peoples. This movement seeks to reverse the globalization process by revitalizing local economies, advocating for local food production, and critically examining trade policies that make people dependent on distant sources.

REFORMIST MOVEMENTS. These movements identify a specific way in which society is organized and targets it to be changed.

Many local officials such as those in Nashville, Tennessee, are pushing a movement to encourage people not to give to panhandlers, arguing that such giving supports the lifestyle (whatever that may be, as panhandlers are very diverse in background and need) that brings people to panhandle. Those against giving argue that panhandlers are often not honest about why they are panhandling, and that some of them are not as poor as they appear. Critics argue that even if they are part of a network of panhandlers, those in it can hardly be called well-off (Klan 2012).

Chris Caldeira

REVOLUTIONARY MOVEMENTS. When people take action to make broad, sweeping, and radical structural changes to society's basic social institutions, they are engaged in revolutionary movements (Benford 1992). The Earth Liberation Front (ELF) is an underground eco-defense movement with no formal leadership or membership. Its members anonymously and autonomously engage in economic sabotage, including property destruction and guerrilla warfare, against those seen as exploiting and destroying the natural environment. ELF members have made news for setting fire to SUVs on dealership parking lots and to equipment at logging companies (Goldman 2007; Gillespie 2008).

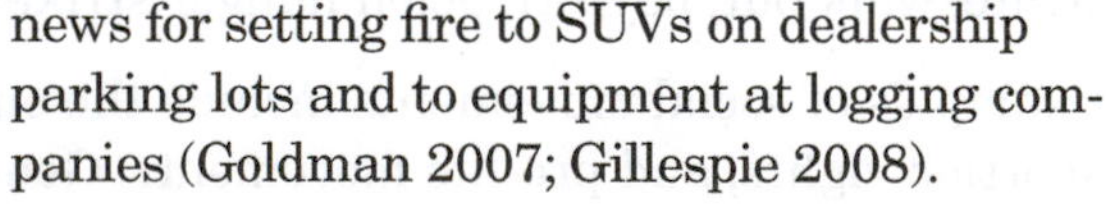

Maj. Jason Dickerman

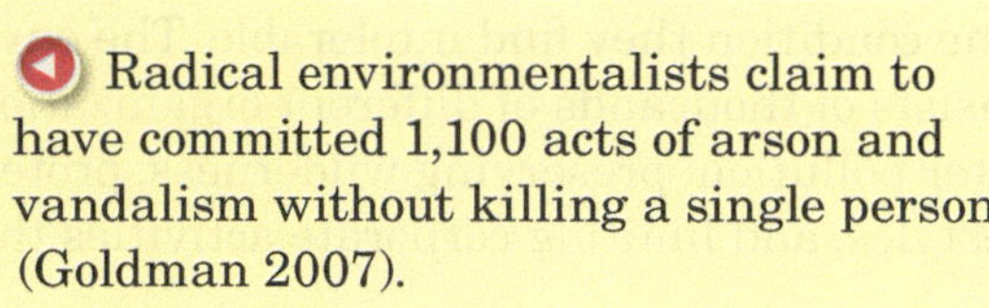

Radical environmentalists claim to have committed 1,100 acts of arson and vandalism without killing a single person (Goldman 2007).

COUNTERREVOLUTIONARY MOVEMENTS. People sometimes join together to return to a social order that once was, but has been changed. Counterrevolutionary movements are responses to revolutionary and reform movements. The Global Warming Petition Project (2010) qualifies as such a movement; it seeks to challenge movements demanding that greenhouse gas emissions be reduced. The Petition Project recruits basic and applied scientists to sign a Global Warming Petition urging the U.S. government to reject international agreements to limit greenhouse gas emissions. The movement's website lists the names of 31,487 basic and applied American scientists who have signed the petition.

On the surface, it would appear that social movements form when enough people feel deprived, either in an objective or a relative sense. Objective deprivation is the condition of those who are the worst off or most disadvantaged—the people with the lowest incomes, the least education, the lowest social status, the fewest job opportunities, and so on. **Relative deprivation** is a social condition that is measured not by objective standards but rather by comparing one group's situation with the situations of groups that are more advantaged. Someone earning an annual income of $100,000 is not deprived in any objective sense. He or she may, however, feel deprived relative to someone making $300,000 or more per year (Theodorson and Theodorson 1969). The research on social movements shows that those who are objectively deprived are less likely than those who are relatively deprived to form or join social movements. The research also shows that people join social movements not to address real or imagined personal deprivation but rather to address larger moral issues, such as mistreatment of animals. The point is that actual deprivation alone cannot explain why people form or join social movements or how social movements take off.

The Life of a Social Movement

Sociologist Ralf Dahrendorf (1973) offers a three-stage model to capture the life of a social movement. Progression from one stage to the next depends on many things. In the first stage, those without power decide to organize against those holding power.

Dahrendorf maintained that it is "immeasurably difficult to trace the path on which a person . . . encounters other people just like himself, and at a certain point . . . [says] 'Let us join hands'" against those in power to change the system (240).

Chris Caldeira

Such encounters occurred in the fall of 2009, when thousands of California college students came together to protest a 32 percent hike in tuition imposed as a result of state budget shortfalls. Often, some event—a tuition hike, a shooting, a toxic emission into the air—makes seemingly disconnected and powerless people aware that they share an interest in changing the system. After the event people dare to complain openly and loudly to one another (Ash 1989). At other times, people organize because they have nothing left to lose. They have reached the point where they do not care anymore about what happens if they speak out (Reich 1989).

In the second stage of conflict, those without power organize, provided they find ways to communicate with one another, they have some freedom to organize, they can garner the necessary resources, and a leader emerges. At the same time, those in power often censor information, restrict resources, and undermine attempts to organize. Resource mobilization theorists maintain that having a core group of sophisticated strategists is key to getting a social movement off the ground. Effective strategists can harness and channel the energies of the disaffected, attract money and supporters, capture media attention, forge alliances with those in power, and develop an organizational structure. Cell phones, text messaging, and the Internet have made organizing easier by allowing interested parties to connect in ways that defy place and time (Lee 2003).

In the third stage of conflict, those seeking change enter into direct conflict with those in power. The capacity of the ruling group to stay in power and the amount and kind of pressure exerted from below them affect the speed and depth of change. The intensity of the conflict can range from heated debate to violent civil war. It depends on many factors, including the belief that change is possible and that those in power can control the conflict. If protestors believe that their voices will eventually be heard, the conflict is unlikely to become violent or revolutionary. If those in power decide that they cannot compromise and they mobilize all their resources to thwart protests, two results are possible. First, the protesters may withdraw from the fray because they believe that the sacrifices will be too great if they continue. Alternatively, the protesters may decide to meet the enemy head-on, in which case the conflict may become violent.

In spite of protests, California college students were unable to prevent hikes in tuition and fees. What factors do you think prevented them from engaging in perpetual protest until their demands were met?

What Do Sociologists See?

This scene represents the third stage of a social movement. Clearly the protestors have come together because they believe something about society needs to be changed. In this third stage those seeking change enter into direct conflict with those in power. The outcome of that confrontation depends on the strength of those in power to control or suppress the conflict and whether protestors believe change is possible. If protestors believe that they can make change, the conflict is likely to escalate. If their sacrifice seems too great, protesters may withdraw. If they believe they have nothing to lose, they may press ahead, in which case the conflict may become violent.

U.S. Air Force photo by Civ/Nan Wylie

Critical Thinking

Is there a contemporary or historical social movement with which you are familiar? Under which of the four types of social movements would it fall?

Key Terms

relative deprivation

social movement

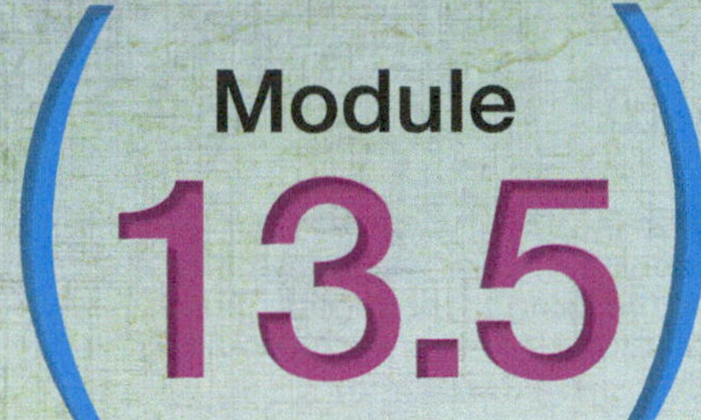

Module 13.5 Aging Societies

Objective

You will learn how those age 65 and over have come to outnumber those age 30 and under.

When your family gets together for a holiday or other special occasion, are there more people under 30 or 65 and over?

Tony Rotundo

There is no historical precedent in the United States, or the world, for such a large proportion of a population moving toward older ages. The theory of the demographic transition offers one explanation for how this came to be.

Theory of the Demographic Transition

Sociologists are interested in the factors that affect population size, growth, and age composition. As you can imagine, births and deaths are key factors. The **theory of the demographic transition** postulates that a country's birth and death rates are linked to its level of industrial or economic development. The demographic transition includes three stages: in stage 1, birth and death rates are both high; in stage 2, death rates decrease, causing population size to increase dramatically; and in stage 3, both birth and death rates drop below 20 per 1,000 persons. Since the theory was put forth in in 1929, a fourth stage has been added (and even a fifth and sixth stage not covered here). The theory centers around a graph depicting historical changes in birth and death rates in Western Europe and the United States, and it helps us understand the forces behind population change, especially the rise of aging populations (see Figure 13.5a).

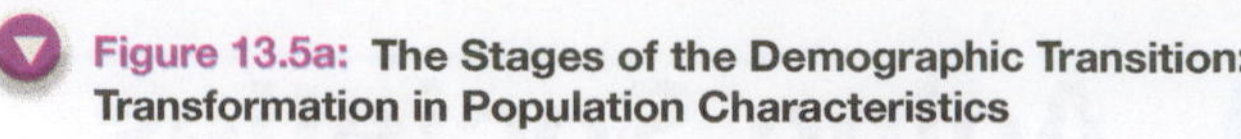
Figure 13.5a: The Stages of the Demographic Transition: Transformation in Population Characteristics

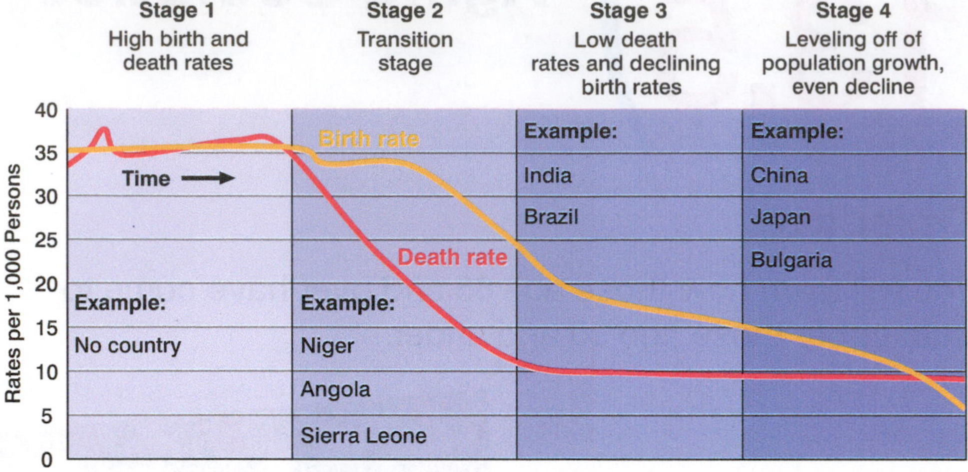

Sources of Data: U.S. Central Intelligence Agency (2014); Thompson (1929); Santosa, Ailiana, et al. (2014)

STAGE 1: HIGH BIRTH AND DEATH RATES. For most of human history—the first 2 to 5 million years—populations grew very slowly, if at all. Demographers speculate that growth until that time was slow because **mortality crises**—violent fluctuations in the death rate caused by war, famine, or epidemics—were a regular feature of life. Stage 1 of the demographic transition is often called the stage of high potential growth: If something happened to cause the death rate to decline—for example, improvements in agriculture, sanitation, or medical care—the population would increase dramatically. In this stage, life is short and brutal; average life expectancy at birth remains short—perhaps 20 to 35 years—with the most vulnerable groups being women of reproductive age, infants, and children younger than age five. It is believed that women gave birth to large numbers of children and families remained small because one of every three infants died before reaching age one, and two of every three died before reaching adulthood.

STAGE 2: TRANSITION. Around 1650, mortality crises became less frequent, and by 1750 the death rate began to decline slowly. This decline was triggered by a complex array of factors associated with the onset of the Industrial Revolution. The two most important factors were (1) increases in the food supply, which improved the nutritional status of the population and increased its ability to resist diseases; and (2) public health and sanitation measures, including the use of cotton to make clothing and new ways of preparing food. Over a 100-year period, the death rate fell from 50 per 1,000 to less than 20 per 1,000, and life expectancy at birth increased to approximately 50 years of age. As the death rate declined, fertility remained high, and thus the **demographic gap**—the difference between birth and death rates—widened, and the population grew explosively. Accompanying the unprecedented growth in population was urbanization.

Around 1880, birth rates began to drop. The decline was not caused by innovations in contraceptive technology, because the methods available in 1880 had been available throughout history. Instead, the decline in fertility seems to have been associated with several other factors. First, the economic value of children declined; children no longer represented a source of cheap labor but rather became an economic liability to parents. Second, with the decline in infant and childhood mortality, women no longer had to bear a large number of children to ensure that a few survived. Third, a change in the status of women gave them greater control over their reproductive life and made childbearing less central to their life.

STAGE 3: LOW DEATH RATES AND DECLINING BIRTH RATES. Around 1930, both birth and death rates fell to less than 20 per 1,000, and the rate of population growth slowed considerably. Life expectancy at birth surpassed 70 years. The remarkable successes in reducing infant, childhood, and maternal mortality placed accidents, homicide, and suicide among the leading causes of death of young people. The risk of dying from infectious diseases declined, allowing those who would have died of infectious diseases in an earlier era to survive into middle age and beyond, at which point they faced an elevated risk of dying from degenerative and environmental diseases such as heart disease, cancer, and strokes. For the first time in history, people age 50 and older accounted for the largest share of deaths. Before stage 3, infants, children, and young women accounted for the largest share (Olshansky and Ault 1986). As death rates decline, disease prevention becomes important. As a result, people become conscious of the link between health and lifestyle factors such as sleep, nutrition, exercise, and bad habits.

STAGE 4: LEVELING OFF OF POPULATION GROWTH, EVEN DECLINE. Since the demographic transition was first proposed, a fourth stage has been added, in which both birth rates and death rates are low. Birth rates drop to a level below that needed to replace those who die. While death rates are low, there is an increase in lifestyle diseases caused by lack of exercise, poor nutrition, and obesity. Birth rates fall below replacement when the average woman has fewer than two children over the course of her reproductive life. Eastern European countries, Italy, and Japan are examples of countries in which this is the case (see Figure 13.5b).

Consider that in Japan, the average woman has 1.12 babies; the death rate is 9.0 per 1,000, and the birth rate is 8.0. The number of people age 70–74 exceeds the population age 5–9 (U.S. Census Bureau 2012e).

Katie Englert

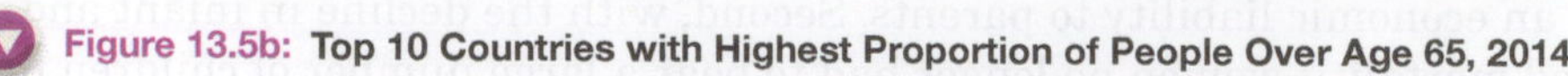

Figure 13.5b: Top 10 Countries with Highest Proportion of People Over Age 65, 2014

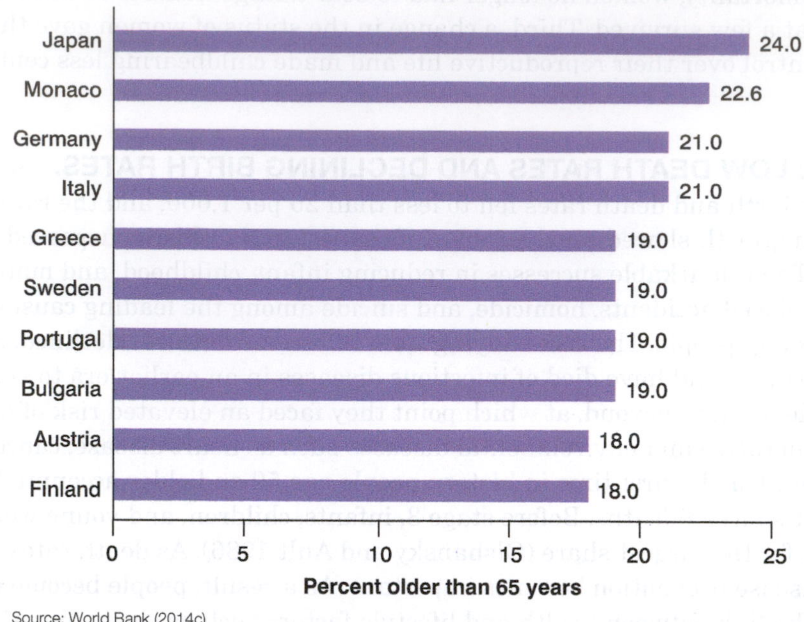

Source: World Bank (2014c)

Developing Countries

The theory of the demographic transition forecasted that developing countries would follow the original three-stage model. But their path has deviated in some fundamental ways. Most countries we call developing were once colonies of today's most industrialized countries. Colonization created economies in developing countries that supported the colonizing countries' industrialization, but it did not put the developing countries on the path that the demographic transition predicts would result in lower birth and death rates.

BIRTH RATES. Birth rates, although still relatively high, are declining in most developing countries. Some important thresholds are associated with declines in fertility. First, less than 50 percent of the labor force is employed in agriculture. (The economic value of children decreases in industrial and urban settings.) Second, at least 50 percent of those between ages 5 and 19 are enrolled in school. (Especially for women, education "widens horizons, sparks hope, changes status concepts, loosens tradition, and reduces infant mortality"; Samuel 1997.) Third, infant mortality is less than 65 per 1,000 live births. (When parents have confidence that their babies and children will survive, they limit the size of their families.) Fourth, 80 percent of the females between ages 15 and 19 are unmarried. (Delayed marriage is important when it is accompanied by delayed sexual activity or protected premarital sex; Berelson 1978.)

DEATH RATES. Death rates in the developing countries have declined at a faster rate than they did in the industrialized countries. That decline has been attributed, in part, to imported Western technologies such as pesticides, fertilizers, immunizations, and antibiotics. Because birth rates were slower to decline, some countries have been caught in a **demographic trap**—the point at which population growth overwhelms the environment's carrying capacity. That is, people's needs for food and shelter exceed the sustainable yield of surrounding forests, grasslands, croplands, or aquifers to the point that people begin to consume the resource base itself (Brown 1987).

URBANIZATION. Urbanization in developing economies differs in several major ways from urbanization in developed economies. For one, in the 18th and 19th centuries, millions of Europeans who were pushed off the land were able to migrate to sparsely populated places such as North America, South America, South Africa, New Zealand, and Australia. If these millions had been forced to make their living in the European cities, the conditions there would have been much worse than they actually were (Light 1983). The problem of urbanization in developing countries is compounded by the fact that many people who migrate to the cities come from some of the most economically precarious sections of their countries. In fact, most rural-to-urban migrants are not pulled into the cities by employment opportunities; rather, they are forced to move there because they have no alternatives. When these migrants come to the cities, they face not only unemployment but also a shortage of housing and a lack of services (electricity, running water, and waste disposal).

One distinguishing characteristic of cities in developing countries is the prevalence of slums and squatter settlements, which are much poorer and larger than even the worst slums in industrialized countries.

Prince Brown, Jr., Courtesy of Joan Ferrante

Sociologists use the term **over-urbanization** to describe a situation in which urban misery—poverty, unemployment, housing shortages, and insufficient infrastructure—is exacerbated by an influx of unskilled, illiterate, and poverty-stricken rural migrants who have come to the cities out of desperation (Dugger 2007).

What Do Sociologists See?

Sociologists see this child as being part of a society where children have economic value. The young girl carrying water and taking care of her younger sibling is contributing to the household economy. She is not a "drain" on resources like children in economies where they consume financial and other resources.

U.S. Army Corps of Engineers

Critical Thinking

Think about your family in terms of the family members who come together for important holiday gatherings. What are their ages? Does the structure of the gathering reflect the aging of American society?

Key Terms

demographic gap

demographic trap

mortality crises

over-urbanization

theory of the demographic transition

Module 13.6 Ageism and the Rise of the Anti-Aging Industry

Objective

You will learn the meaning and origins of ageism.

If you were asked to describe one of your grandparents, what would you say?

Lisa Southwick

Chances are that you would have very positive things to say about your grandparents. The positive experiences most of us have stand in stark contrast to the pervasive negative stereotypes regarding the mental abilities and physical characteristics of older people in general.

Professor Glen Hougan (2007) found that students taking his Design for an Aging Population course held deeply ingrained negative stereotypes about the elderly. Hougan required his students to work in teams to create an aging suit that would simulate some of the physiological challenges and experiences associated with aging. The point of this exercise was to sensitize students to the needs of older populations. However, the suits students created actually did little to stimulate innovative product designs; instead, they supported designs that mirrored and reinforced stereotypes about the elderly.

As Hougan described it, one group of students decided to wear "an 'old man' latex mask that one buys at a joke shop. Their explanation, in all seriousness, was that it enhanced the feeling of being elderly. Another group, using that same principle but a different tactic, incorporated movement restriction devices into clothes they perceived that the elderly wear. Their version of what the elderly wear was a drab, wrinkled, oversized suit jacket and pants one finds in the far reaches of a thrift store" (2007, n.p.).

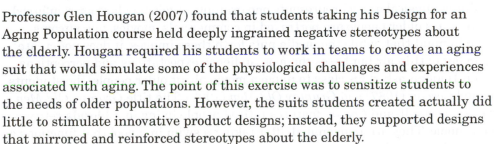

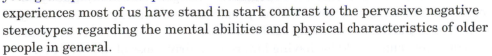

NKU Sociology, Missy Gish

Hougan also asked his students to write down three words to describe the elderly. Seventy-five percent of the words related to being weak, slow, or feeble. For the course to be effective, the professor had to find a strategy to help his students overcome their prejudices and stereotypes. Hougan did this by capitalizing on students' generally positive relationships with their grandparents, asking students to interview them, research their lives, and design something for them.

Ageism

Ageism is systematic prejudice, stereotyping, and/or discrimination on the basis of age. Ageism applies to any age group but is especially common toward those considered elderly. **Institutionalized ageism** occurs when discrimination based on age is viewed as an accepted, established, and even fair way of doing things, but on close analysis, such practices put a certain age group at a disadvantage relative to other age groups. In the context of product design for older people, senior design analyst Gretchen Anderson (2010) maintains that "if we view seniors through the products that are available to them, then they would be viewed as 'cranky, stupid, and tacky.'" The products actually reflect the ageist beliefs held by designers and manufacturers. Anderson offers the example of orthopedic shoes. "The orthopedic shoe, big and bulky, has a style and a color that has not changed in over 40 years." Anderson believes that when most designers "talk about needs of seniors there is a tendency to imagine someone whose eyesight, dexterity, and hearing are so impaired that they are incapable of having an experience."

Of all the -isms—racism, sexism, classism—ageism is the most widespread and the least likely to be challenged. It is the most widespread because everyone, if they live long enough, regardless of race, class, or gender, will eventually be considered "older." Ageism toward older people goes unchallenged on a variety of fronts. The popularity of birthday cards that make fun of, even ridicule, people sometimes as young as 30 for having lowered sex drive, loss of memory, and declining physical appeal represents one example. Ageism is further reinforced by the ways older people are portrayed in commercials, movies, and sitcoms. Keep in mind that, while the media puts forth these images, it does not create them out of thin air but bases them on audience expectations and perceptions. When Hougan asked his class to name a movie about older people, the students named *Cocoon* (1987) and *Grumpy Old Men* (1993). The title *Grumpy Old Men* speaks for itself, and *Cocoon* features retired people living in a Florida retirement home. They are rejuvenated after taking a swim in a pool that functions as a fountain of youth.

Hougan maintains that if we take a long view, we will see that ageism is rooted in a number of social shifts that have gradually eroded the status of the elderly over time. Those shifts include the following:

- Increased literacy rates have diminished older persons' role as authorities and keepers of oral traditions. Literate societies have libraries, bookstores, and now online holdings. Increased literacy is also synonymous with formal schooling. In the 1860s the United States became the first country in the world to embrace mass education, making elementary-level education for all the law. Within 60 years all states had passed compulsory attendance laws. Of course, these laws meant that young people spent their days in an age-segregated environment (i.e., schools) isolated from the influence of other age groups, in particular the elderly.

- With industrialization, technical skills became more valued than lived experiences. Moreover, since technical skills are always in need of update, these steady advances put older people who did not grow up with the latest technologies in a position in which they must find ways to master them.
- The increased mobility that accompanied industrialization dispersed families geographically and as a result weakened connections to older relatives. This is significant for ageism because it is interactions with older people that help dispel stereotypes and other misconceptions about them.
- At one time people worked until they were no longer able, at which point they "retired." As early as 1875 the idea that workers retire and enter a new stage in life took root. In that year the American Express railroad company established the first private pension plan in the United States. In subsequent years and decades, other employers began offering pension plans and governments implemented benefit plans.

This poster is prompting people to "boldly go" online to use Social Security services, including applying for Social Security. In 1956 (for women) and 1961 (for men) the Social Security Act became law, allowing people to retire at age 62 with partial monthly benefits. Retirement age with full monthly benefits ranges from 65 to age 67 depending on year of birth (Sidor 2013). Retirement can be viewed as a systematic process that removes older people from the workforce (a kind of de facto segregation).

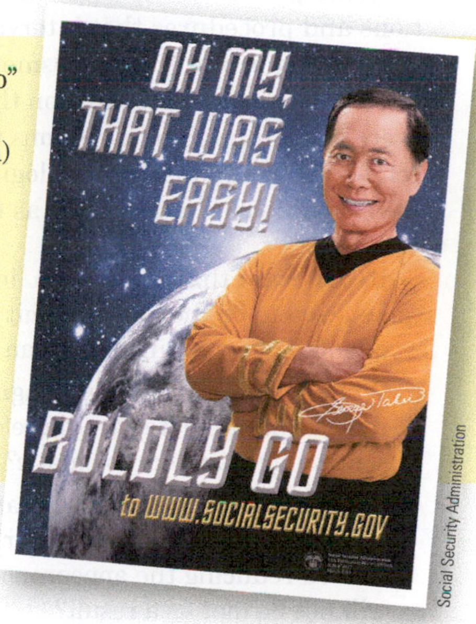

Social Security Administration

- The numbers of age-segregated environments in which people spend much of their day surrounded by people who are their own age (schools, day care, camps, retirement communities, clubs, neighborhoods) are increasing.
- The biological aging process has come to be perceived as a process that moves people closer to the inevitable—death. While this may seem obvious, death was not always something associated with old age.

NKU Sociology, Missy Gish

At one time in the United States death was something people in every age group experienced, but especially those less than 1 year of age. Twenty-one percent of all deaths in 1900 involved infants less than 1 year old. As late as the 1950s, 7 percent of deaths (1 in every 14) involved infants less than 1 (Centers for Disease Control 2014).

- An increasing value is placed not just on youth and beauty but on idealized youth and beauty that can only be achieved through artificial means (surgery, right camera angles, computerized touch-up). When these ideals are juxtaposed against dominant images of old age, it makes aging an especially dreaded process, so much so that an anti-aging industry has emerged.

The Anti-Aging Industry

In the past decade or so there has been an unprecedented increase in the number of products and procedures that are marketed under the label "anti-aging." These products promise to prevent visible signs of aging, prevent physical and mental decline, reverse the biological process of aging, and extend life. There are essentially two branches of what can only be called an anti-aging industry: (1) potions, lotions, and cosmetic treatments that reduce the physical signs of aging (wrinkle-prevention crèmes, hair restoring products); and (2) medical interventions and procedures that intervene to prevent or reverse aging (prescriptions such as Botox and Viagra, cosmetic surgeries) (Vincent et al. 2008). While there is little systematic research on the social consequences of anti-aging products, there are a number of questions that guide the sociological perspective as it relates to this industry. Sociologist John A. Vincent and his colleagues (2008) have outlined such questions as the following:

- In the event that science can find a way to prolong life beyond the 120 years believed to be the natural limit of human life, what are the potential social consequences of extending life for an already aging population?
- What symbolic meanings of aging and old age does the anti-aging industry reflect and promote? Are there other meanings of aging and old age other than something that is problematic?
- How will anti-aging trends change the way we think about the purpose of medicine and the role it plays in our lives? Will medicine redirect limited resources toward reducing the appearance and effects of aging? What other medical needs will not be met as a result?
- Does the anti-aging enterprise intensify ageism and subject those who are aging to "dissection, manipulation and control" (292)?

Lisa Southwick

Do aging and old age have a value, or are they something to be resisted at any cost? When we consider the important role that grandparents and great-grandparents play in people's lives, we see there is value to stepping up and assuming one's place in the life cycle. One could argue that young people need secure older people in their lives who are not offended by that status. How might the anti-aging movement affect that security if older people seek to be young and not act like grandparents or great-grandparents?

What Do Sociologists See?

Sociologists see an advertisement for cosmetic surgery that promises a "natural look." The prevalence of such advertisements suggests that an increasing value is placed, not just on youth and beauty, but on idealized youth and beauty that can only be achieved through artificial means. These kinds of ads, which are part of the anti-aging industry, play a significant role in degrading the status of the elderly and supporting and intensifying ageism.

Rachel Ellison, Courtesy of Joan Ferrante

Critical Thinking

Describe someone considered older who has made an impact on your life. In what ways did their older age contribute to that impact?

Key Terms

ageism

institutionalized ageism

Module

13.7 Changing Environment

Objective

You will learn that environmental sociologists focus on how human activities, especially profit-making activities and overconsumption, affect the quality of life on the planet.

Is the Buddhist monk lifestyle, which rejects accumulation of material wealth, a model of sustainable living?

Chris Caldeira

As we will learn, the answer to this question is that it depends—it depends on where in the world the Buddhist monk lives. To answer this question reference must be made to the carbon footprint. A **carbon footprint** is the impact a person makes on the environment by virtue of his or her lifestyle. That impact is measured in terms of fossil fuels consumed or units of carbon dioxide emitted from burning that fuel. We can think of carbon footprints as primary or secondary. The primary carbon footprint is the total amount of carbon dioxide emitted as the result of someone's direct use of fossil fuels to heat and cool a home, run appliances, power an automobile, and so on. The secondary footprint is the total amount of carbon dioxide emitted to manufacture a product for or deliver a service to that person.

Carbon Footprints

Tim Gutowski (2008a), a professor of mechanical engineering at MIT, and 21 of his students studied 18 different lifestyles in the United States, including a person without a home (homeless), a Buddhist monk, a patient in a coma, and a professional golfer. After extensive interviews they estimated the energy each lifestyle requires. They estimated that none of the 18 lifestyles uses less than 120 gigajoules of energy, even that of a Buddhist monk, whose lifestyle is devoted to simple living, modest dress, and a vegetarian diet. A gigajoule (GJ) is a metric measure of energy consumption. It is a particularly useful measure because it can be applied to different types of energy consumption, such as kilowatts of electricity, liters of heating oil or gasoline, and cubic feet of natural gas. One GJ is equivalent to the energy needed to cook over 2,500 hamburgers or to keep a 60-watt bulb lit for six months (Natural Resources Canada 2010).

Gutowski and his students estimated that the Buddhist monk living in the United States consumed 120 GJ of energy each year and the professional golfer consumed 8,000 GJ. The Buddhist monk consumes about one-third as much energy as the average American (350 GJ) but double the average amount of energy a typical person on the planet consumes (64 GJ). The Buddhist monk's 120 GJ emits 8.5 metric tons of carbon dioxide each year; the professional golfer's lifestyle (8,000 GJ) produces 566 metric tons of carbon emissions.

Gutowski's study (2008b) has important implications: the United States has a "very energy-intensive system" that in effect sets lower limits on how much energy anyone living in the United States uses. If even a Buddhist monk living in the United States uses 120 GJ, then we might conclude that energy use is woven into the fabric of American society (Revkin 2005).

Environment and Society

Environmental sociologists focus on how human activities, especially profit-making activities and overconsumption, affect the quality of life on the planet. Environmental sociologists seek to identify the social, political, economic, and technological factors that contribute to pollution, overconsumption, and waste, which in turn threaten ecosystems, human life, and other species that share the planet (American Sociological Association [ASA] 2012a). This focus differs from classic sociological inquiry that makes humans the center of study. Environmental sociologists seek to uncover the ways in which human activities influence and affect the natural environment (Dunlap and Catton 1979; Buttel and Humphrey 2002).

Much of the focus of environmental sociology has been on explaining how the consumption patterns of the United States and other industrialized countries are connected to environmental destruction. Sociologists associate environmental problems with the global economy's ever-increasing drive for profit. The term **treadmill of production** captures the ceaseless increases in production and, by extension, consumption that are needed to sustain the global economy's success, which is measured by increased profits (Schnaiberg 1980; Gould et al. 2008).

When sociologists think about the cycle of consumption, they don't think about any one consumer, but rather they think about what any one consumer's behavior represents with regard to waste. Consider disposable diapers and the number that any one baby goes through before being potty trained. One estimate puts that number at 8,000 (Environmental Protection Agency 2011). Since the advent of disposable diapers in the 1970s, the average age at which children are potty trained has increased, further increasing the number of diapers any one child uses.

Karl Wiesbaden

This never-ending cycle or treadmill has a devastating impact on the environment because the relentless focus on producing and consuming increases energy consumption, waste, and harmful emissions. Environmental sociologists contend that the damage to the environment is not shared equally; minority and low-income people who consume much less than the wealthy are disproportionately overexposed to environmental pollutants (Stretesky and Lynch 2002; Hooks and Smith 2004; Downey 2005). For example, a disproportionate share of toxic waste dumps and hazardous emissions is found in minority communities within the United States and in poorer nations abroad. This unequal distribution of environmental hazards based on race or socioeconomic status is evidence of **environmental injustice** (Bullard et al. 2007). For example, a study of factors that affect exposure to hazardous waste found that the majority of people living within 1.8 miles of hazardous waste facilities in the United States are minorities (Bullard et al. 2007).

Changing Consumption Habits

In thinking of ways to change consumption habits, we must recognize that energy consumption is woven into the fabric of American society and, to varying degrees, into the fabric of the world's countries. Changing these habits requires a revolution in thinking and behaving. This revolution will not look like the so-called Green Revolution, which to date seems to emphasize consuming green products and taking easy steps to change the way we live, as described by the following headlines: "21 Ways to Save the Earth and Make More Money," "365 Ways to Save the Planet Earth," and "Ten Ways to Green Up Your Sex Life" (Friedman 2008). Our current energy system took more than a hundred years of investments to establish (the age of oil in the United States began in 1901); it may take a hundred years of investments to put a clean energy system in place (Friedman 2008). If we consider the total energy consumed from all fossil fuel sources on a global scale (although it is unequally distributed), humans consume the equivalent of 10 million barrels or 420 million gallons per hour—the United States consumes 84 million gallons of that hourly total (Friedman 2008). This is the scale of demand, which is only expected to increase in coming years.

Chris Caldeira

Environmental sociologists are particularly interested in identifying realistic strategies that countries, communities, and individuals are taking to promote sustainable relationships between humans and their natural environment. Examples of successful strategies can be found from the individual level, such as solar-powered showers, to the country level, where South Korea is one model of recycling and Sweden a model of energy efficiency (Zumbrun 2008; Williamson 2011).

What Do Sociologists See?

Sociologists see a product that was designed to be thrown away after a few uses, or even one use. By one estimate people throw away 2.5 billion disposable razors a year (Pennsylvania Department of Environmental Protection 2014). These products are part of the treadmill of destruction.

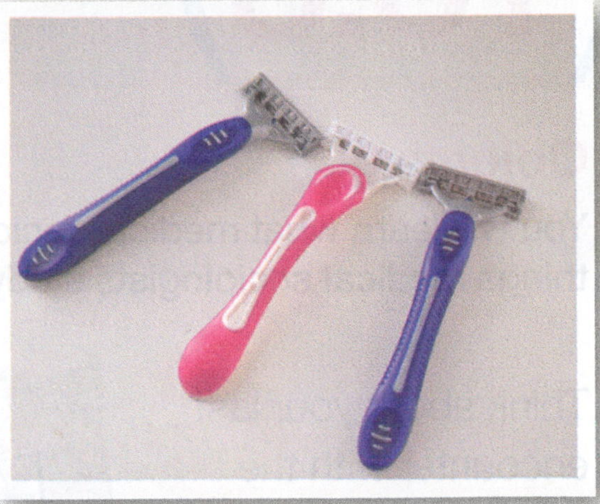

NKU Sociology, Missy Gish

Critical Thinking

Are there areas of your life in which you are guilty of overconsumption? Explain.

Key Terms

carbon footprint

treadmill of production

environmental injustice

Module 13.8 Health Care

Objective

You will learn what medical sociology is and the kinds of things medical sociologists study.

Think about your last encounter with the health care system in the United States. Did you come away with a positive or a negative assessment?

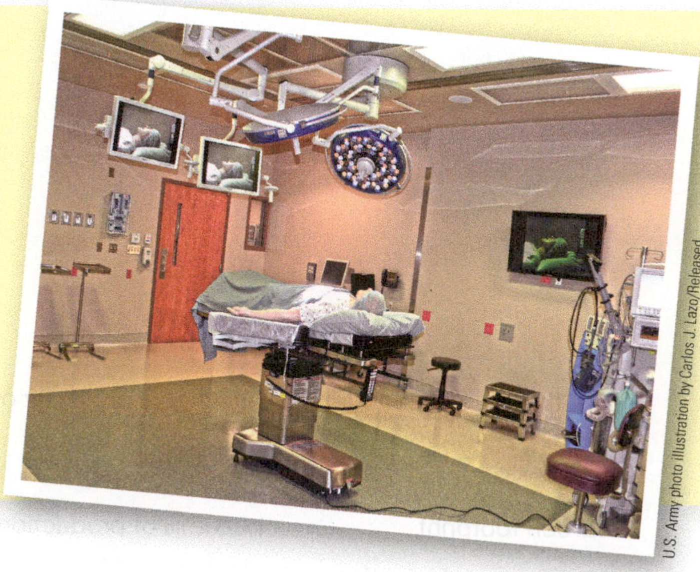

U.S. Army photo illustration by Carlos J. Lazo/Released

According to a nationally representative survey of 15,735 health care consumers in 12 countries (Belgium, Brazil, Canada, China, France, Germany, Luxembourg, Mexico, Portugal, Switzerland, the United Kingdom, and the United States), Americans are among those least satisfied with their system of health care. Almost 40 percent of U.S. respondents gave their system a grade of D or F. Only 22 percent gave the health care system an A (excellent) or B (very good) compared with the 50 percent or more of respondents in Luxembourg (69 percent), Belgium (57 percent), Switzerland (52 percent), France (51 percent), and Canada (50 percent) who gave an A or B grade to their systems (Deloitte 2011b). Why such high dissatisfaction? This is the kind of question medical sociologists seek to answer.

Medical sociologists pay special attention to the social context of health care and give special emphasis to the ways in which health, disease, and illness are defined and experienced. They also study how medical care is organized and delivered, and the relationships among health care providers and other stakeholders.

There are 13,700 sociologists who belong to the ASA. Approximately one in five has teaching or research interests that relate to medical sociology (Scelza et al. 2011; ASA 2012b). As one measure of medical sociology's importance to the discipline, the ASA publishes the *Journal of Health and Social Behavior*. In conjunction with the journal's 50th anniversary, a special issue titled "Reflection on 50 Years of Medical Sociology" was released. Here we review some key findings from that issue that focus on the social dimensions of medicine. The social dimensions are important because health, disease, and illness are not simply biological experiences. They are also social experiences, if only because people give names and assign meaning to biological experiences and establish systems for promoting health and fixing disease and illness.

Key Features of the U.S. System of Health Care

On March 23, 2010, President Barack Obama signed into law the Patient Protection and Affordable Care Act, also referred to as the Affordable Care Act (ACA) or "Obamacare." Regardless of one's opinion about this law, there is no doubt that it mandates a revolutionary overhaul of the rules governing health care in the United States, not seen since Social Security Amendments of 1965, which gave us Medicare and Medicaid. Among other things, the ACA was written with the goals of increasing the quality of health care, making insurance and health care more affordable, reducing the number of uninsured, and expanding coverage to include coverage for preexisting conditions. The Affordable Care Act is one response to the following features of the U.S. system of health care.

1. *The U.S. health care system is the most expensive in the world.* In 2013 the United States spent $2.7 trillion on health care, an amount that is equal to 17.6 percent of its gross domestic product (GDP), or an average of $8,508 per person. Switzerland is next in line, spending 12.3 percent of its GDP, or an average of $5,643 per person (Figure 13.8a). While the United States spends substantially more than other countries, it is not recognized as a leader in health outcomes, most notably life expectancy, infant mortality, and other measures of well-being.

Figure 13.8a: Percent of GDP and Per Capita Spending on Health Care

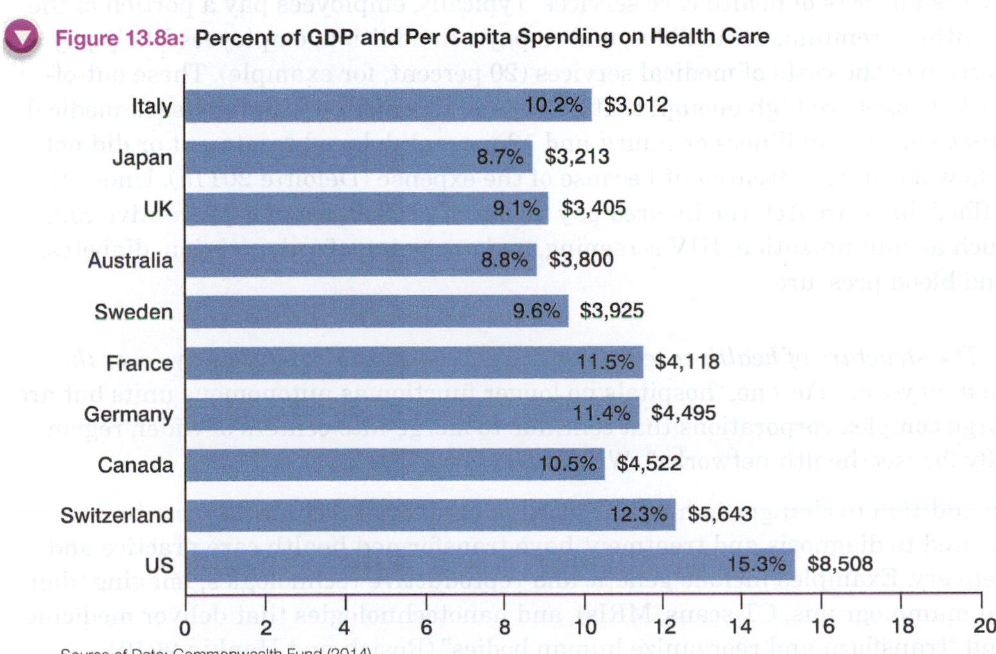

Source of Data: Commonwealth Fund (2014)

2. *When compared with the world's wealthiest countries, the United States is the only country that does not provide some kind of universal health insurance coverage* (Deloitte 2011b; Figure 13.8b). As a result, before the ACA 16.3 percent of the U.S. population (or 41.1 million) was counted as uninsured. At the time of this writing that number was reduced by 7.1 million.

 **Figure 13.8b: Profile of Uninsured Now Eligible for Health Care Under the Affordable Care Act**
There are 41.1 million uninsured Americans who are eligible for coverage through the marketplace. Of these 41.1 million, what percentage are nonelderly? Are male?

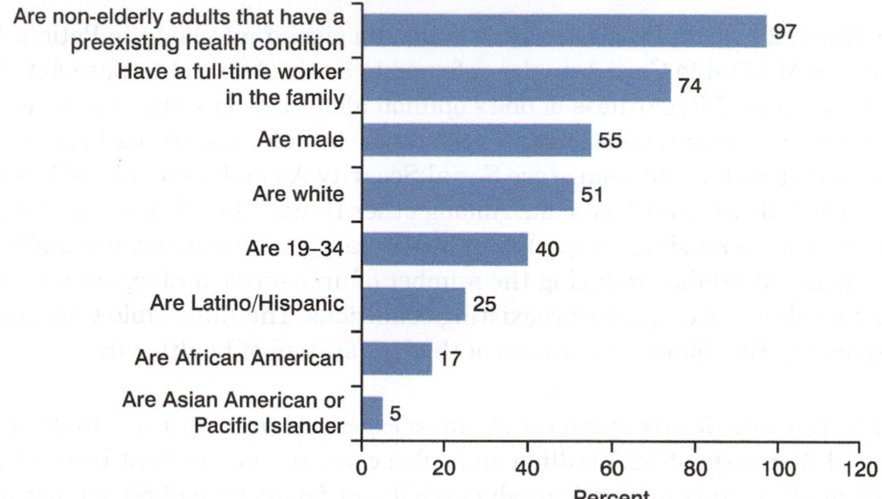

Source: HHS.gov (2014)

3. *Rising and expensive health care costs have made health care unaffordable for many Americans, even those who are insured.* It is important to point out that employer-sponsored health care coverage does not mean that the employer covers all costs of health care services. Typically, employees pay a portion of the monthly premium, deductibles, and co-pays. In addition, employees likely pay a portion of the costs of medical services (20 percent, for example). These out-of-pocket costs are high enough that 25 percent of Americans did not seek medical treatment for an illness or injury, and 19 percent delayed treatment or did not follow a course of treatment because of the expense (Deloitte 2011a). Under the Affordable Care Act, the insured pay no out-of-pocket costs for preventive care such as immunization, HIV screening, and screenings for depression, diabetes, and blood pressure.

4. *The structure of health care delivery has changed in dramatic ways over the past 50 years.* For one, "hospitals no longer function as autonomous units but are large complex corporations that continue to merge into centers of wider, regionally focused health networks" (Wright and Perry 2010).

In addition to changes in hospital reach, a number of technical innovations related to diagnosis and treatment have transformed health care practice and delivery. Examples include genetic and reproductive technologies, imaging (digital mammograms, CT scans, MRIs), and nanotechnologies that deliver medicine and "transform and reorganize human bodies" (Rosich and Hankin 2010).

Transplant and other body part replacement options are among the technologies that have transformed patient treatment and outcomes. An amputee can be fitted with a sophisticated, myoelectric arm that responds to electrical signals from the muscle to which it is attached, enabling patients to perform complex movements connected to personal hygiene.

Lamel Hinton

Because U.S. health care providers get paid for services rendered and not for health care outcomes, there are strong incentives to use the newest technologies because they are considered state-of-the-art and to also overtest and overprescribe treatments (see Figure 13.8c).

Figure 13.8c: Number of MRI Exams and CT Scans Performed per 1,000 Population in Selected Countries, 2009

The charts show the number of people per 1,000 population who received an MRI exam or CT scan in 2009. Which country has the highest rates? Where does the United States rank? One might wonder whether Greece and the United States have older populations, and whether that fact might help explain the high use. All countries listed have aging populations, however, so that is not an explanation. In addition, the average cost of MRI and CT scans are considerably higher in the United States than in other countries. As one example, the average cost of an MRI in the United States is $1,080 versus $280 in France (Klein 2012).

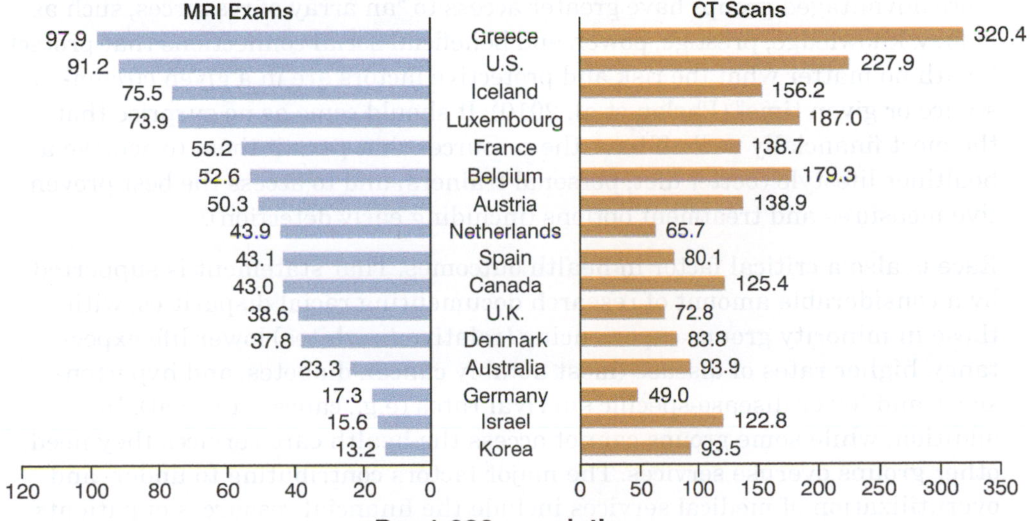

Source of Data: OECD (2011b, slide 25)

One international survey of 100 insurers in 25 countries found that the cost of 22 out of 23 medical services and products ranging from a routine doctor's visit to coronary bypass surgery were highest in the United States. The exception to the rule was cataract surgery, which is highest in Switzerland, followed by the United States. The leading explanation for the high cost of health care was, "Providers largely charge what they can get away with, often offering different prices to different insurers, and an even higher price to the uninsured" (Klein 2012).

5. *Until recently physicians have "been the voice of authority"* (Rosich and Hankin 2010). In other words, physicians controlled the course of treatment and the flow of information related to diagnosis, prognosis, and treatment. That dominance has been undermined in a number of ways. Most notably, other stakeholders—hospitals, pharmaceutical companies, insurance companies, and patients—have taken a visible and more active role in deciding the diagnosis and course of treatment. As a case in point, pharmaceutical companies market drugs directly to consumers. And because patients have access to medical information via the Internet and advertisements, they now assume a larger role in managing preventive care, the course of illness, and treatment.

Insurance companies have also undermined the authority of the doctor by setting the terms of care and treatment. For example, some insurance companies require doctors to handle one health problem per visit so that patients must return a second time (or a third) to address another health concern. This, of course, involves multiple co-payments. Finally, many hospitals assign physicians on staff to care for admitted patients, eliminating the involvement of the patient's primary care and other doctors.

6. *Health inequalities are deeply rooted and persistent.* It is an established fact that those in low-income and disadvantaged populations have worse health outcomes and lower life expectancy than those in advantaged populations. The theory of fundamental causes offers one explanation. According to this theory, health status and socioeconomic status are intertwined so much that those in more advantaged groups have greater access to "an array of resources, such as money, knowledge, prestige, power, and beneficial social connections that protect health no matter what the risk and protective factors are in a given circumstance or given time" (Phelan et al. 2010). It should come as no surprise that the most financially well-off have the resources that permit them to achieve a healthier lifestyle (better diet, personal trainers) and to access the best preventive measures and treatment options (including early detection).

Race is also a critical factor in health outcomes. That statement is supported by a considerable amount of research documenting racial disparities, with those in minority groups experiencing (relative to whites) lower life expectancy, higher rates of disease (most notably cancer, diabetes, and hypertension), and lower disease-specific survival rates (e.g., cancer survival). In addition, while some groups cannot access the health care services they need, other groups overuse services. The major factors contributing to under- and overutilization of medical services include the financial resources of patients (do patients have insurance, does their plan cover a particular treatment or medication?), financial incentives, and the availability of technologies such as MRI and CT scans.

7. *Stressful life events and the quality of social relationships affect health outcomes.* We know without a doubt that stressful events have an impact on a person's mental and physical health. Such events include the death of a spouse or partner, an abusive relationship, unemployment, chronic discrimination, insufficient income to make ends meet, and caretaking responsibilities for someone with a chronic condition. We also know that health is affected by **social relationships**, the nature, number, and quality of the ties that bind people formally and informally to others. Formal ties are relationships people have with those they interact with in the context of their workplace, school, religious group, and other organizations. Informal relationships include those ties that transcend shared membership in an organization—ties with friends, family, and other relatives. We know that the more social support people have to draw upon, the greater the positive impacts on health and lifestyle (Thoits 2010; Umberson 2010).

8. *Disease, illness, and health are "social constructs as well as medical constructs"* (Rosich and Hankin 2010). Disease, health, and illness are not simply biological events; they are social events as well. Perhaps the most dramatic example of this relates to AIDS, which was originally named GRIDS (gay-related immune deficiency syndrome). People notice diseases and give them names and in that process convey meanings about the medical condition. The concept of medicalization helps us see the social processes by which the inevitable problems associated with living and aging become defined as medical problems. Arguably the most vivid examples are from the pharmaceutical industry, which plays an important role in naming physical conditions that can be treated with medicines—not enough eyelashes, menstruation (shortened and fewer cycles), and erectile dysfunction are physical conditions that have been defined as diseases treatable with medication.

When something is medicalized, it becomes something in need of treatment. In this sense a woman's labor to deliver a baby has been medicalized. Rates of caesarean deliveries are considerably higher in some countries such as the United States. These differences in how delivery of a baby is handled offer insights into the social construction of what constitutes risk to baby and mother (see Figure 13.8d). About one-third of women who give birth in Korea, the United States, Australia, and Germany get a C-section. Contrast that to Iceland and the Netherlands, where about 15 percent of women who give birth get a C-section.

Figure 13.8d: Number of C-Sections per 100 Live Births by Selected Countries
As you can see, defining a physical condition as a "disease" shapes the experience of illness, the kind of treatment prescribed, and cost of care. Which country listed has the highest number of C-sections per 100 births? Which country has the lowest rate per 100 births?

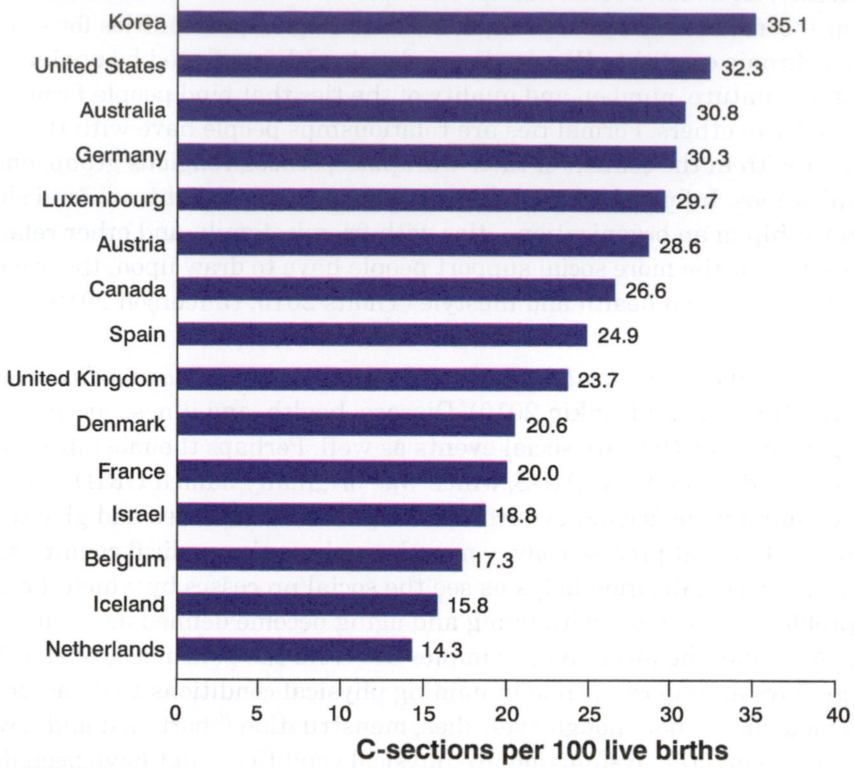

Source of Data: OECD (2011b)

What Do Sociologists See?

Physicians' ability to control the course of treatment and the flow of information related to diagnosis, prognosis, and treatment has been reduced significantly. Other stakeholders such as pharmaceutical companies, insurance companies, and patients have assumed more active, even aggressive roles in directing diagnosis of illness and course of treatment.

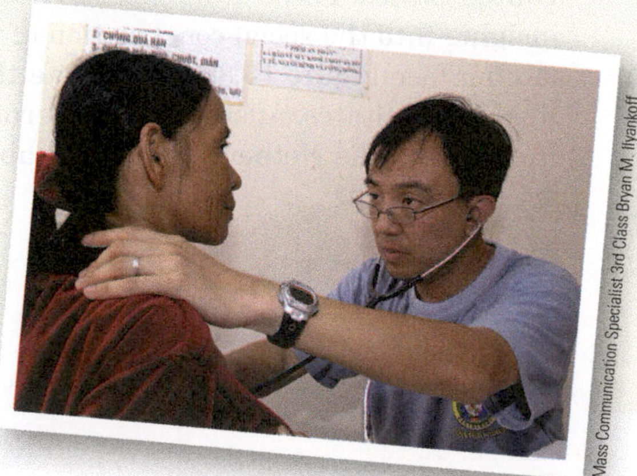

Mass Communication Specialist 3rd Class Bryan M. Ilyankoff

Critical Thinking

Do you know anyone, including yourself, who has been affected by the Affordable Care Act? Describe how they were affected and assess whether the experience is positive, negative, or both.

Key Term

social relationships

Module 13.9 Applying Theory: The Affordable Care Act

Objective

You will learn how sociologists apply each of the four perspectives to assess some dimension of the Affordable Care Act.

Do you have an opinion of the Affordable Care Act? Do you think benefits will outweigh drawbacks?

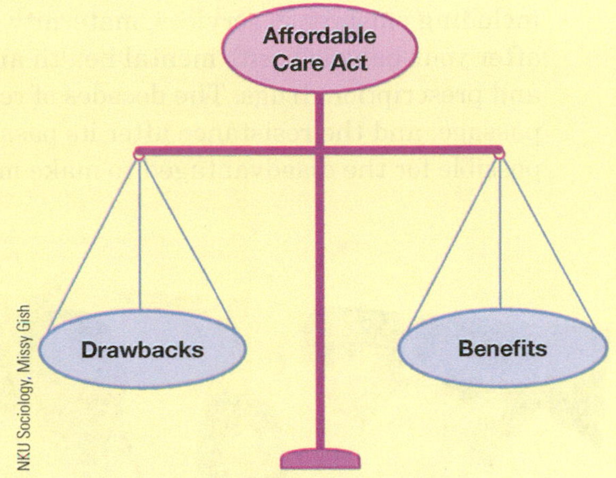

By any measure the Affordable Care Act is revolutionary in its impact on the U.S. system of health care. It will likely be decades before its full impact will be known or understood. The four sociological perspectives offer insights about how to analyze its impact.

Functionalists focus on existing social order and stability. From a functionalist point of view, the Affordable Care Act represents an attempt to correct instability in an existing social order. That instability is related to the high percentage of the population who were uninsured and who could not secure insurance coverage because they had a preexisting condition. Of course, the ACA disrupted the existing order to create a new order. As one example, insurance providers must now spend 80 percent of premium dollars on providing for health care, which means profit margins are trimmed.

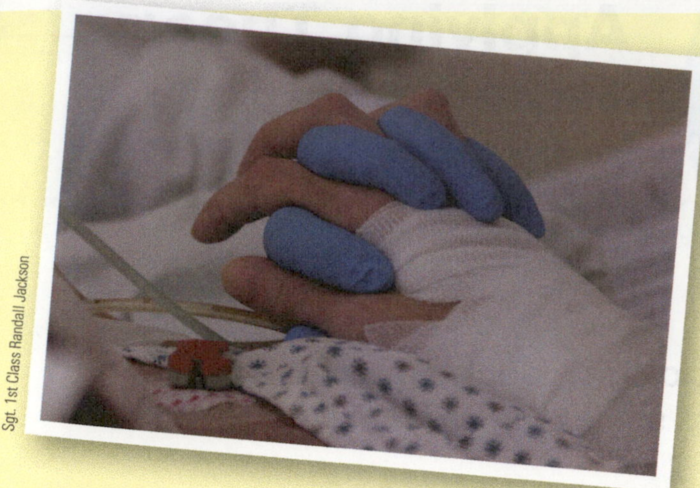
Sgt. 1st Class Randall Jackson

Conflict theorists focus on conflict over scarce and valued resources. The Affordable Care Act gives access to health care to those previously denied coverage because they had a preexisting health condition. Under the health care law, insurers cannot charge more or deny coverage to anyone because of a preexisting condition. Insurance plans must also include health benefits deemed essential in the ACA, including emergency services, maternity and newborn care (care before and after your baby is born), mental health and substance use disorder services, and prescription drugs. The decades of resistance leading up to ACA's eventual passage, and the resistance after its passage, suggest that through conflict it is possible for the disadvantaged to make meaningful gains.

Community Health Centers
Access to care in all 50 States.
HealthCare.gov
Health Insurance Marketplace

HealthCare.gov

Sociologist thinking as symbolic interactionists reflect on how the meaning assigned to the Affordable Care Act varies depending on what it is called. It seems that referring to the plan as Obamacare versus the Affordable Care Act makes a difference in people's opinion about the plan. Symbolic interactionists' thoughts extend to the role the Affordable Care Act plays in opening opportunities for the newly insured to interact with doctors and other health care professionals.

The feminist perspective assesses the Affordable Care Act in terms of how well it addresses gender inequalities in health care. There are a number of areas where women benefit. For one, under this act women cannot be charged more for health insurance because of their gender. "Women gained expanded access to preventive services with no cost-sharing, including mammograms, cervical cancer screenings, prenatal care, flu and pneumonia shots, and regular well-baby and well-child visits. And many health plans now cover additional preventive services with no cost-sharing, including well-woman visits, screening for gestational diabetes, domestic violence screening, breastfeeding supplies and contraceptive services" (HHS.gov 2014).

Being a woman
will no longer be a pre-existing condition, starting in 2014.
HealthCare.gov

Summary: Putting It All Together

Chapter 13

When sociologists study social change they ask: What has changed? What factors trigger change? and What are the consequences of change? The discipline of sociology offers concepts to help us identify and describe revolutionary change (e.g., globalization, rationalization, urbanization, and so on). The triggers include technological innovation, revolutionary ideas, conflict, globalization, the pursuit of profit, and social movements. For example, technological innovations (domestication, the hoe, the plow, mechanization, computer/digitalization) created six broad types of societies, each defined by a signature technology that allowed each society to create varying amounts of surplus wealth. Those six societies are hunting and gathering, pastoral, horticultural, agrarian, industrial, and postindustrial.

Social movements form when there is (1) an actual or imagined condition that enough people find objectionable, (2) a shared belief that something needs to be and can be done about that condition, and (3) an organized effort to attract supporters, articulate a mission, and define a change-making strategy. Usually those involved in social movements work outside the system to advance their cause because the system has failed to address the problem.

Globalization and glocalization shape much of what goes on in our day-to-day lives. The concept of glocalization draws our attention to the intersection of the local and the global. Globalization and glocalization are intertwined because globalization always involves a series of countless glocalizations.

We also focused on three specific areas of social change that count among the most pressing issues of our day—the aging of the world's population, the environment, and health care (and the new health care changes mandated by the Affordable Care Act). With regard to aging, there is no historical precedent in the United States, or the world, for such a large proportion of the population in and moving toward older ages. This dramatic demographic change is complicated by ageism and the rise of the anti-aging industry.

When sociologists study the environment, they focus on how human activities, especially profit making and overconsumption, affect the quality of life on the planet. The ever-increasing need of the economic system to make a profit supports what is known as the treadmill of production (overproduction).

This chapter also considered the characteristics of the U.S. system of health care and changes in the ways medical care is delivered. Characteristics of note include the fact that the U.S. system is the most expensive in the world and that the United States is the only wealthy country that does not provide some form of universal health care coverage. These two characteristics explain in large part the forces that helped push through revolutionary change mandated by the Affordable Care Act.

Glossary

absolute poverty—a situation in which people lack the resources to satisfy the basic needs no person should be without. Absolute poverty is usually expressed as a living condition that falls below a certain threshold or minimum.

absorption assimilation—the process by which a subordinate ethnic, racial, and/or cultural group adapts to the ways of the dominant group, which sets the standards to which they must adjust.

achieved statuses—human-created social categories and characteristics acquired through some combination of personal choice, effort, and ability.

adaptive culture—the part of the culture (nonmaterial) that adjusts to a new product or innovation, specifically to the associated changes that product or innovation promotes.

advantaged—a situation in which the positive symbolic representations and valued resources are disproportionately held by a particular group relative to another group.

ageism—a systematic prejudice, stereotyping, and/or discrimination on the basis of age.

agents of gender socialization—the significant people, groups, and institutions that act to shape our gender identity—whether we identify as male, female, or something else.

agents of primary socialization—family members and caretakers who prepare infants and children to live as family members and to go out into the larger society.

agents of secondary socialization—people other than family and caretakers who expose and teach people of all ages things they need to know to assume a particular role outside the home in the larger society.

agents of socialization—significant people, groups, and institutions that act to shape our sense of self and social identity, help us realize our human capacities, and teach us to negotiate the world in which we live.

aging population—a society in which the percentage of the population that is 65 and older is increasing relative to other age groups.

agrarian society—a society that emerged with the invention of the plow 6,000 years ago, triggering a revolution in agriculture. The plow made it possible to cultivate large fields and increase food production to a level that could support thousands to millions of people, many of whom lived in cities and/or were part of empires.

alienation—a state of being in which humans lose control over the social world they have created and are dominated by the forces of their inventions.

altruistic (suicide)—a state in which the ties attaching the individual to the group are so strong that a person's sense of self cannot be separated from the group.

anomic (suicide)—a state in which the ties attaching the individual to the group are disrupted due to dramatic changes in circumstances.

anomie—a state of cultural chaos resulting from structural strain.

ascribed statuses—human-created social categories and characteristics to which people are assigned and/or with which they identify that are the result of chance in that people exert no effort to obtain them. Birth order, race, sex, and age typically qualify as ascribed statuses.

assimilation—a process by which ethnic, racial, and/or cultural distinctions between groups disappear because one group is absorbed, sometimes by force, into another group's culture or because two cultures blend to form a new culture.

audience reference groups—those watching, listening, or otherwise giving attention to someone who has anticipated that audience's response in preparing what is the focus of attention (e.g., the audience response matters to that someone).

authoritarian government—a form of government in which no separation of powers exists; a single person (a dictator), a group (a family, the military, a single party), or a social class holds all power. No official ideology projects a vision of the "perfect" society or guides the government's political or economic policies.

authority—legitimate power, or power that people believe is just and proper. A leader has authority to the extent that people view him or her as being legitimately entitled to it.

back stage—areas out of sight of an audience, where individuals let their guard down and do things that would be inappropriate or unexpected in a front-stage setting.

beliefs—conceptions that people accept as true concerning how the world operates and the place of the individual in relation to others.

biography—all the events and day-to-day interactions from birth to death that make up a person's life.

biological sex—the physiological, including genetic, characteristics associated with being male or female.

bourgeoisie—the owners of the means of production.

brain drain—the emigration from a country of the most educated and most talented people, including those trained to be hospital managers, nurses, accountants, teachers, engineers, political reformers, and other institution builders.

bureaucracy—a completely rational organization—one that uses the most efficient means to achieve a valued goal.

calculability—a feature of McDonaldization that emphasizes numerical indicators by which customers and the service provider can judge the amount of product and the speed of service (e.g., delivery within 30 minutes).

capitalism—an economic system in which the raw materials and the means of producing and distributing goods and services are privately owned.

carbon footprint—the impact a person makes on the environment by virtue of his or her lifestyle. That impact is measured in terms of fossil fuels consumed or units of carbon dioxide emitted from burning that fuel.

carceral culture—a social arrangement under which the society largely abandons physical and public punishment and replaces it with surveillance to control people's activities and thoughts.

caregivers—those who provide service to people who, because of physical impairment, a chronic condition, or cognitive impairment cannot do certain activities without help.

case studies—objective accounts intended to educate readers about a person, group, or situation.

caste system—a system of stratification in which people are ranked according to ascribed statuses.

censorship—an action taken to prevent information believed to be sensitive, unsuitable, or threatening from reaching some audience.

chance—something not subject to human will, choice, or effort.

charismatic authority—legitimate power that is grounded in exceptional and exemplary personal qualities. Charismatic leaders are obeyed because their followers believe in and are attracted irresistibly to the leader's vision.

choice—the act of choosing from a range of possible behaviors or appearances.

church—a group whose members hold the same beliefs regarding what is considered sacred and profane, share rituals, and gather in body or spirit at agreed-on times to share and reaffirm their commitment to those beliefs and practices.

civil religion—an institutionalized set of beliefs about a nation's past, present, and future and a corresponding set of rituals. These beliefs and rituals can take on a sacred quality and elicit intense feelings of patriotism.

claims makers—those who articulate and promote claims and who tend to gain in some way if the targeted audience accepts their claims as true.

class conflict—an antagonism between exploiting and exploited classes.

class system—a system of stratification in which people are ranked on the basis of their achievements. That ranking is based on merit, talent, ability, or past performance.

coercive organizations—organizations that draw in people who have no choice but to participate.

colonialism—a situation in which a foreign power uses superior military force to impose its political, economic, social, and cultural institutions on an indigenous population in order to control their resources, labor, and markets.

color line—a barrier supported by customs and laws separating nonwhites from whites, especially with regard to their roles in the division of labor. The origin of that line can be traced to the European colonization of Africa.

commercialization of gender ideals—the process of introducing products into the market using advertising campaigns that promise consumers that they will achieve a gender ideal if they buy and use the product.

commodification—a process by which economic value is assigned to things not previously thought of in economic terms such as an idea, a natural resource (water, a view of nature), or a state of being (youth, sexuality).

commodification of sexuality—a process by which companies create products for people to buy with the promise that those products will allow them to express themselves as sexual beings or elicit a sexual response from others.

communitarian utopias—a counterculture in which members withdraw into a separate community in order to live with minimum interference from the larger society, which they view as evil, materialistic, wasteful, or self-centered.

comparison reference groups—groups that provide people with a frame of reference for judging the fairness of a situation in which they find themselves. That judgment is used to justify certain behaviors or assess performance relative to others.

comprehensive dyads—two people who have more than a superficial knowledge of each other's personality and life; they know each other in a variety of ways.

conflict—the major force that drives social change.

conformists—people who have not violated the rules or expectations of a group and are treated accordingly.

conformity (as a response to structural strain)—acceptance of cultural goals and the pursuit of those goals through legitimate means.

content analysis—a method of analysis in which researchers identify themes, sometimes counting the number of times something occurs or specifying categories in which to place observations.

context (with regard to race)—the social setting in which racial and ethnic categories are recognized, created, and challenged.

control—a feature of McDonaldization that emphasizes replacing employee labor with technologies and/or requiring, even demanding, that employees and customers behave in a certain way.

control variables—variables that researchers hold constant in order to focus on just the relationship between the independent variable and the dependent variable.

core economies—the wealthiest, most highly diversified economies in the world, with strong, stable governments.

corporate crime—a crime committed by a corporation as a result of the way that it does business competing with other companies for market share and profits.

countercultures—subcultures that challenge, contradict, or outright reject the values of the mainstream culture of which they are a part.

credential society—a situation in which employers use educational credentials as screening devices for sorting through a pool of largely anonymous applicants.

crime—an act that breaks a law.

cultural anchor—some item of material or nonmaterial culture that elicits broad consensus and support among a diverse membership, even in the face of debate and dissent about its meaning.

cultural capital—a person's nonmaterial resources, including educational credentials, the kinds of knowledge acquired, social skills, and acquired aesthetic tastes.

cultural diffusion—the process by which an idea, an invention, or a way of behaving is borrowed from a foreign source and then adapted to the borrowing people's culture.

cultural diversity—the cultural variety that exists among people who find themselves sharing some physical or virtual space.

cultural particulars—the *specific* practices that distinguish cultures from one another.

cultural relativism—a point of view advocating that a foreign culture not be judged by the standards of a home or other favored culture, and that a behavior or way of thinking must be examined in its cultural context—that is, in terms of that culture's values, norms, beliefs, environmental challenges, and history.

cultural universals—those things all cultures have in common.

culture—the way of life of a people, specifically the shared and human-created strategies for adapting and responding to the social and physical environment.

culture shock—the mental and physical strain that people can experience as they adjust to the ways of a new culture. In particular, newcomers find that many of the behaviors and responses they learned in their home culture, and have come to take for granted, do not apply in the foreign setting.

culture of spectacle—a social arrangement by which punishment for crimes—torture, disfigurement, dismemberment, and execution—is delivered in public settings for all to see.

decolonization—the process of gaining political independence from a colonizing power.

demographic gap—the difference between the birth and death rates.

demographic trap—the point at which population growth overwhelms the environment's ability to support that population.

dependent variable—the behavior to be explained or predicted.

desegregation—the process of ending legally sanctioned racial separation and discrimination, including removing legal barriers to interaction and offering legal guarantees of protection and equal opportunity.

deviance—any behavior or physical appearance that is socially challenged and/or condemned because it departs from the norms and expectations of some group.

differential association—a theory that explains how deviant behavior, especially juvenile delinquency, is learned. This theory states that it is exposure to criminal patterns and isolation from anticriminal influences that put people at risk of turning criminal.

disability—a state of being that society has imposed on those with certain impairments because of how inventions have been designed and social activities have been organized to accommodate the shortcomings of only those considered unimpaired.

disciplinary society—a social arrangement that institutionalizes surveillance and fuels an anxiety about being watched—an anxiety that promotes socially desired behavior.

discrimination—the intentional or unintentional unequal treatment of racial, ethnic, or other group without considering merit, ability, or past performance.

disenchantment—a great spiritual void accompanied by a crisis of meaning in which the natural world becomes less mysterious and revered as it becomes the object of human control and manipulation.

dispositional factors—things that people are believed to control, including personal qualities related to motivation, interest, mood, and level of effort.

division of labor—work that is broken down into specialized tasks, each performed by a different set of workers trained to do that task. The labor and resources needed to manufacture products often come from many locations around the world.

dominant ethnic group—the most advantaged ethnic group in a society; the ethnic group that possesses the greatest access to valued resources, including the power to create and maintain the system that gives it these advantages.

dramaturgical sociology—a sociological perspective that focuses on social interactions emphasizing the ways in which those involved work to create, maintain, dismantle, and present a shared understanding of reality.

dyad—the smallest group, consisting of two people.

economic capital—material resources—wealth, land, money—upon which a person can draw.

economic systems—the social institutions that coordinate human activity to produce, distribute, and consume goods and services.

education—any experiences that train, discipline, and shape the mental and physical potentials of the maturing person.

efficiency—a feature of McDonaldization that emphasizes using the methods that will achieve a desired end in the shortest amount of time.

egoistic (suicide)—a state in which the social ties attaching the individual to the group are weak.

embodied cultural capital—all that has been consciously and unconsciously instilled in a person through the socialization process.

emotional labor—work that requires employees to display or suppress specific emotions and/or manage customer/client emotions.

emotion work—the process by which people consciously manage their feelings by evoking an expected emotional state or suppressing an inappropriate emotional state.

endogamy—norms requiring or encouraging people to choose a partner from the same social category as their own—for example, a partner of the same race, sex, ethnicity, religion, or social class.

environmental injustice—unequal exposure to environmental hazards based on race or socioeconomic status.

essentialism—the belief that men and women are different by divine design and/or nature, and that inequalities between them are natural.

esteem—the reputation that someone occupying a particular status has acquired based on the opinion of those who know and observe him or her.

ethnic cleansing—an extreme form of forced segregation in which a dominant group uses force and intimidation to remove people of a targeted racial or ethnic group from a geographic area, leaving it ethnically pure, or at least free of the targeted group. Ethnic cleansing also involves the destruction of cultural artifacts associated with the targeted groups, such as monuments, cemeteries, and churches.

ethnic group—people who share, believe they share, or are believed by others to share a national origin, a common ancestry, a place of birth, or distinctive social traits (such as religion, style of dress, or language) that set them apart from other ethnic groups.

ethnic renewal—a situation in which someone discovers an ethnic identity, including the process by which an individual takes it upon himself to find, learn about, and claim an ethnic heritage.

ethnocentrism—a point of view in which people use their home culture as the standard for judging the worth of another culture's ways.

ethnomethodology—an investigative and observational approach that focuses on how people make sense of everyday social activities and experiences.

exogamy—norms encouraging or requiring people to choose a partner from a social category other than their own—for example, to choose a partner outside their immediate family who is of the other sex or of another race.

externality costs—hidden costs of using, making, or disposing of a product that are not figured into the price of the product or paid for by the producer. Such costs include those associated with cleaning up the environment and with treating injured and chronically ill workers, consumers, and others.

extreme wealth—the most excessive form of wealth, in which a very small proportion of people in the world have money, material possessions, and other assets (minus liabilities) in such abundance that a small fraction of it, if spent appropriately, could provide adequate food, safe water, sanitation, and basic health care for the 1 billion poorest people on the planet.

façade of legitimacy—an explanation to justify the existing social arrangement that downplays or dismisses charges that the arrangement advantages some and disadvantages others.

false consciousness—with regard to religion, a point of view in which oppressed individuals or groups accept the economic, political, and social arrangements that constrain their chances in life because they are promised compensation for their suffering in the next world.

falsely accused—people who have not broken the rules but are treated as if they have.

family—a social institution that binds people together through blood, marriage, law, and/or social norms.

fatalistic (suicide)—a state in which the ties attaching the individual to the group are so oppressive there is no hope of release.

femininity—traits believed to be characteristic of females.

feminism—a perspective that examines the larger social, economic, and political context in order to understand the position of women in society relative to men and that advocates equal opportunity.

finance aristocracy—according to Marx, that class of people who live in obvious luxury among masses of starving, low-paid, and unemployed workers; includes bankers and stockholders seemingly detached from the world of work. Marx described the finance aristocracy's source of income as "created from nothing—without labor and without creating a product or service to sell in exchange for wealth" gained.

folkways—norms that apply to the mundane aspects or details of daily life.

formal care—caregiving provided, usually for a fee, by credentialed professionals (whether in the person's home or some other facility).

formal curriculum—the content of the various academic subjects—mathematics, science, English, reading, physical education, and so on.

formal dimension (of organizations)—the official, by-the-book way an organization should operate.

formal organizations—coordinating mechanisms that bring together people, resources, and technology and then direct human activity toward achieving a specific outcome such as maintaining order in a community (a police department) or fund-raising (Race for the Cure).

formal sanctions—reactions backed by laws, rules, or policies specifying the conditions under which people should be rewarded or punished for specific behaviors.

front stage—areas visible to an audience, where people feel compelled to present themselves in expected ways.

function—the contribution a part makes to maintain the stability of an existing social order.

fundamentalism—a belief in the timelessness of sacred writings and a belief that such writings apply to all kinds of environments.

games—structured, organized activities that involve more than one person; they are characterized by a number of constraints, such as established roles and rules and a purpose toward which all activity is directed.

gender—the socially created and learned distinctions that specify the physical, behavioral, and mental and emotional traits believed to be characteristic of the recognized sexes, males and females.

gendered—a situation in which institutions have an established pattern of segregating the sexes, empowering one sex and not the other, and/or subordinating one sex relative to the other in ways such that gender systematically shapes the experiences, constraints, and opportunities of the participating men and women.

gender gap—the disparity in opportunities available to men and women relative to each other.

gender ideals—a standard for masculinity or femininity against which real cases can be compared. A gender ideal is at best a caricature, in that it exaggerates the characteristics that are believed to make someone the so-called perfect male or female.

gender identity—the awareness of being a man or woman, of being neither, or something in between (gender identity also involves the ways one chooses to hide or express that identity).

gender role—the cultural norms that guide people in enacting what is considered to be feminine and masculine behavior.

gender stratification—the extent to which opportunities and resources are unequally distributed between men and women.

generalizability—the extent to which researchers' findings can be applied to the larger population of which their sample was a part.

generalized other—a system of expected behaviors and meanings that transcend the people participating. An understanding of the generalized other is achieved by simultaneously and imaginatively relating the self to many others "playing the game."

generation—a cohort composed of people who are born at a particular time in history and/or separated from other age cohorts by time. A generation is distinguished from others by its cultural disposition (dress, language, preferences for songs, activities, entertainment), posture (walk, dance, the way the body is held), access to resources, and socially expected privileges, responsibilities, and duties.

genocide—the calculated and systematic large-scale destruction of a targeted racial or ethnic group that can take the form of killing an ethnic group en masse, inflicting serious bodily or psychological harm, creating intolerable living conditions, preventing births, "diluting" racial or ethnic lines through rape and forced births, or forcibly removing children to live with another group.

gesture—any action that requires people to interpret its meaning before responding. Language is a particularly important gesture because people interpret the meaning of words before they react. In addition to spoken words, gestures also include nonverbal cues, such as tone of voice, inflection, facial expression, posture, and other body movements or positions that convey meaning.

glass ceiling—a barrier that prevents women from rising past a certain level in an organization, especially women who work in male-dominated workplaces and occupations. The term applies to women who have the ability and qualifications to advance but who are not well connected to those who are in a position to advocate for or mentor them.

glass escalator—the largely invisible forces that launch men into positions of power, even within female-dominated occupations, as when management singles out men for special attention and advancement.

global interdependence—a situation in which human interactions and relationships transcend national borders and in which social problems within any one country are shaped by events taking place outside the country—indeed in various parts of the globe.

globalization—the ever-increasing flow of goods, services, money, people, technology, information, and other cultural items across political borders. This flow has become more dense and quick-moving as constraints of space and time break down. As a result of globalization, no longer are people, goods, services, technologies, money, and images fixed to specific geographic locations.

glocalization—the process by which a locality embraces, adapts to, or resists a product, an idea, or a way of behaving that has come to them in the cross-national flow; also the process by which something unique to a locality is launched on a path toward globalization.

government—the organizational structure that directs and coordinates people's involvement in the political activities of a country or some other territory, such as a city, county, or state.

group—two or more people interacting in largely predictable ways who share expectations about their purpose for being. Group members hold statuses and enact roles that relate to the group's purpose.

group think—a phenomenon that occurs when a group under great pressure to take action achieves the illusion of consensus by putting pressure on its members to shut down discussion, dismissing alternative courses of action, suppressing expression of doubt or dissent, dehumanizing those against whom the group is taking action, and/or ignoring the moral consequences of their actions.

habitus—a frame of mind that has internalized the objective reality of society. This objective reality becomes the mental filter that structures people's perceptions, experiences, responses, and actions. It is through the habitus that the social world is understood and that people acquire a sense of place and a point of view that informs how they interpret their own and others' actions.

Hawthorne effect—a phenomenon in which research subjects alter their behavior when they learn they are being observed.

heteronormativity—a normative system that presents the gendered heterosexual nuclear family as the ideal and departures from that system as deviant, even threatening.

hidden curriculum—the teaching method, types of assignments, kinds of tests, tone of the teacher's voice, attitudes of classmates, number of students absent, frequency of teacher absences, and criteria teachers use to assign grades. These so-called extraneous factors convey messages to students not only about the value of the subject but also about the values of society, the place of learning in their lives, and their role in society.

hidden ethnicity—for members of an advantaged ethnic group, a sense of self that is based on no awareness of an ethnic identity because their culture is considered normal, normative, or mainstream.

homophobia—an irrational fear held by some heterosexuals that a same-sex person will make a sexual advance toward them. It also refers to a fear of being in close contact with someone of the same sex.

horticultural societies—societies organized around the use of hand tools such as hoes to work the soil and digging sticks to punch holes in the ground into which seeds are dropped. Horticultural peoples grow crops rather than gather food and employ slash-and-burn technology to make fields for growing crops and grazing animals.

human activity—involves all the things people do with, to, and for one another and what they think and do as a result of others' influence.

human agency—the capacity of individuals to act autonomously, apart from the constraints of social structure.

hunting and gathering societies—societies that do not possess the technology to create surplus wealth; people subsist on wild animals and vegetation and are always on the move, securing food and other subsistence items. A typical hunting and gathering society comprises 45 to 100 members related by blood or marriage.

hydrocarbon society—a society in which the use of fossil fuels shapes virtually every aspect of people's personal and social lives.

hypothesis—a trial prediction about the relationship between the independent and dependent variables. Specifically, the hypothesis predicts how change in an independent variable brings about change in a dependent variable.

I—the active and creative aspect of the self. It is the part of the self that questions the expectations and rules.

ideal type—a deliberate simplification or caricature in that it exaggerates essential traits of something. Ideal does not mean desirable; an ideal is simply a standard against which real cases can be compared.

illegitimate opportunity structures—social settings and arrangements that offer people the opportunity to commit particular types of crime.

impairment—a physical or mental condition that interferes with someone's ability to perform a major life activity that the average person can perform without technical or human assistance or without changing the physical environment around them.

impression management—in social situations, as on a stage, people manage the setting, their dress, their words, and their gestures so that they correspond to an impression they are trying to make.

income—the money a person earns, usually on an annual basis through salary or wages.

independent variable—the variable that explains or predicts the dependent variable.

individual discrimination—behavior that blocks another's opportunities or does harm to life or property.

industrial societies—societies that rely on mechanization or on externally powered machines to subsist. Mechanization allowed humans to produce food, extract resources, and manufacture goods at revolutionary speeds and on an unprecedented scale.

informal care—caregiving that family members, neighbors, and friends provide in a home setting.

informal dimension (of organizations)—any aspect of an organization's operations that departs from the way the organization is officially supposed to operate.

informal sanctions—spontaneous, unofficial expressions of approval not backed by the force of law or official policy.

information explosion—an unprecedented increase in the amount of stored and transmitted data and messages in all media (including electronic, print, radio, and television).

ingroup—the group to which a person belongs, identifies, admires, and/or feels loyalty.

inner-city poor—diverse groups of families and individuals living in the inner city who are "outside the mainstream of the American occupational system and consequently represent the very bottom of the economic hierarchy" (Wilson 1983, 80).

innovation (as a response to structural strain)—the acceptance of cultural goals but the rejection of legitimate means to achieve them.

institutionalized ageism—discrimination based on age that is viewed as an accepted, established, and even fair way of doing things. On close analysis, however, such practices put a certain age group at a disadvantage relative to those in other age groups.

institutionalized cultural capital—anything (material or nonmaterial) recognized as important to success in a particular social setting.

institutionalized discrimination—the established, customary way of doing things in society—the unchallenged laws, rules, policies, and day-to-day practices established by a dominant group that keep minority groups in disadvantaged positions.

institutions—relatively stable and predictable social arrangements created and sustained by people that have emerged over time with the purpose of coordinating human activities to meet some need, such as food, shelter, or clothing. Institutions consist of statuses, roles, and groups.

instrumental rational action—result-oriented behaviors and practices that emphasize the most efficient methods for achieving some valued goal, regardless of the consequences. In the context of an industrial and capitalist society, *efficient* means the most cost-effective and time-saving way to achieve a goal.

integration—a situation in which two or more racial groups interact in what was once a segregated setting; integration may be court-ordered, legally mandated, or the natural outcome of people crossing the "color line" once legal barriers have been removed.

internalization—a process by which people accept as binding learned ways of thinking, appearing, and behaving.

intersectionality—the interconnections among the various categories people occupy, including race, class, gender, sexual orientation, religion, ethnicity, age (generation), nationality, and disability. The various categories people occupy are interlocking and when taken together "cultivate profound differences in our personal biography" and in experiences with others as we move through the world (Collins 2000a, 460).

intersexed—people with some mixture of male and female primary sex characteristics.

involuntary ethnicity—an umbrella ethnic category created by the government or another dominant group to which people from many different cultures and countries are assigned. That category becomes the label by which these diverse peoples are known and with which they are forced to identify.

involuntary minorities—those who did not choose to be a part of a country (nor did their ancestors); rather, they were forced to become part of it through enslavement, conquest, or colonization. Those of Native American, African, Mexican, and Hawaiian descent are examples.

iron cage of irrationality—the process by which supposedly rational systems produce irrationalities.

issue—a societal matter that affects many people and that can only be explained by larger social forces that transcend the individuals affected.

knowledge economy—an economy driven by information-gathering and data-intensive activities that lead to some commercial use; people who succeed in knowledge economies are those with technical expertise, problem-solving skills, analytic skills, and an ability to manage and manipulate data on consumer behavior.

language—a symbol system that assigns meaning to particular sounds, gestures, pictures, or specific combinations of letters to convey meaning.

latent dysfunctions—unanticipated disruptions to the existing social order.

latent functions—a part's unanticipated, unintended, and unrecognized effects on an existing social order.

law—a rule governing conduct created by those in positions of power and enforced by entities given the authority to do so, such as police. A law specifies the prohibited behavior, the categories of people to whom the law applies, and the punishment to be applied to violators.

legal-rational authority—legitimate form of power that derives from a system of impersonal and formal rules that specify the qualifications for occupying an administrative or judicial position; the individual holding that position has the power to command others to act in specific ways.

life chances—a critical set of potential opportunities and advantages, including the chance to survive the first year of life, to grow to a certain height, to receive medical and dental care, to avoid a prison sentence, to graduate from high school, to live a long life, and so on.

life expectancy—the average number of years after birth a person can expect to live.

linguistic relativity hypothesis—the idea that "[n]o two languages are ever sufficiently similar to be considered as representing the same social reality. The worlds in which different societies live are distinct worlds, not merely the same world with different labels attached" (Sapir 1949, 162).

lobbyists—people whose job it is to solicit and persuade state and federal legislators to create legislation and vote for bills that favor the interests of the group they represent. Lobbyists can work for corporations, a private individual, or the public interest.

looking-glass self—the way in which a sense of self develops; specifically, people act as mirrors for one another. We see ourselves reflected in others' real or imagined reactions to our appearance and behaviors. We acquire a sense of self by being sensitive to the appraisals that we perceive others to have of us.

manifest dysfunctions—a part's anticipated disruptions to an existing social order.

manifest functions—a part's anticipated, recognized, or intended effects on maintaining order.

masculinity—traits believed to be characteristic of males.

mass media—forms of communication designed to reach large audiences without direct face-to-face contact between those creating/conveying and receiving messages.

master status—a status that takes on such great importance that it overshadows all other statuses a person occupies. That is, it shapes every aspect of life and dominates social interactions.

material culture—all the physical objects that people have invented or borrowed from other cultures.

McDonaldization of society—a process whereby the principles governing the fast-food industry come to dominate other sectors of the American economy and society and the world. Those principles are (1) efficiency, (2) quantification and calculation, (3) predictability, and (4) control.

me—the social self or the part of the self that is the product of interaction with others and that knows the rules and expectations; the me is the sense of self that emerges out of role-taking experiences.

means of production—the resources such as land, tools, equipment, factories, transportation, and labor that are essential to the production and distribution of goods and services.

mechanical solidarity—a system of social ties based on uniform thinking and behavior.

mechanisms of social control—strategies people use to encourage, often force, others to comply with social norms.

melting pot assimilation—a process by which previously separate groups accept many new behaviors and values from one another, intermarry, procreate, and identify with a blended culture.

military-industrial complex—a relationship between those who declare, fund, and manage wars (the Department of Defense, the office of the president, and Congress) and corporations that make the equipment and supplies needed to wage war.

minority groups—subpopulations within a society that are regarded and treated as inherently different from those in the mainstream. They are systematically excluded (whether consciously or unconsciously) from full participation in society and denied equal access to power, prestige, and wealth.

misandry—sexism directed at men that is so extreme that it involves a hatred of those in that category.

misogyny—sexism directed at women that is so extreme that it involves a hatred of those in that category.

mixed contacts—interactions between stigmatized persons and so-called normals.

modernization—a process of economic, social, and cultural transformation in which a country "evolves" from an underdeveloped to a modern society.

monarchy—a form of government in which the power is in the hands of a leader (known as a monarch) who reigns over a state or territory, usually for life and by hereditary right. Typically, the monarch expects to pass the throne on to someone who is designated as the heir, usually a firstborn son.

moral superiority—the belief that an ingroup's standards represent the only way, leaving no room for negotiation and no tolerance for other ways.

mores—norms that people define as essential to the well-being of a group. People who violate mores are usually punished severely: they may be ostracized, institutionalized, or condemned to die.

mortality crises—violent fluctuations in the death rate caused by war, famine, or epidemics.

mortification—the process by which the self is stripped of all its supports and "shaped and coded" (Goffman 1961).

mystic—a type of counterculture in which members search for "truth and for themselves" (Yinger 1977) and in the process turn inward.

nature—human genetic makeup or biological inheritance.

negatively privileged property class—people completely lacking in skills, property, or employment, or who depend on seasonal or sporadic employment; they constitute the very bottom of the class system.

negative sanctions—expressions of disapproval for violating norms.

negotiated order—the sum of existing and newly negotiated expectations that are part of any social situation.

neocolonialism—a new form of colonialism where more powerful foreign governments and foreign-owned businesses continue to exploit the resources and labor of the postcolonial peoples.

nonmaterial culture—intangible human creations that include beliefs, values, norms, and symbols.

nonparticipant observation—detached watching and listening by a researcher who only observes and does not become part of group life.

nonprejudiced discriminators—fair-weather liberals or people who accept the creed of equal opportunity but discriminate because they simply fail to consider discriminatory consequences or because discriminating gives them some advantage.

nonprejudiced nondiscriminators—all-weather liberals or people who accept the creed of equal opportunity, and their conduct conforms to that creed.

normative reference groups—groups that provide people with norms that they draw upon or consider when evaluating a behavior or a course of action.

norms—rules and expectations for the way people are supposed to behave, feel, and appear in a particular social situation.

nurture—the interaction experiences that make up every person's life, or more generally, the social environment.

objectified cultural capital—physical and material objects that a person owns outright or has direct access to.

observation—a research strategy that involves watching, listening to, and recording behavior and conversations in context as they happen.

oligarchy—rule by the few, or the concentration of decision-making power in the hands of a few persons who hold the top positions in a hierarchy.

operationalized—a variable is operationalized when the researcher gives clear, precise instructions about how to observe or measure it.

organic solidarity—a system of social ties founded on interdependence, specialization, and cooperation.

outgroup—any group to which a person does not belong. Obviously, one person's ingroup is another person's outgroup.

over-urbanization—a situation in which urban misery—poverty, unemployment, housing shortages, and insufficient infrastructure—is exacerbated by an influx of unskilled, illiterate, and poverty-stricken rural migrants who have come to the cities out of desperation.

panethnicity—a broad catchall ethnic category in which people with distinct histories, cultures, languages, and identities are lumped together and viewed as belonging to that category (e.g., Hispanic/Latino).

panopticon—philosopher Jeremy Bentham's design for the most efficient and rational prison—the perfect prison. (*Pan* means a complete view and *optic* means seeing.)

participant observation—a method of research in which the researchers join a group, interact directly with those they are studying, assume a role critical to the group's purpose, and/or live in a community under study.

pastoral societies—societies that rely on domesticated herd animals to subsist. Pastoralism was adopted by people living in deserts and other regions in which vegetation was limited. Domesticating animals allowed people to produce surplus wealth or more milk and meat than they needed to subsist.

patriarchy—an arrangement in which men have systematic power over women in public and private (family) life.

peer group—people who are approximately the same age, participate in the same day-to-day activities, and share a similar overall social status in society.

peer pressure—instances in which people feel directly or indirectly pressured to engage in behavior that meets the approval and expectations of peers and/or to fit in with what peers are doing. That pressure may be to smoke (or not smoke) cigarettes, to drink (or not drink) alcohol, and to engage (or not engage) in sexual activity.

penalties—constraints on a person's opportunities and choices, as well as the price paid for engaging in certain activities, appearances, or choices deemed inappropriate for someone in a particular category.

peripheral economies—economies built around a few commodities or even a single commodity, such as coffee, peanuts, or tobacco, or on a natural resource, such as oil, tin, copper, or zinc.

play—voluntary, spontaneous activity with few or no formal rules.

pluralism—a situation in which different racial and ethnic groups coexist in harmony; have equal social standing; maintain their unique cultural ties, communities, and identities; and participate in the economic and political life of the larger society. These groups also possess an allegiance to the country in which they live and its way of life.

political action committees (PACs)—special-interest groups that raise money to be donated to the political candidates who seem most likely to support their economic, social, and/or political needs and interests. There are more than 4,500 registered PACs.

political parties—organizations that try to acquire power to influence social action. Parties are organized to represent people of a certain class or social status or with certain interests.

political system—the institution that regulates the access to and use of power to control access to scarce and valued resources and to make laws, policies, and decisions that affect others' life chances.

positive checks—events that increase deaths, including natural disasters, epidemics of infectious and parasitic diseases, war, and famine.

positively privileged property class—those who monopolize the purchase of the highest-priced consumer goods, have access to the most socially advantageous kinds of education, control the highest executive positions, own the means of production, and live on income from property and other investments.

positive sanctions—expressions of approval for complying with norms.

positivism—the belief that valid knowledge about the world can be derived only by using the scientific method.

postindustrial societies—societies that rely on the intellectual technologies of telecommunications and computers. These signature technologies encompass four interdependent revolutionary innovations: (1) electronics, (2) miniaturization, (3) digitalization, and (4) software.

power—the probability that an individual can achieve his or her will, even against opposition.

power elite—those few people who occupy such lofty positions in the social structure of leading institutions that their decisions affect millions, even billions, of people worldwide.

predestination—the belief that God has foreordained all things, including the salvation or damnation of individual souls.

predictability—a principle of McDonaldization that emphasizes the expectation that a service or product will be the same no matter where in the world or when (time of year, time of day) it is purchased.

prejudice—a rigid and, more often than not, unfavorable judgment about a category of people that is applied to anyone who belongs to that category.

prejudiced discriminators—active bigots or people who reject the notion of equal opportunity and profess a moral right, even a duty, to discriminate. They derive significant social and psychological gains from the conviction that anyone from their racial or ethnic group is superior to other such groups.

prejudiced nondiscriminators—timid bigots or people who reject the creed of equal opportunity but refrain from discrimination, primarily because they fear possible sanctions or being labeled as racists. Timid bigots rarely express their true opinions about racial and ethnic groups, and often use code words such as *inner city* or *those people* to camouflage their true feelings.

primary group—a group characterized by strong emotional ties among members who feel an allegiance to one another.

primary sector—economic activities that generate or extract raw materials from the natural environment. Mining, fishing, growing crops, raising livestock, drilling for oil, and planting and harvesting forest products are examples.

primary sex characteristics—the anatomical traits essential to reproduction.

prison-industrial complex—the corporations and agencies with an economic stake in building and supplying goods and services to correctional facilities. This stake fuels an ongoing "need" for prisoners so that these companies can maintain or increase profit margins. These private corporations represent a significant lobbying force shaping legislation and correctional policy.

privileges—special taken-for-granted advantages, immunities, and benefits enjoyed by a dominant group relative to other groups.

profane—everything that is not considered sacred, including things opposed to the sacred (such as the unholy, the irreverent, and the blasphemous) and things that stand apart from the sacred (such as the ordinary, the commonplace, the unconsecrated, and the bodily).

proletariat—those individuals who must sell their labor to the bourgeoisie.

pure deviants—people who have broken the rules and are caught, punished, and labeled as outsiders.

race—human-created or -constructed categories that have come to assume great social importance.

racial common sense—ideas and assumptions people hold in common about race or a racial group believed to be so obvious or natural that their validity need not be questioned.

racial formation—a theoretical viewpoint in which race is presented, not as a concrete biological category, but as a product of the system of racial classification.

racism—a set of beliefs that uses biological or innate factors to explain and justify inequalities between racial and ethnic groups.

radical activist—a type of counterculture in which members preach, create, or demand a new order with new obligations to others.

random—a sampling method in which every person or case in theory has an equal chance of being selected.

rationalization—a process in which thought and action rooted in emotion, superstition, respect for mysterious forces, or tradition is replaced by instrumental rational action or means-to-ends thinking.

rebellion (as a response to structural strain)—the rejection of both the valued goals and the legitimate means of attaining them. Rebellion involves establishing a new set of goals and means of obtaining them.

redlining—systematic and institutionalized practices that deny, limit, or increase the cost of services to neighborhoods because residents are low-income and/or minority. Redlining can involve financial services (loans, checking accounts, credit cards, mortgages), insurance, health care, and grocery stores.

reentry shock—culture shock that can be experienced upon returning to a home culture after living in another culture.

reference group—any group whose standards people take into account when evaluating something about themselves or others, whether it be personal achievements, aspirations in life, or individual circumstances.

relative deprivation—a social condition that is measured not by objective standards but rather by comparing one group's situation with the situations of groups who are more advantaged.

relative poverty—a situation that is measured not by some essential minimum but rather by comparing a particular situation against an average or advantaged situation.

reliability—a standard for assessing an operational definition that emphasizes the ability of a measure to yield consistent results.

religion—a system of shared rituals and beliefs about the sacred and the profane that bind together a community of worshipers.

representative democracy—a form of government in which power is vested in citizens who vote into office those candidates they believe can best represent their interests.

research design—a plan for deciding who or what to study and the method of gathering data.

research methods—various techniques that sociologists and other investigators use to formulate and answer meaningful questions and to collect, analyze, and interpret data.

resocialization—an interactive process during which the affected party reconstructs his or her identity, and by which he or she renegotiates relationships with significant others who must also adjust to the changing person and circumstances.

retreatism (as a response to structural strain)—the rejection of both culturally valued goals and the legitimate means of achieving them.

reverse ethnocentrism—a viewpoint that regards a home culture as inferior to an idealized foreign culture.

ritualism (as a response to structural strain)—the rejection of the cultural goals but a rigid adherence to legitimate means society has in place to achieve them.

rituals—rules that govern how people behave in the presence of the sacred. These rules may take the form of instructions detailing the appropriate day(s) and occasions for worship, acceptable dress, and wording of chants, songs, and prayers.

role—the behavior expected of a status in relation to another status—for example, the role of brother in relation to sister; the role of physician in relation to patient.

role conflict—a predicament in which the roles associated with two or more distinct statuses that a person holds conflict in some way.

role expectations—norms about how a role should be enacted relative to other statuses.

role performance—the actual behavior of the person occupying a role.

role-set—the various role relationships with which someone occupying a status is involved.

role strain—a predicament in which there are contradictory or conflicting role expectations associated with a single status.

role-taking—imaginatively stepping into another person's shoes to view and evaluate the self.

routine—the usual ways of thinking and doing things.

sacred—everything that is regarded as extraordinary and that inspires in believers deep and absorbing sentiments of awe, respect, mystery, and reverence. Sacred things may include objects, living creatures, elements of nature, places, states of consciousness, holy days, ceremonies, and other activities.

samples—a subset of cases from a larger population of interest.

sanctions—reactions of approval or disapproval to behavior that departs from or conforms to group norms.

schooling—a program of formal, systematic instruction that takes place primarily in classrooms but also includes extracurricular activities and out-of-classroom assignments.

scientific method—a carefully planned research process with the goal of generating observations and data that can be verified by others.

scientific racism—the use of faulty science to support systems of racial rankings and theories of social and cultural progress that placed whites in the most advanced ranks and stage of human evolution.

secondary groups—groups that consist of two or more people who interact for a specific purpose. Secondary group relationships are confined to a particular setting and specific tasks. Members relate to each other in terms of specific roles.

secondary sector—economic activities that transform raw materials from the primary sector into manufactured goods such as computers and cars.

secondary sex characteristics—physical traits not essential to reproduction such as breast development, quality of voice, distribution of facial and body hair, and skeletal form, that supposedly result from the action of so-called male (androgen) and female (estrogen) hormones.

secondary sources or archival data—third-party data that have been collected for a purpose not related to the research study.

secret deviants—people who have broken the rules but whose violation goes unnoticed or, if it is noticed, no sanctions are applied.

secularization—a process by which religious influences on thought and behavior are gradually removed or reduced. More specifically, it is a process by which some element of society, once part of a religious sphere, separates from its religious or spiritual connection or influences.

segmentalized dyad—a two-person group in which the parties know little about each other's personality and personal life, and what they do know is confined to a specific situation, such as the classroom, a hair salon, or other specialized setting.

segregation—the physical and/or social separation of people by race or ethnicity. It may be legally enforced (de jure) or socially enforced without the support of laws (de facto).

selective forgetting—a process by which people forget, dismiss, or fail to pass on to their children an ancestral connection to one or more ethnicities.

selective perception—a situation in which prejudiced persons notice only the behaviors that support their stereotypes and then use those observations to support the stereotypes they hold.

self-administered survey—a set of questions that respondents read and answer.

self-fulfilling prophecy—a point of view that begins with a false definition of a situation that is assumed to be accurate. People behave as if that definition were true so that the misguided behavior produces responses that confirm the false definition.

self-referent terms—terms to distinguish the self (including I, me, mine, first name, and last name) and to specify the statuses one holds in society (athlete, doctor, child, and so on).

semiperipheral economies—economies characterized by moderate wealth (but extreme inequality) and a moderately diverse system of production and consumption. Semiperipheral economies exploit peripheral economies and are in turn exploited by core economies.

sense of self—self-knowledge derived from stepping outside the self and seeing it from another's point of view and also imagining the effects one's words and actions have on others.

sex—a distinction based on primary sex characteristics.

sexism—the belief that one sex—and by extension, one gender—is innately superior to another, justifying unequal treatment of the sexes.

sexuality—all the ways people experience and express themselves as sexual beings. The study of sexuality considers the range of social activities, behaviors, and thoughts that generate sexual sensations and experiences and that allow for sexual expression.

sexual orientation—"an enduring pattern of emotional, romantic, and/or sexual attractions to men, women, or both sexes. Sexual orientation also refers to a person's sense of identity based on those attractions, related behaviors, and membership in a community of others who share those attractions" (American Psychiatric Association 2009).

sexual scripts—responses and behaviors that people learn, in much the same way that actors learn lines for a play, to guide them in sexual activities and encounters. These scripts are gendered in that males and females learn different scripts about the sex-appropriate responses and behavioral choices open to them in specific situations.

significant others—people or characters (such as cartoon characters, a parent, or the family pet) who are important in a child's life, in that they greatly influence the child's self-evaluation and way of behaving.

significant symbols—gestures that convey the same meaning to the persons transmitting them and receiving them; gestures or sounds that must be interpreted before a response is made.

situational factors—things believed to be outside a person's control—such as the weather, bad luck, and another's incompetence.

social action—actions people take in response to others.

social change—any significant alteration, modification, or transformation in the way social activities and human relationships are organized.

social class—a person's overall economic and social status in a system of social stratification.

social emotions—feelings that we experience as we relate to other people, such as empathy, grief, love, guilt, jealousy, and embarrassment.

social facts—collectively imposed ways of thinking, feeling, or behaving that have "the remarkable property of existing outside the consciousness of the individual" (Durkheim 1901, 51).

social forces—anything human-created that influences, pressures, or pushes people to behave and think in specified ways.

social inequality—the unequal access to and distribution of income, wealth, and other valued resources.

social interaction—everyday encounters in which people communicate, interpret, and respond to each other's words and actions.

socialism—an economic system in which raw materials and the means of producing and distributing goods and services are collectively owned. That is, public ownership—rather than private ownership—is an essential characteristic of this system.

socialization—the lifelong process by which people learn the ways of the society in which they live. More specifically, it is the process by which humans acquire a sense of self or a social identity, develop their human capacities, learn the culture(s) of the society in which they live, and learn expectations for behavior.

social mobility—movement from one social class to another.

social movement—a phenomenon in which a substantial number of people organize to make a change, resist a change, or undo a change to some area of society.

social network—a web of closely to loosely knit social relationships linking people to one another.

social order—the way people have organized interaction and other activities to achieve some valued goal.

social prestige—a level of respect or admiration for a status apart from any person who happens to occupy it.

social relationships—the nature, number, and quality of the ties that bind people formally and informally to others.

social reproduction—the process by which teachers and staff unwittingly perpetuate inequalities of the larger society and the inequalities children bring with them to the classroom by the way they treat students and organize their academic experiences.

social status—a human-created and defined position in society such as female, teenager, patient, retiree, sister, homosexual, and heterosexual.

social stratification—the systematic process of categorizing and ranking people on a scale of social worth where one's ranking affects life chances in unequal ways.

social structure—a largely invisible system that broadly shapes and coordinates human behavior and activity in noticeable and predictable ways.

sociological imagination—a perspective that allows us to consider how outside forces, especially time in history and place (geographic location), shape life stories or biographies.

sociological perspective—a conceptual framework for thinking about and explaining how human activities are organized and/or how people relate to one another and respond to their surroundings.

sociology—the scientific study of human activity in society.

solidarity—the system of social ties that acts as a cement connecting people to one another and to the wider society.

special-interest groups—groups consisting of people who share an interest in a particular economic, political, or other social issue and who form an organization or join an existing organization to influence public opinion and government policy.

status group—an amorphous group of persons held together by virtue of a lifestyle that has come to be "expected of all those who wish to belong to the circle" and by the level of social esteem and honor others accord them (Weber 1948, 187).

status set—all the statuses any one person assumes.

status symbols—visible markers of economic and social position and rank.

status value—a situation in which people who possess one characteristic are regarded and treated as more valuable or worthy than people who possess other characteristics.

stereotypes—generalizations about people who belong to a particular category that do not change even in the face of contradictory evidence. Stereotypes give holders an illusion that they know the other group and that they possess the right to control images of the other group.

stigma—an attribute that is deeply discrediting.

structural constraints—the established and customary rules, policies, and day-to-day practices that affect a person's life chances.

structural strain—a situation in which there is an imbalance between culturally valued goals and the legitimate means to obtain them. An imbalance exists when (1) the sole focus is on achieving valued goals by any means necessary, (2) people are unsure whether following the legitimate means will lead to success, or (3) there are not enough legitimate opportunities to satisfy demand.

subcultures—groups that share in certain parts of the mainstream culture but have distinctive values, norms, beliefs, symbols, language, and/or material culture that set them apart in some way.

subsist—to meet basic needs for human survival.

surplus wealth—a situation in which the amount of available food items and other products exceed that which is required to subsist.

surveillance—a mechanism of social control that involves monitoring the movements, conversations, and associations of those believed to be or about to be engaged in some wrongdoing and then intervening at appropriate moments.

symbols—anything (a word, an object, a sound, a feeling, an odor, a gesture, an idea) to which people assign a name and a meaning.

sympathetic knowledge—firsthand knowledge gained by living and working among those being studied.

system of oppression—a system that empowers and privileges some categories of people while disempowering other categories. The act of disempowering includes marginalizing, silencing, or subordinating another.

technology—knowledge, tools, applications, and other inventions used in ways that allow people to adapt to and exercise control over their surroundings. Technology can be material and/or conceptual.

tertiary sector—economic activities related to delivering services, such as health care or entertainment, and to creating and distributing information, such as books or data.

theocracy—a form of government in which political authority rests in the hands of religious leaders or a theologically trained elite group. The primary purpose of a theocracy is to uphold divine laws in its policies and practices. Thus, there is no legal separation of church and state. Government policies and laws correspond to religious principles and laws.

theory of the demographic transition—a theoretical perspective that postulates that a country's birth and death rates are linked to its level of industrial or economic development and that the so-called developing countries will eventually achieve birth and death rates like those of Western European and North American countries.

this-worldly asceticism—a belief that people are instruments of divine will and that God determines and directs their activities. Calvinists glorified God when they accepted a task assigned to them, carried it out in an exemplary and disciplined fashion, and did not indulge in the fruits of their labor.

Thomas Theorem—a theory stating that before people take action or respond to a situation, they decide on and attach a meaning to the situation, which then affects their response. In other words, if people "define situations as real they are real in their consequences" (Thomas and Thomas 1928, 572).

tipping point—a situation in which what was once a rare practice or event snowballs into something dramatically more common.

total fertility—the average number of children that a woman bears in her lifetime.

total institutions—settings in which people surrender control of their lives, voluntarily or involuntarily, thereby submitting to the authority of an administrative staff and undergoing a program of resocialization cut off from the rest of society.

totalitarianism—a form of government characterized by (1) a single ruling party led by a dictator, (2) an unchallenged official ideology that defines a vision of the "perfect" society and the means to achieve that vision, (3) a system of social control that suppresses dissent, and (4) centralized control over the media and the economy.

tracking—also known as ability grouping; a sifting and sorting mechanism by which students are assigned to separate instructional groups within a single classroom; programs such as college preparatory versus general studies; or advanced placement, honors, or remedial classes.

traditional authority—a form of power grounded in the sanctity of time-honored norms that govern how someone comes to hold a powerful position, such as chief, king, queen, or emperor. Usually the person inherits that position by virtue of being born into a family that has held power for some time.

transgender—the label applied to those who feel that their inner sense of being a man or woman does not match their anatomical sex, so they behave and/or dress to actualize their gender identity.

transnational corporation—a corporation that has operations in more than one country; sometimes referred to as a multinational or global corporation.

treadmill of production—a term used to describe the ceaseless increases in production and, by extension, consumption that are needed to sustain the global economy's success, which is measured by increased profits.

triad—a three-person group that is sociologically significant because a third person added to a two-person group (a dyad) significantly alters the pattern of interaction between them.

troubles—individual problems or difficulties that are caused by personal shortcomings related to motivation, attitude, ability, character, or judgment. The resolution of a trouble lies in changing the person in some way.

trust—the taken-for-granted assumption that in a given social encounter others share the same expectations and definitions of the situation and that they will act to meet those expectations.

typificatory schemes—systematic mental frameworks that allow people to place what they observe into preexisting social categories with essential characteristics.

tyranny of the normal—a point of view that measures differences against what is thought to be normal and that assumes those with impairments fall short in other ways.

urbanization—a transformative process in which people migrate from rural to urban areas and change the ways they use land, interact, and earn a living.

urban underclass—diverse groups of families and individuals residing in the inner city who are on the fringes of the American occupational system and as a result are in the most disadvantaged position of the economic hierarchy.

utilitarian organizations—organizations that draw in those seeking to achieve some desired goal in exchange for money. That goal may be to earn an income as an employee, to acquire a skill by enrolling in a special program, or to purchase a desired product at a department store.

validity—a standard by which operational definitions are assessed and that focuses on the extent to which a measure accurately represents what it is intended to measure.

values—general, shared conceptions of what is good, right, desirable, or important.

variable—any behavior or characteristic that consists of more than one category.

virtual integration—a perception that racial integration exists derived from simply seeing other racial groups on television and in advertisements; it gives "the sensation of having meaningful, repeated contact with other racial groups without actually having it" (Lynch 2007).

voluntary organizations—organizations that draw in people who give time, talent, or money to address a human and community need or to achieve some other not-for-profit goal.

wealth—the combined value of a person's income and other material assets such as stocks, real estate, and savings, minus debt.

welfare state—a term that applies to an economic system that is a hybrid of capitalism and socialism.

white-collar crime—crimes committed by people whose position of respectability and high social status allows them the opportunity to do so in the course of doing their jobs.

witch hunt—campaign to identify, investigate, and correct behavior that has been defined as undermining a group. The targeted behavior is rarely the real cause but is addressed to make a problem appear as if it is being managed.

References

8Asians.com. 2009. "Oh, You Crazy Test Studying Asians Trying to Get into College." http://www.8asians.com/2009/06/16/oh-you-crazy-test-studying-asians-trying-to-get-into-college/.

Abercrombie, Nicholas, Stephen Hill, and Bryan S. Turner. 1988. *The Penguin Dictionary of Sociology.* New York: Penguin.

Abercrombie, Nicholas, and Bryan S. Turner. 1978. "The Dominant Ideology Thesis." *British Journal of Sociology* 29(2): 149–170.

Aboulafia, Mitchell. 2012 "George Herbert Mead." *The Stanford Encyclopedia of Philosophy,* edited by Edward N. Zalta. http://plato.stanford.edu/archives/sum2012/entries/mead/.

Addams, Jane. 1910. *Twenty Years at Hull-House: With Autobiographical Notes.* New York: Macmillan.

———. 1912. *A New Conscience and an Ancient Evil.* Urbana: University of Illinois Press.

Advameg, Inc. 2007. "How Products Are Made." http://www.madehow.com/Volume-2/Bread.html.

Aldrich, Howard E., and Peter V. Marsden. 1988. "Environments and Organizations." In *Handbook of Sociology,* edited by N. J. Smelser, 361–392. Newbury Park, CA: Sage.

Allen, Walter, and Angie Y. Chung. 2000. "Your Blues Ain't Like My Blues: Race, Ethnicity and Social Inequality in America." *Contemporary Sociology* 29(6): 796–805.

Alvarez, Lizette, and Robbie Brown. 2011. "Student's Death Turns Spotlight on Hazing." *New York Times,* November 30. http://www.nytimes.com/2011/12/01/us/florida-am-university-students-death-turns-spotlight-on-hazing.html?pagewanted=all.

American Psychiatric Association (APA). 2009. "What Is Sexual Orientation?" http://www.apa.org/topics/sorientation.html.

American Sociological Association. 2012a. "Environment and Technology." www.asanet.org/sections/environment_awards.cfm.

———. 2012b. "Section Membership Counts." http://www.asanet.org/sections/CountsLastFiveYears.cfm.

Anderson, Gretchen. 2010. Quoted in Glen Hougan, "What Has Design Got to Do with It?" March 28. http://theconference.ca/what-has-design-got-to-do-with-it.

Angier, Natalie. 2002. "The Most Compassionate Conservative." *New York Times Book Review,* October 27: 8–9.

Anspach, Renee R. 1987. "Prognostic Conflict in Life-and-Death Decisions: The Organization as an Ecology of Knowledge." *Journal of Health and Social Behavior* 28(3): 215–231.

Anonymous (NKU student). 2012. Student Interview Paper (anonymous). Soc 100: Introduction to Sociology, taught by Joan Ferrante, Fall Semester.

Appelbaum, Eileen, and John Schmitt. 2009. "Review Article: Low Wage Work in High Income Countries: Labor-Market Institutions and Business Strategy in U.S. and Europe." *Human Relations* 62(12): 1907–1934.

Appelrouth, Scott, and Laura D. Edles. 2007. *Sociological Theory in the Contemporary Era.* Thousand Oaks, CA: Pine Forge Press.

Archdiocese of Cincinnati. 2014. "Elementary School Program." http://www.catholiccincinnati.org/ministries-offices/cise-catholic-inner-city-schools-education-fund/what-we-do/elementary-school-program/.

Arenson, Karen W. 2006. "Can't Complete High School? Just Go Right Along to College." *New York Times,* May 30: A1.

Aron, R. 1969. Quoted in *The Sociology of Max Weber,* by Julien Freund. New York: Random House.

Ash, Timothy Garton. 1989. *The Uses of Adversity: Essays on the Fate of Central Europe.* New York: Random House.

Ashenmiller, Bevin. 2011. "The Effect of Bottle Laws on Income: New Empirical Results." *American Economic Review* 101(3): 60–64.

Asian American Hotel Owners Association. 2009. "AAHOA History." http://www.aahoa.com/Content/NavigationMenu/AboutUs/History/default.htm.

Asian Studies Center. 2013. "Afghanistan Culture." Windows on Asia. http://asia.isp.msu.edu/wbwoa/central_asia/afghanistan/culture.htm.

Associated Press. 2006. "Country Singer Carrie Underwood Graduates." *USA Today,* May 7. http://www.usatoday.com/life/people/2006-05-07-underwood_x.htm.

———. 2009. "Cowboy Churches Rope in New Christians." *Faith* on msnbc.com, January 8. http://www.msnbc.msn.com/id/28567031/ns/us_news-faith/t/cowboy-churches-rope-new-christians/.

Aud, S., W. Hussar, G. Kena, K. Bianco, et al. 2011. *The Condition of Education 2011* (NCES 2011-033). U.S. Department of Education, National Center for Education Statistics. Washington, DC: U.S. Government Printing Office. http://nces.ed.gov/programs/coe/pdf/coe_rmc.pdf.

Bailey, Stanley. 2008. "Unmixing for Race Making in Brazil." *American*

Journal of Sociology 114(3): 577–614.

Baker, Peter. 2009. "Obama Presses for Action on Credit Cards." *New York Times,* May 14. www.nytimes.com/2009/05/15/us/politics/15obamacnd.html?ref=business.

Banker, Carter. 2011. "Jeans as a Global Phenomenon." *Johns Hopkins News-Letter,* February 3. http://www.jhunewsletter.com/2011/02/03/jeans-as-a-global-phenomenon-51318/.

Barkley, Charles. 2006. "Who Is Afraid of a Large Black Man?" *Today Books*. http://www.today.com/id/7338308/ns/today-today_books/t/whos-afraid-charles-barkley/#.U3oTBC8mRJM.

Barnet, Richard J. 1990. "Reflections: Defining the Moment." *The New Yorker,* July 16: 45–60.

Barrio, Federico. 1988. "History and Perspective of the Maquiladora Industry in Mexico." In *Mexico In-Bond Industry,* edited by T. P. Lee, 7–13. Tiber, Mexico: ASI.

Barrionuevo, Alexai. 2007. "Globalization in Every Loaf: Ingredients Come from All Over, but Are They Safe?" *New York Times,* June 25: C1.

Barton, Len. 1991. "Sociology, Disability Studies and Education: Some Observations." In *Disability Reader: Social Science Perspectives,* edited by T. Shakespeare, Ch. 4. London, UK: Continuum International Publishing Group.

Bates, Karen Grigsby. 2012. "Nailing the American Dream, with Polish." NPR Special Series: *American Dreams Then and Now*. http://www.npr.org/2012/06/14/154852394/with-polish-vietnamese-immigrant-community-thrives.

Bauman, Zygmunt. 2005. *Liquid Life*. Cambridge, UK: Polity.

Baumgartner-Papageorgiou, Alice. 1982. *My Daddy Might Have Loved Me: Student Perceptions of Differences between Being Male and Being Female.* Denver: Institute for Equality in Education.

Bauwens, Daan. 2013. "Japan Values Women Less—As It Needs Them More." Inter Press Service News Agency, January. http://www.ipsnews.net/2013/01/japan-values-women-less-as-it-needs-them-more/.

Bearman, Peter, and Peter Hedström, eds. 2009. "Self-Fulfilling Prophecies." *The Oxford Handbook of Analytical Sociology*. Oxford, UK: Oxford University Press. http://users.ox.ac.uk/~sfos0060/prophecies.shtml.

Becker, Howard S. 1963. *Outsiders: Studies in the Sociology of Deviance.* New York: Free Press.

Becker, Howard, and Ruth Hill Useem. 1942. "Sociological Analysis of the Dyad." *American Sociological Review* 7(1): 13–26.

Beilin, H. 1992. "Piaget's Enduring Contribution to Developmental Psychology." *Developmental Psychology* 28: 191–204.

Bell, Daniel. 1976. "Welcome to the Post-Industrial Society." *Physics Today,* February: 46–48.

———.1999. *The Coming of Post-Industrial Society.* New York: Basic Books.

Bem, Sandra Lipsitz. 1993. *The Lenses of Gender: Transforming the Debate on Sexual Inequality.* Binghamton, NY: Vail-Ballou.

Ben-Ari, Nirit, and Ernest Harsch. 2005. "Sexual Violence, an 'Invisible War Crime.'" *African Renewal* 18(4) (January). http://www.un.org/en/africarenewal/vol18no4/184sierraleone.htm.

Ben-David, Amith, and Yoav Lavee. 1992. "Families in the Sealed Room: Interaction Patterns of Israeli Families during SCUD Missile Attacks." *Family Process* 31(1): 35–44.

Benford, Robert D. 1992. "Social Movements." In *Encyclopedia of Sociology,* edited by E. F. Borgatta and M. L. Borgatta. New York: Macmillan.

Berelson, Bernard. 1978. "Prospects and Programs for Fertility Reduction: What? Where?" *Population and Development Review* 4: 579–616.

Berger, Peter L. 1963. *Invitation to Sociology: A Humanistic Perspective.* New York: Anchor.

Berger, Peter, and Thomas Luckmann. 1966. *The Social Construction of Reality.* Garden City, NY: Anchor.

Berners-Lee, Tim. 1996. "The World Wide Web: Past, Present and Future." http://www.w3.org/People/Berners-Lee/1996/ppf.html.

Bernstein, Mary. 2002. "Identities and Politics: Toward a Historical Understanding of the Lesbian and Gay Movement." *Social Science History* 26(Fall): 3.

Best, Joel. 1989. *Images of Issues: Typifying Contemporary Social Problems.* New York: Aldine de Gruyter.

Bianchi, Suzanne M., and Melissa A. Milkie. 2010. "Work and Family Research in the First Decade of the 21st Century." *Journal of Marriage and the Family* 72(June): 705–725.

Biblarz, Timothy J. and Evren Savci. 2010. "Lesbian, Gay, Bisexual, and Transgender Families." *Journal of Marriage and Family* 72(3): 480–497.

Bieltz, Brandon. 2012. "Denim Day to Be Celebrated at Fort Meade." U.S. Army. Retrieved April 19, 2012, from http://www.army.mil/article/78170/Denim_Day_to_be_celebrated_at_Fort_Meade/.

Birdsall, Nancy, and Arvind Subramanian. 2004. "Saving Iraq from Its Oil." *Foreign Affairs* 83(4): 77–89.

Blackman, Tim. 2013. "Care Robots for Supermarket Shelf: A Product Gap in Assistive Technologies." *Ageing and Society* 33(5): 763–781. ww.ncbi.nlm.nih.gov/pmc/articles/PMC3665953/.

Blood, Peter R., ed. 2001. *Afghanistan: A Country Study.* Washington, DC: GPO for the Library of Congress. http://countrystudies.us/afghanistan/index.htm.

Blount, Heather. 2013. "Starbucks Introduces Cross-Channel, Multi-Brand Loyalty Program." Sunbelt FS Website, March 21. http://www.sunbeltfoodservice.com/2013/03/21/starbucks-introduces-cross-channel-multi-brand-loyalty-program.

Blumer, Herbert. 1969. *Symbolic Interactionism: Perspective and Method.* Berkeley: University of California Press.

Boisey Swim Team. 2014. "Fees, Fund Raising, and Other Charges for 2014." http://www.boiseyswimteam.org/SubTabGeneric.jsp?team=srsbyst&_stabid_=15318.

Bologna, Sergio. 2008. "Marx as Correspondent of the *New York Daily Tribune*, 1856–57." *Money and Crisis.* http://wildcat-www.de/en/material/cs13bolo.htm.

Boone, Jon. 2009. "'Worse Than the Taliban': New Law Rolls Back

Rights for Afghan Women." *The Guardian,* May 31. http://www.guardian.co.uk/world/2009/mar/31/hamid-karzai-afghanistan-law.

Boudon, Raymond, and François Bourricaud. 1989. *A Critical Dictionary of Sociology,* selected and translated by P. Hamilton. Chicago: University of Chicago Press.

Boulding, Elise. 1976. *The Underside of History.* Boulder, CO: Westview.

Bourdieu, Pierre. 1984. *Distinction: A Social Critique of the Judgement of Taste,* translated by Richard Nice. Cambridge, MA: Harvard University Press.

———. 1986. "The Forms of Capital." In *Handbook of Theory and Research for the Sociology of Education,* edited by J. Richardson, 241–258. New York: Greenwood.

Boxer, Barbara (U.S. Senator). 2007. "Historical Timeline for Women's History." http://boxer.senate.gov/.

Bradsher, Keith. 2006. "Ending Tariffs Is Only the Start." *New York Times,* February 28: C1.

Brennan, Christine. 2013. "Colleges Doing Poorly Hiring Women Coaches." *USA Today,* December 19. http://www.usatoday.com/story/sports/college/2013/12/18/christine-brennan-women-college-coaches/4118067/.

Brewer, Marilynn B. 1999. "The Psychology of Prejudice: Ingroup Love or Outgroup Hate." *Journal of Social Issues,* Fall. http://findarticles.com/p/articles/mi_m0341/is_3_55/ai_58549254/.

Britton, Dana. 2000. "The Epistemology of the Gendered Organization." *Gender and Society* 14(3): 418–434.

Brookings Institution. 2012. "Tax Facts Historical Capital Gains." http://www.taxpolicycenter.org/taxfacts/displayafact.cfm?Docid=161.

Brooks, David. 2011. *The Social Animal.* New York: Random House.

Brown, David. 2010. "A Mother's Education Has a Huge Effect on a Child's Health." *Washington Post,* September 16. http://www.washingtonpost.com/wp-dyn/content/article/2010/09/16/AR2010091606384.html.

Brown, Lester R. 1987. "Analyzing the Demographic Trap." *State of the World 1987: A Worldwatch Institute Report on Progress Toward a Sustainable Society.* New York: Norton.

Buckley, Alan D. 1998. "Comparing Political Systems." http://homepage.smc.edu/buckley_alan/.

Bullard, Robert, Paul Mohai, Robin Saha, and Beverly Wright. 2007. "Toxic Wastes and Race at Twenty: 1987–2007." http://www.ucc.org/assets/pdfs/toxic20.pdf.

Bullock, Alan. 1977. "Democracy." In *The Harper Dictionary of Modern Thought,* edited by A. Bullock, S. Trombley, and B. Eadie, 211–212. New York: Harper & Row.

Bumiller, Elisabeth. 2009. "Rare Source of Attack on 'Don't Ask, Don't Tell.'" *New York Times,* September 30. http://www.nytimes.com/2009/10/01/us/01pentagon.html.

Burke, James. 1978. *Connections.* Boston: Little, Brown.

Bush, George H. W. 1991. Address before a Joint Session of the Congress on the State of the Union (January 29). http://www.gpo.gov:80/fdsys/pkg/PPP-1991-book1/pdf/PPP-1991-book1-doc-pg74.pdf.

Bush, George W. 2003. "President Bush Announces Major Combat Operations in Iraq Have Ended." Remarks by the president from the USS *Abraham Lincoln* at sea off the coast of San Diego. http://web.archive.org/web/20041020235150/http://www.whitehouse.gov/news/releases/2003/05/iraq/20030501-15.html.

Business-in-Asia.com. 2012. "What Impact Vietnam's New WTO Membership Will Have and What Vietnam Has Committed." Runckel & Associates Consultants. http://www.business-in-asia.com/wto_vietnam_impacts.html.

Business Insider. 2013. "The 50 Most Expensive Private High Schools in the United States." (September 4). http://www.businessinsider.com/most-expensive-private-high-schools-us-2013-8?op=1.

Buttel, Frederick H., and Craig R. Humphiturey. 2002. "Sociological Theory and the Natural Environment." In *Handbook of Environmental Sociology,* edited by Riley E. Dunlap and William Michelson, 33–69. Westport, CT: Greenwood Press.

Cahill, Betsy, and Eve Adams. 1997. "An Exploratory Study of Early Childhood Teachers' Attitudes toward Gender Roles." *Sex Roles: A Journal of Research* 36(7–8): 517–530.

Campbell, April. 2011. "Afghan Youth Taekwondo Team Shows ISAF Kickin' Good Time." http://www.army.mil/article/67826/Afghan_youth_Taekwondo_team_shows_ISAF_Kickin_good_time/.

Campbell, E. Q. 1964. "The Internalization of Norms." *Sociometry* 27: 391–412.

Caplan, Lionel. 1987. "Introduction: Popular Conceptions of Fundamentalism." In *Studies in Religious Fundamentalism,* edited by L. Caplan, 1–24. Albany: State University of New York Press.

Carver, Terrell. 1987. *A Marx Dictionary.* Totowa, NJ: Barnes and Noble.

CBS News. 2013. "Study Shows 70 Percent of Americans Take Prescription Drugs" (June 20). http://www.cbsnews.com/news/study-shows-70-percent-of-americans-take-prescription-drugs/.

CBS News Polls. 2005. "Women's Movement Worthwhile." http://www.cbsnews.com/stories/2005/10/22/opinion/polls/main965224.shtml.

Center for Responsive Politics. 2009. "Lobbying—Top Spenders." http://www.opensecrets.org/lobby/top.php?showYear=2009&indexType=s.

———. 2014. "Super PACs." http://www.opensecrets.org/pacs/superpacs.php?cycle=2014.

Centers for Disease Control. 2011a. "Health Disparities and Inequalities Report." MMWR, January 14, 60(Supplement). www.cdc.gov/mmwr/pdf/other/su6001.pdf.

———. 2011b. "Sexual Behavior, Sexual Attraction, and Sexual Identity in the United States: Data from the 2006–2008 National Survey of Family Growth." National Health Statistics Report 36, March 3. http://www.cdc.gov/nchs/data/nhsr/nhsr036.pdf.

———. 2013. Suicide Prevention. http://www.cdc.gov/ViolencePrevention/suicide/index.html.

———. 2014. "Life Expectancy." http://www.cdc.gov/nchs/fastats/life-expectancy.htm.

Chaliand, Gerard, and Jean-Pierre Rageau. 1995. *The Penguin Atlas of Diasporas.* New York: Penguin Group.

Chambliss, William. 1974. "The State, the Law, and the Definition of Behavior as Criminal or Delinquent." In *Handbook of Criminology,* edited by D. Glaser, 7–44. Chicago: Rand McNally.

Chan, Sewell. 2010. "World Trade Organization Upholds American Tariffs on Tires from China." *New York Times,* December 13. http://www.nytimes.com/2010/12/14/business/global/14trade.html.

Chance, James. 2002. "Tomorrow the World." *New York Review of Books,* November 21: 33–35.

Chanda, Nayan. 2003. "The New Leviathans: An Atlas of Multinationals Throws Unusual Light on Globalization." Review of *Global Inc.: An Atlas of the Multinational Corporation,* by Medard Gabel and Henry Bruner, in *YaleGlobal Online,* November 12. http://yaleglobal.yale.edu.

Chang, Leslie. 2009. *Factory Girls: From Village to City in a Changing China.* New York: Spiegel & Grau.

Charney, William, ed. 2010. *Handbook of Modern Hospital Safety* (2nd ed.). Boca Raton, FL: Taylor and Francis.

Che, Yi, Akiko Hayashi, and Joseph Tobin. 2007. "Lessons from China and Japan for Preschool Practice in the United States." *Journal of College Education* 40(1): 7–12.

Chehabi, H. E., and Juan J. Linz, eds. 1998. *Sultanistic Regimes.* Baltimore and London: Johns Hopkins University Press.

Chen, David W. 2004. "What Is a Life Worth?" *New York Times,* June 20: 42K.

ChildStats.gov. 2012. "Family and Social Environment." *America's Children: Key National Indicators of Well-Being, 2011.* http://www.childstats.gov/americaschildren/tables.asp.

———. 2013. "Secure Parental Employment." *America's Children: Key National Indicators of Well-Being, 2013.* http://www.childstats.gov/americaschildren/eco2.asp.

Christian, Julie, Richard Bagozzi, Dominic Abrams, and Harriet Rosenthal. 2012. "Social Influence in Newly Formed Groups: The Roles of Personal and Social Intentions, Group Norms, and Social Identity." *Personality and Individual Differences* 52: 255–260.

Churchill, Winston. 1947. "Speech, House of Commons, 11 November 1947." In *Winston S. Churchill: His Complete Speeches, 1897–1963,* edited by Robert Rhodes James, Vol. 7, 7566. New York: Chelsea House.

Clark, Don. 2010. "Leadership and Developing Diversity." http://www.nwlink.com/~donclark/leader/diverse.html.

Clark, Kirsten. 2014. "Cowboy Church Puts Unique Brand on Worship." *Louisville Courier Journal*, May 14. http://www.courier-journal.com/story/news/local/oldham/2014/05/16/cowboy-church-puts-unique-brand-worship/2205056.

CNN Fortune Global 500. 2014. http://money.cnn.com/magazines/fortune/global500/index.html.

CNN Money. 2013. "Top 20 Paid CEOs," April 8. http://money.cnn.com/gallery/news/companies/2013/04/08/executive-pay.

Coca-Cola. 2014. "Coca-Cola History." http://www.worldofcoca-cola.com/coca-cola-facts/coca-cola-history/.

Cole, Roberta L. 2002. "Manifest Destiny Adapted for 1990s War Discourse: Mission and Destiny Intertwined." *Sociology of Religion* 63(4): 403–426.

Collins, Elizabeth M. 2011. "From Hospital to Hollywood." *Soldier's Magazine,* January 26. http://www.army.mil/article/50855/From_hospital_to_Hollywood_a_Soldier_s_story/.

Collins, Patricia Hill. 2000a. *Black Feminist Thought.* Boston: Unwin Hyman.

———. 2000b. *Black Feminist Thought: Knowledge, Consciousness and the Politics of Empowerment* (2nd ed.). New York: Routledge.

———. 2006. *From Black Power to Hip Hop: Racism, Nationalism, and Feminism.* Philadelphia: Temple.

Collins, Randall. 1961. "Conflict Theory of Educational Stratification." In *Schools and Society: A Sociological Approach to Education,* edited by Jeanne H. Ballantine and Joan Z. Spade, 34–40. Thousand Oaks, CA; Sage.

———. 1971. "A Conflict Theory of Sexual Stratification." *Social Problems* 19(1): 3–21.

Connidis, Ingrid Arnet, and Julie Ann McMullin. 2002. "Sociological Ambivalence and Family Ties: A Critical Perspective." *Journal of Marriage and the Family* 64(3): 558–567.

Cook-Lynn, Elizabeth. 2008. "Deadliest Enemies: Law and the Making of Race Relations on and off Rosebud Reservation, and: Not Without Our Consent: Lakota Resistance to Termination, 1950–59." *Wicazo Sa Review* 23(1): 155–158. http://muse.jhu.edu/login?uri=/journals/wicazo_sa_review/v023/23.1cook-lynn.html.

Cooley, Charles Horton. 1909. *Social Organization.* New York: Scribner's.

———. 1961. "The Social Self." In *Theories of Society: Foundations of Modern Sociological Theory,* edited by T. Parsons, E. Shils, K. D. Naegele, and J. R. Pitts, 822–828. New York: Free Press.

Coonan, Clifford. 2008. "Western Demand Drains Philippines of 85 Percent of Its Trained Nurses." *Independent News* (UK), January 25. http://www.independent.co.uk/news/.

Coren, Stanley. 2011. "Do We Treat Dogs the Same Way as Children in Our Modern Families?" *Psychology Today,* May 2. http://www.psychologytoday.com/blog/canine-corner/201105/do-we-treat-dogs-the-same-way-children-in-our-modern-families.

Cornell, Stephen, and Douglas Hartmann. 2007. *Ethnicity and Race: Making Identities in a Changing World* (2nd ed.). Thousand Oaks, CA: Pine Forge.

Corsaro, William A., and Laura Fingerson. 2003. "Development and Socialization in Childhood." In *Handbook of Social Psychology,* edited by J. Delamater, 125–156. New York: Kluwer Academic/Plenum.

Coser, Lewis A. 1973. "Social Conflict and the Theory of Social Change." In *Social Change: Sources, Patterns, and Consequences,* edited by E. Etzioni-Halevy and A. Etzioni, 114–122. New York: Basic Books.

Cote, James. 1997. "A Social History of Youth in Samoa: Religion, Capitalism, and Cultural Disenfranchisement." *International Journal of Comparative Sociology* 38(3–4): 217.

Coutu, Walter. 1951. "Role-Playing vs. Role-Taking: An Appeal for Clarification." *American Sociological Review* 16(2), 180–187.

CowboyChurch.Net. 2014. "Directories." http://www.cowboychurch.net/list-your-ministry.html.

Crapanzano, Vincent. 1985. *Waiting: The Whites of South Africa.* New York: Random House.

Culbertson, Howard. 2006. "Why Is Ethnocentrism Bad?" Southern Nazarene University. http://home.snu.edu/~HCULBERT/ethno.htm.

Cuzzort, R. P., and E. W. King. 2002. *20th Century Social Thought.* New York: Harcourt.

D'Andrade, Roy. 1984. "Cultural Meaning Systems." In *Culture Theory,* edited by R. A. Shweder and R. A. LeVine, 88–119. Cambridge: Cambridge University Press.

Dahrendorf, Ralf. 1973. "Toward a Theory of Social Conflict." In *Social Change: Sources, Patterns, and Consequences* (2nd ed.), edited by E. Etzioni-Halevy and A. Etzioni, 100–113. New York: Basic Books.

Daily Mail Online. 2011. "The End of the Line" (April 25). http://www.dailymail.co.uk/sciencetech/article-1380383/Worlds-typewriter-factory-ends-production-Godrej-Boyce-closes-doors.html.

Daly, Kerry J. 1992. "Toward a Formal Theory of Interactive Socialization." *Qualitative Sociology* 15(44): 395–417.

Dao, James. 2013. "In Debate over Military Sexual Assault, Men Are Overlooked Victims." *New York Times,* June 23. http://www.nytimes.com/2013/06/24/us/in-debate-over-military-sexual-assault-men-are-overlooked-victims.html?pagewanted=all&_r=0.

DARPA. 2012. "DARPA Researchers Design Eye-Enhancing Virtual Reality Contact Lenses." http://www.darpa.mil/NewsEvents/Releases/2012/01/31.aspx.

———. 2014. "Legged Squad Support System (LS3)." Retrieved April 9, 2014, from http://www.darpa.mil/Our_Work/TTO/Programs/Legged_Squad_Support_System_%28LS3%29.aspx.

Davis, F. James. 1978. *Minority-Dominant Relations: A Sociological Analysis.* Arlington Heights, IL: AHM.

Davis, Jeffry. 2104. http://www.wheaton.edu/Academics/Faculty/D/Jeffry-Davis/guidance/Why-Read-Books-at-All.

Davis, Kingsley. 1940. "Extreme Isolation of a Child." *American Journal of Sociology* 45: 554–565.

———. 1947. "Final Note on a Case of Extreme Isolation." *American Journal of Sociology* 3(5): 432–437.

———. 1984. "Wives and Work: The Sex Role Revolution and Its Consequences." *Population and Development Review* 10(3): 397–417.

Davis, Kingsley, and Wilbert E. Moore. 1945. "Some Principles of Stratification." In *Ideological Theory: A Book of Readings,* edited by L. A. Coser and B. Rosenberg, 413–445. New York: Macmillan.

Davis, Nancy J. 2005. "Taking Sex Seriously: Challenges in Teaching about Sexuality." *Teaching Sociology* 33(1): 16–31.

Davis, Shannon, Theodore N. Greenstein, and Jennifer P. Gerteisen Marks. 2007. "Effects of Union Type on Division of Household Labor: Do Cohabiting Men Really Perform More Housework?" *Journal of Family Issues* 28: 1246–1272.

Day, Thomas. 2009. "About Caregiving." http://www.longtermcarelink.net/eldercare/caregiving.htm#caregivers.

Defense Contract Management Agency. 2014. "About Us." http://www.dcma.mil/about.cfm.

Defense News. 2014. "Defense News Top 100 for 2013." http://special.defensenews.com/top-100/charts/rank_2013.php.

Defense of Marriage Act of 1996, Public Law 104-199, 104th Cong. (September 21, 1996). http://frwebgate.access.gpo.gov/cgi-bin/getdoc.cgi?dbname=104_cong_public_laws&docid=f:publ199.104.pdf.

Dell Social Innovation Challenge. 2014. "2013 Winners." http://www.dellchallenge.org/about/2013-winners.

Deloitte. 2011a. "2011 Survey of Health Care Consumers in the United States: Key Findings, Strategic Implications." http://www.deloitte.com/assets/Dcom-UnitedStates/Local%20Assets/Documents/US_CHS_2011ConsumerSurveyinUS_062111.pdf.

———. 2011b. "2011 Survey of Heath Care Consumers Global Report: Key Findings, Strategic Implementations." http://www.deloitte.com/assets/Dcom-UnitedStates/Local%20Assets/Documents/US_CHS_2011ConsumerSurveyGlobal_062111.pdf.

Dennis, Kingsley. 2008. "Viewpoint: Keeping a Close Watch: The Rise of Self-Surveillance and the Threat of Digital Exposure." *Sociological Review* 56(3): 347–357.

Der Spiegel. 2003. Quoted in "Europe Seems to Hear Echoes of Empires Past," by Richard Bernstein. *New York Times,* April 14. http://www.nytimes.com/.

Dewhurst, Christopher J., and Ronald R. Gordon. 1993. Quoted in "How Many Sexes Are There?" *New York Times,* March 12: A15.

Dhar, Aarti. 2012. "Blow to Misuse of Sex Determination Tests." *The Hindu*. http://www.thehindu.com/health/policy-and-issues/article2801466.ece.

Diamond, Jared. 1992. Quoted in "A Brief History of Human Society: The Origin and Role of Emotion in Social Life: 2001 Presidential Address," by D. S. Massey. *American Sociological Review* 67(1): 1–29.

Dixson, Mark. 2013. "Public Education Finances." U.S. Bureau of the Census. http://www2.census.gov/govs/school/11f33pub.pdf.

Doane, Ashley W. 1997. "Dominant Group Ethnic Identity in the United States: The Role of 'Hidden' Ethnicity in Intergroup Relations." *Sociological Quarterly* 38(3): 375–397.

Doob, Christopher B. 2013. *Social Inequality and Social Stratification in US Society*. Upper Saddle River, NJ: Pearson Education, Inc.

Downey, Liam. 2005. "Assessing Environmental Inequality: How the Conclusions We Draw Vary According to the Definitions We Employ." *Sociological Spectrum* 25(3): 349–369.

Dreger, Alice. 2009. "The Sex of Athletes: One Issue, Many Variables." *New York Times,* October 24. http://www.nytimes.com/2009/10/25/sports/25intersex.html?_r=.

Dreifus, Claudia. 2005. "Declaring with Clarity, When Gender Is Ambiguous." *New York Times,* May 31: D2.

DuBois, W. E. B. 1919 [1970]. "Reconstruction and Africa." In *W. E. B. DuBois: A Reader,* 372–381. New York: Harper & Row.

Dugger, Cecilia. 2005. "Study Finds Small Developing Lands Hit Hardest by 'Brain Drain.'" *New York Times,* October 23: A10.

———. 2007. "UN Predicts Urban Population Explosion." *New York Times,* June 28: A6.

Duhigg, Charles, and Keith Bradsher. 2012. "How U.S. Lost Out on iPhone Work." *New York Times,* January 22: A1+.

Duncan, David Ewing. 2013. "How Long Do You Want to Live?" *New York Times*, August 25. http://www.nytimes.com/2012/08/26/sunday-review/how-long-do-you-want-to-live.html?_r=0.

Dunlap, Riley E., and William R. Catton Jr. 1979. "Environmental Sociology." *Annual Review of Sociology* 5: 243–273.

Durkheim, Émile. 1897. *Suicide.* New York: Free Press.

———. 1901. *The Rules of Sociological Method and Selected Texts on Sociology and Its Method,* edited by S. Lukes, translated by W. D. Halls. Repr., New York: Free Press, 1982.

———. 1915. *The Elementary Forms of the Religious Life* (5th ed.), translated by J. W. Swain. Repr., New York: Macmillan, 1964.

———. 1933. *The Division of Labor in Society,* translated by G. Simpson. Repr., New York: Free Press, 1964.

———. 1961. "On the Learning of Discipline." In *Theories of Society: Foundations of Modern Sociological Theory,* Vol. 2., edited by T. Parsons, E. Shils, K. D. Naegele, 860–865. New York: Free Press.

Ebersole, Luke. 1967. "Sacred." In *A Dictionary of the Social Sciences,* edited by J. Gould and W. L. Kolb. New York: UNESCO.

Eberstadt, Nicholas. 2012. "The Islamic World's Quit Revolution." *Foreign Policy* (March 9). http://www.foreignpolicy.com/articles/2012/03/09/the_islamic_worlds_quiet_revolution?page=full.

Eckstein, Susan, and Thanh-Nghi Nguyen. 2011. "The Making and Transnationalization of an Ethnic Niche: Vietnamese Manicurist." *International Migration Review* 45(3): 639–674.

The Economist. 2011. "Japan's Cramming Schools: Testing Times." (December 31). http://www.economist.com/node/21542222.

———. 2012. "Daily Chart Kings of the Carnivores" (April 30). http://www.economist.com/blogs/graphicdetail/2012/04/daily-chart-17.

Education Encyclopedia. 2009. "Early Childhood Education: International Context." http://www.answers.com/topic/early-childhood-education-international-context.

Ehrenreich, Barbara. 2001. *Nickel and Dimed.* New York: Holt.

Eisenhart, Corrine. 2011. "Why Do Taiwanese Children Excel at Math?" *Phi Delta Kappan,* September. http://www.academia.edu/987689/Why_do_Taiwanese_Children_Excel_at_Math.

Ellis, Blake. 2011. "Average Student Loan Debt Tops $25,000." CNN Money, November 3. http://money.cnn.com/2011/11/03/pf/student_loan_debt/index.htm.

Emirbayer, Mustafa, and Ann Mische. 1998. "What Is Agency?" *American Journal of Sociology* 103(4): 962–1023.

Engelman, Robert. 2013. Quoted in "Global Meat Production and Consumption Continue to Rise." World Watch Institute. http://www.worldwatch.org/global-meat-production-and-consumption-continue-rise.

Engels, Friedrich. 1884. *The Origin of the Family, Private Property, and the State,* edited by E. B. Leacock. New York: International Publisher.

Ennis, Sharon R., Marays Rios-Vargas, and Nora G. Albert. 2011. "The Hispanic Population: 2010." *2010 Census Briefs,* May 11. http://2010.census.gov/2010census/data/.

Environmental Protection Agency. 2011. "10 Fast Facts on Recycling" (updated October 7). http://www.epa.gov/reg3wcmd/solidwasterecyclingfacts.htm.

Environmental Working Group. 2011. "2011 Farm Subsidy Data Base." http://farm.ewg.org/top_recips.php?fips=00000&progcode=totalfarm®ionname=theUnited States (accessed March 29, 2012).

Epstein, Cynthia Fuchs. 2006. "Great Divides: The Cultural, Cognitive, and Social Bases of the Global Subordination of Women." *American Sociological Review* 72(February): 1–22.

Erikson, Kai T. 1966. *Wayward Puritans.* New York: Wiley.

Esposito, John L. 1986. "Islam in the Politics of the Middle East." *Current History* (February): 53–57, 81.

Esser, James K. 1998. "Alive and Well After 25 Years: A Review of Groupthink Research." *Organizational Behavior and Human Decision Processes* 73(2/3): 116–141.

Etzioni, Amitai. 1975. *Comparative Analysis of Complex Organizations.* New York: Free Press of Glencoe.

European Social Innovation Competition. 2014. "Semi-Finalist Number 4." http://socialinnovationcompetition.eu/396/.

Evans, Pam. 1991. Quoted in "What Is Prejudice as It Relates to Disability Anti-Discrimination Law?" by David Reubain. *Disability Rights Education and Defense Fund.* http://www.dredf.org/international/paper_ruebain.html.

Export.gov. 2012. "Maquiladoras: Frequently Asked Questions." NAFTA, Mexico. http://infoserv2.ita.doc.gov/ticwebsite/naftaweb.nsf/504ca249c786e20f85256284006da7ab/ab04e16e721ef690852566ff005ece25!OpenDocument.

Eyerman, Ron, and Bryan S. Turner. 1998. "Outline of a Theory of Generations." *European Journal of Social Theory* 1(1): 91–106.

Family Caregiver Alliance. 2014. "Selected Caregiver Statistics." https://caregiver.org/selected-caregiver-statistics.

Federal Procurement Data System. 2012. "Top 100 Contractors Report." Fiscal Year 2011. https://www.fpds.gov/fpdsng_cms/index.php/reports.

Fein, Melvyn L. 2011. "On Loss and Losing. A Crucial Nexus." Listed in *Selected Works of Melvyn L. Fein.* http://works.bepress.com/melvyn_fein/.

Feinberg, Lynne, Susan C. Reinhard, Ari Housar, and Rita Choula. 2011. "Valuing the Invaluable: The

Growing Contributions and Cost of Family Care Giving." AARP Public Policy Institute. http://assets.aarp.org/rgcenter/ppi/ltc/i51-caregiving.pdf.

Felluga, Dino. 2009. "Modules on Foucault: On Panoptic and Carceral Society." *Introductory Guide to Critical Theory.* http://www.cla.purdue.edu/academic/engl/theory/.

Forbes. 2010. "America's Best Prep Schools" (April 29). http://www.forbes.com/2010/04/29/best-prep-schools-2010-opinions-private-education.html.

———. 2011. "The 10 Top-Earning American Idols." http://www.forbes.com/pictures/eeel45ekei/1-carrie-underwood-20-million/ (accessed February 28, 2011).

———. 2013. "Americas Highest Paid Chief Executives, 2012." http://www.forbes.com/lists/2012/12/ceo-compensation-12_rank.html.

———. 2014a. "The 50 Philanthropists Who Have Given Away the Most Money." http://www.forbes.com/sites/randalllane/2013/11/18/the-50-philanthropists-who-have-given-away-the-most-money/.

———. 2104b. "The World's Billionaires." http://www.forbes.com/billionaires/list/#tab:overall.

———. 2014c. "Without Much Fanfare, Apple Has Sold Its 500 Millionth iPhone." http://www.forbes.com/sites/markrogowsky/2014/03/25/without-much-fanfare-apple-has-sold-its-500-millionth-iphone/.

Foucault, Michel. 1977. *Discipline and Punish: The Birth of the Prison.* Paris: Gallimard.

Frank, Robert. 2013. "Where the Wealthy Get Their Income." *Inside Wealth.* http://www.cnbc.com/id/100958750.

Fraser, Laura. 2002. "The Islands Where Boys Grow Up to Be Girls." *Marie Claire* (December): 72-79.

Freedom House. 2008. "Yemen." http://www.freedomhouse.org/country/yemen.

Freund, Julien. 1968. *The Sociology of Max Weber.* New York: Random House.

Friedman, Thomas. 2008. *Hot, Flat, and Crowded: Why We Need a Green Revolution—and How It Can Renew America.* New York: Farrar, Strauss, Giroux.

Frontline. 2009. "Drowning in Electronics." www.pbs.org/frontlineworld/stories/ghana804/resources/ewaste.html.

Fry, Richard T. 2007. "The Changing Racial and Ethnic Composition of U.S. Public Schools." Pew Hispanic Center Report, August 30. http://www.pewhispanic.org/files/reports/79.pdf.

Gabel, Medard, and Henry Bruner. 2003. *Global Inc.: An Atlas of the Multinational Corporation.* New York: New Press.

Galginaitis, Carol. 2007. "The Lessons of Loss." http://www.clf.uua.org/betweensundays/earlychildhood/LessonsLoss.html.

Gans, Herbert. 1972. "The Positive Functions of Poverty." *American Journal of Sociology* 78: 275–289.

———. 1979. "Symbolic Ethnicity: The Future of Ethnic Groups and Cultures in America." *Race and Ethnicity* 2(1): 1–20.

Garb, Frances. 1991. "Secondary Sex Characteristics." In *Women's Studies Encyclopedia,* Vol. 1: *Views from the Sciences,* edited by H. Tierney, 326–327. New York: Bedrick.

García-Ordaz, Mercedes, Rocío Carrasco-Carrasco, and Francisco José Martínez-López. 2013. "Should We Program Robotic Emotions from the Gender Perspective?" *International Journal of Robotics Applications and Technologies* 1(1): 1–13.

Garfinkel, Harold. 1967. *Studies in Ethnomethodology.* Malden, MA: Blackwell.

Geertz, Clifford. 1984. *American Anthropologist* 86(2): 263–278.

Gendered Innovations. 2014. "Sex and Gender Methods for Research." http://genderedinnovations.stanford.edu/.

Gerth, Hans, and C. Wright Mills. 1954. *Character and Social Structure: The Psychology of Social Institutions.* London: Routledge & Kegan Paul.

Ghaziani, Amin, and Delia Baldassarri. 2011. "Cultural Anchors and the Organization of Differences: A Multi-Method Analysis of LGBT Marches on Washington." *American Sociological Review* 76(2): 179–206.

Gibbs, Jack P. 1965. "Norms: The Problem of Definition and Classification." *American Journal of Sociology* 60: 586–594.

Gibson, David. 2004. "What Did Jesus Really Look Like?" *New York Times,* February 21: A17.

Gillespie, Elizabeth M. 2008. "Earth Liberation Front Believed Responsible for Arson at Multimillion Dollar Seattle Homes." *Huffington Post,* March 3. http://www.huffingtonpost.com/2008/03/03/earth-liberation-front-be_n_89651.html.

Glanz, James. 2012. "The Cloud Factories: Power, Pollution and the Internet." *New York Times,* September 22. http://www.nytimes.com/2012/09/23/technology/data-centers-waste-vast-amounts-of-energy-belying-industry-image.html?pagewanted=all&_r=0.

The Globalist. 2014. "Just the Facts: CEOs and the Rest of Us." http://www.theglobalist.com/just-facts-ceos-rest-us/.

Global Warming Petition Project. 2010. http://www.petitionproject.org/.

Goffman, Erving. 1959. *The Presentation of Self in Everyday Life.* New York: Anchor.

———. 1961. *Asylums: Essays on the Social Situation of Mental Patients and Other Inmates.* New York: Anchor.

Goldman, Russell. 2007. "Environmentalists Classified as Terrorists Get Stiff Sentences." ABC News, May 25. http://www.abcnews.go.com/.

Goldstein, Fred. 2009. "GM Restructuring Will Deepen Capitalist Crisis." *Workers World,* May 3. www.workers.org/2009/us/gm_restructuring_0507.

Google. 2014. "Our Mission. Frequently Asked Questions." https://investor.google.com/corporate/faq.html.

Gordon, John Steele. 1989. "When Our Ancestors Became Us." *American Heritage* (December): 106–221.

Gordon, Milton M. 1978. *Human Nature, Class and Ethnicity.* New York: Oxford University Press.

Gordon, Steven. 1981. "The Sociology of Sentiments and Emotion." In *Social Psychology: Sociological*

Perspectives, edited by M. Rosenberg and R. H. Turner, 562–592. New York: Basic Books.

Gould, Kenneth A., David N. Pellow, and Allan Schnaiberg. 2008. *The Treadmill of Production: Injustice and Unsustainability in the Global Economy.* Boulder, CO: Paradigm Publishers.

Grameen Bank. 2005. "Grameen Bank at a Glance." http://www.grameen-info.org.

———. 2014. "At a Glance." http://www.grameen-info.org/index.php?option=com_content&task=view&id=178&Itemid=182.

Granovetter, Mark. 1973. "The Strength of Weak Ties." *American Journal of Sociology* 76(6): 1360–1380.

Greenaway, Peter. 2009. "Afterword." In *Ladies and Gents: Public Toilets and Gender*, edited by Olga Gershenson and Barbara Penner. Philadelphia: Temple University Press.

Greenberg, Jack. 2001. Quoted in "McAtlas Shrugged: An FP Interview with Jack Greenberg." *Foreign Policy* (May/June): 26+.

Grindeland, Sherry. 2014. "Sammamish Firefighter Climbs to Honor Her Sister." *Sammamish Review,* March 16. http://sammamishreview.com/2014/03/16/sammamish-firefighter-climbs-to-honor-her-sister.

Gross, Terry. 2009. "Fighting America's 'Financial Oligarchy.'" *Fresh Air,* April 15. http://www.npr.org/templates/transcript/transcript.php?storyId=103122382.

Guđmundsson, Hans Kristján. 2008. "The Gender Challenge in Research Funding—Assessing the European National Scenes: Iceland." http://ec.europa.eu/research/science-society/document_library/pdf_06/iceland-research-funding_en.pdf.

Gutowski, Timothy (and students). 2008a. "Environmental Life Style Analysis (ELSA)." IEEE International Symposium on Electronics and the Environment, May 19–20, San Francisco. http://web.mit.edu/ebm/www/Publications/ELSA%20IEEE%202008.pdf.

———. 2008b. Quoted in "Reducing Your Carbon Footprint." *Congressional Quarterly Researcher* 18(42): 985–1008.

Hacker, Andrew. 1971. "Power to Do What?" In *The New Sociology: Essays in Social Science and Social Theory in Honor of C. Wright Mills,* edited by I. L. Horowitz, 134–146. New York: Oxford University Press.

Hall, Kenji. 2009. "No Outcry about CEO Pay in Japan." *Bloomsberg Businessweek,* February 10. http://www.businessweek.com/globalbiz/content/feb2009/gb20090210_949408.htm.

Hallinan, M. T. 1988. "Equality of Educational Opportunity." In *Annual Review of Sociology,* Vol. 14, edited by W. R. Scott and J. Blake, 249–268. Palo Alto, CA: Annual Reviews.

Hamington, Maurice. 2007. "Jane Addams." *Stanford Encyclopedia of Philosophy* (online ed.). http://plato.stanford.edu/entries/addams-jane/.

———. 2010. "Jane Addams." In *Stanford Encyclopedia of Philosophy*, edited by Edward N. Zalta. http://plato.stanford.edu/archives/sum2010/entries/addams-jane/.

Haney Lopez, Ian F. 1994. "The Social Construction of Race: Some Observations on Illusion, Fabrication, and Choice." *Harvard Civil Rights—Civil Liberties Law Review* 29: 39–53.

Hannerz, Ulf. 1992. *Cultural Complexity: Studies in the Social Organization of Meaning.* New York: Columbia University Press.

Hansen, Ronald J. 2013. "University of Phoenix Parent to Lay Off 500 Workers on Declining Enrollment." AZCentral, October 22. http://www.azcentral.com/business/news/articles/20131022university-of-phoenix-expected-to-announce-layoffs.html.

Harris, Gardiner. 2003. "If the Shoe Won't Fit, Fix the Foot? Popular Surgery Raises Concern." *New York Times,* December 7: 24.

Harris-Kojetin, Laurin, Manisha Sengupta, Eunice Park-Lee, and Roberto Valverde. 2013. "Long-Term Care Services in the United States: 2013 Overview." National Center for Health Care Statistics. http://www.cdc.gov/nchs/data/nsltcp/long_term_care_services_2013.pdf.

Harrison, Lily. 2013. "Toby Keith, Taylor Swift and Carrie Underwood Top Forbes' Highest-Paid Country Music Acts List." Music Channel, July 1. http://www.eonline.com/news/435350/toby-keith-taylor-swift-and-carrie-underwood-top-forbes-highest-paid-country-music-acts-list.

Harvard Pediatric. 2010. "Violent Video Games and Young People." *Harvard Mental Health Letter,* October. http://www.health.harvard.edu/newsletters/Harvard_Mental_Health_Letter/2010/October/violent-video-games-and-young-people.

Hasbro Toys. 2010. "G.I. Joe: Chronology." http://www.hasbro.com/gijoe/en_US/.

Hatfield, William. 2006. "White Men Can Jump." In *Sociology: A Global Perspective* (6th edition), edited by Joan Ferrante, 309–310. Belmont, CA: Wadsworth.

Heater, Brian. 2012. "Robotic Butlers, Bartenders and Receptionists at Carnegie Mellon." Engaget, October 24. http://www.engadget.com/2012/10/24/robotic-butlers-bartenders-and-receptionists-at-carnegie-mellon/.

Hedges, Chris. 2002. *War Is the Force That Gives Us Meaning.* New York: Anchor Books/Doubleday.

Henrik, Kreutz. 2000. "A Pragmatic Theory of Action—Collective Action as Theory of the Result of Individual Harmful Addiction." Abstract. http://www.mtapti.hu/.

Henry, S. 2009. "The Social Construction of Crime." In *21st Century Criminology: A Reference Handbook*, edited by J. Miller, 296–305. Thousand Oaks, CA: Sage. http://www.sagepub.com/haganintrocrim8e/study/chapter/handbooks/42347_1.2.pdf.

Hersh, Seymour. 2004. "Annals of National Security: Torture at Abu Ghraib." *New Yorker,* May 10. http://www.newyorker.com/archive/2004/05/10/040510fa_fact.

HHS.gov. 2014. "State by State." http://www.hhs.gov/healthcare/facts/bystate/statebystate.html.

Hill, Patrice. 2013. "Top 1 Percent Reap 90 Percent of Income Gains Since Obama Took Office." *Washington Post,* January 13. http://www.washingtontimes.com/news/2014/jan/13/obama-economic-policies-fail-to-turn-trends-hurtin/?page=all.

Hippocratic Oath. 1943. *Hippocratic Oath: Text, Translation, and Interpretation,* translated from the Greek by Ludwig Edelstein. Baltimore: Johns Hopkins Press.

Hira, Ron, and Anil Hira. 2008. *Outsourcing America.* New York: American Management Association.

Hirsch, E. D., Jr., Joseph F. Kett, and James Trefil. 1993. *The Dictionary of Cultural Literacy.* New York: Houghton Mifflin.

History Learning Site. 2014. "Vietnamese Boat People." Retrieved February 7, 2014, from http://www.historylearningsite.co.uk/vietnam_boat_people.htm.

Hochschild, Adam. 1998. *King Leopold's Ghost.* New York: Houghton Mifflin.

Hochschild, Arlie. 2003a. *The Commercialization of Intimate Life: Notes from Home and Work.* San Francisco and Los Angeles: University of California Press.

———. 2003b. *The Managed Heart: Commercialization of Human Feeling.* Berkeley: University of California Press.

Hoffman, Reid, and Ben Casnocha. 2012. *The Start Up of You.* New York: Crown.

Holloway, Thomas H. 1996. "Immigration." In *Encyclopedia of Latin American History and Culture,* edited by B. A. Tenenbaum, 239–242. New York: Scribner's.

Holmquist, K. 2008. "Jeans Blues." *The Irish Times,* February 9: 7.

Holt, Steve. 2013. "Bolivia: A Country with No McDonald's." *Take Part,* June 5. http://www.takepart.com/article/2013/06/05/country-no-mcdonalds.

Hooks, Gregory, and Chad L. Smith. 2004. "The Treadmill of Destruction: National Sacrifice Areas and Native Americans." *American Sociological Review* 69(4): 558–575.

Hougan, Glen. 2007. "Ageism and Design: Connecting to Your Grandparents." http://www.idsa.org/sites/default/files/Hougan-Ageism_and_Design.pdf.

Hubpages. 2014. "How to Build a House Made from PET Plastic Bottles." Retrieved February 10, 2014, from http://izzym.hubpages.com/hub/How-to-Build-a-House-Made-from-Plastic-Bottles.

Huffington Post. 2011. "World's Last Typewriter Factory Closes Its Doors (Update)" (updated June 26). http://www.huffingtonpost.com/2011/04/26/worlds-last-typewriter-factory-closes_n_853670.html.

Hughes, Everett C. 1984. *The Sociological Eye: Selected Papers.* New Brunswick, NJ: Transaction.

Hull, Anne. 2005. "In Rural Texas, Blessings and Culture Shock." *Washington Post,* September 11: A01.

Humes, Karen R., Nicholas A. Jones, and Roberto R. Ramirez. 2011. "Overview of Race and Hispanic Origin: 2010." *2010 Census Briefs,* March. http://www.census.gov/prod/cen2010/briefs/c2010br-02.pdf.

Hyslop, Noreen. 2011. "Cowboy Church Movement Gaining Popularity." *Tulsa (OK) World,* November 26. Reprinted from "'Cowboy Church' Movement Gains Popularity in Southwest," by Noreen Hyslop, *Dexter Daily Statesman.* http://www.tulsaworld.com/news/article.aspx?subjectid=18&articleid=20111126_11_A22_ULNSer484407.

Independent Lens. 2013 (retrieved). "Two Spirits." http://www.pbs.org/independentlens/two-spirits/film.html.

Innocence Project. 2014. "Frequently Asked Questions." http://www.innocenceproject.org/about/FAQs.php.

International Centre for Prison Studies. 2012. "Entire World-Prison Population Rates per 100,000 of the National Population." http://www.prisonstudies.org/info/worldbrief/wpb_stats.php?area=all&category=wb_poprate (accessed March 29, 2012).

International Forum on Globalization. 2010. http://www.ifg.org/about.htm.

———. 2014. "About Us." http://ifg.org/about/.

International Labour Organization. 2010. *Maternity at Work: A Review of National Legislation* (2nd ed.). Geneva, Switzerland: United Nations Statistics Division.

International Labour Union. 2009. "Recovering from the Crisis: A Global Jobs Pact." http://www.ilo.org/public/libdoc/ilo/2009/109B09_101_engl.pdf.

Intersex Society of North America. 2014. "How Common Is Intersexed?" http://www.isna.org/faq/frequency.

IRS. 2014. "IRS Criminal Investigation Combats Identity Theft Refund Fraud." http://www.irs.gov/uac/Newsroom/IRS-Criminal-Investigation-Combats-Identity-Theft-Refund-Fraud.

Isaacs, Julia B. 2011. "Economic Mobility of Families Across Generations." Executive Summary Economic Mobility Project. Pew Charitable Trusts. http://www.pewtrusts.org/en/archived-projects/economic-mobility-project.

Jaffe, Eric. 2013. "How Free Transit Works in the United States." The Atlantic Cities, March 6. http://www.theatlanticcities.com/commute/2013/03/how-free-transit-works-united-states/4887/.

James Madison College. 2014. "Comparative Cultures and Politics." http://jmc.msu.edu/major/ccp.

Janis, Irving L. 1972. *Victims of Groupthink.* Boston: Houghton Mifflin.

JapanReference. 2014. "Sumo." Retrieved January 9, 2014, from http://www.jref.com/japan/culture/sumo.shtml.

Japan Times. 2003. "Japan: A Developing Country in Terms of Gender Equality" (June 14). http://www.japantimes.co.jp.

Jargon, Julie. 2013. "At McDonald's, Salads Just Don't Sell." *Wall Street Journal*, October 18. http://online.wsj.com/news/articles/SB10001424052702304384104579139871559464960.

Jerome, Richard. 1995. "Suspect Confessions." *New York Times Magazine,* August 13: 28–31.

Johansson, S. Ryan. 1987. "Status Anxiety and Demographic Contraction of Privileged Populations." *Population and Development Review* 13(3): 439–470.

Johnson, Anjelah. 2014. "Nail Salon." Comedy Central: *Funny Videos*. http://www.youtube.com/watch?v=SsWrY77o77o.

Johnson, Jim. 2013. "US Plastic Bottle Recycling Up 6.2 Percent in 2012." *Plastic News,* November 13. http://www.plasticsnews.com/article/20131107/NEWS/131109940/us-plastic-bottle-recycling-up-6-2-percent-in-2012.

Johnson, M. Alex. 2013. "Incandescent Bulb Ban Just One of New Year's New Laws." NBC News,

December 31. http://www.cnbc.com/id/101303132.

Johnston, Norman. 2009. "Prison Reform in Pennsylvania." Pennsylvania Prison Society. http://www.prisonsociety.org/about/history.shtml.

Jones, Sharyn. 2011. *Food and Gender in Fiji*. Lanham, MD: Lexington Books.

Josephson Institute. 2013. "The Ethics of American Youth." http://charactercounts.org/programs/reportcard/.

Judt, Tony. 2004. "Dreams of Empire." *New York Review of Books,* November 4. http://www.nybooks.com/.

Julian, Tiffany A. and Robert A. Kominski. 2011. "Education and Synthetic Work-Life Earnings Estimates." American Community Survey Reports, ACS-14. U.S. Census Bureau, Washington, DC.

Kagan, Jerome. 1989. *Unstable Ideas: Temperament, Cognition, and Self.* Cambridge, MA: Harvard University Press.

Kaiser, Robert G. 2005. "In Finland's Footsteps: If We're So Rich and Smart, Why Aren't We More Like Them?" *Washington Post,* August 7: B01.

Kaiser Family Foundation. 2010. "Generation M^2: Media in the Lives of 8- to 18-Year-Olds" (January 20). http://www.kff.org/entmedia/mh012010pkg.cfm.

Kane, Emily W. 2012. *The Gender Trap: Parents and the Pitfalls of Raising Boys and Girls*. New York: New York University Press.

Kang, K. Connie. 1995. *Home Was the Land of Morning Calm: A Saga of a Korean-American Family.* Reading, MA: Addison-Wesley.

Karen, David. 2005. "No Child Left Behind? Sociology Ignored!" *Sociology of Education* 78(2): 165–169.

Karmadonov, Oleg A. 2014. "Sociology of a Changing Society." http://web.ceu.hu/crc/Syllabi/alumni/sociology/karmadonov.html.

Katzenbach, Jon, Fredrick Beckett, Steven Dichter, Marc Feigen, et al. 1997. *Real Change Leaders: How You Can Create Growth and High Performance at Your Company.* New York: Crown.

Kellner, Douglas. 2002. *Sociological Theory* 20(3): 285–305.

Kemper, Theodore. 1968. "Reference Groups, Socialization and Achievement." *American Sociological Review* 33(1): 31–45.

Kilcullen, John. 1996. "Marx on Capitalism." http://philpapers.org/rec/KILMOC.

Kipnis, Laura. 2004. "The State of the Unions: Should This Marriage Be Saved?" *New York Times,* January 25: 25.

Kittaka, Louise George. 2013. "Juku: An Unnecessary Evil or Vital Steppingstone to Success?" *Japan Times*, March 5 http://www.japantimes.co.jp/community/2013/03/05/issues/juku-an-unnecessary-evil-or-vital-steppingstone-to-success/#.U4DTHC8mRJM.

Kivisto, Peter, and Dan Pittman. 2007. "Goffman's Dramaturgical Sociology: Personal Sales and Service in a Commodified World." In *Illuminating Social Life: Classical and Contemporary Theory,* 271–290. Thousand Oaks, CA: Pine Forge Press.

Klan, Ashley. 2012. "Donating to Panhandlers Not the Answer." GoLocal: Worchester, July 4. http://www.golocalworcester.com/news/donating-to-panhandlers-not-the-answer/.

Klapp, Orrin E. 1986. *Overload and Boredom: Essays on the Quality of Life in an Information Society.* New York: Greenwood.

Klein, Ezra. 2012. *Wonkblog* (blog). "Why an MRI Costs $1,080 in America and $280 in France." *Washington Post,* March 3. http://www.washingtonpost.com/blogs/ezra-klein/post/why-an-mri-costs-1080-in-america-and-280-in-france/2011/08/25/gIQAVHztoR_blog.html.

Kochhar, Rakesh, and Rich Morin. 2014. "Despite Recovery, Fewer Americans Identify as Middle Class." Pew Research Center. Retrieved January 27, 2014, from http://www.pewresearch.org/fact-tank/2014/01/27/despite-recovery-fewer-americans-identify-as-middle-class/.

Koehler, Nancy. 1986. "Re-Entry Shock." In *Cross-Culture Re-Entry: A Book of Readings,* 89–94. Abilene, TX: Abilene Christian University Press.

Kotbi, Nabil. 2009. Quoted in "Clinic Treats Emotional Toll of Recession," by C. Ferrette. *The Journal News* (NY), February 28. http://www.lohud.com.

Kroeber, A. L., and C. Kluckhohn. 1952. *Culture: A Critical Review of Concepts and Definitions.* Cambridge, MA: Peabody Museum.

Kuhn, Thomas. 1975. *The Structure of Scientific Revolutions.* Chicago: University of Chicago.

Kurdek, Lawrence A. 2005. "What Do We Know about Gay and Lesbian Couples?" *Current Directions in Psychological Science* 14(5): 251–254.

LaFraniere, Sharon. 2009. "No Detail Is Overlooked as China Prepares to Celebrate." *The New York Times,* September 29: A4.

Lam, Peng Er. 2009. "Declining Fertility Rates in Japan: An Ageing Crisis Ahead." *East Asia* 26: 177–190.

Lamb, David. 1987. *The Arabs: Journeys beyond the Mirage.* New York: Random House.

Lauby, Jennifer, and Oded Stark. 1988. "Individual Migration as a Family Strategy: Young Women in the Philippines." *Population Studies* 42(3): 473–486. http://www.jstor.org/stable/2174408.

Laughlin, Lynda. 2013. "Who's Minding the Kids? Child Care Arrangements: Spring 2011." Household Economic Studies, U.S. Bureau of the Census. http://www.census.gov/prod/2013pubs/p70-135.pdf.

Layton, Lyndsey. 2012. "High School Graduation Rate Rises in U.S." *Washington Post,* March 19. http://www.washingtonpost.com/local/education/high-school-graduation-rate-rises-in-us/2012/03/16/gIQAxZ9rLS_story.html.

Lechner, Frank J. 1989. "Fundamentalism Revisited." *Society* (January/February): 51–59.

Lee, Fang. 2014. "Where Have All the Lobbyists Gone? What's Going On?" *The Nation*, March 10. http://www.thenation.com/article/178460/shadow-lobbying-complex.

Lee, Jennifer. 2003. "Critical Mass: How Protesters Mobilized So Many and So Nimbly." *New York Times,* February 23. http://www.nyt.com/.

Lee, Min Kyung, Jodi Forlizzi, Sara Kiesler, Paul Rybski, John Antanitis, and Sarun Savetsila. 2012a. "Personalization in HRI: A

Longitudinal Field Experiment." 7th ACM/IEEE International Conference on Human–Robot Interaction, pp. 319–326.

Lee, Min Kyung, Sara Kiesler, Jodi Forlizzi, and Paul Rybski. 2012b. "Ripple Effects of an Embedded Social Agent: A Field Study of a Social Robot in the Workplace." Proceedings at CHI'12, May 5–10, 2012, Austin, TX.

Lee, Sejun. 2013. "Korean, 'Brandeis-Style.'" *Brandeis Now,* January 7. http://www.brandeis.edu/now/2013/october/kpop.html.

Lehrman, Sally. 1997. "WO: Forget Men Are from Mars, Women Are from Venus. Gender." *Stanford Today* (May/June): 47.

Leidner, Robin. 1993. *Fast Food, Fast Talk: Service Work and the Routinization of Everyday Life.* Berkeley: University of California Press.

Leland, John. 2008. "More Men Take the Lead Role in Caring for Elderly Parents." *New York Times,* November 28. http://www.nytimes.com/2008/11/29/us/29sons.html.

Lencioni, Patrick. 2007. *Three Signs of a Miserable Job.* San Francisco: Jossey-Bass.

Lepkowski, Wil. 1985. "Chemical Safety in Developing Countries: The Lessons of Bhopal." *Chemical and Engineering News* 63: 9–14.

Levine, Linda. 2012a. "The Distribution of Household Income and the Middle Class." Congressional Research Service, November 13. http://www.fas.org/sgp/crs/misc/RS20811.pdf.

———. 2012b. "Offshoring (or Offshore Outsourcing) and Job Loss among U.S. Workers." Congressional Research Service. https://www.fas.org/sgp/crs/misc/RL32292.pdf.

Levi-Strauss and Company. 2013. Annual Report, 2012. http://levistrauss.com/lsco/ar2012/two-page.php?pg=016&zl=3.

Lewis, M. Paul. 1998. "Marx's Stock Resurges on 150-Year Tip." *New York Times,* June 27: A17.

Li, Dingjun, P. L. Patrick Rau, and Ye Li. 2010. "A Cross-Cultural Study: Effect of Robot Appearance and Task." *International Journal of Social Robotics* 2: 175–186.

Light, Ivan. 1983. *Cities in World Perspective.* New York: Macmillan.

Lilly, Amanda. 2009. "A Guide to China's Ethnic Groups." Washington Post.com, July 8. http://www.washingtonpost.com/wp-dyn/content/article/2009/07/08/AR2009070802718.html.

Lincoln, C. Eric, and Lawrence H. Mamiya. 1990. *The Black Church in the African American Experience.* Durham, NC: Duke University Press.

Linton, Ralph. 1936. *The Study of Man: An Introduction.* New York: Appleton-Century-Crofts.

Lloyd, Janice. 2011. "Apples Top Most Pesticide-Contaminated List." *USA Today,* June 13. http://yourlife.usatoday.com/ntness-food/safety/story/2011/06/Apples-top-list-of-produce-contaminated-with-pesticides/48332000/1.

Lockard, C. Brett, and Michael Wolf. 2012. "Employment Outlook: 2010-2020." *Monthly Labor Review* (January). http://www.bls.gov/opub/mlr/2012/01/art5full.pdf.

Lockheed Martin. 2012. "Who We Are." http://www.lockheedmartin.com/us/who-we-are.html.

———. 2014. "Who We Are." http://www.lockheedmartin.com/us/who-we-are.html.

Lorber, Judith. 2005. "Night to His Day: The Social Construction of Gender." In *The Spirit of Sociology: A Reader,* edited by R. Matson, 292–305. New York: Penguin.

Lourenço, Orlando, and Armando Machado. 1996. "In Defense of Piaget's Theory: A Reply to 10 Common Criticisms." *Psychological Review* 103(1): 143–164.

Love, David A. 2007. "Walgreens Suit Shows Employment Discrimination Still a Problem." Progressive Media Project, March 13. http://www.progressive.org/media_mplove031307.

Luhby, Tami. 2011. "Income Inequality in America Worsening Wealth Inequality by Race." *CNN Money,* June 21. http://money.cnn.com/2012/06/21/news/economy/wealth-gap-race/.

Lyall, Sarah. 2007. "Gay Britons Serve in Military with Little Fuss, as Predicted Discord Does Not Occur." *New York Times,* April 27: A14.

Lynch, Michael. 2007. Review of *By the Color of Our Skin: The Illusion of Integration and the Reality of Race,* by Leonard Steinhorn and Barbara Diggs-Brown. http://findarticles.com/p/articles/mi_m1568/is_7_31/ai_57815517/.

Mahmood, Cynthia K., and Sharon L. Armstrong. 1992. "Do Ethnic Groups Exist? A Cognitive Perspective on the Concept of Cultures." *Ethnology* 31(1): 1–14.

Marcus, Steven. 1998. "Marx's Masterpiece at 150." *New York Times Book Review,* April 26: 39.

Marger, Martin. 1991. *Race and Ethnic Relations: American and Global Perspectives.* Belmont, CA: Wadsworth.

———. 1992. *Race and Ethnic Relations.* Belmont, CA: Wadsworth.

Markus, Hazel, and Paula Nurius. 1986. "Possible Selves." *American Psychologist* 41(9): 954–969.

Martin, Patricia Yancey. 2004. "Gender as Social Institution." *Social Forces* 82(4): 1249–1273.

Martinez, J. R. 2004. Quoted in "Seeing Past the Scars of Battle," by T. Dwyer. *Washington Post,* October 24. http://www.washingtonpost.com/wp-dyn/articles/A57280-2004Oct23.html.

———. 2011. "Biography for J. R. Martinez." *The Internet Movie Data Base.* http://www.imdb.com/name/nm3210615/bio (accessed December 30, 2011).

Marx, Karl. 1843. Introduction to "A Contribution to the Critique of Hegel's Philosophy of Right." In *Collected Works,* Vol. 3, edited by J. O'Malley, translated by A. Jolin and J. O'Malley. New York: Cambridge University Press, 1970. http://www.marxists.org/archive/marx/works/1843/critique-hpr/intro.htm.

———. 1844. *Economic and Philosophic Manuscripts of 1844,* translated by Richard Hooker. http://www.marxists.org/archive/marx/works/1844/manuscripts/labour.htm.

———. 1845. *The German Ideology,* Part IA: *Idealism and Materialism.* http://www.marxists.org/archive/marx/works/1845/german-ideology/ch01a.htm.

———. 1856. *Class Struggles in France, 1848–1850.* http://www.marxists.org/archive/marx/works/1850/class-struggles-france/index.htm.

———. 1888. "The Class Struggle." In *Theories of Society,* edited by T. Parsons, E. Shils, K. D. Naegele, and

J. R. Pitts, 529–535. New York: Free Press, 1961.

Marx, Karl, and Friedrich Engels. 1848. *Manifesto of the Communist Party.* http://www.marxists.org/archive/marx/works/1848/communist-manifesto/ch01.htm.

Massey, Douglas S. 2002. "A Brief History of Human Society: The Origin and Role of Emotion in Social Life: 2001 Presidential Address." *American Sociological Review* 67(1): 1–29.

Mattel. 2010. "Barbie: The History." http://corporate.mattel.com/about-us/history/default.aspxp.

Mayor, Susan. 2005. "BMA Calls on G8 Governments to Address 'Brain Drain.'" *British Medical Journal* 330(7506): 1466.

McDonald's. 2014. "About Us." http://www.mcdonalds.com/us/en/home.html.

McIntosh, Peggy. 1992. "White Privilege and Male Privilege: A Personal Account of Coming to See Correspondences through Work in Women's Studies." In *Race, Class and Gender: An Anthology,* edited by M. L. Andersen and P. H. Collins, 70–81. Belmont, CA: Wadsworth.

Mead, George Herbert. 1934. *Mind, Self and Society.* Chicago: University of Chicago Press.

Meeks. Larry. 2011. "White Woman Scared of Black Men." https://www.creators.com/advice/ethnically-speaking-larry-meeks/white-woman-scared-of-black-men.html.

Merton, Robert K. 1938. "Social Structure and Anomie." *American Sociological Review* 3(5): 672–682.

———. 1957a. "The Role-Set: Problems in Sociological Theory." *British Journal of Sociology* 8: 106–120.

———. 1957b. *Social Theory and Social Structure.* Glencoe, IL: Free Press.

———. 1976. "Discrimination and the American Creed". In *Sociological Ambivalence and Other Essays,* 189–216. New York: Free Press.

———. 1997. "On the Evolving Synthesis of Differential Association and Anomie Theory: A Perspective from the Sociology of Science." *Criminology* 35(3): 517–525.

Michels, Robert. 1962. *Political Parties,* translated by E. Paul and C. Paul. New York: Dover.

Migration Policy Institute. 2007. "New Report Probes Efforts to Protect Migrant Workers." http://www.migrationpolicy.org/news.

Mihesuah, Devan A. 2008. *Big Bend Luck.* Bangor, ME: Booklocker.com.

Milgram, Stanley. 1974. *Obedience to Authority: An Experimental View.* New York: Harper & Row.

———. 1987. "Obedience." In *The Oxford Companion to the Mind,* edited by R. L. Gregory, 566–568. Oxford, UK: Oxford University Press.

Military Times. 2014. "Some Fast Food Outlets Closing on Military Bases: New Federal Wage Rules May Be Factor, Sources Say" (March 17). http://www.militarytimes.com/article/20140317/NEWS/303170027/Some-fast-food-outlets-closing-military-bases.

Miller, Daniel, and Sophie Woodward. 2014. "A Manifesto for the Study of Denim." http://www.ucl.ac.uk/anthropology/people/academic_staff/d_miller/mil-12.

Mills, C. Wright. 1956. *The Power Elite.* New York: Oxford University Press.

———. 1959. *The Sociological Imagination.* New York: Oxford.

———. 1963. "The Structure of Power in American Society." In *Power, Politics and People: The Collected Essays of C. Wright Mills,* edited by I. L. Horowitz, 23–38. New York: Oxford University Press.

Mills, Janet Lee. 1985. "Body Language Speaks Louder Than Words." *Horizons* (February): 8–12.

Mitten, David. 2001. Quoted in "Harvard's Muslims Grieving, Wary," by Ken Gewertz. *Harvard Gazette,* September 20. http://news.harvard.edu/gazette/2001/09.20/29-muslims.html.

Monnier, Christine. 2009. "What Is Social Structure?" *PB Works.* https://globalsociology.pbworks.com/w/page/14711305/What%20is%20Social%20Structure.

Montague, James. 2011. "A First in Cup Qualifying for a Player and a Team." *New York Times,* November 25. http://www.nytimes.com/2011/11/26/sports/soccer/jonny-saelua-transgender-player-helps-american-samoa-to-first-international-soccer-win.html?_r=0.

Morrow, David J. 1996. "Trials of Human Guinea Pigs." *New York Times,* May 8: C1.

Moss, Michael. 2003. "False Terrorism Tips to F.B.I. Uproot the Lives of Suspects." *New York Times,* June 19: A1.

Muller, Jerry Z. 2013. "Capitalism and Inequality: What the Right and Left Got Wrong." *Foreign Affairs* (March/April): 30–51.

Murdock, George P. 1945. "The Common Denominator of Cultures." In *The Science of Man in the World Crisis,* edited by Ralph Linton, 123–142. New York: Columbia.

Murphy, Dean E. 2004. "Imagining Life without Illegal Immigrants." *New York Times,* January 11: 16WK.

Murphy, S.L., J. Xu and K.D. Kochanek. 2012. "Deaths: Preliminary Data for 2010." *National Vital Statistics Reports* 60(4) (January 11). http://www.cdc.gov/nchs/data/nvsr/nvsr60/nvsr60_04.pdf.

Naofusa, Hirai. 1999. "Traditional Cultures and Modernization: Several Problems in the Case of Japan." Institute for Japanese Culture and Classics. http://www2.kokugakuin.ac.jp.

National Archives. 2014. "Teaching with Documents: An Act of Courage. The Arrest Records of Rosa Parks." http://www.archives.gov/education/lessons/rosa-parks/.

National Association of Home Builders. 2007. http://www.nahb.org/.

———. 2011. "NAHB Study: New Homes in 2015 Will Be Smaller, Greener and More Casual." http://www.nahb.org/news_details.aspx?newsID=12244.

National Center for Education Statistics. 2011a. *The Condition of Education 2011.* "Programs and Courses: Undergraduate Fields of Study," Table A-40-1. http://nces.ed.gov/programs/coe/tables/table-fsu-1.asp.

———. 2011b. "Post-Secondary Education." *Digest of Educational Statistics: 2010,* Chap. 3. http://nces.ed.gov/pubs2011/2011015_3a.pdf.

———. 2013a. "Fast Facts: Highest Enrollment." http://nces.ed.gov/fastfacts/display.asp?id=74.

———. 2013b. "Public School Graduates and Dropouts from

the Common Core of Data: School Year 2009–10." http://nces.ed.gov/pubsearch/pubsinfo.asp?pubid=2013309rev.

———. 2013c. "Digest of Education Statistics, 2012." NCES 2014-015, Table 5. http://nces.ed.gov/fastfacts/display.asp?id=84.

National Council for Crime Prevention in Sweden. 1985. *Crime and Criminal Policy in Sweden.* Report No. 19. Stockholm: Liber Distribution.

National Park Service. 2009. "We Shall Overcome: Historic Places of the Civil Rights Movement." http://www.nps.gov/nr/travel/civilrights/.

National Public Radio. 2013. "90 Cents: The Cost of Safely-Made Jeans in Bangladesh" (July 10). http://hereandnow.wbur.org/2013/07/10/bangladesh-factory-safety.

The Nation's Report Card. 2011. "Reading Report." http://www.nationsreportcard.gov.

Natural Resources Canada. 2010. "Gigajoule (GJ)." http://oee.nrcan.gc.ca/commercial/technical-info/tools/gigajoule-definition.cfm.

Nature. 2007. "Holy Cow: Hinduism's Sacred Animal." PBS, August 12. http://www.pbs.org/wnet/nature/holycow/hinduism.html.

Newport, Sally F. 2000. "Early Childhood Care, Work, and Family in Japan: Trends in a Society of Smaller Families." *Childhood Education* 77(2): 68+.

New York Daily News. 2013. "How Long Can You Wear High Heels before Pain Kicks In? An Hour and Six Minutes, Women Say" (June 13). http://www.nydailynews.com/life-style/long-high-heels-start-hurt-article-1.1368068.

Nicholson, Peter. 2008. Quoted in "More Men Take the Lead Role in Caring for Elderly Parents." *New York Times,* November 28. http://www.nytimes.com/2008/11/29/us/29sons.html.

Norris, Floyd. 2012. "Manufacturing Is Surprising Bright Spot in U.S. Economy." *New York Times,* January 5. http://www.nytimes.com/2012/01/06/business/us-manufacturing-is-a-bright-spot-for-the-economy.html.

North Carolina Department of Corrections. 2009. "Greene Correctional Institution." http://www.doc.state.nc.us/DOP/prisons/greene.htm.

Northern Kentucky University. 2014. "Financial Aid and Scholarship Information." http://coehs.nku.edu/content/dam/coehs/docs/departments/0809freshNKU_scholar.pdf.

Norton, Michael, and Dan Ariely. 2011. "Building a Better America: One Wealth Quintile at a Time." *Perspectives on Psychological Science* 6: 9–12. www.people.hbs.edu/mnorton/norton%20ariely%20in%20press.pdf.

Nottingham, Elizabeth K. 1971. *Religion: A Sociological View.* New York: Random House.

Novas, Himilce. 1994. "What's in a Name?" In *Everything You Need to Know about Latino History,* 2–4. New York: Dutton Signet.

NPR. 2012. "'Life behind the Lobby of Indian-American Motels" (May 30). http://www.npr.org/2012/06/02/153988290/life-behind-the-lobby-indian-american-motel-owners.

Oakes, Jeannie. 1986a. "Keeping Track. Part 1: The Policy and Practice of Curriculum Inequality." *Phi Delta Kappan* 67(September): 12–17.

———. 1986b. "Keeping Track. Part 2: Curriculum Inequality and School Reform." *Phi Delta Kappan* 67(October): 148–154.

Obadina, Tunde. 2000. "The Myth of Neo-Colonialism." *African Economic Analysis.* www.afbis.com.

Obama, Barack. 2004. *Dreams from My Father.* New York: Crown.

———. 2009. Quoted in "Michelle Obama May Have White Slave-Owner Ancestor." *The Guardian,* October 9. http://www.guardian.co.uk/world/2009/oct/09/michelle-obama-white-slave-owner.

———. 2011. Transcript: "Osama bin Laden Killed: Barack Obama's Speech in Full." *The Telegraph* (UK), May 2. http://www.telegraph.co.uk/news/worldnews/barackobama/8487354/Osama-bin-Laden-killed-Barack-Obamas-speech-in-full.html.

———. 2014. "Statement by the President on the Death of Nelson Mandela." http://www.whitehouse.gov/the-press-office/2013/12/05/statement-president-death-nelson-mandela.

Oberschall, Anthony. 2000. "Utopian Visions: Engaged Sociologies for the 21st Century." *Contemporary Sociology* 29(1): 1–13.

OECD.. 2011a. *Education at a Glance 2011.* http://www.oecd.org/.

———. 2011b. *Health at a Glance 2011: OECD Indicators,* released November 23. http://www.oecd.org/dataoecd/24/8/49084488.pdf.

———. 2013. "PISA 2012 Results in Focus: What 15-Year-Olds Know and What They Can Do with What They Know." http://www.oecd.org/pisa/keyfindings/pisa-2012-results-overview.pdf.

———. 2014a. "Employment Policies and Data." http://www.oecd.org/employment/emp/onlineoecdemploymentdatabase.htm#earndisp.

———. 2014b. "Real Minimum Wage Laws." StatExtracts. http://stats.oecd.org/Index.aspx?DatasetCode=U_D_D.

Ogawa, Naohiro, Robert D. Retherford, and Rikiya Matsukura. 2009. "Japan's Declining Fertility and Policy Responses." In *Ultra-Low Fertility in Pacific Asia,* edited by G. Jones, P. T. Straughan, and A. Chan, 40–72. New York: Routledge.

Ogbu, John U. 1990. "Minority Status and Literacy in Comparative Prospective." *Daedalus* 119(2): 141–168.

Ogburn, William F. 1968. "Cultural Lag as Theory." In *Culture and Social Change* (2nd ed.), edited by O. D. Duncan, 86–95. Chicago: University of Chicago Press.

Okello, K. L. 2013. "A Survey of the Productivity and Functions of Goats in Uganda." Food and Agricultural Organization of the United Nations. Retrieved January 7, 2013, from http://www.fao.org/wairdocs/ilri/x5464b/x5464b16.htm.

Olshansky, S. Jay, and A. Brian Ault. 1986. "The Fourth Stage of the Epidemiologic Transition: The Age of Delayed Degenerative Diseases." *Milbank Quarterly* 64(3): 355–391.

Omi, Michael, and Howard Winant. 1986. *Racial Formation in the United States: From the 1960s to the 1980s.* New York: Routledge & Kegan Paul.

———. 2002. "Racial Formation". In *Race Critical Theories,* edited by Philomena Essed and David T. Goldberg, 123–145. London: Blackwell.

The Oregonian. 2011. "2011 Oregon High School Graduation Rates." http://schools.oregonlive.com.

Orfield, John Kucsera, and Genevieve Siegel-Hawley. 2012. "E Pluribus . . . Separation: Deepening Double Segregation for More Students." Civil Rights Project, September 19. http://civilrightsproject.ucla.edu/research/k-12-education/integration-and-diversity/mlk-national/e-pluribus...separation-deepening-double-segregation-for-more-students.

Page, Clarence. 1996. *Showing My Colors: Impolite Essays on Race and Identity*. New York: HarperCollins.

Painter, Nell Irvin. 2010. *History of White People*. New York: Norton.

Parke, Ross D. 2013. "Challenges to the Ideal Family Form." http://media.wiley.com/product_data/excerpt/90/04706744/0470674490-5.pdf.

Parker, Kim, and Wendy Wang. 2013. "Modern Parenthood: Roles of Moms and Dads Converge as They Balance Work and Family." Pew Research: Social and Demographic Trends (March 14). http://www.pewsocialtrends.org/2013/03/14/modern-parenthood-roles-of-moms-and-dads-converge-as-they-balance-work-and-family/.

Pennsylvania Department of Environmental Protection. 2014. "Waste Reduction in the Home." http://www.dep.state.pa.us/dep/deputate/airwaste/wm/RECYCLE/facts/reduce.htm.

Petrich, Marisa. 2011. "Sikh Soldier Answers Lifelong Calling to Serve." U.S. Army Homepage, June 2. http://www.army.mil/article/58866/.

Pew Charitable Trusts. 2013. "Payday Lending in America, Report 2." http://www.pewstates.org/uploadedFiles/PCS_Assets/2013/Pew_Choosing_Borrowing_Payday_Feb2013.pdf.

Pew Research Center. 2011a. "For Millennials, Parenthood Trumps Marriage." http://www.pewsocialtrends.org/2011/03/09/iii-millennials-attitudes-about-marriage/.

———. 2011b. "Is College Worth It?" (May 15). http://www.pewsocialtrends.org/2011/05/15/is-college-worth-it/.

Phelan, Jo C., Bruce G. Link, and Parisa Tehranifar. 2010. "Social Conditions as Fundamental Causes of Health Inequalities: Theory, Evidence, and Policy Implications." *Journal of Health and Social Behavior* 51(1, Suppl.): S28–S40.

Pizzitutti, Francesco , Carlos F. Mena, and Stephen J. Walsh. 2104. "Modelling Tourism in the Galapagos Islands: An Agent-Based Model Approach." *Journal of Artificial Societies and Social Simulation* 17(1): 14. http://jasss.soc.surrey.ac.uk/17/1/14.html.

Porter, Eduardo. 2013. "In Public Education, the Edge Still Goes to the Rich." *New York Times*, November 5. http://www.nytimes.com/2013/11/06/business/a-rich-childs-edge-in-public-education.html.

Powell, Walter W., and Kaisa Snellman. 2004. *Annual Review of Sociology* 30: 199–220. http://www.stanford.edu/group/song/papers/powell_snellman.pdf.

Proudhon, Pierre-Joseph. 1847. "The Philosophy of Misery." www.marxists.org/reference/subject/economics/proudhon/philosophy/intro.htm.

Proweller, Amira. 1998. *Constructing Female Identities: Meaning Making in an Upper Middle Class Youth Culture.* Albany: State University of New York.

Rabin, Roni Caryn. 2008. "TV Ads Contribute to Childhood Obesity, Economists Say." *New York Times,* November 20. http://www.nytimes.com/2008/11/21/health/research/21obesity.html.

Radelet, Michael L., Hugo Adam Bedau, and Constance Putnam. 1994. *In Spite of Innocence: Erroneous Convictions in Capital Cases.* Boston: Northeastern University Press.

Raz, Guy. 2010. "A Bottled-Water Drama in Fiji." NPR, December 1. http://www.npr.org/2010/12/01/131733493/A-Bottled-Water-Drama-In-Fiji.

Redfield, Robert. 1962. "The Universally Human and the Culturally Variable." In *Human Nature and the Study of Society: The Papers of Robert Redfield,* Vol. 1, edited by M. P. Redfield, 439–453. Chicago: University of Chicago Press.

Reich, Jens. 1989. Quoted in "People of the Year." *Newsweek,* December 25: 18–25.

Reporters without Borders. 2010. "Web 2.0 versus Control 2.0" (March 18). http://en.rsf.org/web-2-0-versus-control-2-0-18-03-2010,36697.

Research and Markets. 2014. "Denim Jeans: Global Strategic Business Report." Retrieved January 11, 2014, from http://globenewswire.com/news-release/2013/10/25/583751/10054269/en/Denim-Jeans-Global-Strategic-Business-Report.html.

Reuters. 2014. "U.S. Lifts Ban Blocking BP from New Government Contracts" (May 13). http://www.reuters.com/article/2014/03/13/us-bp-usa-contracts-idUSBREA2C24E20140313.

Revkin, Andrew C. 2005. "On Climate Change, a Change of Thinking." *New York Times,* December 4. http://www.nytimes.com/2005/12/04/weekinreview/04revkin.html.

Rhode, Deborah L., and Christopher J. Walker. 2008. "Gender Equity in College Athletics: Women Coaches as a Case Study." *Stanford Journal of Civil Rights and Civil Liberties* 4: 1–50.

Ritzer, George. 1993. *The McDonaldization of Society.* Newbury Park, CA: Pine Forge Press.

———. 2008. *The McDonaldization of Society* (5th ed.). Thousand Oaks, CA: Pine Forge Press.

Robertson, Ian. 1988. *Sociology* (3rd ed.). New York: Worth.

Robertson, Roland. 1987. "Economics and Religion." *The Encyclopedia of Religion.* New York: Macmillan.

The Rock Club. 2014. "About Climbing." http://climbrockclub.com/getting-started/about-climbing/.

Rodney, Walter. 1973. *How Europe Underdeveloped Africa.* Dar-Es-Salaam: Bogle-L'Ouverture Publications, London and Tanzanian Publishing House. http://www.marxists.org/subject/africa/rodney-walter/how-europe/index.htm.

Roethlisberger, F. J., and William J. Dickson. 1939. *Management and the Worker.* Cambridge, MA: Harvard University Press.

Rokeach, Milton. 1973. *The Nature of Human Values.* New York: Free Press.

Rosich, Katherine J., and Janet R. Hankin. 2010. "Executive Summary: What Do We Know? Key Findings from 50 Years of Medical Sociology." *Journal of Health and Social Behavior* 51(1, Suppl.): S1–S9.

Rostow, W. W. 1960. *The Stages of Economic Growth: A Non-Communist Manifesto.* Cambridge, UK: Cambridge University Press.

Rotundo, Tony. 2014. Personal Correspondence (January 8).

Royal Geographical Society. 2014. "Who Wants to Be a Billionaire? What Do Billionaires Do?" http://www.rgs.org/OurWork/Schools/Teaching+resources/Key+Stage+3+resources/Who+wants+to+be+a+billionaire/What+do+billionaires+do.htm.

Samuel, John. 1997. "World Population and Development: Retrospect and Prospects." *Development Express.* Ottawa: Canadian International Development Agency.

Sandia National Laboratories. 2014. "Sandia National Laboratories Presents Fully Integrated Lab-on-a-Chip Development." http://www.sandia.gov/media/NewsRel/NR2000/labchip.htm.

Santosa, Ailiana, et al. 2014. "The Development and Experience of Epidemiological Transition Theory over Four Decades: A Systematic Review." *Global Health Action* 7: 23574

Sapir, Edward. 1949. "Selected Writings of Edward Sapir." In *Language, Culture and Personality,* edited by D. G. Mandelbaum. Berkeley: University of California Press.

Satterly, D. J. 1987. "Jean Piaget (1896–1980)." In *The Oxford Companion to the Mind,* edited by R. I. Gregory, 621–622. Oxford, UK: Oxford University Press.

Satya. 2014. "Where Do All the Feathers Go? The *Satya* Interview with Walter Schmidt." http://www.satyamag.com/feb06/schmidt.html.

Save the Children. 2007. "State of the World's Mothers." http://www.savethechildren.org/site/c.8rKLIXMGIpI4E/b.6153061/k.A0BD/Publications.htm.

———. 2014. "State of the World's Mothers." http://www.savethechildren.org/atf/cf/%7B9def2ebe-10ae-432c-9bd0-df91d2eba74a%7D/SOWM_2014.PDF.

Scelza, Janene, Roberta Spalter-Roth, and Olga Mayorova. 2011. "A Decade of Change: ASA Membership from 2000–2010." http://www.asanet.org/images/research/docs/pdf/2010_asa_membership_brief.pdf.

Schilt, Kristen, and Catherine Connell. 2007. "Do Workplace Gender Transitions Make Gender Trouble?" *Gender, Work and Organization* 14(6): 596–618.

Schmalz, Jeffrey. 1993. "From Midshipman to Gay-Rights Advocate." *New York Times,* February 4: B1+.

Schmid, Randolph. 1997. "'Multiracial' Category Rejected for Census Forms." *Chicago Sun-Times,* July 9.

Schmitt, John. 2012. "Low Wage Lessons." Center for Economic and Policy Research, January. http://www.cepr.net/documents/publications/low-wage-2012-01.pdf.

Schnaiberg, Allan. 1980. *The Environment: From Surplus to Scarcity.* New York: Oxford University Press.

Schuster Institute for Investigative Journalism. 2014. "Ubiquitous Palm Oil: Products with Palm Oil & Its Derivatives." http://www.schusterinstituteinvestigations.org/#!products-with-palm-oil/c1z3e.

Schwartz, Barry. 2004. *The Paradox of Choice*. New York: Harper.

Scully, Matthew. 2002. *Dominion: The Power of Man, the Suffering of Animals, and the Call to Mercy.* New York: St. Martin's.

September 11 Victim Compensation Fund. 2001. "Calculating the Losses." *New York Times,* December 21: B6.

Shattuck, Rachel M., and Rose M. Kreider. 2013. "Social and Economic Characteristics of Currently Unmarried Women With a Recent Birth: 2011." http://www.census.gov/prod/2013pubs/acs-21.pdf.

Shen, Hua, and Adrian Ziderman. 2009. "Student Loans Repayment and Recovery: International Comparisons." *Higher Education* 57(3): 315–333.

Shiao, Jiannbin Lee, and Mia H. Tuan. 2008. "Korean Adoptees and the Social Context of Ethnic Exploration." *American Journal of Sociology* 113(4): 10023–10066.

Shoup, Anna. 2006. "Afghanistan and the War on Terror: The Taliban." *Online Newshour.* http://www.pbs.org/newshour/indepth_coverage/asia/afghanistan/keyplayers/taliban.html.

Sidor, Gary. 2013. "Fact Sheet: The Social Security Retirement Age." Congressional Research Brief, January 24. http://www.fas.org/sgp/crs/misc/R41962.pdf.

Sierra Club. 2007. "Sprawl Overview." http://www.sierraclub.org/sprawl/overview/.

Simmel, Georg. 1950. *The Sociology of Georg Simmel,* translated, edited, and with an introduction by Kurt H. Wolff. New York: Free Press.

Simpson, Richard L. 1956. "A Modification of the Functional Theory of Social Stratification." *Social Forces* 35:132–137.

SIPRI. 2012. *SIPRI Yearbook 2011.* http://www.sipri.org/yearbook/2011.

Slobodchikoff, Con. 2011. "New Language Discovered: Prairiedogese." Interview with Con Slobodchikoff by Jad Abumrad and Robert Krulwich, National Public Radio, January 11. http://www.npr.org/2011/01/20/132650631/new-language-discovered-prairiedogese.

Smart, Ninian. 1976. *The Religious Experience of Mankind.* New York: Scribner's.

Smith, Adam. 1776. *An Inquiry into the Nature and Causes of the Wealth of Nations.* www2.hn.psu.edu/faculty/jmanis/adam-smith/Wealth-Nations.pdf.

Smith, Nate. 2012. "Guest Blog: Can I Keep Chickens in My Backyard?" *The Survivalist Blog,* May 14. http://www.thesurvivalistblog.net/can-i-keep-chickens-in-my-backyard/.

Smith, Vicki. 2010. "Enhancing Employability: Human, Cultural and Social Capital in an Era of Turbulent Unpredictability." *Human Relations* 63(2): 279–300.

Sobie, Jan Hipkins. 1986. "The Cultural Shock of Coming Home Again." In *The Cultural Transition: Human Experience and Social Transformation in the Third World and Japan,* edited by M. I. White and S. Pollack, 95–102. Boston: Routledge & Kegan.

Social Issues Reference. 2009. "Preschool-History and Demographics." http://social.jrank.org/pages/519/Preschool-History-Demographics.html.

Solaro, Erin. 2010. "Introductions: An Unabashed Feminist Writes about Women in the Military." POV, *Regarding War.* http://www.pbs.org/pov/regardingwar/conversations/women-and-war/introductions-an-unabashed-feminist-writes-about-women-in-the-military.php.

Son, Eugene. 1998. "G.I. Joe—A Real American FAQ." www.eugeneson.com/.

Sowell, Thomas. 1981. *Ethnic America: A History.* New York: Basic Books.

Sparks, Dinah, and John Malkus. 2013. "First-Year Undergraduate Remedial Coursetaking: 1999–2000, 2003–04, 2007–08." Statistics in Brief. National Center of Education Statistics, U.S. Department of Education. http://nces.ed.gov/pubs2013/2013013.pdf.

Spector, Malcolm, and J. I. Kitsuse. 1977. *Constructing Social Problems.* Menlo Park, CA: Cummings.

Spitz, Rene A. 1951. "The Psychogenic Diseases in Infancy: An Attempt at Their Etiologic Classification." In *The Psychoanalytic Study of the Child,* Vol. 27, edited by R. S. Eissler and A. Freud, 255–278. New York: Quadrangle.

Spreitzer, Elmer A. 1971. "Organizational Goals and Patterns of Informal Organizations." *Journal of Health and Social Behavior* 12(1): 73–75.

Stanger, Melissa, and Peter Jacobs. 2013. "The 50 Most Expensive Private High Schools in America." *Business Insider,* September 4. http://www.businessinsider.com/most-expensive-private-high-schools-us-2013-8?op=1.

Starbucks. 2013. "About Us: Company Profile." http://www.starbucks.com/about-us/company-information.

Stein, Arlene. 1989. "Three Models of Sexuality: Drives, Identities, and Practices." *Sociological Theory* 7(1): 1–13.

Stinson Beach County Water District. 2014. "Drought Alert." http://stinson-beach-cwd.dst.ca.us/downloads/drought_alert_2014.pdf.

Stoller, Gary. 2014. "Conn. Governor Signs Bill Raising State Minimum Wage." *USA Today.* http://www.usatoday.com/story/news/nation/2014/03/27/connecticut-boost-minimum-wage/6973715/.

Stretesky, Paul B., and Michael J. Lynch. 2002. "Environmental Hazards and School Segregation in Hillsborough County, Florida, 1987–1999." *Sociological Quarterly* 43(4): 553–573.

Stub, Holger. 1982. *The Social Consequences of Long Life.* Springfield, IL: Thomas.

Sugrue, Thomas. 2007. "Stamping Out Detroit Postmark Irks 'Michigan Metro' Residents." *Bloomberg News,* October 30.

Sumner, William Graham. 1907. *Folkways.* Boston: Ginn.

Susan G. Komen. 2014. "About Us." http://ww5.komen.org/AboutUs/AboutUs.html.

Sutherland, Edwin H., and Donald R. Cressey. 1978. *Principles of Criminology* (10th ed.). Philadelphia: Lippincott.

Takahashi, Junko. 1999. "Century of Change: Marriage Sheds Its Traditional Shackles." *Japan Times,* December 13. http://www.japantimes.co.jp.

Tampa Bay Times. 2012. "Subsidies and Tariffs: Sweetening the Pot." March 29. http://www.tampabay.com/specials/2008/graphics/sugar-prices/.

Telles, Edward E. 2004. *Race in Another America: The Significance of Skin Color in Brazil.* Princeton, NJ: Princeton University Press.

Thanh Nien News. 2006. "Cars Mean Traffic Jams in Vietnam's Southern Metro." http://www.thanhniennews.com/2006/Pages/20061171234220.aspx.

Theodorson, George A., and Achilles G. Theodorson. 1969. *A Modern Dictionary of Sociology.* New York: Harper.

Thoits, Peggy A. 2010. "Stress and Health: Major Findings and Policy Implications." *Journal of Health and Social Behavior* 51: S41–S53.

Thomas, William I. 1923. *The Unadjusted Girl.* Boston: Little, Brown.

Thomas, William I., and Dorothy Swain Thomas. 1928. *The Child in America.* Repr., New York: Johnson, 1970.

Thompson, Warren S. 1929." Population." *American Journal of Sociology* 34(6): 959-975.

Tilghman, Andrew. 2012. "Pentagon Opens More Military Jobs to Women." *Army Times,* February 9. http://www.armytimes.com/news/2012/02/military-ban-on-women-lifted-for-1-percent-of-military-jobs-020912w/.

Timmerman, Kelsey. 2012. *Where Am I Wearing? A Global Tour to the Countries, Factories, and People That Make Our Clothes.* New York: John Wiley and Sons.

Toro, Luis Angel. 1995. "'A People Distinct from Others': Race and Identity in Federal Indian Law and the Hispanic Classification in OMB Directive No. 15." *Texas Tech Law Review* 26: 1219–1274.

Traina, Cristina. 2005. "Touch on Trial: Power and Right to Physical Affection." *Journal of the Society of Christian Ethics* 25(1): 3–34.

Tulenko, Kate. 2010. "Countries without Doctors?" *Foreign Policy* (June 11). www.foreignpolicy.com/articles/2010/06/11/countries_without_doctors.

Tuller, David. 2004. "Gentleman, Start Your Engines?" *New York Times,* June 21: E1.

Tumin, Melvin M. 1953. "Some Principles of Stratification: A Critical Analysis." *American Sociological Review* 18: 387–394.

Turce, Nick. 2013. "AFRICO"s Gigantic 'Small Footprint.'" TomDispatch.com, September 5. http://www.tomdispatch.com/blog/175743/tomgram%3A_nick_turse,_africom%27s_gigantic_%22small_footprint%22.

Tyler, Greg. 2011. "Salaries Go Up Even When Income Doesn't." *Sport Digest,* April 11. http://thesportdigest.com/2011/04/salaries-go-up-even-when-income-doesn%E2%80%99t/.

Umberson, Debra. 2010. "Social Relationships and Health: A Flashpoint for Health Policy." *Journal of Health and Social Behavior* 51: S54–S66.

UNICEF. 2013. "Afghanistan." http://www.unicef.org/infobycountry/afghanistan_statistics.html.

United Nations. 1948. "Convention on the Punishment and Prevention of the Crime of Genocide." http://www.ess.uwe.ac.uk/documents/gncnvntn.htm.

———. 2006. *Beyond Scarcity: Power, Poverty, and the Global Water Crisis.* Human Development Report. http://hdr.undp.org/hdr2006/.

———. 2010a. "Beyond the Midpoint." http://content.undp.org/go/newsroom/publications/poverty-reduction/poverty-website/mdgs/beyond-the-midpoint.en.

———. 2010b. "We Can End Poverty 2015." www.un.org/millenniumgoals.

———. 2012. "UN Data." http://data.un.org/.

———. 2013. "Afghanistan." Human Development Report. http://hdr.undp.org/sites/default/files/Country-Profiles/AFG.pdf.

United Nations Development Programme. 2011. "Towards Human Resilience: Sustaining MDG Progress in an Age of Economic Uncertainty" (October 13). http://www.undp.org/content/undp/en/home/librarypage/poverty-reduction/inclusive_development/towards_human_resiliencesustainingmdgprogressinanageofeconomicun/.

United Nations Millennium Development Project. 2013. "We Can End Poverty: Millennium Development Goals and Beyond 2015." http://www.un.org/millenniumgoals/reports.shtml.

U.S. Air Force. 2012. "Young Girl's Curiosity of Science Leads to Becoming First African-American Woman in Space." http://media.dma.mil/2012/Feb/07/2000180508/-1/-1/0/120207-F-BZ728-002.JPG.

U.S. Army. 2012. "TAMC Head, Neck Docs Use Robot to Increase Patients' Quality of Life" (January 31). http://www.army.mil/media/233611.

U.S. Bureau of Labor Statistics. 2012. "National Occupational Employment and Wage Estimates, United States." http://www.bls.gov/oes/current/oes_nat.htm#29-0000.

———. 2013a. "Employed Persons by Detailed Occupation, Sex, Race, and Hispanic or Latino Ethnicity." http://www.bls.gov/cps/cpsaat11.pdf.

———. 2013b. "Highlights of Women's Earnings in 2012." http://www.bls.gov/cps/cpswom2012.pdf.

———. 2013c. "Women in the Labor Force: A Databook." http://www.bls.gov/cps/wlf-databook-2012.pdf.

———. 2013d. "Employment by Major Industry Sector." http://www.bls.gov/emp/ep_table_201.htm.

———. 2014a. "Fastest Declining Occupations." http://www.bls.gov/emp/ep_table_105.htm.

———. 2014b. "Pharmacy Technicians." http://www.bls.gov/oes/current/oes292052.htm.

———. 2014c. "Wives Who Earn More Than Their Husbands, 1987–2012." http://www.bls.gov/cps/wives_earn_more.htm.

U.S. Census Bureau. 2002. *Demographic Trends in the 20th Century,* by F. Hobbs and N. Stoops. Census 2000 Special Reports, Series CENSR-4. Washington, DC: U.S. Government Printing Office.

———. 2011. *Overview of Race and Hispanic Origin: 2010.* http://www.census.gov/prod/cen2010/briefs/c2010br-02.pdf.

———. 2012a. "2012 Census of Governments." http://www.census.gov/govs/.

———. 2012b. "Who's Minding the Kids? Child Care Arrangements: Spring 2010." http://www.census.gov/hhes/childcare/data/sipp/2010/tables.html.

———. 2012c. "Selected Social Characteristics in the United States, 2012: American Community Survey 1-Year Estimates." http://factfinder2.census.gov/faces/nav/jsf/pages/searchresults.xhtml?refresh=t.

———. 2013a. "America's Families and Living Arrangements: 2012." https://www.census.gov/prod/2013pubs/p20-570.pdf.

———. 2013b. "Asian Alone by Separate Groups." *2012 American Community Survey 1-Year Estimates.* http://factfinder2.census.gov/faces/nav/jsf/pages/searchresults.xhtml?refresh=t.

———. 2013c. "Supplemental Measure of Poverty Remains Unchanged" (November 6). https://www.census.gov/newsroom/releases/archives/poverty/cb13-183.html.

———. 2013d. "The Diversifying Electorate—Voting Rates by Race and Hispanic Origin in 2012 (and Other Recent Elections)." Current Population Survey, May 13. http://www.census.gov/prod/2013pubs/p20-568.pdf.

———. 2014a. "Country Rank." http://www.census.gov/population/international/data/countryrank/rank.php.

———. 2014b. "International Data Base." http://www.census.gov/population/international/data/idb/informationGateway.php.

U.S. Central Intelligence Agency. 2012. *World Factbook.* https://www.cia.gov/library/publications/the-world-factbook/.

———. 2014. *World Factbook.* https://www.cia.gov/library/publications/the-world-factbook/geos/us.html.

U.S. Department of Agriculture. 2011a. "Rural Income, Poverty, and Welfare: Poverty Geography." Briefing Rooms, September 17. http://www.ers.usda.gov/Briefing/IncomePovertyWelfare/PovertyGeography.htm.

———. 2011b. "Rural Labor and Education: Farm Labor." http://www.ers.usda.gov/Briefing/LaborAndEducation/Farm Labor.htm.

———. 2013. "Parents Projected to Spend $241,080 to Raise a Child Born in 2012, According to USDA Report." http://www.usda.gov/wps/portal/usda/usdahome?contentid=2013/08/0160.xml.

———. 2014. "Farm Labor." http://www.ers.usda.gov/topics/farm-economy/farm-labor/background.aspx#wages.

U.S. Department of Criminal Justice Statistics. 2010. "Prisoners under the Jurisdiction of the Federal Bureau of Prisons, by Adjudication Status, Type of Offense, and Sentence Length, 1990, 2000, and 2009a." http://www.albany.edu/sourcebook/pdf/t600232009.pdf.

U.S. Department of Defense. 1990. "DOD Directive 1332.14." In *Gays in Uniform: The Pentagon's Secret Reports,* edited by K. Dyer, 19. Boston: Alyson.

———. 2013. "Contracts: Defense Logistics Agency" (December 11). http://www.defense.gov/contracts/contract.aspx?contractid=5182.

U.S. Department of Education. 2010. *Digest of Education Statistics.* http://nces.ed.gov/programs/digest/.

———. 2012. "Percentage of Students at or above Selected National Assessment of Educational Progress (NAEP) Reading Achievement Levels, by Grade and Selected Student Characteristics: Selected Years, 1998 through 2011." *Digest of Education Statistics.* http://

nces.ed.gov/programs/digest/d12/tables/dt12_143.asp.

———. 2013a. "Characteristics of Private Schools in the United States: Results from the 2011–12 Private School Universe Survey." http://nces.ed.gov/pubs2013/2013316.pdf.

———. 2013b. "Table 216.60. Number and Percentage Distribution of Public School Students Eligible for Free or Reduced-Price Lunch, by School Level, Locale, and Student Race/Ethnicity: 2011–12." *Digest of Education Statistics*. http://nces.ed.gov/programs/digest/d13/tables/dt13_216.60.asp.

U.S. Navy. 2014. Photo Gallery: Destroyers. http://www.navy.mil/viewGalleries.asp.

———. 2014. "2014 Poverty Guidelines." http://aspe.hhs.gov/poverty/14poverty.cfm.

U.S. Department of Justice. 2007. "Criminal Offenders Statistics." http://bjs.ojp.usdoj.gov/.

———. 2013. "For Second Consecutive Year Violent and Property Crime Rates Increased in 2012." http://www.bjs.gov/content/pub/press/cv12pr.cfm.

U.S. Department of Labor, Bureau of Labor Statistics. 2011. "College Enrollment and Work Activity of 2010 High School Graduates." Economic News Release, April 8. http://www.bls.gov/news.release/hsgec.nr0.htm.

———. 2012a. "The Employment Situation—January 2012." News Release, April 6. http://www.bls.gov/news.release/empsit.nr0.htm.

———. 2012b. "Labor Force Statistics from the Current Population Survey." http://www.bls.gov/cps/tables.htm#weekearn.

———. 2013. "Employment by Summary Education and Training Assignment, 2012 and Projected 2022." http://www.bls.gov/emp/ep_table_education_summary.htm.

———. 2014a. "Earnings and Unemployment Rate by Level of Educational Attainment." http://www.bls.gov/emp/ep_chart_001.htm.

———. 2014b. "Occupations with Most Job Growth." http://www.bls.gov/emp/ep_table_104.htm.

———. 2014c. "College Enrollment and Work Activity of High School Graduates News Release." Bureau of Labor Statisics (April 22). http://www.bls.gov/news.release/archives/hsgec_04222014.htm.

———. 2014d. "Median Weekly Earning of Full-Time and Salaried Workers Age 25 Years and Over." http://www.bls.gov/news.release/pdf/wkyeng.pdf.

U.S. General Accounting Office. 2004. "Defense of Marriage Act: Update to Prior Report." http://www.gao.gov/new.items/d04353r.pdf.

U.S. Military.com. 2014. "Foreign Born Troops Sworn in as Citizens." http://www.usmilitary.com/2571/foreign-born-troops-sworn-in-as-citizens.

U.S. Navy. 2014. Photo Gallery: Destroyers. http://www.navy.mil/viewGalleries.asp.

U.S. Office of Management and Budget. 1997. "Revisions to the Standards for Classification of Federal Data on Race and Ethnicity." *Federal Register Notice,* October 30. http://www.whitehouse.gov/omb/fedreg/1997standards.html.

University of Kentucky (UK) News. 2003 "UK Offers Drive-Thru Flu Shots." http://news.uky.edu.

University of Michigan Health System. 2010. "Television and Children" (updated August 2010). http://www.med.umich.eduyourchild/topics/tv.htm.

UPS. 2014. "Worldwide Facts." http://www.ups.com/content/us/en/about/facts/worldwide.html.

USA Today. 2014a. "Coaches' Salaries." http://www.usatoday.com/sports/college/salaries/ncaab/coach/.

———. 2014b. "Top 10 Internet-Censored Countries" (February 5). http://www.usatoday.com/story/news/world/2014/02/05/top-ten-internet-censors/5222385/.

VanderLaan, D. P., D. L. Forrester, L. J. Petterson, and P. L. Vasey. 2012. "Offspring Production among the Extended Relatives of Samoan Men and *Fa'afafine*." *PLoS ONE* 7(4). http://www.plosone.org/article/info%3Adoi%2F10.1371%2Fjournal.pone.0036088.

Van Evera, Stephen. 1990. "The Case against Intervention." *Atlantic Monthly* (July): 72–80.

Varadarajan, Tunku. 1999. "A Patel Motel Cartel?" *New York Times,* July 4. http://www.nytimes.com/1999/07/04/magazine/a-patel-motel-cartel.html?pagewanted=all.

Verkuyten, Maykel. 2005. "Ethnic Group Identification and Group Evaluation among Minority and Majority Groups." *Journal of Personality and Social Psychology* 88(1): 121–138.

Vincent, John A., Emmanuelle Tulle, and John Bond. 2008. "The Anti-Aging Enterprise: Science, Knowledge, Expertise, Rhetoric, and Values." *Journal of Aging Studies* 22(4): 291–376.

Wacquant, Loïc J. D. 1989. "The Ghetto, the State, and the New Capitalist Economy." *Dissent* (Fall): 508–520.

Wallerstein, Immanuel. 1984. *The Politics of the World-Economy: The States, the Movements and the Civilizations.* New York: Cambridge University Press.

———. 1990. "Culture as the Ideological Battleground of the Modern World-System." *Theory, Culture, and Society* 7: 31–55.

Walmsley, Roy. 2002. "Global Incarceration and Prison Trends." United Nations Office on Drugs and Crime. http://www.unodc.org/pdf/crime/forum/forum3_Art3.pdf.

Wang, Wendy. 2012. "The Rise of Intermarriage Rates, Characteristics Vary by Race and Gender." Pew Research Social and Demographic Trends, February 16. http://www.pewsocialtrends.org/2012/02/16/the-rise-of-intermarriage/.

Warner, Melanie. 2006. "Salads or No, Cheap Burgers Revive McDonald's." *New York Times,* April 19. http://www.nytimes.com/.

Waters, Mary C. 1990. *Ethnic Options: Choosing Identities in America.* Berkeley: University of California Press.

———. 1994. "Ethnic and Racial Identities of Second Generation Black Immigrants in New York City." *International Migration Review* 28(4): 795–820.

Weber, Max. 1922a. *Economy and Society,* Vol. 2, edited by Guenther Roth and Claus Wittich, translated by Ephraim Fischof, 1978. Berkeley: University of California Press.

———. 1922b. *The Sociology of Religion,* translated by E. Fischoff. Boston: Beacon.

———. 1925. *Economy and Society,* Part III, Chap. 6, 650–678. http://www.faculty.rsu.edu/~felwell/Theory Web/readings/WeberBurform.html #Weber.

———. 1947. *The Theory of Social and Economic Organization,* edited and translated by A. M. Henderson and T. Parsons. New York: Macmillan.

———. 1948. "Class, Status, and Party." In *Essays from Max Weber,* edited by H. Gerth and C. W. Mills. New York: Routledge & Kegan Paul.

———. 1958. *The Protestant Ethic and the Spirit of Capitalism* (5th ed.), translated by T. Parsons. New York: Scribner's.

———. 1994. *Weber: Political Writings* (Cambridge Texts in the History of Political Thought), edited by Peter Lassman, translated by Ronald Speirs, xvi. Cambridge, UK: Cambridge University Press.

Wellman, Barry, and Keith Hampton. 1999. "Living Networked On and Off Line." *Contemporary Sociology* 28(6): 648–654.

Wells, Amy Stuart, and Jeannie Oakes. 1996. "Potential Pitfalls of Systematic Reform: Early Lessons from Research on Detracking." *Sociology of Education* 69(extra issue): 135–143.

West, Amanda, and Linda Allin. 2010. "Chancing Your Arm: The Meaning of Risk in Rock Climbing." *Sport in Society* 13: 7–8. http://www.tandfonline.com/doi/abs/10.1080/17430431003780245#.Us4CgfYmRJM.

Whorf, Benjamin. 1956. *Language, Thought, and Reality: Selected Writings of Benjamin Lee Whorf,* edited by J. B. Carroll. Cambridge: Technology Press of MIT.

Widmer E. D. (2010). *Family* CONFIGURATIONS: A STRUCTURAL APPROACH OF FAMILY DIVERSITY. London: Ashgate Publishing.

Wikipedia. 2104. "All-Time Olympic Games Medal Table." http://en.wikipedia.org/wiki/All-time_Olympic_Games_medal_table.

Williams, Terry. 1989. *The Cocaine Kids: The Inside Story of a Teenage Drug Ring.* Reading, MA: Addison-Wesley.

Williamson, Lucy. 2011. "South Korea's Enthusiasm for Recycling." BBC News, June 9. http://news.bbc.co.uk/2/hi/programmes/from_our_own_correspondent/9508181.stm.

Wilson, Matthew J., et al. 2011. "NCAA Division I Men's Basketball Coaching Contracts." *Journal of Issues of Intercollegiate Athletics* 4: 396–410.

Wilson, Reid. 2014. "Washington, Colorado Move to Ban Kids from Tanning Beds." *Washington Post,* February 21. http://www.washingtonpost.com/blogs/govbeat/wp/2014/02/21/washington-colorado-move-to-ban-kids-from-tanning-beds/.

Wilson, William Julius. 1983. "The Urban Underclass: Inner-City Dislocations." *Society* 21: 80–86.

———. 1987. *The Truly Disadvantaged.* Chicago: University of Chicago Press.

Wirth, Louis. 1945. "The Problem of Minority Groups." In *The Science of Man,* edited by R. Linton, 347–372. New York: Columbia University Press.

Wiseman, Rosalind. 2003. *Queen Bees and Wannabes.* New York: Three Rivers Press.

Witte, Thomas E. 2005. "The Biggest Man on the Field." *Sportshooter,* October 31. http://www.sportsshooter.com/news/1477.

Wooddell, George, and Jacques Henry. 2005. "The Advantage of a Focus on Advantage: A Note on Teaching Minority Groups." *Teaching Sociology* 33(3): 301–309.

World Bank. 2008. *World Development Indicators 2008.* Washington, DC: World Bank.

———. 2009. "Understanding Poverty." http://web.worldbank.org/.

———. 2010. "World Bank Updates Poverty Estimates for the Developing World" (February 17). http://web.worldbank.org/WBSITE/EXTERNAL/NEWS/0,,contentMDK:21882162~pagePK:34370~piPK:34424~theSitePK:4607,00.html.

———. 2011a. "Gini Index." http://data.worldbank.org/indicator/SI.POV.GINI.

———. 2011b. "National Oil Companies and Value Creation." Working Paper No. 218. http://siteresources.worldbank.org/INTOGMC/Resources/9780821388310.pdf.

———. 2014a. "Adolescent Fertility Rate (Births per 1,000 Women Ages 15–19)." http://data.worldbank.org/indicator/SP.ADO.TFRT.

———. 2014b. "Percentage of Women Age 15–49 Using Any Kind of Contraception." http://search.worldbank.org/data?qterm=childless+women&language=EN&op=&os=20.

———. 2014c. "Population Ages 65 and Above (% of Total)." http://data.worldbank.org/indicator/SP.POP.65UP.TO.ZS.

———. 2014d. "Poverty Overview." http://www.worldbank.org/en/topic/poverty/overview.

World Economic Forum. 2011. *The Global Gender Gap Report 2011: Rankings and Scores.* http://www.weforum.org/reports/global-gender-gap-report-2011 World Watch Institute.

———. 2013. *The Global Gender Gap Report 2013.* http://www.weforum.org/issues/global-gender-gap.

World Health Organization. 2013. "Key Facts: Malaria." www.who.int/mediacentre/factsheets/fs094/en/index.html.

———. 2014. "Trends in Maternal Mortality: 1990 to 2013." http://apps.who.int/iris/bitstream/10665/112682/2/9789241507226_eng.pdf?ua=1.

World Socialist Movement. 2014. "'Imagine' by John Lennon." http://www.worldsocialism.org/articles/imagine_by_john_lennon.php.

Wright, Eric R., and Brea L. Perry. 2010. "Medical Sociology and Health Services Research: Past Accomplishments and Future Policy Challenges." *Journal of Health and Social Behavior* 51(November): S107–S119.

Wright, Erik Olin. 2004. "Social Class." In *Encyclopedia of Sociological Theory,* edited by George Ritzer. Thousand Oaks, CA: Sage.

Yamada, Masahiro. 2000. "The Growing Crop of Spoiled Singles." *Japan Echo* 27(3). http://www.japanecho.com.

Yamamoto, Noriko, and Margaret I. Wallhagen. 1997. "The Continuation of Family Caregiving in Japan." *Journal of Health and Social Behavior* 38(June): 164–176.

Yavorsky, J. E., and L. Sayer. 2013. 'Doing Fear': The Influence of Hetero-Femininity on (Trans) women's Fears of Victimization." *Sociological Quarterly* 54: 511–533. doi: 10.1111/tsq.12038.

Yinger, J. Milton. 1977. "Presidential Address: Countercultures and Social Change." *American Sociological Review* 42(6): 833–853.

Young, T. R. 1975. "Karl Marx and Alienation: The Contributions of Karl Marx to Social Psychology." *Humboldt Journal of Social Relations* 2(2): 26–33.

YouTube. 2014. "Top 10 Most Viewed Videos."

Zelditch, Morris. 1964. "Family, Marriage, and Kinship." In *Handbook of Modern Sociology,* edited by Robert E. L. Faris, 680–733. Chicago: Rand McNally.

Zuboff, Shoshana. 1988. *In the Age of the Smart Machine: The Future of Work and Power.* New York: Basic Books.

Zumbrun, Joshua. 2008. "In Pictures: Ten Most Energy-Efficient Countries." *Forbes,* July 7. http://www.forbes.com/2008/07/03/energy-efficiency-japan-biz-energy_cx_jz_0707efficiency_countries.html.

Index